AF877420

KEY TOPICS IN
NUCLEAR STRUCTURE

KEY TOPICS IN NUCLEAR STRUCTURE

Proceedings of the 8th International Spring Seminar on Nuclear Physics

Paestum, Italy 23—27 May 2004

edited by

Aldo Covello

Dipartimento di Scienze Fisiche
Università di Napoli Federico II

World Scientific

NEW JERSEY • LONDON • SINGAPORE • BEIJING • SHANGHAI • HONG KONG • TAIPEI • CHENNAI

Published by

World Scientific Publishing Co. Pte. Ltd.

5 Toh Tuck Link, Singapore 596224

USA office: 27 Warren Street, Suite 401-402, Hackensack, NJ 07601

UK office: 57 Shelton Street, Covent Garden, London WC2H 9HE

British Library Cataloguing-in-Publication Data
A catalogue record for this book is available from the British Library.

KEY TOPICS IN NUCLEAR STRUCTURE
Proceedings of the 8th International Spring Seminar on Nuclear Physics

Copyright © 2005 by World Scientific Publishing Co. Pte. Ltd.

All rights reserved. This book, or parts thereof, may not be reproduced in any form or by any means, electronic or mechanical, including photocopying, recording or any information storage and retrieval system now known or to be invented, without written permission from the Publisher.

For photocopying of material in this volume, please pay a copying fee through the Copyright Clearance Center, Inc., 222 Rosewood Drive, Danvers, MA 01923, USA. In this case permission to photocopy is not required from the publisher.

ISBN 981-256-093-9

Printed in Singapore by World Scientific Printers (S) Pte Ltd

LOCAL ORGANIZING COMMITTEE

A. Covello, Seminar Chairman
A. Gargano, Scientific Secretary
F. Andreozzi
L. Coraggio
N. Itaco
N. Lo Iudice
G. La Rana
A. Porrino

INTERNATIONAL ADVISORY COMMITTEE

G. de Angelis (*Legnaro*)
N. Benczer-Koller (*Rutgers*)
F. Catara (*Catania*)
R. Donangelo (*Rio de Janeiro*)
G. Dracoulis (*Canberra*)
A. Faessler (*Tübingen*)
B. Fornal (*Kraków*)
W. Gelletly (*Surrey*)
H. Grawe (*GSI*)
F. Iachello (*Yale*)
R. Julin (*Jyväskylä*)
R. Machleidt (*Idaho*)
Jie Meng (*Beijing*)
E. Moya de Guerra (*Madrid*)
T. Otsuka (*Tokyo*)
J. Pinston (*Grenoble*)
A. V. Ramayya (*Vanderbilt*)
D. Schwalm (*Heidelberg*)
E. Vigezzi (*Milano*)
V. Voronov (*Dubna*)
R. Wyss (*Stockholm*)
A. Zuker (*Strasbourg*)

SPONSORS OF THE SEMINAR

Università di Napoli "Federico II"
Istituto Nazionale di Fisica Nucleare

HOST TO THE SEMINAR

Dipartimento di Scienze Fisiche, Università "Federico II"

FOREWORD

The Eighth International Spring Seminar on Nuclear Physics was held in Paestum from May 23 to May 27, 2004. This Seminar was the eighth in a series of topical meetings to be held every two or three years in the Naples area. The series began with the Sorrento meeting in 1986 and continued with the Capri meeting in 1988, the Ischia meeting in 1990, the Amalfi meeting in 1992, the Ravello meeting in 1995, the S. Agata meeting in 1998, and the Maiori meeting in 2001. The aim of this eighth meeting was to discuss recent advances and new perspectives in nuclear structure experiment and theory.

Nuclear structure studies of exotic nuclei are currently being performed in several laboratories where beams of rare isotopes are available. Meanwhile the development of new facilities, which will provide high-intensity beams, is in progress or under discussion in Europe, Asia and North America. At this meeting we had a very lively and comprehensive overview of this fascinating field and of future scenarios thanks to the participation of leaders of the most important projects.

Besides the great impetus toward the exploration of nuclear regions far away from stability, new exciting results of spectroscopic studies conducted with stable beams have been reported, reminding us that these "traditional" beams coupled with high-efficiency detectors are still, and will remain for a long time, a very valuable tool to advance our knowledge of nuclear structure.

Nuclear structure theory is setting new frontiers. On the one hand, the experimental studies of nuclei far from stability are fostering theoretical studies of possible changes of nuclear structure. On the other hand, sustained efforts are being made to understand the properties of nuclei in terms of the basic interactions between the constituents. This means that a truly microscopic theory of nuclear structure is on the way.

As usual, the program of the meeting consisted of general talks and of more specialized seminars, the latter including most of the contributions submitted by participants. The speakers covered five main topics: i) Present and Future of Nuclear Structure with Rare Isotope Beams; ii) Nuclear Forces and Nuclear Structure; iii) The Role of Shell Model in the Understanding of Nuclear Structure; iv) Collective Aspects of Nuclear Structure; v) Special Topics. This volume contains the invited papers and all oral and poster contributions considered relevant to the Seminar according to the judgment of the Advisory Committee. The actual program of the Seminar is also included to give an idea of how it was organized. We are confident that the high quality of both invited and contributed papers collected in these Proceedings will be appreciated by the nuclear physics community.

As was the case for most of the previous Seminars, the Paestum Seminar too ended with a Round Table Discussion on the theme "Trends and Perspectives in Nuclear Structure". C. Baktash, F. Iachello, E. Rehm, D. Schwalm, I. Talmi and A. Zuker kindly agreed to be on the panel and their remarks were essential in bringing about the active involvement of the audience.

As compared with the previous conferences, this Seminar had a larger number of participants, about 100 coming from some 20 countries. While the Greek temples in Paestum are certainly an important attraction, we would like to see in this increased participation a gratifying sign of the vitality of nuclear structure research and of this Seminar. For those liking statistics, let me mention that more than 50% of the present participants attended one or more of the previous Seminars, which is just in line with the tradition of these meetings.

Unfortunately, this foreword ends with a note of sorrow. Soon after her participation in the Paestum Conference, our young colleague Milena Serra passed away unexpectedly in Tokyo, where she held a postdoctoral appointment at the University of Tokyo. I had just met her in Paestum and could never suspect that our acquaintance would have not gone beyond the few days of the Conference. To my great regret, it has not been possible to have in this volume a written record of the excellent contribution she presented as a poster session paper at our meeting. The loss of Milena Serra is a loss to our community and it is on behalf of all of us that I wish to dedicate these Proceedings to her memory.

I gratefully acknowledge the financial support of the University of Naples Federico II, the Istituto Nazionale di Fisica Nucleare and the Dipartimento di Scienze Fisiche who made the Seminar possible. I also acknowledge the support provided in various ways by the Dipartimento di Scienze Fisiche which acted as host to the Seminar.

Aldo Covello

CONTENTS

Section II
NUCLEAR FORCES AND NUCLEAR STRUCTURE

Section III
THE ROLE OF SHELL MODEL IN THE UNDERSTANDING OF NUCLEAR STRUCTURE

8th INTERNATIONAL SPRING SEMINAR ON NUCLEAR PHYSICS

KEY TOPICS IN NUCLEAR STRUCTURE

PAESTUM, MAY 23–27, 2004

PROGRAM

Sunday, May 23

9:00 Opening address

A. Covello, Seminar Chairman

Chairperson: **K. Kemper** (*Tallahassee*)

9:15 **A. C. Shotter** (*TRIUMF*): Radioactive Beams at TRIUMF - Current Situation - Future Prospectives

9:45 **C. K. Gelbke** (*Michigan*): Rare Isotope Research Capabilities at the NSCL Today and at RIA in the Future

10:15 **K. E. Rehm** (*Argonne*): Experiments with Radioactive Ion Beams at ATLAS - Present Status and Future Plans

10:45 Coffee break

11:15 **C. Baktash** (*Oak Ridge*): Nuclear Structure Studies near Doubly-Magic ^{132}Sn

11:45 **N. Aoi** (*RIKEN*): RIKEN RI Beam Factory

12:15 Session close

Chairperson: **C. Fahlander** (*Lund*)

15:00 **T. Aumann** (*GSI*): Prospects with Rare Isotope Beams at the International Facility for Antiprotons and Ion Research (FAIR)

15:30 **R. F. Casten** (*Yale*): The Evolution of Structure in Exotic Nuclei

16:00 **N. Benczer-Koller** (*Rutgers*): Nuclei-Far-From Stability: Magnetic Moment of the 2_1^+ State in ^{76}Kr

16:30 Coffee break

17:00 **J. Meng** (*Beijing*): New Effective Interactions, New Symmetry, Exotic Phenomena and Mass Limit in Atomic Nuclei

17:30 **L. S. Ferreira** (*Lisboa*): Nuclear Structure beyond the Proton Drip-Line

17:50 **P. Ring** (*München*): Relativistic Quasi-Particle RPA and New Collective Modes in Nuclei Far from Stability

18:10 **M. Tomaselli** (*GSI*): Microscopic Correlations in Nuclear Structure Calculations*

18:30 **M. V. Stoitsov** (*Tennessee*): Particle-Number Projected HFB Method with Skyrme Forces and Delta Pairing*

18:50 **R. Id Betan** (*Stockholm*): Complex Shell Model with Anti-Bound States*

19:10 Session close

Monday, May 24

Chairperson: **F. Catara** (*Catania*)

9:00 **T. T. S. Kuo** (*Stony Brook*): Studies of Phase-Shift Equivalent Low-Momentum NN potentials

9:30 **S. C. Pieper** (*Argonne*): Quantum Monte Carlo Calculations of Light Nuclei

10:00 **S. Fujii** (*Tokyo*): Cutoff Momentum Dependence of the Low-Momentum Interaction on the Nuclear Binding Energy*

10:20 **M. Hjorth-Jensen** (*Oslo*): Coupled Cluster Approaches to Nuclei, Ground States and Excited States

10:50 Coffee break

11:20 **B. R. Barrett** (*Tucson*): Ab Initio Large-Basis No-Core Shell Model

11:50 **A. Zuker** (*Strasbourg*): Monopoles for Pedestrians

12:20 **T. Otsuka** (*Tokyo*): Chaos and Symmetry

12:50 **A. Leviatan** (*Jerusalem*): Relativistic Pseudospin Symmetry as a Supersymmetric Pattern in Nuclei*

13:10 Session Close

Chairperson: **I. Talmi** (*Rehovot*)

15:00 **A. Covello** (*Napoli*): Nuclear Structure Calculations with Modern Nucleon-Nucleon Potentials

15:30 **R. Broda** (*Kraków*): Yrast States in Neutron-Rich Nuclei from the ^{48}Ca Region Studied in Deep-Inelastic HI Reactions

16:00 **J. A. Pinston** (*Grenoble*): Neutron Rich In and Cd Close to the Magic ^{132}Sn

16:30 Coffee break

17:00 **L. Corradi** (*Legnaro*): Multinucleon Transfer Reactions Studied with Large Solid Angle Spectrometers

17:30 **K. P. Lieb** (*Göttingen*): Pair Breaking in a Shears Band of ^{104}In*

17:50 **M. Górska** (*GSI*): Structure of the ^{100}Sn Region Based on a Core-Excited $E4$ Isomer in ^{98}Cd

18:10 Session close

Tuesday, May 24

Chairperson: **P. J. Daly** (*Purdue*)

9:00 **R. Julin** (*Jyväskylä*): Recent Results from Spectroscopic Studies of Exotic Heavy Nuclei at JYFL

9:30 **H. Mach** (*Uppsala*): Testing Shell Model on Exotic Nuclei at ^{135}Sb

10:00 **B. Fornal** (*Kraków*): Yrast Structure of Neutron-Rich N=30-34 Nuclei - Shell Closure at N=32

10:30 Coffee break

11:00 **L. Zamick** (*Rutgers*): Simplified Expressions for Pair Transfer, Especially N=Z Nuclei

11:20 **F. Andreozzi** (*Napoli*): A New Algorithm for Large Scale Shell Model Calculations*

11:40 **J. P. Draayer** (*Baton Rouge*): Extended Pairing Model for Well-Deformed Nuclei*

12:00 **G. Royer** (*Nantes*): Potential Barriers in the Fusion Like Deformation Path*

12:20 Session close

Wednesday, May 26

Chairperson: **G. de Angelis** (*Legnaro*)

9:00 **F. Iachello** (*Yale*): Quantum Phase Transitions in Nuclei

9:30 **D. J. Rowe** (*Toronto*): Quasi-Dynamical Symmetry in the Approach to a Second Order Phase Transition

9:50 **G. Graw** (*München*): Transfer Reactions and Structure of Heavy Nuclei

10:10 **D. Bonatsos** (*Attiki*): Critical Point Symmetry for the Prolate to Oblate Shape Phase Transition*

10:30 Coffee break

11:00 **F. Palumbo** (*Fracati*): Bosonization and Interacting Boson Model

11:30 **A. I. Georgieva** (*Sofia*): Distribution of Collective States within the Framework of Symplectic Symmetries*

11:50 **U. Kneissl** (*Stuttgart*): Dipole Strength Distributions in ^{124,126,128,130,132,134,136}Xe: a Systematic Study in the Mass Region of a Nuclear Shape Transition

12:20 **N. D. Dang** (*RIKEN*): Pairing Effects on the Giant Dipole Resonance Width at Low Temperature*

12:40 **A. Maj** (*Kraków*): Collectivity in Light Nuclei and the GDR*

13:00 Session close

Chairperson: **S. Lunardi** (*Padova*)

15:00 **V. Abrosimov** (*Kiev*): Unified Semiclassical Approach to Isoscalar Collective Modes in Heavy Nuclei*

15:20 **E. Vigezzi** (*Milano*): Pairing Correlations beyond Mean Field

15:40 **M. Gai** (*Yale*): Soft Dipole Excitation near Threshold*

16:00 **E. G. Lanza** (*Catania*): Microscopic Description of Multiple Giant Resonance in Heavy Ion Collisions*

16:20 Coffee break

16:50 **J. H. Hamilton** (*Vanderbilt*): Shape Changes and Triaxiality in Neutron Rich Y, Nb, Mo, Tc, Rh and Ag Nuclei

17:20 **A. V. Ramayya** (*Vanderbilt*): Nuclear Band Structures in $^{93-97}$Sr and Half-Life Measurements

17:40 **E. Jones** (*Vanderbilt*): Shifted Identical Bands from Pt to Pb*

18:00 Session close

Thursday, May 27

Chairperson: **P. Sona** (*Firenze*)

9:00 **D. Goutte** (*GANIL*): The SPIRAL2 Project at GANIL

9:30 **D. Schwalm** (*Heidelberg*): First Experiments with REX-ISOLDE and MINIBALL

10:00 **R. Wyss** (*Stockholm*): Selected Aspects of Collectivity in N=Z Nuclei and the Neutron Deficient Pb-Region

10:30 **N. Lo Iudice** (*Napoli*): Microscopic Study of Low-Lying 0^+ States in Deformed Nuclei

11:00 Coffee break

11:30 **Ch. Stoyanov** (*Sofia*): Microscopyc Study of Low-Lying States in ^{92}Zr*

11:50 **V. V. Voronov** (*Dubna*): Finite Rank Approximation for Nuclear Structure Calculations with Skyrme Interactions

12:10 **A. I. Vdovin** (*Dubna*): Gamma Transitions between Configurations "Quasiparticle $\otimes$ Phonon"*

12:30 **J. Kvasil** (*Praha*): Collective Modes in Fast Rotating Nuclei

12:50 **J. N. Ginocchio** (*Los Alamos*): Test of Pseudospin Symmetry in Deformed Nuclei

13:10 Session close

Chairperson: **G. Pisent** (*Padova*)

15:000 **M. Mutterer** (*Darmstadt*): Recent Experiments on Particle-Accompanied Fission

15:20 **J. M. G. Gómez** (*Madrid*): Chaos and $1/f$ Noise in Nuclear Spectra*

15:40 **V. Zelevinski** (*Michigan*): Super-Radiance:
from Nuclear Physics to Pentaquarks

16:00 **R. A. Broglia** (*Milano*): Solving the Protein Folding Problem
with the Help of Concepts Emerging from the Study
of the Superfluid-Normal Phase Transition in Atomic Nuclei

16:20 Coffee break

Chairperson: **A. Covello** (*Napoli*)

16:50 Round Table Discussion:
Trends and Perspectives in Nuclear Structure

**C. Baktash, F. Iachello, E. Rehm, D. Schwalm,
I. Talmi, A. Zuker**

* Contributed paper

SECTION I

PRESENT AND FUTURE OF NUCLEAR STRUCTURE WITH RARE ISOTOPE BEAMS

KEY TOPICS IN NUCLEAR STRUCTURE
Proceedings of the 8th Int. Spring Seminar on Nuclear Physics
edited by Aldo Covello
© 2005 World Scientific Publishing Co.

RADIOACTIVE BEAMS AT TRIUMF

A. C. SHOTTER
TRIUMF
4004 Wesbrook Mall
Vancouver, BC, Canada, V6T 2A3
E-mail: ashotter@triumf.ca

Nuclear physics research is evolving into a new era. Much of our knowledge concerning nuclear matter to date has been derived from experiments involving reactions between stable projectile and target nuclei. However, new technical developments now enable intense beams of radioactive nuclei to be used as projectiles. The use of such beams enables the nuclear landscape to be investigated over a much wider range of neutron to proton ratios. Not only is this important for nuclear physics, it is also of great importance for nuclear astrophysics, use of the nucleus as a probe for fundamental symmetries, as well as using such beams to probe the atomic structure of new materials. This paper will describe progress at TRIUMF concerning the production of radioactive beams.

1. Introduction

The TRIUMF laboratory is Canada's national laboratory for particle and nuclear physics. The laboratory has facilities and interests that span various areas of subatomic physics. In the last few years, the laboratory has been developing a purpose-built facility (ISAC, Isotope Separation and Acceleration) for the production of intense radioactive ion beams (RIB). The production of RIB's is based on the ISOL method, and uses the spallation reaction initiated by 500 MeV protons on various target materials. The RIB's so produced are used for a variety of research problems; this paper will concern the use of these beams for nuclear physics and nuclear astrophysics investigations.

2. The ISAC Facility

The TRIUMF laboratory facilities are based on a suite of five cyclotron accelerators. These cyclotrons are used to service a range of activities ranging from pure particle and nuclear physics research to medical applications. The ISAC facility uses one of the beams from the main 500 MeV cyclotron.

This cyclotron accelerates H^- ions, so it can simultaneously provide several beams of different intensities and energies to a variety of target stations. The ISAC facility, Figure 1, is based on the ISOL method and uses a beam of up to 100 μA, 500 MeV protons from this cyclotron. The beam is transported into a purpose-built target area where it is directed onto specially constructed targets.

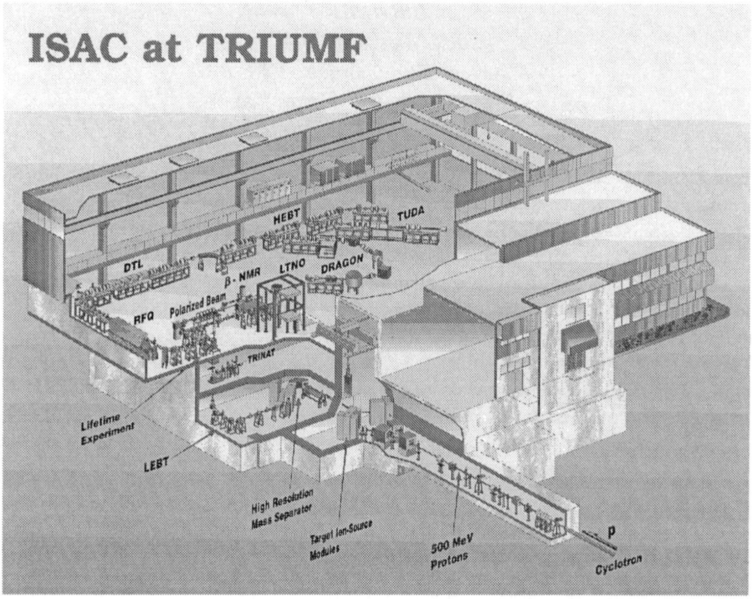

Figure 1. The ISAC radioactive beam facility at TRIUMF.

The resulting spallation reaction produces a variety of radioactive isotopes. The trick then is to extract these isotopes from the target, ionize them, mass separate them, select the appropriate isotope and then deliver this isotope beam to the experimenter. Since this is an online system, isotopes can be delivered to the experimenter with lifetimes as low as tens of milliseconds.

Due to the high radioactivity produced in the target, handling of the targets and their associated ancillary equipment has to be done remotely. The need to do this in a highly shielded area is one of the main costs associated with the ISAC facility, as indeed it will be for any high-powered ISOL facility. To increase the flexibility of the facility, there are two target stations, one in use, and the other in waiting or in maintenance mode.

The ISAC facility has been operating up to the present time using a surface ion source. However, an ECR source has now been installed, and there is progress towards the development of a laser ion source. This will ensure a wide variety of unstable isotopes of different elements can be produced as pure isotope beams.

The most important factor in any ISOL facility is the composition and construction and mode of operation of the isotope production target. The ISAC target is designed to take up to 100 μA of a 500 MeV proton beam. The target material may come as a powder, pellets or compressed composite discs. The target is 1.8 cm in diameter and can be up to 19 cm long. Control of the target temperature is very important for efficient release of spallation produced isotopes. Generally for low intensity proton beams, the target must be externally heated, while for higher intensity proton beams, the target must be cooled. The isotopes produced by the target are first ionized in an appropriate ion source and then mass selected by a high resolution mass spectrometer. The ions leaving this spectrometer will have an energy of 2 keV per mass unit. These ions may then either be delivered to the experimenter as is, or be further accelerated. Generally, for nuclear astrophysics purposes, the ions are accelerated; this is undertaken by the use of an RFQ accelerator section, followed by electronic stripping before further acceleration through a DTL accelerator. The final ion energy can be between 0.15 to 1.8 MeV/u. Different target materials are used depending on the particular isotope that needs to be produced. In this way a whole variety of RIB isotopes ranging in mass from A=8 to 160 have been produced for a range of experiments with an intensity over the range of 10^3 to 10^{11} particles per second. More details can be found on the TRIUMF web site at http://www.triumf.ca/people/marik/homepage.html.

The present ISAC facility, designated ISAC-I, although capable of producing a wide range of isotopes in the KeV/u range, can presently only accelerate ions to the MeV/u range for A<30. To increase the range of ions that can be accelerated and to increase the acceleration energy, a new post-accelerator is under construction. The new accelerator complex, designated ISAC-II, will be capable of acceleration of RIB's ions up to A=160 and to an energy of 6.5 MeV/u. The type of post accelerator will be a superconducting linear accelerator. In addition to the accelerator, a charge state booster is being developed so that higher ion masses can be accelerated through the initial RFQ accelerator that has a restriction of A/q<30. The current plans are to have first beams accelerated to 4.3 MeV/u through this new accelerator towards the end of 2005.

3. The Experimental Program at ISAC

The physics areas of interest to users of the ISAC facility are nuclear astrophysics, nuclear structure physics, fundamental symmetries, atomic physics, and condensed matter. A layout of the experimental area is shown in Figure 2.

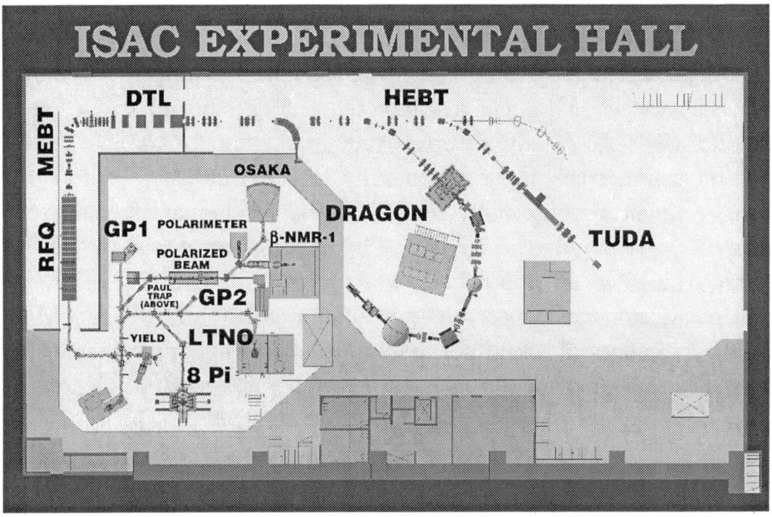

Figure 2. ISAC: Experimental equipment layout.

As main sequence stars evolve, energy is produced primarily through nuclear reactions involving stable nuclei. In contrast, for cataclysmic events such as nova, supernova and X-ray bursters where prodigious outbursts of energy can occur on the time scale of seconds, the relevant nuclear reactions can involve very exotic unstable nuclei. Nuclear astrophysics concerns the study of the relevant nuclear reactions that drive these stellar engines; this study naturally falls into two groups of reactions: those that involve stable nuclei and those involving unstable nuclei. One of the main motivations for the development of RIB facilities is to learn about the nuclear reactions that drive explosive stellar events.

ISAC-I provides an ideal facility to measure some of the key reactions that drive nova and X-ray bursters (hot CNO, Ne-Na cycles, rp process, etc.) The two major facilities designed for nuclear astrophysics experiments are: 1) DRAGON[1] (Detector of Recoils And Gammas Of Nuclear Reactions) a recoil mass separator and associated windowless gas target built to measure the rates of proton and alpha radiative capture reactions (see Figure 3); and 2) TUDA[2] (TRIUMF UK Detector Array) an array of double sided silicon strip detectors located in a general reaction chamber specially designed for low signal counting in high background environments, see Figure 4. The TUDA facility was built to study resonant reactions complementary to DRAGON and transfer reactions (e.g. (α,p) and (p,α)) associated with explosive hydrogen and helium burning. Recent examples of these are associated with the important nova reactions ^{21}Na (p,γ) ^{22}Mg and ^{22}Na (p,p) ^{22}Na (see Ref. [3,4]).

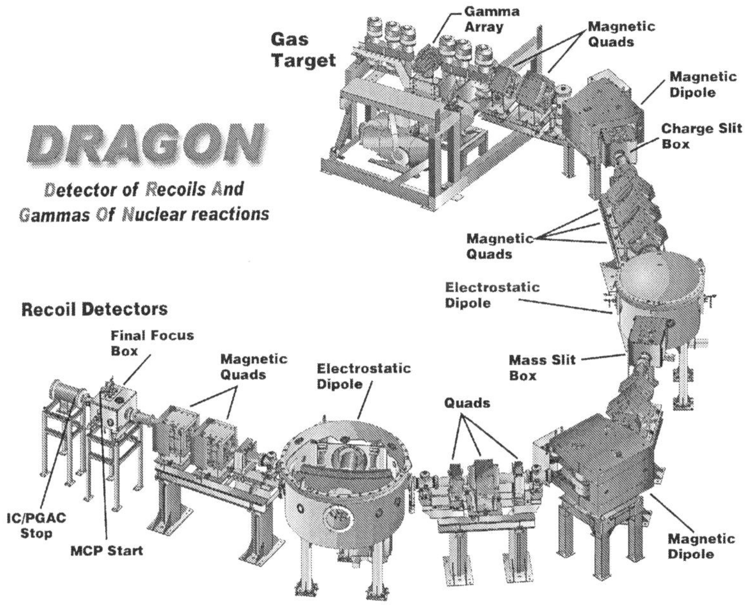

Figure 3. The DRAGON Recoil Spectrometer.

A wide range of facilities has also been developed to utilize the unaccelerated radioactive beams from ISAC-I. These include:

- TRIUMF Neutral Atom Trap (TRINAT) facility for precision tests of the electroweak Standard Model;
- Low Temperature Nuclear Orientation facility (LTNO) set up for measuring nuclear ground state moments;
- A 4π gas proportional counter and moving tape collector system for high precision lifetime measurements of superallowed β emitters;

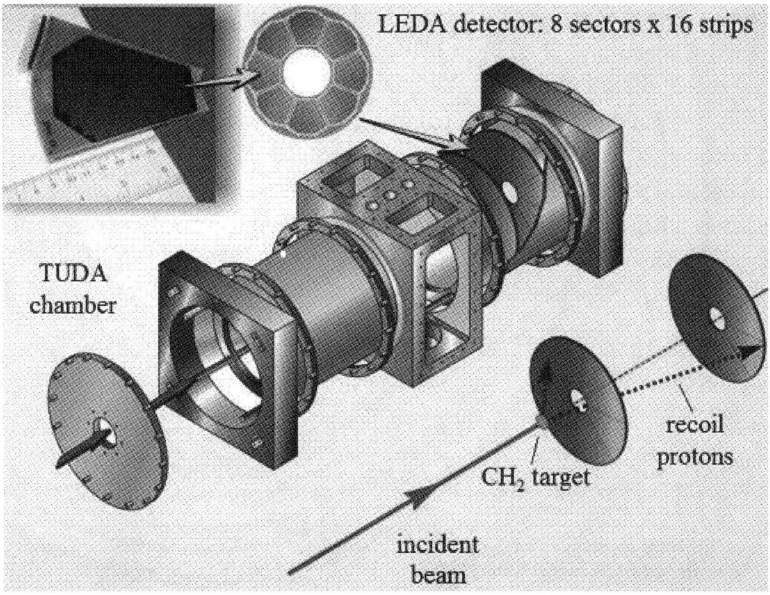

Figure 4. The TUDA facility.

- The 8π γ-ray spectrometer, an array of 20 Compton suppressed HPGe detectors and associated Scintillating Electron Positron Tagging Array (SCEPTAR) and moving tape collector for β-decay studies of exotic nuclei;
- A β-detected nuclear magnetic resonance (β-NMR) facility including a radioactive beam polarizer using an in-flight optical pumping technique for studying materials and phenomena of interest in condensed matter science;
- An array of plastic scintillator detectors (provided by Osaka University) for neutron time of flight measurements in β-n-γ studies of polarized ^{11}Li.

The studies undertaken with these facilities address nuclear structure, fundamental symmetries, and condensed matter research.

The goal of nuclear physics research with the radioactive beam is to understand the structure and dynamics of atomic nuclei. Our present knowledge is based on the relatively small number of isotopes near the valley of stability that can be produced in reactions between beams and targets consisting of stable nuclei. The radioactive ion beams provided by the ISAC facility open broad new regions of the nuclear chart to experimental investigations, providing physicists with a unique opportunity to study nuclear systems with extreme proton to neutron ratios. The limiting conditions of bound nuclei and the new structures that exist near these boundaries are being explored.

ISAC provides access to this new physics, which cannot be studied with stable beams and targets. Of particular interest are the loosely bound nuclei near the limits of nuclear existence, which are expected to exhibit new shell structure and new collective modes. The knowledge gained by studying exotic nuclei offers the promise to guide the development of a unified theory of this many-body quantum system of strongly interacting particles.

The nucleus provides a "laboratory" that can be exploited to enhance tests of discrete symmetries and tests of the Standard Model of particle physics. The wide range of exotic nuclei produced with high intensity at ISAC provides a unique opportunity to search for new physics beyond the electro-weak Standard Model. First of all, a more complete investigation of superallowed Fermi β-decay of the appropriate nuclei far from stability provides definitive information on the apparent non-unitarity of the Cabibbo-Kobayashi-Maskawa (CKM) matrix. Secondly, ISAC will allow for an order-of-magnitude improvement on the current limits of the CP-violating atomic Electric Dipole Moment (EDM) and further constrain the sources of CP-violation responsible for baryogenesis in early universe. Precision measurements of the β-decay of unpolarized and polarized trapped atoms can greatly improve the sensitivity of searches for non V-A contributions to the weak interaction. Finally, measurements of parity violation in radioactive Fr atoms would provide a much more sensitive probe of deviations from the Standard Model than Cs atoms since the electron-nucleus interactions scale approximately as Z^3.

The technique of beta-detected nuclear magnetic resonance (β-NMR) has been developed at ISAC as a new tool for research in the fields of materials science and condensed matter physics. The method utilizes polarized radioactive ions to investigate the local magnetic and electrostatic properties of solids. A unique feature of the ISAC β-NMR facility is a high voltage platform that enables implantation of the radioactive ions at controlled depths. This opens up new applications for β-NMR as a sensitive probe of ultra thin films, nanostructures, interfaces and dilute impurities. Recently, the intense spin-polarized beams of $^8Li^+$, that are available at ISAC, have been used to demonstrate the high sensitivity of β-NMR.

4. Look to the Future

When ISAC-II is operational it can be used to address a range of nuclear physics and nuclear astrophysics problems associated with heavy unstable isotopes. This will be particularly relevant for the heavy nuclei up to mass 160. For these studies, a variety of well-established experimental techniques will be used, but they will need to be specially adapted for radioactive beam experiments. For ISAC-II there is a need to build new particle detection arrays, high efficiency γ-ray detector arrays and magnetic spectrometers. Good progress is being made to fulfill these needs.

A new generation of γ-ray spectrometer, TIGRESS, is under construction, which will comprise up to 16 large volume segmented hyper-pure clover-type germanium detectors. A new recoil spectrometer, EMMA, has been designed. This spectrometer will detect exotic heavy ion products of fusion-evaporation reactions. This spectrometer will be used with the TIGRESS to detect the prompt reaction of γ-rays. In addition to the above, another major facility is being constructed to enable high precision mass measurements of short-lived nuclei. This facility, TITAN, is based on a penning trap but using highly charged ions produced by an EBIT (electron beam ion trap). Precision of Δ m/m $\sim 1.10^{-8}$ should be possible for ions of lifetimes less than 10 ms.

One of the main problems with any ISOL facility is that generally only one isotope beam can be used at any time. Also, since the beam intensity is low, experiments can take a long time to complete. Expensive apparatus not currently being used with a particular RIB is therefore remaining idle; methods should always be sought to increase this efficiency. For this reason, investigations are being undertaken at TRIUMF to increase the number of simultaneous RIB's that can be used at any time. In this respect, the ISAC driver, the H^- cyclotron, can simultaneously produce two high intensity proton beams; each beam can then irradiate separate targets, therefore producing two separate radioactive beams.

References

1. D. A. Hutcheon *et al.*, *Nucl. Instrum. Methods,* **A49**, 190 (2003).
2. T. Davinson *et al.*, *Nucl. Instrum. Methods,* **A454**, 350 (2000).
3. J. M. D'Auria *et al.*, *Phys. Rev.* **C69**, 065803 (2004).
4. C. Ruiz *et al.*, *Phys. Rev.* **C65**, 042801 (2002).

KEY TOPICS IN NUCLEAR STRUCTURE
Proceedings of the 8th Int. Spring Seminar on Nuclear Physics
edited by Aldo Covello
© 2005 World Scientific Publishing Co.

EXPERIMENTS WITH RADIOACTIVE ION BEAMS AT ATLAS - PRESENT STATUS AND FUTURE PLANS

K. E. REHM
Argonne National Laboratory
Physics Division
9700 South Cass Av.
Argonne, IL 60439, USA
E-mail: rehm@anl.gov

During the last 10 years many experiments with radioactive ion beams have been performed at the superconducting linear accelerator ATLAS at Argonne National Laboratory. The production methods employ the two-accelerator technique for long-lived isotopes or the in-flight technique for nuclei with shorter half-lives. Experiments with radioactive beams ranging in mass from ^{6}He ($T_{1/2}$=0.8s) to ^{56}Ni ($T_{1/2}$=6.1d) have been performed so far. The experiments cover studies in nuclear structure and reactions, in nuclear astrophysics, especially in explosive nuclear synthesis, as well as in neutrino physics, which are relevant in connection with recent results from the large neutrino detectors at SNO and SUPERKAMIOKANDE.

1. Introduction

The availability of beams of unstable nuclei has opened many opportunities in nuclear physics. Although the intensities of theses beams are still many orders of magnitude smaller than those available at today's stable beam accelerators, the unique properties of some of these isotopes have allowed important first measurements which previously could not be performed.

In nuclear astrophysics the reaction flows in novae, supernovae or X-ray bursts follow paths located on the proton or neutron-rich side of the valley of stability far away from stable nuclei. The rates for most of the reactions have so far been inferred from theoretical calculations, from symmetry arguments or by using indirect techniques. Experiments with radioactive beams have for the first time allowed direct measurements of astrophysical rates for a few of the critical reactions.

In the field of nuclear reactions the properties of exotic nuclei provide new information which is hard to obtain from experiments with stable isotopes. Examples are the influence of weakly bound neutrons or protons on the reaction mechanism and the influence of the isospin degree of freedom on fusion reactions.

Experiments in nuclear structure also employ the special properties of some of these nuclei. Due to the curvature of the valley of stability in the N-Z plane no stable doubly-closed shell nucleus exists between ^{48}Ca and ^{208}Pb. Several of these 'magic' nuclei which are crucial for a better understanding of the single-particle structure (e.g. ^{56,78}Ni or ^{100,132}Sn) can be studied for the first time.

In this contribution I will describe two recent experiments with radioactive beams from ATLAS which provided new information relevant to nuclear structure. The first one, performed to measure the ^{8}B neutrino spectrum[1] also provides improved information for the first excited 2^+ state in ^{8}Be. The second example describes a high-precision measurement of the charge-radius of the 'halo'-nucleus ^{6}He[2].

2. Experimental Details

The radioactive beam program at ATLAS which started 10 years ago with a ^{18}F beam has now produced more than 15 different ion beams covering the mass region between ^{6}He and ^{56}Ni[3]. The techniques used to produce these beams at ATLAS are modifications of the standard ISOL and fragmentation methods.

In the so-called two-accelerator method, used for the production of longer-lived isotopes ($T_{1/2} \geq 2$ h), the isotope of interest is produced in small amounts at a production accelerator, converted into a suitable chemical form, and then batches of the material are transferred into the ion source of ATLAS for use in the experiments. An advantage of this technique is the possibility to produce beams of elements which, because of their chemical properties (e.g. high chemical reactivity, refractory nature), are difficult to produce by the standard ISOL approach. With this technique beams of ^{18}F ($T_{1/2}$=110m), ^{44}Ti ($T_{1/2}$=59y) and ^{56}Ni ($T_{1/2}$=6.1d) have been produced at ATLAS so far.

The In-Flight production technique, used at ATLAS for the production of short-lived isotopes, can be considered as a low-energy version of the fragmentation method. It utilizes inverse reactions such as d(^{7}Li,^{6}He)^{3}He, or ^{3}He(^{6}Li,^{8}B)n with high intensity ^{7,6}Li beams to produce the unstable nuclei. The choice of inverse kinematics, i.e. bombarding a lighter target with a heavier beam, results in a forward focusing of the secondary reaction products, similar to the standard fragmentation method. The reaction products are then directed to the secondary target through an ion-optical transport system. The beam quality can be improved by including a resonator in the beam transport system[3]. Advantages of the in-flight method compared to the standard ISOL technique are the independence from the chemical properties and the fast transport times (typically less than 1 μs).

3. Experimental Results

3.1. *The 2⁺ state in ⁸Be*

There is considerable uncertainty in the literature about the energy and the width of the first excited state in ^{8}Be. The ^{8}B nucleus (J^π=2^+, $t_{1/2}$=770±3 msec) decays by an allowed β^+ transition through a broad resonance in ^{8}Be, which is described in the R-matrix formalism as a series of overlapping 2^+ nuclear resonances. A comparison of the 1988 Ajzenberg-Selove compilation[4] with the most recent analysis from TUNL[5] shows that the location of the 2^+ state is known with an accuracy of only 300 keV while the width of the state experienced a 10% change between 1988 and 2002.

Table 1. Excitation energy and width of the first excited 2^+ state in ^{8}Be, obtained from various compilations and experiments.

$E(2^+)$ [keV]	$\Gamma(2^+)$ [keV]	source
3040±300	1500±20	1988 Ref.[4]
3060±300	1370±70	2002 Ref.[5]
3024±13	1426±32	Ref.[6]
3012±7	1382±19	Ref.[7]

The large error bar for the energy of the 2^+ state can be traced back to uncertainties in the previous measurements of the ^{8}B decay[8]. In these experiments ^{8}B was produced with a ^{3}He beam via the ^{6}Li(^{3}He,n)^{8}B reaction. The ^{8}B ions were stopped in the ^{6}Li target which was subsequently rotated in front of a Si detector. Energy deposition in the detector was measured and corrections were made for energy losses in the catcher foil and detector dead layer. The first experiments, by Farmer and Class[9], De Braeckeleer and Wright[10], and Wilkinson and Alburger[11], observed the singles α-spectrum. In the later experiment by Ortiz et al[13] the two alpha particles were measured in coincidence. Data from these measurements differ from each other by energy offsets up to ± 100 keV, an effect attributed to systematic problems with detector calibrations and energy loss in the dead layers of the Si detectors[8].

A recent experiment at ATLAS[1] was designed to minimize systematic uncertainties which may have affected the previous α measurements. A beam of 27 MeV ^{8}B ions produced with the In-Flight technique was implanted into the center of a planar Si detector, which eliminated energy loss corrections due to catcher foils and detector dead layers and allowed the energy deposited by both α- particles to be observed with a single detector. Beta-particle energy deposition was minimized by the use of a thin (91 μm) Si detector, sufficient to stop the most energetic α-particles, and by the requirement of a β coincidence in a plastic scintillation detector to define the angles of the emitted betas. For calibration, ^{20}Na ions were implanted into the same detector immediately before the ^{8}B measurement. The β-decay of ^{20}Na proceeds 20% of the time to α-unstable levels in ^{20}Ne and provides lines in the region of the ^{8}B α-spectrum peak.

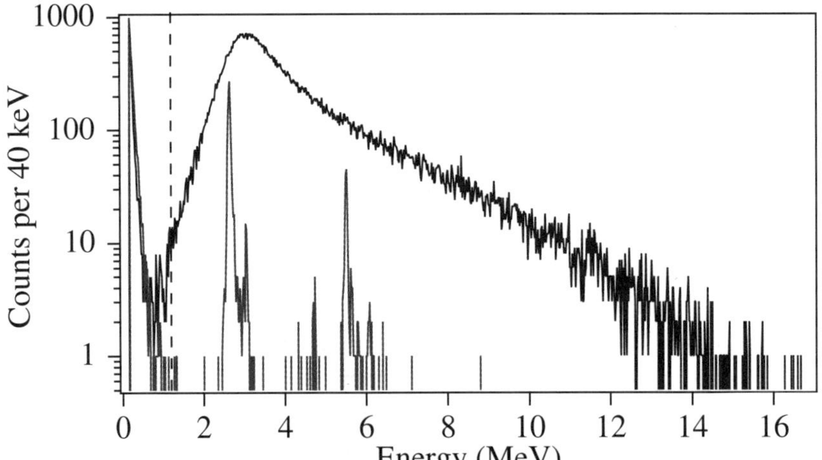

Figure 1. The measured ^{8}B β-delayed alpha spectrum (blue) shown with the β-delayed alpha lines from the ^{20}Na calibration measurement(red).

The ^{6}Li beam was cycled (1.5 sec on/1.5 sec off) and data taken only during the beam-off cycle. With an average ^{6}Li current of 60 pnA about 3 ^{8}B ions/sec were implanted. Energy signals and the relative timing between the Si and β detectors were recorded, as well as the timing of the Si signal with respect to the beam-off cycle. Over three days, 4.5×10^5 ^8B events were observed, 16% of which were in coincidence with a β event in the scintillator.

The α spectrum from this measurement is given in Fig. 1 (blue line) in comparison with the lines (shown in red) obtained from the ^{20}Na calibration experiment. Analyzing the α spectrum data within the R-matrix formalism[12] gave the parameters which are summarized in the last row of Table 1[7]. These values are in excellent agreement with the results obtained from an $\alpha - \alpha$ phase shift analysis[6] and should help in reducing the experimental uncertainties for this state.

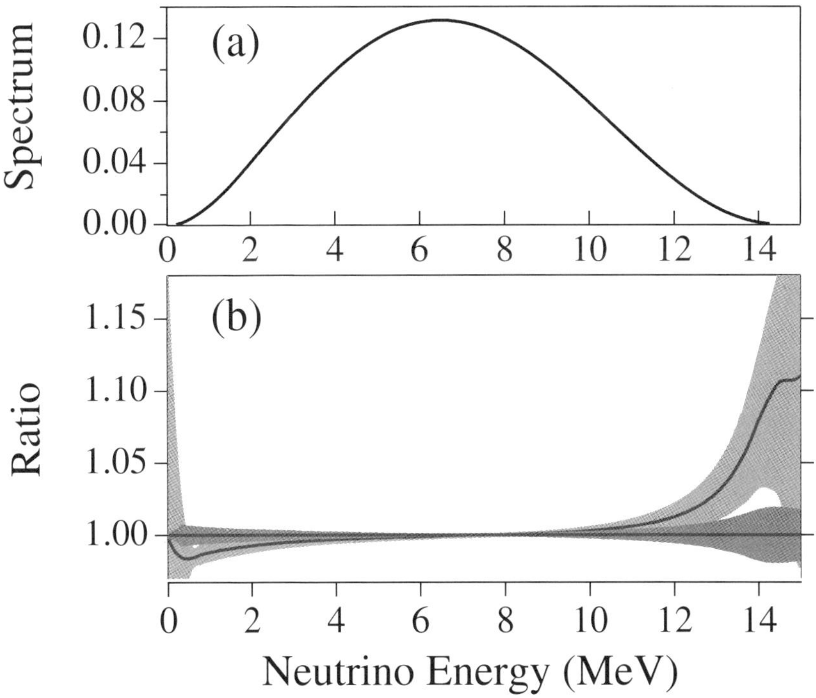

Figure 2. (a) Normalized neutrino spectrum deduced from the experiment. (b) The red line is the ratio between the neutrino spectrum recommended in Ref. 13 to the spectrum obtained in this work. The blue line represents the spectrum deduced here. The bands indicate the $\pm1\sigma$ experimental uncertainties.

This experiment also provided a new ^{8}B neutrino spectrum, presented in Fig.2. This spectrum is an important input parameter for the interpretation of the experiments that detect energetic neutrinos from the Sun. More details about this part of the experiment can be found in Ref.[1].

3.2. *The Charge Radius of ^{6}He*

Some of the most basic observables of a nucleus are its mass and size. While for stable nuclei many high precision data for both variables exist high accuracy measurements for unstable nuclei are still challenging. The nucleus ^{6}He is of particular interest, because two of its neutrons are loosely bound and form a 'halo' with considerably larger radial extent than the alpha-particle core. The charge distribution will reveal this halo character because the motion of the center-of-mass reflects the radial extent of the neutrons as well as the correlations between them.

We have performed a high precision measurement of the isotope shift (IS) of ^{6}He with respect to ^{4}He. Including QED and relativistic corrections the IS is calculated to be[14]:

$$IS(2^3S_1 - 2^3P_2) = 34473.625(13) + 1.210(< r_c^2 >_{He4} - < r_c^2 >_{He6})MHz$$

From this equation and the rms radius of ^{4}He obtained from muonic ^{4}He atoms[15] the charge radius of ^{6}He can be calculated.

The ^{6}He atoms were produced in a hot graphite target via the ^{12}C(^{7}Li,^{6}He)^{13}N reaction with a 60 MeV ^{7}Li beam from the ATLAS accelerator at Argonne National Laboratory. Neutral ^{6}He atoms diffused out of the target, were excited in an RF discharge into the metastable 2^3S_1 level and transferred into a magneto-optical trap (MOT). For the detection and spectroscopy of a trapped atom the (2^3S_1 - 2^3P_2) transition at a wavelength of 389 nm was chosen. The stability of the setup was tested extensively by performing laser spectroscopy on ^{4}He atoms. Based on the weighted average of 18 independent measurements, the isotope shift between ^{6}He and ^{4}He on the 2^3S_1 - 2^3P_2 transition was determined to be 43194.772±0.056 MHz. From this value a point-proton radius of ^{6}He of 1.912 4±0.018 fm was derived.

In Fig. 3 the experimental and theoretical values of the point-proton radius of ^{6}He are compared. The two earlier experimental values, extracted from nuclear collision experiments, require a description of the interaction and a model for the nucleon distribution and are all considerably less ac-

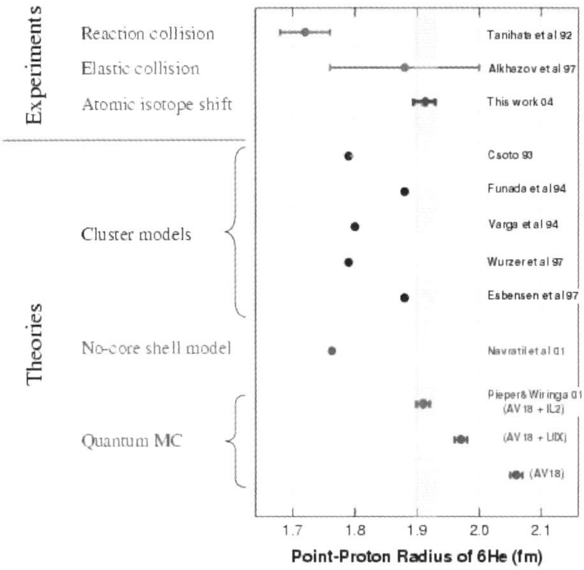

Figure 3. Experimental and theoretical values for the ^{6}He charge radius. The experimental and theoretical values were taken from (top to bottom) Refs.[16,17,2,18,19,20,21,22,23,24], respectively.

curate. The value obtained via the isotope shift represents the first model-independent determination. Among the theoretical calculations some of the cluster model calculations agree within 2% with the experimental value. The same holds for the ab-initio calculation based on the AV18 two-body potential and the IL2 three body force[24,25]. More details about this experiment may be found in Refs.[26,2]

4. R&D at ANL towards a future RIA Facility

The experiments described in the previous section were all performed with very low beam intensities. Considerable improvements can be achieved when beams from the next generation radioactive beam facilities will become available. In the US the Rare Isotope Accelerator (RIA) has obtained the highest priority for new construction in the latest Long Range Plan of the Nuclear Science Advisory Committee. The enhanced capabilities at

RIA come from a combination of factors:

• A powerful superconducting linear accelerator capable of accelerating all stable ions from hydrogen to uranium to energies above 400 MeV/u.

• The use of a variety of production techniques optimized to achieve the highest yields for each isotope.

• A very efficient post acceleration scheme based on the superconducting linac technology with a very low q/m injector.

To achieve these goals a vigorous R&D program is presently going on at numerous laboratories[27]. A few of the topics studied at Argonne National Laboratory are briefly outlined in the following paragraphs.

4.1. *Multiple Charge State Acceleration*

The primary beam intensity requirements at RIA specify a beam power of at least 100 kW for beams of all masses. While this is achievable with existing ion source technologies for lighter-mass beams, heavier beams (e.g. uranium) require the simultaneous acceleration of multiple charge states through the driver accelerator. A concept for this technique has been developed at Argonne and was subsequently tested at the existing ATLAS facility[28]. The multi-charge state operation is made feasible by the large transverse and longitudinal acceptance, which can be obtained in a linac with superconducting cavities. This not only provides a substantial increase in beam current, but also enables the use of two stripper sections in the driver accelerator, thus reducing the length of the linac required. This concept has now been accepted as the basis of the RIA driver accelerator.

4.2. *Liquid Lithium Target Technology*

The high intensity of the heavy ion beams produced by the RIA driver linac results in very high energy densities that solid materials can not survive. For these cases a liquid lithium target has been proposed as a solution. A windowless lithium jet target with a dimension of 5mm x 10 mm has been developed at Argonne[29]. For a 200 kW uranium beam the energy deposition is about 2 MW/cm^3. Calculations have shown that a high intensity electron beam can produce a similar energy deposition in lithium. Because the high energy density can be the main cause for a jet disturbance a test experiment with a liquid lithium jet and a high power electron beam was performed. A 20 mA electron beam with an energy of 1 MeV from a high power Dynamitron accelerator with a beam spot of 1 mm in diameter was focused on a 5 mm x 10 mm flowing lithium jet target. The energy deposition under this condition simulates the one expected for a 200 kW uranium beam. Fig. 4 shows a picture of the liquid lithium target interacting with the high power electron beam. No disturbance of the lithium jet through

Figure 4. Liquid lithium jet target bombarded with a 15 kW electron beam (See Ref.28 for details).

the high energy deposition of the electron beam was observed in these test experiments.

5. Acknowledgments

The experiments described in this contribution were performed in collaboration with I. Ahmad, K. Bailey, J. P. Greene, A. Heinz, D. Henderson, R. J. Holt, R. V. F. Janssens, C. L. Jiang, Z.-T. Lu, E. F. Moore, G. Mukherjee, P. Mueller, T. P. O'Connor, R. C. Pardo, T. Pennington, G. Savard, J. P. Schiffer, D. Seweryniak, X. D. Tang, L.-B. Wang and G. Zinkann (ANL), M. Paul (Jerusalem, Israel), W. T. Winter and S. J. Freedman (LBNL) and G. Drake (Windsor, Ontario). This work was supported by the US D.O.E. Office of Nuclear Physics, under contract No. W-31-109-ENG-38.

References

1. W. T. Winter et al., *Phys. Rev. Lett.* **91**, 252501 (2003).
2. L.-B. Wang et al., to be publ.,
3. B. Harss et al., *Rev. Sci. Instrum.* **71**, 380 (2000).
4. F. Ajzenberg-Selove, *Nucl. Phys.* **A506**, 1 (1988).
5. J. H. Kelley et al., *Nucl. Phys.* to be publ., www.tunl.duke.edu/nucldata.
6. E. K. Warburton, *Phys. Rev.* **C33**, 303 (1986) and references therein.
7. W. T. Winter, S. Freedman, K. E. Rehm and J. P. Schiffer, to be publ.
8. J. N. Bahcall et al., *Phys. Rev.* **C54**, 411 (1996).
9. B. J. Farmer and C. M. Class, *Nucl. Phys.* **15**, 626 (1960).
10. L. De Braeckeleer et al., *Phys. Rev.* **C51**, 2778 (1995).
11. D. H. Wilkinson and D. E. Alburger, *Phys. Rev. Lett.* **26**, 1127 (1971).
12. F. C. Barker, *Aust. Journ. Phys.* **22**, 293 (1969).
13. C. E. Ortiz et al., *Phys. Rev. Lett.* **85**, 2909 (2000).
14. G. W. F. Drake, *Nucl. Phys.* **737c**, 25 (2004).
15. E. Borie and G. A. Rinker, *Phys. Rev.* **A18**, 324 (1978).
16. I. Tanihata et al., *Phys. Lett.* **B289**, 261 (1992).
17. G. D. Alkhazov et al., *Phys. Rev. Lett.* **78**, 2313 (1997).
18. A. Csoto, *Phys. Rev.* **C48**, 165 (1993).
19. S. Funada, H. Kameyama and Y. Sakurai, *Nucl. Phys.* **A575**, 93 (1994).
20. K. Varga, Y. Suzuki, and Y. Ohbayasi, *Phys. Rev.* **C50**, 189 (1994).
21. J. Wurzer and H. M. Hofmann, *Phys. Rev.* **C55**, 688 (1997).
22. H. Esbensen, G. F. Bertsch and K. Hencken, *Phys. Rev.* **C56**, 3054 (1997).
23. P. Navratil et al., *Phys. Rev. Lett.* **87**, 172502 (2001).
24. S. C. Pieper and R. B. Wiringa, *Ann. Rev. Nucl. Part. Sci.* **51**, 53 (2001).
25. The authors of Ref.24 have recently noticed a possible problem in the extraction of rms radii values and associated uncertainties from the QMC calculations and are currently making further investigations.
26. P. Mueller et al., *Nucl. Instr. Meth.* **B204**, 536 (2003).
27. J. Nolen, *Nucl. Phys.* **A734**, 661 (2004).
28. P. Ostroumov et al., *Phys. Rev. Lett.* **86**, 2798 (2001).
29. C. B. Reed, J. A. Nolen, J. R. Specht, V. J. Novick and P. Plotkin, *Nucl. Phys.* subm.

KEY TOPICS IN NUCLEAR STRUCTURE
Proceedings of the 8th Int. Spring Seminar on Nuclear Physics
edited by Aldo Covello
© 2005 World Scientific Publishing Co.

FIRST EXPERIMENTS WITH REX-ISOLDE AND MINIBALL

D. SCHWALM

Max-Planck Institute for Nuclear Physics
Saupferch-Eckweg
D-69117 Heidelberg, Germany
E-mail: d.schwalm@mpi-hd.mpg.de
for the REX-ISOLDE and MINIBALL-Collaboration

The Radioactive beam EXperiment (REX) at ISOLDE (CERN) was conceived as a pilot experiment to demonstrate a novel technique to bunch, charge breed and accelerate radioactive ions produced by ISOL facilities. The experiment was successfully commissioned in 2002 and promoted to a CERN user facility at the end of 2003. First measurements were performed at 2.2 MeV/u in summer 2003 using the MINIBALL array, which consists of 24 individually encapsulated, 6-fold segmented high-purity Ge-detectors and allows to detect de-excitation γ-rays with high efficiency *and* high energy resolution. The REX-ISOLDE facility together with the MINIBALL array offers new and unique possibilities to study collective and single-particle properties of nuclei far off stability with standard nuclear physics tools such as (safe) Coulomb excitation and single-nucleon transfer reactions in inverse kinematics, as well as with other well established experimental methods. Due to the current energy range of REX-ISOLDE (2004: 3.0 MeV/u) the present experimental programme is concentrated on light to medium heavy, neutron rich nuclei. After an introduction to REX-ISOLDE and MINIBALL, the high potential and physics opportunities offered by the new facility are exemplified by first results obtained for Mg isotopes around the "island of inversion" at N = 20.

1. REX-ISOLDE

REX-ISOLDE is a first-generation Radioactive Ion Beam (RIB) facility[1,2], developed and realized by a European collaboration at CERN, to study the structure of nuclei far-off stability and to demonstrate a novel and efficient technique for accelerating exotic ions; it takes advantage of the availability of $\approx$ 800 radioisotopes from about 70 elements delivered as 1^+ ions by the online mass separators at the PS-ISOLDE[3]. These ions are first accumulated, cooled and bunched in a Penning trap (REX-TRAP[4]) and then injected into an Electron Beam Ion Source (REX-EBIS[5]), where their charge state is increased by bombarding them with an intense electron beam. While the EBIS is breeding the ions to higher charge states such that the resulting A/q allows an effective acceleration, the time is used by the trap to accumulate ions for the next EBIS filling. This novel and effective concept of a charge state breeder has been successfully taken into operation in 2001. The beam extracted from the EBIS is then purified from restgas contaminations by a mass separator[6] and finally injected into a short linear accelerator (REX-LINAC[1] consisting of an RFQ, an IH-structure, and three 7-gap resonators) and accelerated up to energies of 2.2 MeV/u. In summer 2004 an additional 9-gap IH-structure[2] will be available, which will raise the final beam energy to 3.1 MeV/u. After a momentum analysis with a dipole magnet the beam is guided to one of the two target stations. A schematic view of REX-ISOLDE is shown in Fig. 1.

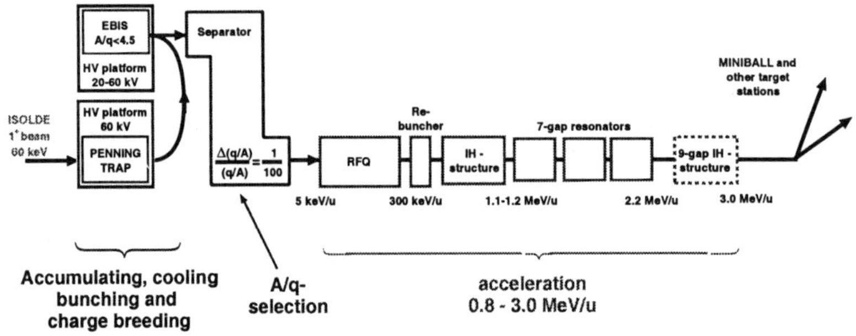

Figure 1. Schematic view of REX-ISOLDE (Final energies of 3.0 MeV will be available in summer 2004)

1.1. *The Charge State Breeder*

Following the different components of the charge state breeder defined by REX-TRAP, REX-EBIS, and the achromatic A/q separator, the singly charged ions of 60 keV are first retarded by the REX-TRAP platform potential of nearly 60 keV and injected continuously into the trap, where they are finally slowed down in the buffer gas (typical gas pressures are 10^{-3} mbar), accumulated and cooled. Besides buffer gas cooling sideband cooling techniques are employed[7] to further improve the transversal phase space of the accumulated ion cloud. After an accumulation time of several milliseconds the ions are extracted from the trap in a short bunch, re-accelerated to 60 keV and transfered to REX-EBIS. The transfer line provides differential pumping to decouple the buffer gas of REX-TRAP from the REX-EBIS vacuum, which is required to be in the range of 10^{-11} mbar to assure a proper functioning of the source and to minimise rest gas contaminations. The transmission through REX-TRAP depends on the number of ions accumulated in the source; for less than 10^5 ions per bunch it amounts to about 40%, declining to 10% for 10^7 ions per bunch due to space charge effects.

The ion bunch transfered from REX-TRAP is retarded again by the platform potential of REX-EBIS and then trapped and ionized by the electron beam. The time $t_b(q)$ needed to breed charge states of $q \geq A/4.5$, required by REX-LINAC, depends on the electron energy and is proportional to the inverse of the electron beam density. For $^{30}Mg^{7+}$, e.g., it takes about 15 ms at an electron energy of 5 keV and a current density of 150 A/cm^2, and about 20% of the injected $^{30}Mg^{1+}$ ions can be extracted as $q = 7^+$ ions. The number of ions REX-EBIS can handle per breeding cycle is expected to be $10^{10}/q$ and is thus not limiting the intensity of REX-ISOLDE.

After the charge breeding the ions are extracted as $\approx$ 100 μs long bunches and injected via the S-shaped separator into the RFQ of the LINAC. The measured $(q/A)/\Delta(q/A)$ resolution of the separator is about 100 at a transmission of 80%. Due to the required low injection energy of 5 keV/u the EBIS platform potential has to be lowered before the extraction. While this can be achieved during the breeding time, the platform has to be upcharged again before the next filling of the source can take place. As this takes about 5 ms, the cycling time of the charge breeder - and thus also the accumulation time of REX-TRAP - is determined by 5 ms+$t_b(q)$; accordingly, for the case of $^{30}Mg^{7+}$ the cycling time was 20 ms. As the minimum charge state required for further acceleration is proportional to A, the cycling times will increase strongly for heavier ions. To reduce the breeding times for these ions, which will finally limit the intensities of the radioactive beams available at REX-ISOLDE, in particular the current density of the electron beam of REX-EBIS has to be increased and the number of ions that can be accumulated in REX-TRAP has to be optimised; a corresponding upgrade program has already been started[2].

Cycling times of the charge breeder of $\geq$ 20 ms and ion pulse lengths of $\approx$ 100 μs are well within the operation parameters of the REX-LINAC, which allows repetition rates of 1-100 Hz and a duty factor of up to 10%.

1.2. *The REX-LINAC*

The linear accelerator of REX-ISOLDE consists of three (2004 of four, see below) different types of resonator structures to provide the required acceleration voltages in a most efficient way and to meet the requirements of the experiments for variable beam energies. The RFQ and IH-structure take the extracted ions from 5 keV/u to intermediate energies of 0.3 MeV/u and 1.2 MeV/u, respectively, from where they can be post accelerated or decelerated by three 7-gap resonators to variable energies between 0.8 MeV/u and 2.2 MeV/u. Due to the charge breeder the total length of the LINAC amounts only to 9 m (2004: 12 m). The RFQ, IH, and 7-gap structures operate at 101.28 MHz, i.e. half of the CERN proton linac frequency.

The 4-rod RFQ is mainly a blue copy of the RFQ[8], which was developed and built for the High Current Injector[9] at the MPI-K in Heidelberg. It employs the MPI-K solution for constructing the modulated rods and for securing their effective cooling. The rf quadrupole field of the structure provides transversal focusing while the modulation performs a smooth bunching of the injected macropulse into micropulses of 10 ns distance and accelerates the ions. In order to secure efficient adiabatic bunching and optimum output emittances, a rather low injection energy of 5 keV/u was chosen.

The Interdigital-H-type(IH) resonator[1] containing a drift-tube structure and a magnetic quadrupole triplet lens is a derivative of a structure developed by GSI for the Pb-Linac at CERN[10]. A new feature is the possibility to decrease the final energy from 1.2 and 1.1 MeV/u to be able to decelerate the beam down to 0.8 MeV/u using the subsequent 7-gap resonators.

The three 7-gap resonators[11] are similar to those developed for the High Current Injector of the MPI-K in Heidelberg[9]. These special types of spiral resonators are designed and optimised for synchronous velocities of 5.4, 6.0, and 6.6% of the velocity of light and compromise between maximum acceleration voltage and maximum flexibility in the transit time factor. The 7-gap resonators allow to vary the final energy of the beam continuously between 0.8 and 2.2 MeV/u.

Several light ion beams have been successfully accelerated to 2.2 MeV/u and transported through the analysing magnet to the target, including the radioactive species ^{9}Li,$^{24-26}$Na and ^{30}Mg. The transmission efficiency through the LINAC to the target was about 80%, which results in an overall efficiency of REX-ISOLDE (defined by the number of ions reaching the target as compared to the number of 1^{+}-ions in front of REX-TRAP) of

$\approx 5\%$. The energy spread of the beam is about 1.5%, while the width of the micropulses is expected to be of the order of a few ns.

The heaviest radioactive ion that has been accelerated so far is ^{88}Kr, however, the final energy of 2.2 MeV/u is below the Coulomb barrier for all targets. To expand the useful mass range of REX-ISOLDE, an upgrade program has been launched. As a first step an additional 9-gap IH-cavity has been inserted in the drift space between the last 7-gap resonator and the bending magnet, which will be ready in summer 2004 and allows final energies of 3.1 MeV/u. The new resonator operates at an eigenfrequency of 202.56 MHz and is a redesign of a prototype of an IH-structure planned for the MAFF accelerator[12]. Together with a foreseen improvement in breeding times of the EBIS, radioactive ions with masses up to $A \approx 100$ will then be available for nuclear and astrophysical as well as solid state experiments.

2. The MINIBALL Array

One of the main tools in nuclear structure studies at energies around the Coulomb barrier has been high resolution γ-ray spectroscopy. To adapt this tool to investigations at rare isotope facilities two eminent problems have to be solved: Due to the very low intensities available it is important to detect γ-rays with high efficiency and thus also with a large solid angle coverage. Consequently it is desirable to use large volume, high resolution Ge detectors close to the target. On the other hand, standard low energy reactions like Coulomb excitation and few nucleon transfer reactions have to be performed in inverse reaction kinematic, which results in large recoil velocities of about 5% c of the γ-decaying nuclei and thus in large Doppler-shifts and -spreads. For this reason it would be important to keep the solid angle subtended by each Ge detector small enough to achieve a good angular resolution so that the Doppler spreads can be kept small and Doppler shift corrections can be made.

The requirement of high efficiency and high granularity at reasonable costs cannot be reconciled using conventional Ge detectors. This has forced us to develop and realize the MINIBALL array[13,14], which is based on the new technology of position sensitive Ge detectors. MINIBALL is the first fully operational spectrometer which uses segmented Ge detectors and pulse shape analysis to determine the entry position of the γ-ray into the detector.

MINIBALL consists out of 24 6-fold segmented, individually encapsulated Ge detectors[15], which are mounted in 8 triple cryostats. The segmentation is achieved by subdividing the outer contact as shown in fig. 2.

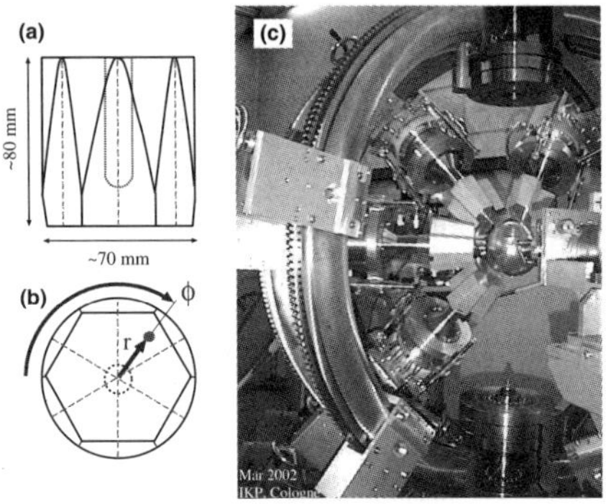

Figure 2. (a) Side and (b) front view of the 6-fold segmented Ge detector together with the definition of r and ϕ of the main interaction point. The dividing lines of the outer contact are shown by the dashed lines. (c) The MINIBALL frame with four of the eight triple clusters mounted.

While the current signal at the core electrode is proportional to the energy left by the γ-ray in the crystal as in a conventional detector, the current signal seen by the segment contact allows to measure the fraction of the energy that was deposited in the triangular shaped segment and thereby to determine the segment in which the highest energy deposition (main interaction) took place. The current signals of the core and the 6 segments are integrated by preamplifiers and digitised by the 40 MHz Flash ADCs of the XIA DGF-4C module[16]. All data processing is from then on digital, first in a field programmable gate array (FPGA) and then in a digital signal processor (DSP), which also performs the user specific pulse shape analysis. As shown in fig. 2c the triple clusters are mounted in a versatile frame which allows the realisation of various detector configurations. In the most dense package (detector-target distance 11 cm) the full energy peak efficiency is about 10% at 1.3 MeV.

The granularity of the detector is increased by determining on an event-by-event basis the (r, ϕ)-coordinates of the main interaction (MI) of the γ-ray in the crystal; simulations have shown[17,18] that the MI point is a reasonable approximation of the first interaction point needed to reconstruct the original direction of the γ-ray. In fact, according to our simulations the mean transversal distance of the MI from the original γ-ray direction is < 5 mm for 350 keV γ-rays and decreases for smaller and higher energies. The distance r of the MI point from the central core of the detector and its azimuthal angle ϕ (see fig. 2) are determined in the following way[18,19]:

(i) r is proportional to the drift time of the electrons from the MI point to the central contact, which is deduced from the steepest-slope time t_{ss} of the core signal - see fig. 3.

(ii) ϕ is given by $\phi = n \cdot 60^0 + \varphi$, where n is the segment number containing the largest energy fraction and φ depends linearly on $\log(|Q_{n-1}/Q_{n+1}|)$, where Q_{n-1} and Q_{n+1} are the heights of the induced charge signals in the two neighbouring segments which are caused by the drifting electrons and holes in segment n, as illustrated in fig. 4.

The core and segment energies, the starting times of the signals, the steepest slope times, as well as the induced charges are determined in real time by the FPGA and DSP of the DGF-4C module[20] and stored for later off-line calibration and analyses. Under the ideal condition present in the measurements shown in fig.s 3,4 position resolutions of ± 5 mm can be achieved for r and the arc length $(r \cdot \phi)$, which corresponds to a granularity increase of a factor of ≈ 100 as compared to an unsegmented detector. For a triple cluster and a target detector distance as small as 11 cm one has to take into account that the detector axes are no longer pointing to the target spot but are slightly misaligned. Nevertheless, the gain in granularity that

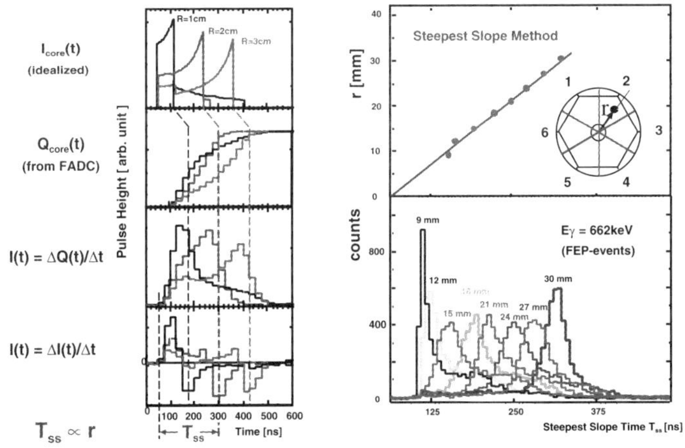

Figure 3. The steepest slope method. Left: The distance r is proportional to the drift time of the electrons from the main interaction point to the core. Their arrival at the core causes a steep drop of the current pulse at T_{ss}, which can be deduced from the second derivative of the digitised charge pulse. Right, bottom panel: Measured steepest slope times when irradiating the detector perpendicular to the detector surface at different positions r with a well collimated γ-source of 662 keV. Only full-energy-peak (FEP) events are analysed (from Ref. 19). Right, top panel: Calibration curve.

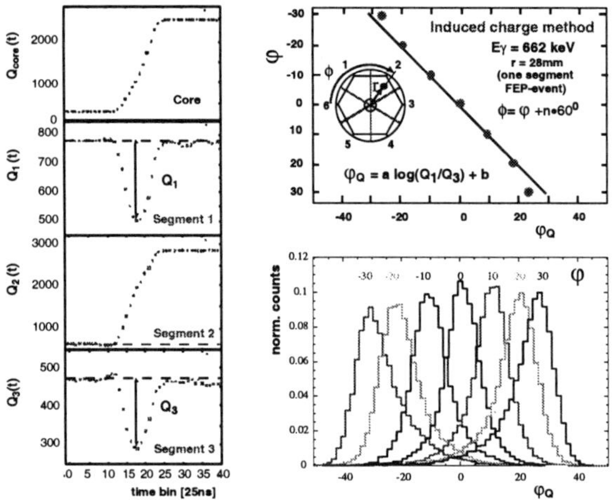

Figure 4. The induced charge method. Left: Measured signal traces of a 662 keV full-energy-peak (FEP) event, where the full energy was deposited in segment 2. The moving charges in segment 2 induce transient charge signals on the contacts of segment 1 and 3 of height Q_1 and Q_3 (For events where also energy is deposited in one or both segments neighbouring the main interaction segment, the induced charge pulse heights Q can still be deduced employing a simple algorithm). Right, bottom panel: Measured values $\varphi_Q = a \log |Q_1/Q_3| + b$ for fixed values of a and b when irradiating the detector perpendicular to the detector surface at $r = 2.8$ cm and various φ angles with a well collimated γ-source of 662 keV (from Ref. 18). Right, top panel: Final calibration curve.

can be achieved by a careful calibration is still larger than 50; in fact, it could be demonstrated by observing a 2.2 MeV γ-ray at 90^0 with respect to the γ-emitting nucleus of 5.6% c, that the energy resolution increases from 35 keV, obtained by considering only the individual positions of the three cluster detectors in the Doppler correction, to about 5 keV FWHM by exploiting the position sensitivity[19].

The MINIBALL set-up is supplemented by a double-sided Si strip detector[21] in form of a CD with 24x4 sector strips and 16 annual strips subdivided in four quadrants, which allows to detect reaction products under forward angles between about 17^0 to 53^0, thus supplying the necessary additional information needed to complete the Doppler shift correction. For the detection of light particles, the 500μm thick detector can be backed by another Si detector to provide a ΔE-E telescope for particle identification.

3. First Experiments

The experimental programme with REX-ISOLDE and MINIBALL started - after some preliminary runs performed already during the commissioning of the facility in 2002 - during the summer campaign 2003 with an investigation of neutron-rich Na and Mg isotopes[22].

This region of the nuclear chart around the magic neutron number $N = 20$ and has attracted considerable experimental and theoretical interest since the discovery of Thibault *et al*[23] in 1975 that the Na isotopes ^{31}Na and ^{32}Na are more tightly bound than predicted by s-d shell model calculations. This was later explained[24] by noticing that for $Z < 14$ and neutron numbers around 20 the $f_{7/2}$ orbit intrudes into the s-d shell at large deformations, which eventually leads to strongly prolate deformed ground-states of these neutron-rich nuclei. But even though the later discovery of a low lying first excited 2^+-state in ^{32}Mg and the observation of large $B(E2, 0^+ \rightarrow 2^+)$-values in ^{30,32,34}Mg by intermediate energy Coulomb excitation experiments[25,26,27,28] have corroborated the large deformation of these nuclei, experimental information on the structure of nuclei in this "island of inversion" is still rather meager. Ground state properties and level energies are so far mainly known from β-decay studies only; thus the resulting level schemes are unlikely to be complete, and only with a few exceptions are spins and parities of the excited states known experimentally. Moreover, the data on the $B(E2)$-values of ^{30,32,34}Mg are very imprecise and measurements of different laboratories disagree by as much as a factor of two (see also fig. 7), indicating that systematic errors in these experiments are not completely understood. The experimental validation of theoretical results obtained from modern nuclear structure calculations attempting to describe nuclei at large isospins is thus severely hampered by the lack of precise experimental data on the single particle and collective properties of these nuclei.

Preliminary results obtained with a beam of ^{30}Mg, which contains 4 more neutrons than the heaviest stable Mg isotope ^{26}Mg, will be used in the following to exemplify the new opportunities opened up by REX-ISOLDE and MINIBALL for a systematic study of properties of low lying states in radioactive isotopes using standard nuclear physics tool like inverse (d,p) reactions or "safe" Coulomb excitation. For these investigations, the ^{30}Mg atoms, which have a halflife of only 335 ms, were ionised to ^{30}Mg^{1+} in the laser ion source of ISOLDE to avoid isobaric impurities in the extracted ion beam, in particular of the more copiously produced ^{30}Al. They were then charge bred to $q = 7^+$ and accelerated to 2.25 MeV/u. The intensity of the ^{30}Mg beam at the target was about $2 \cdot 10^4\,\mathrm{s}^{-1}$; laser-on/laser-off measurements did not reveal any beam impurities within the accuracy of the measurement (<10%).

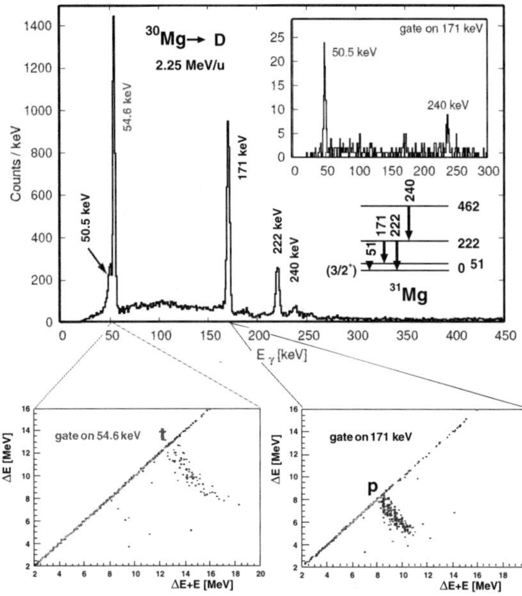

Figure 5. Upper panel: γ-spectrum following the reaction of ^{30}Mg with Deuterium. The spectrum was taken in coincidence with a charged particle in the Si telescope at a beam current of $\approx 2 \cdot 10^4\,\mathrm{s}^{-1}$ in a 3 day run. Insert: $\gamma-\gamma$-coincidence spectrum with a gate on the 171 keV transition. Lower panels: $\Delta E - E$ plots of the Si telescope data with gates on the 54.6 keV and 171 keV transition, respectively.

Fig. 5 shows the low energy part of a γ-ray spectrum observed when bombarding a deuterated polyethylene target (C_3D_6) of about 1 mg/cm^2 for about 3 days. The marked lines are due to the 1n-transfer and 1n-pickup reaction D(^{30}Mg,^{31}Mg)p and D(^{30}Mg,^{29}Mg)t, respectively. Note that the spectrum, which was recorded in coincidence with a charged particle in the Si detector, is almost background free despite of the radioactivity of the projectile and its daughter products, and that it can be measured down to $\approx$ 50 keV due to the very low threshold that can be achieved with the DFG-4C module. The observed γ-lines can be easily attributed to one of the two reaction channels by setting gates on the line and displaying the ΔE vs. E information provided by the Si telescope (see lower panels of fig. 5): While the strong line at 54.5 keV is found to be due to the decay of ^{29}Mg populated in the pickup reaction, the other four marked lines in fig. 5 clearly belong to ^{31}Mg populated in the 1n-transfer channel. This is further corroborated by the $\gamma - \gamma$-coincidence spectrum obtained by gating on the 171 keV line (insert in fig. 5), which is - due to the high FEP efficiency of MINIBALL - still intense enough to observe and identify even the weak 240 keV line. To gain information on the multipolarity of the transitions and thus on the spins of the involved states, the angular distributions of these lines are presently being analysed together with the γ-lines which can be attributed to the decay of higher lying states in ^{31}Mg. Moreover, the energies of the entry states in ^{31}Mg can be deduced by analysing the charged particle energies and their directions.

To Coulomb excite the lowest lying 2^+ state of ^{30}Mg at 1482 keV, a 1 mg/cm^2 a thick natural Nickel target was bombarded and γ-rays from the deexcitation of the projectile as well as of the nickel isotopes were observed in coincidence with either the projectile or the recoiling target nuclei using the Si-detector. The detected Mg projectiles, which belong to cm scattering angles of $22^0 < \theta_{cm} < 74^0$, could be easily distinguished from the detected Ni recoils, corresponding to scattering angles of $78^0 < \theta_{cm} < 150^0$, by means of their higher kinetic energies at the same laboratory scattering angle. In fig. 6 the γ-spectra resulting from a 3 day run are displayed separately for the two θ_{cm}-regions and for the two possible Doppler shift corrections, i.e. assuming that the γ-ray was emitted form the projectile (fig. 6a,b) or from the recoiling target nucleus (fig. 6c,d), respectively. While the γ-spectra measured in coincidence with forward scattered ^{30}Mg (fig. 6a,c) only display transitions from the decay of the projectile and the dominant target isotopes ^{58,60,62}Ni, the spectra taken in coincidence with backscattered ^{30}Mg show strong lines from the decay of the odd nickel isotopes ^{59,61}Ni, obviously produced by an 1n-pickup by ^{58,60}Ni (fig. 6a,c). While this might be surprising at first glance because even for $\theta_{cm} = 150^0$ the distance between the two nuclear surfaces is still larger than $d_{surf} = 3.7$ fm[a], the strong transfer is symptomatic for the long tail of the neutron density distributions of neutron rich nuclei. Analysing only the spectra measured in coincidence with forward scattered ^{30}Mg, which correspond to *safe*[29]

[a]The surface distance d_{surf} is defined by $d_{surf} = D(\theta_{cm}) - R_{proj} - R_{targ}$ with R[fm]= $1.28A^{1/3} - 0.76$ and $D(\theta_{cm})$ denoting the distance of closest approach[29].

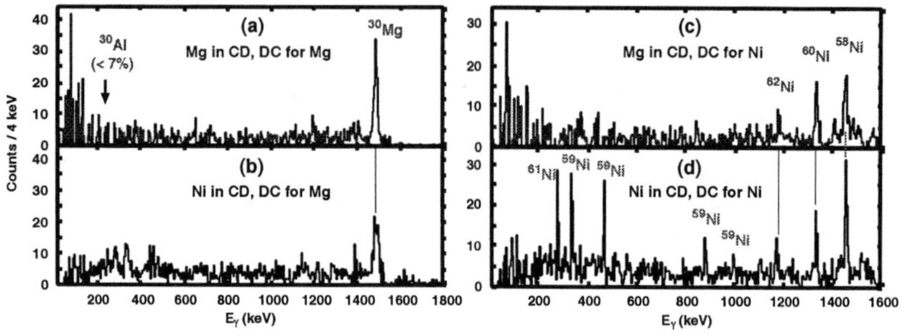

Figure 6. γ-spectra observed in the bombardment of a ^{nat}Ni target with a ^{30}Mg beam of 2.25 MeV/u and an intensity of $2 \cdot 10^4\,\mathrm{s}^{-1}$. The data were taken in a 3-days run requiring coincidences with a projectile or a recoiling target nucleus in the CD-like Si detector. Doppler corrected spectra assuming the γ- emitting nucleus to be Mg-like are shown in (a) when a projectile and in (b) when a target nucleus was detected in the CD detector. In (c) and (d) the corresponding spectra are displayed assuming target-like nuclei to emit the observed γ-rays.

surface distances of $7.1\,\mathrm{fm} < d_{surf} < 26.6\,\mathrm{fm}$, the $B(E2, 0^+ \to 2^+)$-value of ^{30}Mg could be extracted relative to the well known $B(E2, 0^+ \to 2^+)$-values of the three Ni isotopes[30] from the intensity of the Mg- and Ni-lines using well established Coulomb excitation codes. The analysis results in a preliminary value of $B(E2, 0^+ \to 2^+) = 201^{+43}_{-23}\,\mathrm{e}^2\mathrm{fm}^4$ (ref. 31). While the lower error is statistical only, the upper error was increased to allow for possible isobaric beam impurities of up to 10%. To ensure this upper limit, several tests were performed beside the already mentioned laser-on/laser-off measurements by looking for time dependences of the Mg to Ni intensity ratio with respect to the proton pulse reaching PS-ISOLDE, exploiting the short lifetime of ^{30}Mg, and with respect to the beginning of the REX-EBIS bunch, as well as searching for excitations of isobaric impurities such e.g. ^{30}Al.

Our present $B(E2, 0^+ \to 2^+)$-value is displayed together with the values measured previously for the heavier Mg isotopes in fig. 7. Note, that the MSU[26] and the GANIL[27] results for ^{30}Mg are larger than our value by about 50% and 100%, respectively. The reason for these discrepances are not yet understood; however, while the present result is based on the well established nuclear structure tool of "safe" Coulomb excitations and is moreover expected to be rather insensitive to systematic experimental errors due to the relative measurement of the projectile to the target excitation, the previous results are deduced from measurements using inter-

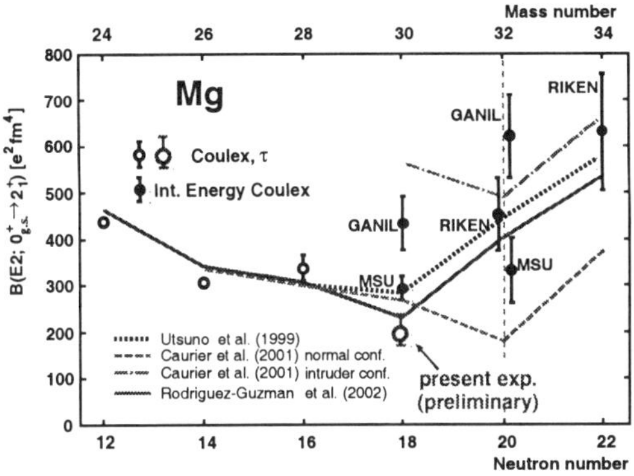

Figure 7. Experimental and theoretical $B(E2, 0^+ \rightarrow 2^+)$ values for the even Mg isotopes with $24 \leq A \leq 34$ including the present result obtained at REX-ISOLDE/MINIBALL by Niedermaier et al. [31]. The experimental values for ^{24,26,28}Mg obtained via safe Coulomb excitation and lifetime measurements are taken from ref. [30], those deduced from Intermediate Energy Coulomb excitation measurements are from ref. [25, 26, 27, 28]. The theoretical results are from ref. [32, 33, 34]

mediate beam energies of about 50 MeV/u, which require large corrections to account for nuclear contributions to the excitation of the 2^+-state and for the feeding from higher lying states, which can only be obtained from rough theoretical estimates. Fig. 7 also compares the experimental values to the results of recent theoretical investigations; it is obvious that precise $B(E2, 0^+ \rightarrow 2^+)$ values are needed to judge the quality and the predictive power of these calculations concerning the deformation of these nuclei. This is in particular true for the $B(E2, 0^+ \rightarrow 2^+)$ value of ^{32}Mg, which we intend to remeasure during the experimental campaign in summer 2004.

4. Outlook

While nuclear structure investigations with and without MINIBALL will play a key role in the future research programme at REX-ISOLDE, the facility also provides new access to intriguing problems of nuclear astrophysics and solid state physics. With the already installed upgrade to 3.0 MeV/u and the presently discussed extension of the accelerator and the experimental equipment, REX-ISOLDE is heading towards an exiting new area of rare isotope research.

Acknowledgements

The author would like to thank all members of the REX-ISOLDE and MINIBALL community for the very fruitful collaboration over many years.

References

1. D. Habs et al., *Hyp. Int.* **129**, 43 (2000).
2. O. Kester et al., *Nucl. Instr. Meth.* **B204**, 20 (2003).
3. E. Kugler, *Hyp. Int.* **129**, 23 (2000).
4. F. Ames et al., *AIP Conf. Proc.* **455**, 927 (1998).
5. F. Wenander, Lic. thesis, Chalmers University of Technology, Sweden (1998), CERN-OPEN-2000-320.
6. R. Rao et al., *Nucl. Instr. Meth.* **A427**, 170 (1999).
7. F. Ames et al., *AIP Conf. Proc.* **606**, 609 (2002).
8. M. Madert et al., *Nucl. Instr. Meth.* **B139**, 437 (1998).
9. R. von Hahn et al., Proc. EPAC2000, Vienna, 596 (2000).
10. D. Warner et al., CERN 93-01 (1993).
11. H. Podlech et al., Proc. EPAC2000, Vienna, 572 (2000).
12. O. Kester et al., *Nucl. Instr. Meth.* **B139**, 28 (1998), and Proc. EPAC2000, Vienna, 827 (2000).
13. J. Eberth et al., *Prog. Part.Nucl. Phys.* **46**, 389 (2001).
14. J. Eberth et al., in: *Frontiers of Nuclear Structure*, ed.s P. Fallon and R. Clark, CP **656**, 349 (2003).
15. The segmented detectors were supplied by EURISYS, Strasbourg, France
16. The DGF-4C modules were supplied by XIA, Newmark(CA), USA
17. Ch. Gund, Diploma thesis, University Heidelberg (1996)
18. Ch. Gund, Dissertation, University Heidelberg (2000)
19. D. Weisshaar, Dissertation, University Cologne (2002)
20. M. Lauer, Dissertation, University Heidelberg (2004)
21. A.N. Ostrowski et al., *Nucl. Instr. Meth.* **A480**, 448 (2002).
22. H. Scheit et al., arXiv:nucl-ex/0401023v1, and to be publ. in *Nucl. Phys. A*.
23. C. Thibauld et al., *Phys. Rev.* **C12**, 644 (1975).
24. X. Campi et al., *Nucl. Phys.* **A251**, 193 (1975).
25. T. Motobayashi et al., *Phys. Lett.* **B346**, 9 (1995).
26. B.V. Pritychenko et al., *Phys. Lett.* **B461**, 322 (1999).
27. V. Criste et al., *Phys. Lett.* **B514**, 233 (2001).
28. H. Iwasaki et al., *Phys. Lett.* **B552**, 227 (2001).
29. D. Schwalm, in: *Probing the Nuclear Paradigm with Heavy Ion Reaction*, ed.s R.A. Broglia, P. Kienle, P.F. Bortignon, World Scientific, p.1 (1994).
30. S. Raman et al., *Atom. Nucl. Data* **78**, 1 (2001).
31. O.Niedermaier et al., to be published.
32. Y. Utsuno et al., *Phys. Rev.* **C60**, 054315 (1999).
33. E. Caurier et al., *Nucl. Phys.* **A693**, 374 (2001).
34. R. Rodriguez-Guzman et al., *Nucl. Phys.* **A709**, 201 (2002).

KEY TOPICS IN NUCLEAR STRUCTURE
Proceedings of the 8th Int. Spring Seminar on Nuclear Physics
edited by Aldo Covello
© 2005 World Scientific Publishing Co.

PROSPECTS WITH RARE ISOTOPE BEAMS AT THE INTERNATIONAL FACILITY FOR ANTIPROTONS AND ION RESEARCH (FAIR)

THOMAS AUMANN
Gesellschaft für Schwerionenforschung mbH (GSI),
Planckstraße 1,
D-64291 Darmstadt, Germany
E-mail: t.aumann@gsi.de

The experimental programme and associated instrumentation for the radioactive-beam facility at the future FAIR facility at GSI is briefly discussed.

1. Overview

A major new facility for ion and antiproton research FAIR is planned at GSI, which will provide high-intensity primary beams for a variety of different experiments enabling a broad range of physics goals to be addressed. The science areas covered include i) the structure of nuclei far from stability and astrophysics, ii) hadron spectroscopy and hadronic matter, iii) compressed nuclear matter, iv) high energy density in bulk matter, v) quantum electrodynamics, strong fields, and ion-matter interactions. The science cases as well as the facility concept and layout is worked out by the international community and is described in detail in the Conceptual Design Report (CDR)[1] of the facility. The concept is based on a double-ring structure SIS100/300, which utilizes the existing UNILAC/SIS18 accelerators as an injector. A schematic layout of the facility is displayed in Figure 1, showing in addition several storage rings and experimental areas. Two aspects of the concept are most important to overcome the present intensity limit given by the fact that the space-charge limit is reached in the synchrotron SIS18: i) First, a faster cycling of the synchrotron of 3 Hz compared to the present situation of 0.3 Hz yields an order of magnitude increase in average intensity, and (ii) the acceleration of uranium ions with charge state 28^+ instead of 73^+ will increase the space-charge limit by one order of magnitude. In order to reach the same beam energy (1.5 to 2 GeV/u) while

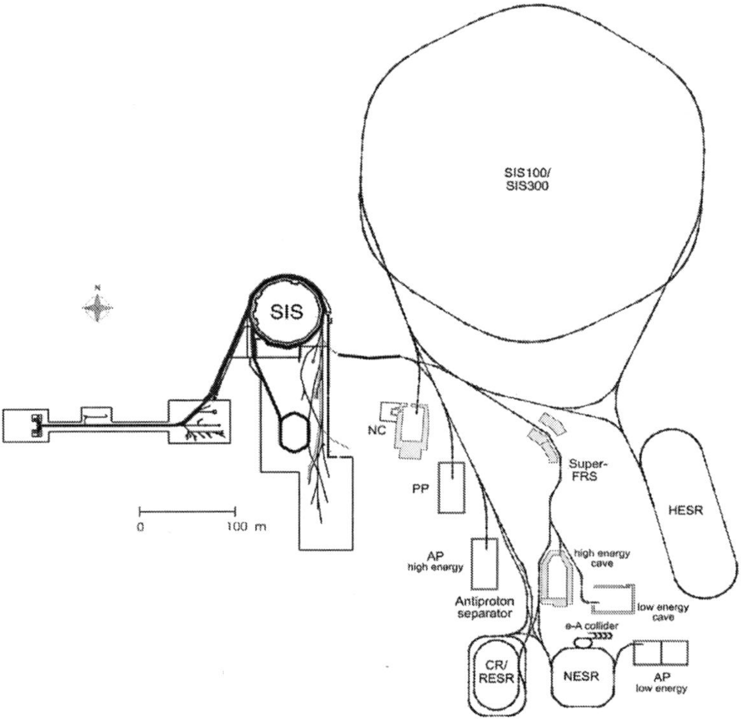

Figure 1. Schematic layout of the FAIR facility.

accelerating a lower charge-state, a synchrotron with higher magnetic rigidity (100 Tm) is needed, which is provided by the SIS100 ring. An average intensity of 10^{12} ions/s is reached, e.g., for uranium beams at 1.5 GeV/u. The second ring, SIS300, serves either as a stretcher ring to provide slow extracted high-duty cycle beams, e.g., for experiments with short-lived nuclei. Or, alternatively, the two synchrotrons may be used to accelerate ions with higher charge state to higher beam energies (with lower intensity) up to about 30 GeV/u as required, e.g., for the nucleus-nucleus collisions programme. The rigidity of 100 Tm does also allow accelerating protons up to 30 GeV, which is optimum for anti-proton production. The system can provide both pulsed beams, e.g., for injection into storage rings or the plasma physics experiments, and continuous high-duty cycle beams as required for external target experiments.

2. The radioactive beam facility

For the production of secondary beams of short-lived nuclei by fragmentation and fission, the key requirements are high intensity and beam energies of up to 1.5 GeV/u. The latter is important due to several reasons: Firstly, an efficient transport of the secondary beams, in particular those produced by uranium fission, is achievable with reasonable sized magnetic separators only by making use of the kinematic forward focussing due to the high beam energy. Secondly, the $B\rho - \Delta E - B\rho$ separation method allows clean isotopic separation and identification only for one ionic charge state, i.e., fully ionized fragments, which can be reached for heavy ions ($Z \approx 80$) only for energies above one GeV/u. Besides the requirements due to the production and separation method, high energies of the secondary beams are also advantageous due to experimental and physics reasons, among those the possibility of using thick targets, high-acceptance measurements, and a clean separation of reaction mechanisms and spectroscopic information to be deduced. Figure 2 shows a schematic layout of the rare-isotope facility including a super-conducting fragment recoil separator Super-FRS[2], and three experimental areas, the low-energy branch, high-energy branch, and ring branch.

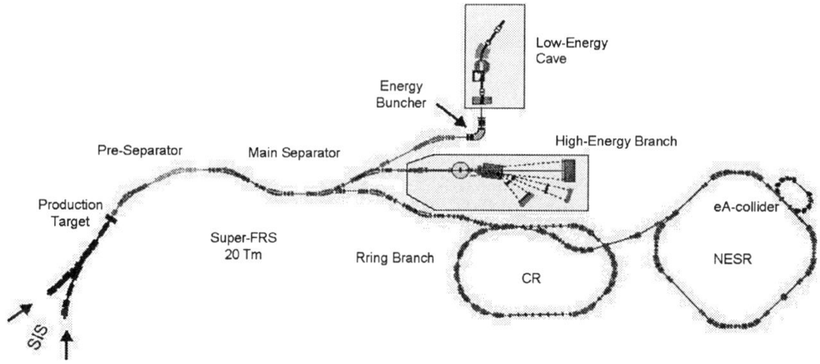

Figure 2. Schematic drawing of the Super-FRS and the three experimental areas, the low-energy branch, high-energy branch, and ring branch.

The Super-FRS is optimized for efficient transport of fission fragments implying a rather large acceptancce of 5% in momentum and ± 40 mrad and ± 20 mrad angular acceptance in the horizontal and vertical planes, respectively. Although the acceptance is largely increased compared to the present FRS, an ion-optical resolving power of 1500 has been retained (for 40 π mm mrad). Besides the resolution, an additional pre-separator ensures low background and clean isotopic separation of the beams of interest. The calculated transmission for beams of neutron-rich medium mass nuclei produced via fission of uranium with primary beam energy of 1.5 GeV/u amounts to about 50%. The gain in transmission compared to the present separator is, as an example, a factor of 30 for the ^{78}Ni region. Together with the gain in primary beam intensity, intensities for separated rare-isotope beams will increase by significant more than three orders of magnitude.

In order to achieve a broad and at the same time in-depth insight into the structure of exotic nuclei, it is indispensable to utilize different experimental techniques to measure similar or related observables. Different probes and/or different beam energies are needed in order to optimize the sensitivity to particular nuclear structure observables. Three experimental areas are foreseen behind the Super-FRS, one housing a variety of experiments with low-energy and stopped beams, a high-energy reaction setup, and a storage and cooler ring complex including electron and antiproton colliders. In the following, a few aspects of the different branches will be briefly outlined. This discussion can not be comprehensive, a more complete and detailed description of the proposed experiments can be found in the Letters of Intent of the NUSTAR collaboration [3].

3. Experiments with slowed-down and stopped beams

For experiments requiring low-energy beams or implanted ions, the high-energy radioactive beams have to be slowed down. The key instrument for these experiments is an energy-focusing device[4], which reduces the energy spread of the secondary beams delivered by the Super-FRS. The principle is sketched in Figure 3. It consists of a high-resolution dispersive separator stage in combination with a set of profiled energy degraders. After the monoenergetic degrader, the momentum spread is greatly reduced allowing, e.g., the ions to be stopped in a 1m long gas cell, from which they can be extracted for further manipulation and experiments. This method combines the advantages of in-flight separation (no limitations on lifetime and chemistry) with the ISOL type experimental methods. For gamma spectroscopy studies, beams can be slowed down also to intermediate energies (100 MeV/u) or Coulomb barrier energies. Table 1 gives an overview on the planned experimental programme.

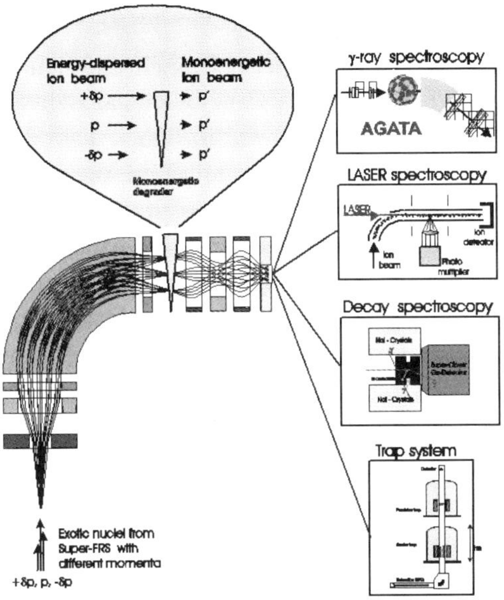

Figure 3. Schematic view of the experimental setups for experiments with low-energy and stopped radioactive beams.

Table 1. Experimental opportunities at the low-energy branch[3].

Field	Experimental method	Subject
Nuclear structure and astrophysics	Multiple Coulex, direct reactions, fusion evaporation Intermediate energy Coulex, fragmentation	B(E2), lifetime, moments, spins, high spin structure, single particle structure, dynamical properties B(E2), lifetime, moments, spins, shape evolution, shell structure
Nuclear structure and astrophysics	α-, β-, γ-, particle-decay spectroscopy	Half-lifes, spins, nuclear moments, GT strength, isomer decay, β-decay branching ratios, β-delayed neutrons
Atomic physics methods applied to nuclear structure	Laser spectroscopy	Nuclear charge radii Magnetic dipole and electric quadrupole moments
Precision experiments	Ion trapping, charge breeding, and spectroscopy	Nuclear binding energies (mass measurements, Q-values)
Tests of the Standard Model	Atom trapping (MOT)	Half-life measurements β-decay branching ratios β-ν correlations
Nuclear structure and antimatter	RIB+pbar trapping	Charge radii, skin- and halo distributions

4. Scattering experiments with high-energy rare-ion beams

The instrumentation at the high-energy branch is designed for experiments using directly the high-energy secondary beams as delivered from the separator with magnetic rigidities up to 20 Tm. The R3B[5] experimental configuration is based on a concept similar to the existing LAND reaction setup[6] at GSI introducing substantial improvement with respect to resolution and an extended detection scheme, which comprises the additional detection of light (target-like) recoil particles and a high-resolution fragment spectrometer. The setup, which is schematically depicted in Figure 4, foresees two operation modes, one for large-acceptance measurements of heavy fragments and light charged particles (left bend in Figure 4), and alternatively for high-resolution momentum measurements ($\Delta p/p \approx 10^{-4}$) using a magnetic spectrometer (right bend in Figure 4). The experimental configuration is suitable for kinematical complete measurements for a wide variety of scattering experiments, i.e., such as heavy-ion induced electromagnetic excitation, knockout and breakup reactions, or light-ion (in)elastic and quasi-free scattering in inverse kinematics, thus enabling a broad physics programme with rare-isotope beams to be performed. Table 2 gives an overview on the planned experimental programme. More details can be found in the Letter of Intent of the R^{3}B collaboration [5,3].

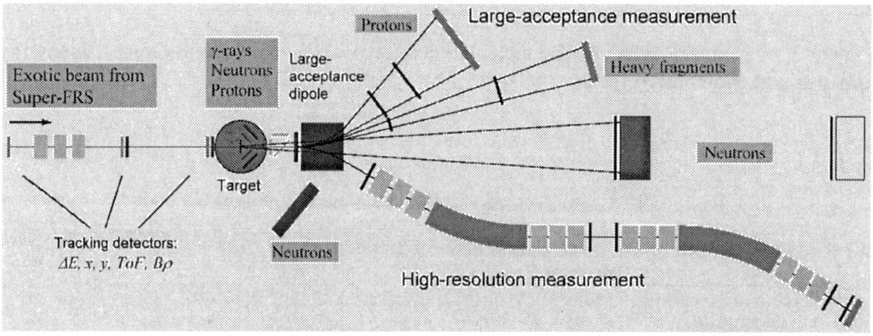

Figure 4. Schematic view of the experimental setup for scattering experiments with relativistic radioactive beams comprising γ-ray and target recoil detection, a large-acceptance dipole magnet, a high-resolution magnetic spectrometer, neutron and light-charged particle detectors, and a variety of heavy-ion detectors.

Table 2. Reactions with high-energy beams and corresponding achievable information.

Reaction type	*Physics goals*
Total-absorption measurements	Nuclear matter radii, halo and skin structures
Elastic p scattering	Nuclear matter densities, halo and skin structures
Knockout	Shell structure, valence-nucleon wave function, many-particle decay channels unbound states, nuclear resonances beyond the drip lines
Quasi-free scattering	Single-particle spectral functions, shell-occupation probabilities, nucleon-nucleon correlations, cluster structures
Heavy-ion induced electromagnetic excitation	Low-lying transition strength, single-particle structure, astrophysical S factor, soft coherent modes, low-lying resonances in the continuum, giant dipole (quadrupole) strength
Charge-exchange reactions	Gamow-Teller strength, soft excitation modes, spin-dipole resonance, neutron skin thickness
Fission	Shell structure, dynamical properties
Spallation	Reaction mechanism, astrophysics, applications: nuclear-waste transmutation, neutron spallation sources
Projectile fragmentation and multifragmentation	Equation-of-state, thermal instabilities, structural phenomena in excited nuclei, γ-spectroscopy of exotic nuclei

5. Experiments with stored and cooled beams

The third branch of the Super-FRS serves a storage- and cooler-ring complex. A schematic layout is shown in the left part of Figure 5. Fragment pulses as short as 50 ns are injected into a Collector Ring (CR) with large acceptance, where fast stochastic pre-cooling is applied. A momentum spread of $\Delta p/p \approx 10^{-4}$ is achieved within less than 500 ms. The CR may also be used for mass measurements of short-lived nuclei applying ToF measurements in the isochronous mode[7]. In the New Experimental Storage Ring NESR, fragments can be further cooled by electron cooling. Several experiments with exotic nuclei are foreseen in the NESR including mass and

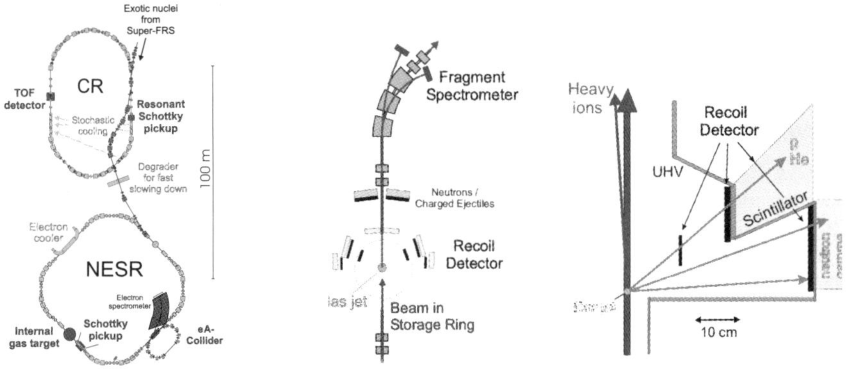

Figure 5. Left: Schematic layout of the storage ring complex. Right: Detection scheme for reaction studies at the internal target in the NESR.

life-time measurements applying the Schottky method[7], as well as electron scattering experiments at an intersecting electron storage ring (eA collider), which allow for the first time studying ground state and transition form factors of unstable nuclei with a pure electromagnetic probe. The eA collider may as well be used to facilitate collision experiments between stored ions and antiproton beams. From such measurements, information on rms radii for proton and neutron ground state densities can be obtained. Light ion scattering can be studied at the internal gas target (see right part of Fig. 5). The advantage of performing such experiments in a storage ring arises mainly from the fact that thin targets can be used while gaining the luminosity by the revolution frequency of 1 MHz. This allows to measure reactions at low-momentum transfer implying (in inverse kinematics) low-energy recoils. Table 3 gives an overview on the different reactions and the nuclear structure information that can be obtained.

Table 3. Nuclear structure information from scattering off light nuclei [3].

Method (reactions)	**Physical observables**	**Related effects in exotic nuclei**
elastic scattering (p,p); (^{4}He,^{4}He);	nuclear matter radii and distributions	halo; neutron skin; central density; optical potential
inelastic scattering (p,p');($\bar{p}$,p'); (^{4}He,^{4}He')	surface collective states; electric giant resonances; isovector magnetic excitation for (p,p'); analyzing powers	bulk properties in N-Z asymmetric matter; proton/neutron deformation; nuclear compressibility; threshold strength; soft modes
charge exchange (p,n); (d,^{2}He); (^{3}He,t)	spin-isospin excitations; Gamow-Teller; Spin-dipole resonance	(stellar) weak interaction rates; spin excitations; neutron skin
transfer reaction (p,d); (d,^{3}He); (p,t); (d,p)	Spectroscopic factors; Single particle (hole) states; Pair transfer	single-particle structure; spin-orbit; pairing interaction
quasi-free scattering (p,2p);(p,np); (p,p ^{4}He)	single-particle spectral function; cluster knockout	single-particle structure; nucleon-nucleon and cluster correlations; in-medium interactions

References

1. *An International Accelerator Facility for Beams of Ions and Antiprotons*, Conceptual Design Report, GSI (2001), http://www.gsi.de/GSI-Future/cdr/.
2. H. Geissel *et al.*, Nucl. Instr. and Meth. in Phys. Res. B **204** (2003) 71.
3. Letters of Intent of the NUSTAR collaboration at FAIR (April 2004), http://www.gsi.de/forschung/kp/kp2/nustar_e.html/.
4. C. Scheidenberger *et al.*, Nucl. Instr. and Meth. B **204** (2003) 119.
5. Reactions with Relativistic Radioactive Beams (R3B), http://www-land.gsi.de/r3b/.
6. R. Palit *et al.*, Phys. Rev. C **68** (2003) 034318.
7. H. Geissel, H. Wollnik, Nucl. Phys. A **693** (2001) 19.

KEY TOPICS IN NUCLEAR STRUCTURE
Proceedings of the 8th Int. Spring Seminar on Nuclear Physics
edited by Aldo Covello
© 2005 World Scientific Publishing Co.

THE SPIRAL 2 PROJECT AT GANIL

Dominique GOUTTE
(GANIL)
and the
SPIRAL2 APD-Group

The SPIRAL2 project is now close to the end of the Detailed Design Study (APD) phase. The baseline project, as well as possible extensions, was presented at various meetings and an intermediate report has been published, showing the technical solutions chosen, as well as the tentative time schedule and construction costs. The reference project, including the baseline project and the addition of a number of extensions, was discussed with the physics group and presented at the Steering Committee, held on the 5th of February 2004. A decision for the beginning of the construction in 2005 could be taken before the end of this year.

1- Physics objectives and specifications

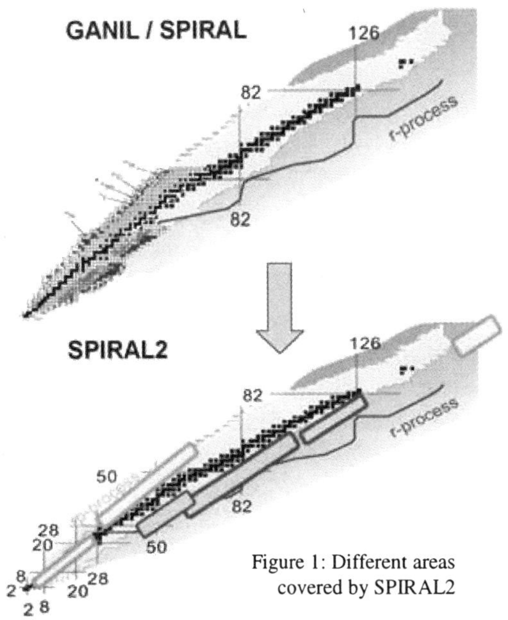

Figure 1: Different areas covered by SPIRAL2

The SPIRAL 2 project aims at delivering high intensities of rare isotope beams for a much wide part of the nuclear chart than accessible today. Several production methods will be used as the fission induced by fast neutrons in a uranium target or by direct bombardment of the fissile material, or via fusion-evaporation with unstable beams or heavy ion beams. This will allow to cover broad areas of the nuclear chart as shown in Figure 1.

In addition to fundamental research in nuclear physics, the SPIRAL 2 facility will also offer a high performance multidisciplinary tool, especially in fields of science requiring high fluxes of neutron, such as material sciences, atomic, plasma and surface physics. In particular, it will provide a high neutron flux with an adequate energy spectrum for the study of the behaviour of material under irradiation, of interest for future fusion experimental facilities and fusion reactors.

The list of specifications resulting from the physics needs are summarized below:

Driver and primary beams

The driver must deliver deuterons up to an energy of 40 MeV with a beam current up to 5 mA and heavy ions with beam currents up to 1 mA. It will be optimised in energy for ions of mass-to-charge ratio A/q=3, resulting in an output energy of about 14 MeV/u. It will also be able to accelerate ions of mass-to-charge ratio A/q=6. The beam energy will be adjustable between the maximal energy and as low as the RFQ output energy. The layout of the facility takes account of the possible future increase in energy up to 100 MeV/u. A fast chopper is required for some physics experiments to select one bunch out of a few hundred to a few thousand.

Production hall

The production rate of the radioactive beams produced by neutron-induced fission of an uranium target from a deuteron beam bombarding a carbon converter, must be higher than 10^{13} fissions/s. The use of high-density targets could allow us to reach an upper limit of $\sim 10^{14}$ fissions/s. This fission rate of 10^{14} fissions/s has been used for all safety and radiation-protection-related calculations. The converter has to withstand a maximum beam power of 200 kW. Without the use of a converter, the primary beam will consist of deuterons or other species (such as $^{3,4}He$) and the maximum power is limited by the most restricting condition, such that the induced activity remains below the activity induced by 10^{14} fissions/s obtained with the converter method and the power deposited in the target is lower than the maximum power that the target can withstand (presently estimated to about 6 kW for a UC_x target). Different thick targets will replace the uranium target for fusion evaporation reactions with stable ion beams. Different types of ion sources will be studied in order to get the best efficiency for the selected ion specie. A mass separator must deliver at least two independent beams, with mass resolution of 250. An identification station is essential for the control of the desired specie output.

The isotopes will be bred to higher charge states by means of an ECR charge breeder prior to post-acceleration.

Experimental area

Without post-acceleration, the secondary beams will be transported to the low-energy experimental hall (LIRAT). After post-acceleration in the existing CIME cyclotron, the secondary beams will be transported to the existing experimental area at GANIL. New direct beam transfer lines will allow the direct delivery of beams out of CIME to the existing caves G1/G2. For the study of fusion evaporation reactions with the in-flight method, the high-intensity stable ion beams from the linac will be transported to a new experimental hall.

Use of neutrons for other applications

The possibility of material irradiation studies, using the large neutron flux, especially for the study of the behaviour of materials considered for future fusion machines (ITER, DEMO), has to be investigated.
Room should be left for a possible installation of pulsed neutron beam experiments, including an experimental hall and a ~10 m long neutron line to be used for neutron-TOF like experiments.

2- Reference project

The schematic layout is shown in Fig. 2. The facility can be divided into 4 main areas: accelerator driver, production station (including converter, target and ion source), secondary beam transfer lines and high energy RIB beam lines.

The **reference project** contains all fundamental parameters for RIB and heavy ions production (e.g.,. beam power, fission rate, etc) :
- the entire driver delivering a deuteron beam at the design power, and a heavy ion beam (q/A=1/3) at intermediate intensity (state-of-the-art)
- one station for RIB production and the purchase of 2 plugs equipped with converter, target & ECR ion source
- the secondary beamlines including the separator, the charge-breeder and the transport to the CIME cyclotron
- the experimental hall for stable heavy ions
- the RIB transfer beamlines to the Low Energy experimental area (LIRAT)
- the direct beamlines to G1/G2 caves
- one location reserved for a possible irradiation station

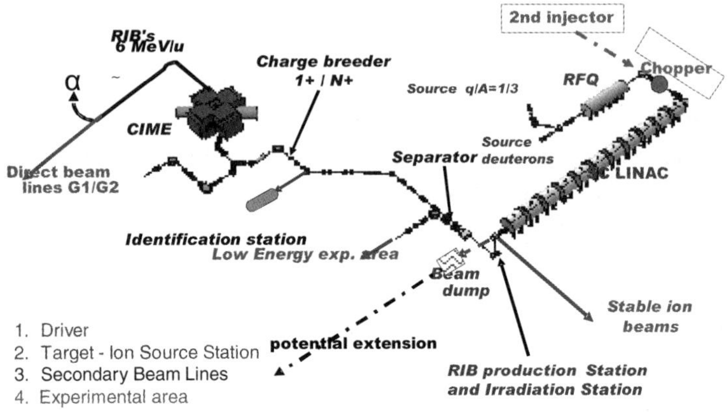

Figure 2: Schematic layout : accelerator driver (green) ; RIB production and irradiation stations (red) secondary beam transfer lines (blue) ; high-energy RIB beam lines (violet).

In addition to this reference project, **technical improvements** that will be purchased in the course of the operation of the facility, and **extensions** that can be installed at a later stage, and the possible **future upgrades** were considered. Among these, an extension uo to 100 MeV/A of the linear accelerator, a second injector (A/Q=6), ……

3 - Accelerator driver

The driver must accelerate beams of high power (200 kW deuteron beam power), different ion species and mass-to-charge ratios (deuterons as well as heavier ions with mass-to-charge ratio A/q=3 and up to A/q=6 at a later stage) with high output-energy flexibility (from 40 MeV deuteron energy down to

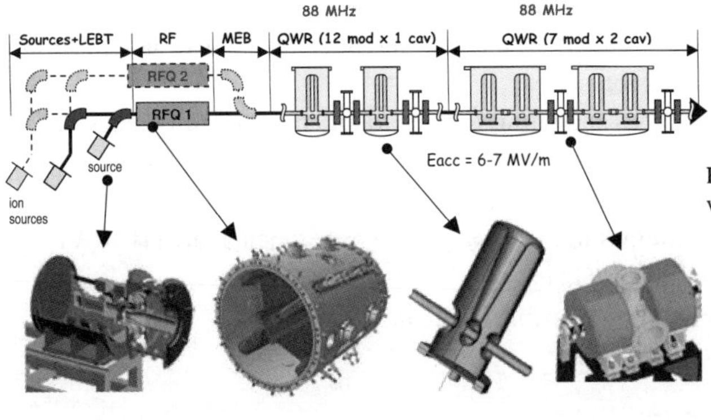

Figure 3: Schematic view of the accelerator

energies, as low as that at the RFQ exit). The concept of "Independently Phased Superconducting Linac" has been chosen because it provides safe continuous wave (CW) operation and high flexibility in the acceleration of different ion species and charge-to-mass ratios. A schematic view of the accelerator is shown in Fig. 3.

Injector

One injector will deliver both kinds of ion beams at the required energy of 0.75 MeV/u.

The deuteron source (5 mA) is developed either from a "downgraded version" of the high intensity SILHI-type source or from a version of the Micro-Phoenix source. The deuteron beam will be injected into the RFQ cavity through a bending magnet, thus allowing us to get rid of other ion species (i.e. D_2^+,D_3^+). First beam measurements of the Micro-Phoenix source have shown an ***Figure*** horizontal rms emittance lower than $\varepsilon_{norm} = 0.1$ π.mm.mrad for a deuteron current of 5.4 mA. Beam characterisation of the SILHI-type source are scheduled from April 2004.

For the A/q=3 ions, the present state-of-the-art of ECR sources produces 1 mA for O^{6+} and 0.3 mA for Ar^{12+}. Such a source can be installed as the initial source and is part of the baseline project. High confinement fields (Br ~ 2-3 T) and high frequency (f > 28 GHz) will be required to increase the ion beam currents. The A-Phoenix source, based on the combination of permanent and high temperature superconducting magnets will permit us to reach the highest intensities for noble gases and should be ready before 2008. For the production of metallic ions (Ni, Cr, etc), development will be carried out on existing sources at GANIL.

The RFQ cavity must bunch and accelerate the beam to the required energy with a high transmission to allow for hands-on maintenance. Different technologies at 88 MHz were studied and the four-vane structure was finally chosen because the RF power consumption is the lowest (~ 150 kW) and the team has much experience on this type of structure. The design is based on a full mechanical assembly without brazing and the vanes are dismountable. The critical issues are the RF joints and the displacement of the vane tips in operation. A 1 meter long prototype has been ordered and first tests are planned before the end of 2004.

A second injector for heavier ions (A/q=6), including a new ion source and a second RFQ cavity, is planned to feed ions into the MEBT system (Medium Energy Beam Transport) but is considered as an extension and will be installed later. The beam coming from both ion sources can then be independently transferred either to the 1st RFQ or to the 2nd RFQ (dashed lines of the source area in Fig. 3).

In order to satisfy the physics request, a fast chopper has to be inserted in the MEBT line to select one bunch from N = 10^3 to 10^5 bunches (for physics of solids and atomic physics) or from N = 10^2 bunches (for nuclear physics). This device, which needs significant R&D effort owing to the small rise time required (less than 8 ns), can be installed later as an extension.

Superconducting linac

The choice of short superconducting cavities, exhibiting very wide velocity acceptance in comparison with long multi-cell structures, allows the optimisation of the output energy for each ion specie by re-adjusting the individual RF phases. Two types of superconducting cavities were considered, Quarter-Wave Resonators (QWR) and Half-Wave Resonators (HWR), and several different frequency scenarios have been studied:

88 MHz for the whole linac

88 and 176 MHz for the low- and high-energy parts, respectively

176 MHz for the whole linac (with intermediate IH-structure)

The single frequency scheme at 176 MHz has been ruled out because it required an intermediate IH-structure to boost the RFQ output energy. Although the dual-frequency scheme was first preferred, the use of 2 families of QWR resonators at 88 MHz (β=0.07 and β=0.12) was finally adopted for the following reasons:

- the total number of cavities is lower (26 cavities instead of 36)
- there is no frequency jump which would require longitudinal matching
- the cavity aperture is potentially larger
- the frequency is identical for all RF sources
- a slight cost reduction

All initial reservations about this choice (cavity fabrication which could be difficult owing to the large diameter, steering effects which could not be compensated for, the 100 MeV/u linac extension which could necessitate an intermediate frequency before 352 MHz) have now been removed.

In addition, the focusing by means of room-temperature quadrupoles, instead of superconducting solenoids, resulting in one cryostat per focusing lattice, has been chosen. Despite a slight cost increase but still lower than the one of the dual frequency scheme, this arrangement offers many advantages: the residual magnetic field of solenoids close to resonators cannot be sufficiently lowered, the cryostats are much simpler, the cavity and magnet alignment is much easier, the space available for diagnostics is larger and the linac tuning is simplified.

A realistic accelerating field of 6-7 MV/m was chosen because the resulting maximum peak fields (*Epk* < 40 MV/m, *Bpk* < 80 mT) can be achieved without too much effort by using well-tried methods developed in the last ten years, such as high-pressure rinsing, high-purity niobium and clean conditions. Furthermore, free room has been left at the end of the linac to allow for the insertion of two additional high-β cryomodules should the field gradient in operation be lower than the specifications. Two cavities, one of each family, have been differently designed (removable botton plate with Nb/Ti flange for the low-β cavity and welded bottom plate with two ports for high-pressure rinsing for the high-β cavity) and will be tested before the end of 2004.
The design of the input coupler, as well as the RF power test bench, is in progress. The RF generator will be purchased in 2004 and power tests are planned in 2005.
Lastly, beam dynamics calculations with space charge forces, 3-D field maps of resonators and simple "one-to-one" trajectory correction, have shown very low emittance growths. Systematic start-to-end simulations including all combined effects, such as field and alignment errors, are in progress.

High energy beam transport

From the linac exit, the beam will be transported either straight to a beam dump (10-20% of full beam power) or to a new experimental hall for stable ions, or down to the RIB production station. In the latter case, a non-linear magnetic lens has been added in order to provide a uniform beam distribution at the converter location. The layout also allows for a possible 100 MeV/u linac extension by removing the beam dump and different solutions using 3 or only 2 frequencies (88 and 352 MHz) have been found to accelerate the various ion species with a total linac length of ~100 m.

4 - RIB production station

In order to provide against radiation and contamination, the "plug" technology, developed at TRIUMF (Canada) was chosen and adapted to the RIB production system of SPIRAL 2.
The production plug (Fig. 4) comprises essentially:

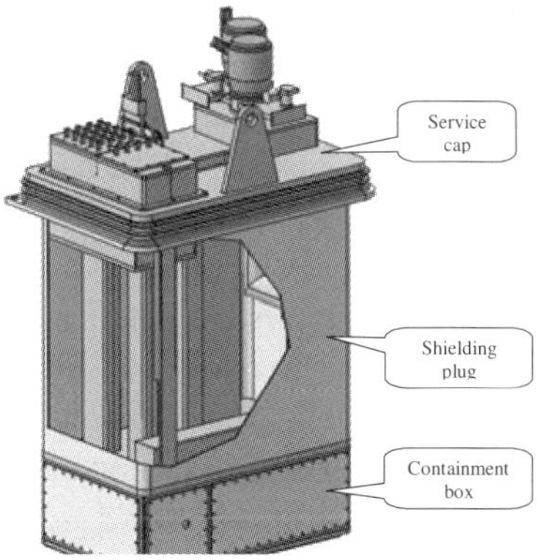

Figure 4: Schematic view of the production plug

- containment tanks for the converter, the target and the ion source
- shielding for biological protection against radiation
- a service cap for the ancillary equipment (e.g. pumping system, motors for converter, valves, etc)

After target irradiation and enough cooling time, the plug will be isolated by valves and disconnected from all external supplies. It can be then remotely transported to a shielded bunker. After a few months of storage, the plug will be transported into a hot cell for maintenance (disassembly and replacement of components) by means of remote hand-operated manipulators. A minimum of two production plugs, one in place and one in preparation, will be needed to ensure an acceptable RIB production time.

The entire plug will be insulated and raised to high potential (up to 60 kV) and will be mounted within a vacuum-tight tank. The other solution, which consists of raising only the essential equipment to high voltage, has been ruled out because the free space between converter and target either would be too small to sustain the high voltage, or would require an electrical contact on the rotating carbon wheel.

Converter

The converter, which has to withstand a maximum incident beam power of 200 kW, is a carbon wheel, rotating at a sufficiently high speed (a few hundreds rpm) so as to distribute the temperature uniformly along the circumference. The beam impinges horizontally onto the rim of the wheel, composed of individual graphite tiles. Based on the experience of such carbon wheels developed at PSI, the maximum temperature has been fixed at 1750°C in order to limit the evaporation, resulting in a diameter of 1 m.

Targets

The neutron-rich isotopes are produced by fission of a depleted uranium carbide target. A low-density target, based on the technology used at PARRNE and ISOLDE, has been designed to reach at least 10^{13} fissions/s. A high-density target permitting us to reach 10^{14} fissions/s is under study, in collaboration with the Gatchina and Legnaro laboratories.

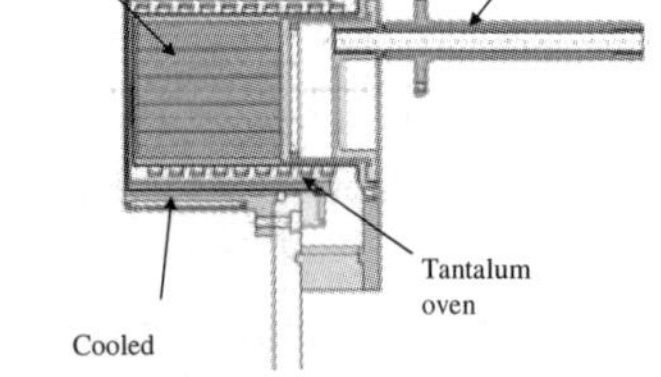

Figure 5: UCx target with the oven

The geometry of the low-density target has been optimised by taking into account the distribution of the incoming neutron flux and the effusion process. A tantalum oven has been designed to stabilize the target temperature around 2200°C to allow efficient diffusion (Fig. 5).

Ion sources

Different types of ion sources have to be coupled to the target to cover the largest range of radioactive isotope beams. Two ion sources - very likely ECR and thermo-ionisation sources - are included in the reference project, the other sources being purchased in the course of the operation of the facility..

Secondary beam transport lines

The secondary beam lines have to transport the radioactive beams either to a new low-energy physics area (LIRAT) or to the CIME cyclotron. Direct beam lines to G1/G2 caves from CIME would extend the capability of GANIL to deliver simultaneous beams, from both the existing GANIL cyclotrons and from SPIRAL 2.

The radioactive isotopes extracted from the ion source are collected and then mass selected through a separator. The beams of different ion species are split up, some to the charge breeder prior to post-acceleration in the existing CIME cyclotron, and some to the low-energy experimental hall. In addition, the high-energy beams should be as pure as possible.

The section between ion source and separator is short and included in the plugs. For the magnetic separator, a BRAMA ("Broad-Range Acceptance Mass Analyser") type solution, based on sliding electrostatic deflectors has been

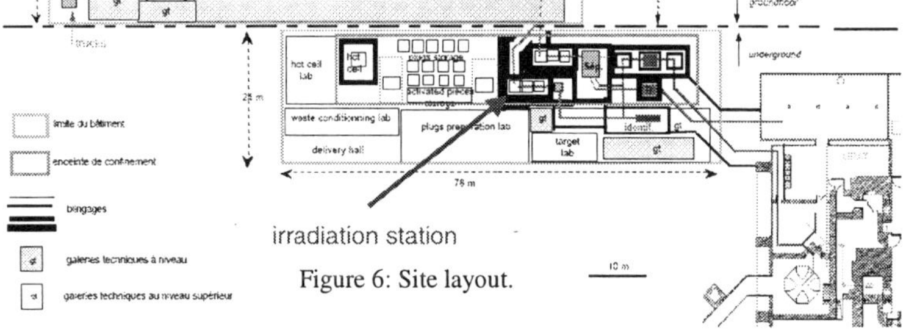

Figure 6: Site layout.

selected as reference solution for its compactness and intrinsic ability to switch the required mass to the required channel.

The charge breeder is the "Phoenix booster", developed at LPSC/Grenoble. The system includes a double Einzel lens for injection matching and up- and downstream bending magnets for beam cleaning.

6 - Infrastructure and conventional facilities

The infrastructure and conventional facilities were designed around the reference project in such a way that all future extensions can be implemented (such as the 100 MeV/u linac and an experimental hall for pulsed neutrons). The site layout is shown in Fig. 6.

The concrete shielding needed for biological protection is included in the planned production hall, mainly around the production station and downstream components, as the separator and charge breeder. In the same way, the driver accelerator is surrounded by concrete shielding for radiation protection, with a thickness dependent on the location along the linac (from 30 cm around the deuteron source to 1.5 m at the high-energy end).

Although safety aspects have been taken into account in the design of the production building, detailed studies by nuclear engineering companies will be shortly launched. These studies include also the nuclear ventilation system, the hot cell design and the nuclear waste management system.

7 - Project schedule and cost

The construction of the SPIRAL 2 facility is planned to take place over 5 years including 6 months for sequential commissioning of the accelerator, the RIB production station and the chain of the beam transport lines up to the experimental areas. Therefore, assuming that the decision of construction is taken before the end of this year and that all required authorisations will been obtained from the safety authorities at the right time the first beams to the experimental area could start at the end of 2009. During the last year of the construction period, the beam power would increase in steps, up to the specified 200 kW for deuterons to be delivered to the target and ion-source station.

The construction costs - i.e. all the costs during the 5-year construction period of the SPIRAL 2 reference facility amount to 115 Million Euro at 2003 prices

This figure includes not only capital investments for all components necessary for the SPIRAL 2 reference facility, but also all the manpower required (in-house and subcontracted staff) at the various stages of the project: i.e. detailed design, procurement, construction, testing, installation and commissioning.

KEY TOPICS IN NUCLEAR STRUCTURE
Proceedings of the 8th Int. Spring Seminar on Nuclear Physics
edited by Aldo Covello
© 2005 World Scientific Publishing Co.

THE EVOLUTION OF STRUCTURE IN EXOTIC NUCLEI

R. F. CASTEN
Wright Nuclear Structure Laboratory, Yale University
New Haven, Connecticut, 06520-8124, USA
E-mail: rick@riviera.physics.yale.edu

With the increasing likelihood that major next–generation radioactive beam facilities will be built in the not–too–far future, interest in exotic nuclei has become even more intense. Of course, major foci of this interest concern the frontiers, namely extremely proton rich nuclei, nuclei near the neutron drip line, and the heaviest nuclei that can exist. However, another major area of interest will be the evolution of structure between the known nuclei and those furthest from stability. Recent work has revealed several new and important facets of this structural evolution. These concern phase transitional behavior and its nature; the expected locus and the theoretical description of nuclei in phase transitional regions; new examples of evolutionary trajectories with implications for the description of nuclei in terms of regular and chaotic behavior; and new empirical signatures for such behavior. Several of these aspects of nuclei far from stability will be discussed.

1. Introduction

With the expectation that major new facilities for the study of exotic nuclei will be constructed and operational in the U.S., Europe and Japan within the next decade, interest in the properties of these nuclei is growing rapidly. There are four frontiers to this field. Three of them are frequently discussed, namely proton rich nuclei, where key issues involve the study of nuclei "beyond" the drip line, and the special properties of N$\sim$Z nuclei; neutron rich nuclei, where an abundance of interest lies, including such topics as exotic topologies (halo and n–skin nuclei), the physics of weakly bound quantal systems, the effects of coupling to the continuum, and changes in residual interactions and the underlying shell structure; and the heaviest nuclei that can exist, where the essential challenge is to understand the microscopic origins of their binding through the occupation of specific quantum states by the outermost nucleons.

However, there is a fourth frontier, somewhat less discussed, but equally important and, ironically, more accessible, namely, the evolution of structure between the bounding limits of nuclear existence. Access to large reaches of exotic nuclei—in particular, long isochains—will give us an unprecedented opportunity to study structural changes as a function of nucleon number. It will even be possible in special cases, such as Ni, to gain glimpses into the behavior of a given element in four different major neutron shells. Development of structure from magic or near–magic to deformed, and back to magic or near–magic, will shed light both on the emergence of collective modes and on the role of various microscopic residual interactions.

In this paper, we will discuss two aspects of structural evolution that reflect diverse but complementary facets of it. Both represent important topics for study in the new arena of nuclei that will become available in the next generations of exotic beam facilities. One is completely new in this context, while the other is an important recent development that has had significant impact in the last few years and has already become a major new research area. The latter topic, of course, refers to phase transitions in the equilibrium nuclear shape as a function of nucleon number, and the proposal and discovery of empirical examples of the new critical point symmetries X(5) and E(5). The other topic concerns the issue of regularity and chaos in nuclear spectra. The discussion is based primarily on Refs. [1–6].

2. Shape/Phase Transitions in Finite Nuclei and Their Locus

Phase transitions in infinite systems exhibit an abrupt change in some observables (the order parameter, usually related to the symmetry of the system) as a function of another observable, the control parameter. In finite systems, this change in properties is more gradual but one can nevertheless speak in terms of (muted) phase transitional behavior. Typically, in nuclear shape–related phase transitions the order parameter is the nuclear deformation while the control parameter is either nucleon number or some surrogate for it, such as a model parameter that determines the equilibrium shape. First order phase transitions include phase coexistence in which both spherical and deformed minima in the energy surface exist. The critical point occurs where they cross and the equilibrium deformation changes discontinuously. Traditionally such transitional regions have been the most difficult to describe as they involve competing degrees of freedom.

Recently, however, Iachello [1] introduced a new, extraordinarily simple, model of such a situation, called X(5), whose basic ansatz is illustrated in Fig. 1. The figure shows a set of successive (as a function of, say, neutron number) energy surfaces. The curve labelled 3 and drawn in bold has degenerate spherical and deformed minima, with a (small) barrier between them. Iachello simulates this by ignoring the barrier and taking a simple (infinite) square well. The solutions are Bessel functions of irrational order. The predictions for energies and B(E2) values are fixed and completely parameter free except for scale. When X(5) was proposed it was immediately identified empirically in ^{152}Sm [2] and ^{150}Nd [3]. Figure 2 shows a comparison [2] of the empirical level scheme of ^{152}Sm with X(5). The agreement is excellent, in particular the two signature characteristics of X(5), an $R_{4/2}$ value of ~3 and an $R_{0/2} \equiv E(0_2^+)/E(2_1^+)$ value of ~5.6. The discrepancies in the scale of interband B(E2) values and in the energy scale for the 0_2^+ sequence have been discussed and are not viewed as serious.

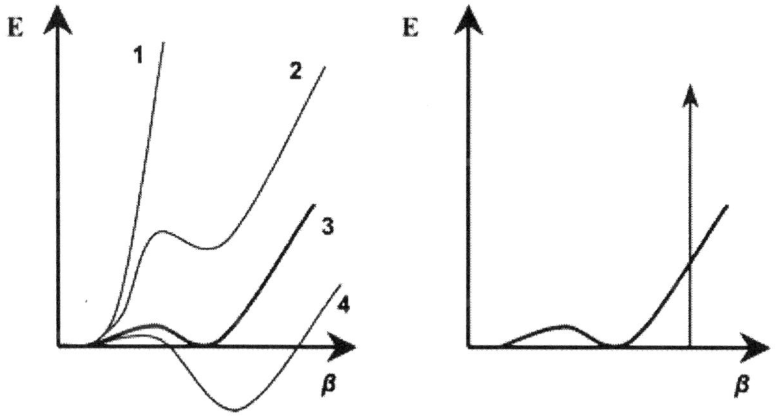

Figure 1. Energy surfaces in a shape transitional region. The curve on the right is the same as the one in bold on the left and corresponds to the phase transitional point. The square well on the right is the X(5) ansatz.

Several other N=90 nuclei are also close in structure to X(5). To seek still further examples, we used the P–Factor [7,8], with a value P $= N_pN_n/[N_p + N_n]$ ~5, to estimate the locus of the onset of deformation. This value is chosen because it is found [7] empirically that nuclei in all regions of medium and heavy nuclei develop quadrupole deformation when N_pN_n is on the order of five times $N_p + N_n$. In the rare earth region, one such nucleus is ^{162}Yb [4]. The previously existing scheme for this nucleus, however, had a low lying 0^+ state with $R_{0/2}$ ~3.9, in disagreement with

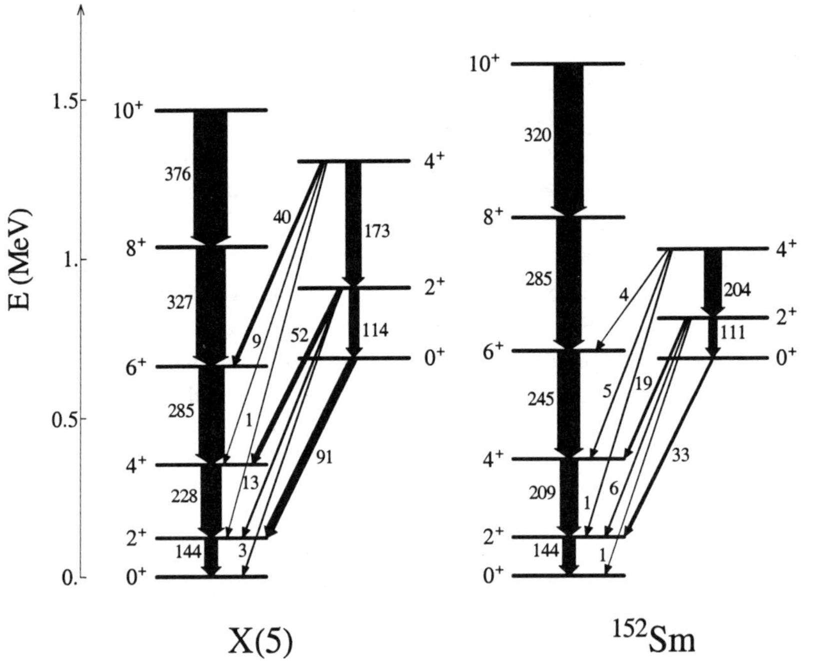

Figure 2. Comparison of the predictions of X(5) with the low lying level scheme of ^{152}Sm. From Ref. [2].

X(5). At Yale, we therefore carried out extensive β–decay studies of ^{162}Yb and showed that this 0^+ state, in fact, does not exist. The new lowest 0_2^+ level (previously 0_3^+), gives $R_{0/2}$ ~5.4 and the rest of the ^{162}Yb yrast and yrare energies are quite close to X(5). The yrast B(E2) values, however, are lower than in X(5) and new experiments are being carried out to assess if the existing literature values are correct.

An important question arises as to whether phase transitional behavior occurs elsewhere in the nuclear chart. Certainly, the A=100 region offers an obvious candidate, but, here, structural evolution is so rapid that no single nucleus appears to reflect X(5): Due to the discontinuous changes in structure with *integer* nucleon number, the critical point, and therefore spectra resembling X(5), are effectively by–passed. This makes it all the more important to seek out additional regions of first order phase transitional behavior elsewhere. To do this, we again use the P–Factor, and, using standard shell and subshell magic numbers, sketch the locus of P–5 values. This is done for a large segment of the nuclear chart in Fig. 3.

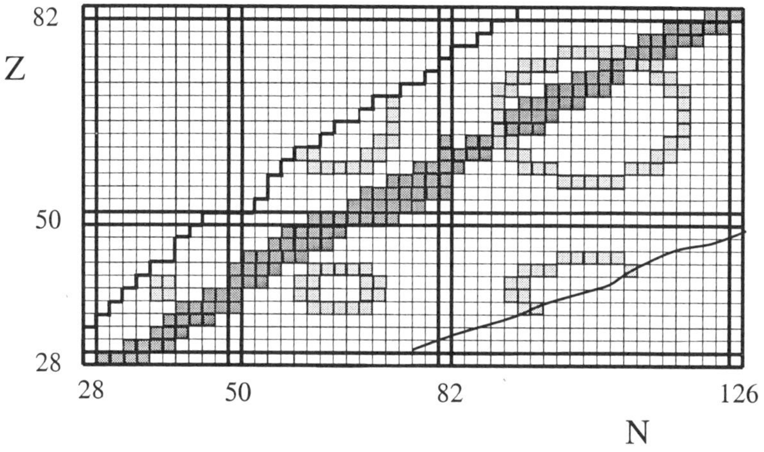

Figure 3. Locus of P ∼5 in a large region of the nuclear chart. The grey boxes are stable nuclei and the slanting contours at upper left and lower right are estimates of the proton and neutron drip lines. The loci of P ∼5 are indicated. Based on Ref. [9].

Interestingly, most of the candidate regions are located off the valley of stability and will have to be sought in future experiments with exotic beams. Many of these regions in Fig. 3 should be accessible. Key experimental techniques will be Coulomb excitation in inverse kinematics and β–decay. The focus in first experiments should be the yrast energies, the B(E2) values for the $2_1^+ \rightarrow 0_1^+$ and $4_1^+ \rightarrow 2_1^+$ transitions and the energies of the 0_2^+, 2_2^+ and 4_2^+ levels. This figure shows that the advent of new ideas about shape transitional behavior has led in turn to new interest in structural evolution in exotic nuclei and to new regions in which to seek and to study critical point behavior.

3. Mapping the Symmetry Triangle

One interesting by–product of the discovery of empirical examples of X(5) has been a new attempt [5] to locate rare earth nuclei in the symmetry triangle [10] of the IBA where the vertices correspond to the dynamical symmetries of the IBA, U(5), SU(3), and O(6). The reason has to do with the 0_2^+ states, which have historically been difficult to understand in most models and which, consequently, have tended to be ignored in fitting phenomenological models to the data. With the success of X(5), which derives largely from the prediction of the properties of the 0_2^+ sequence of states, it was decided to re–do the extensive set of IBA fits in Ref. [11], but to require that the 0_2^+ also be well reproduced. The standard two–term IBA Hamiltonian H $= \epsilon n_d + \kappa Q \cdot Q$ was used, with parameters ϵ/κ and χ. Excellent fits were obtained, typically characterized by larger ϵ values than before. As a result, the trajectories of structural evolution across the rare earth region were significantly altered, especially for the Gd, Yb and Hf isotopic chains. The new trajectories [5] are summarized in Fig. 4. [See Ref. [5] for the detailed parameter values, for which Fig. 4 is a schematic proxy.]

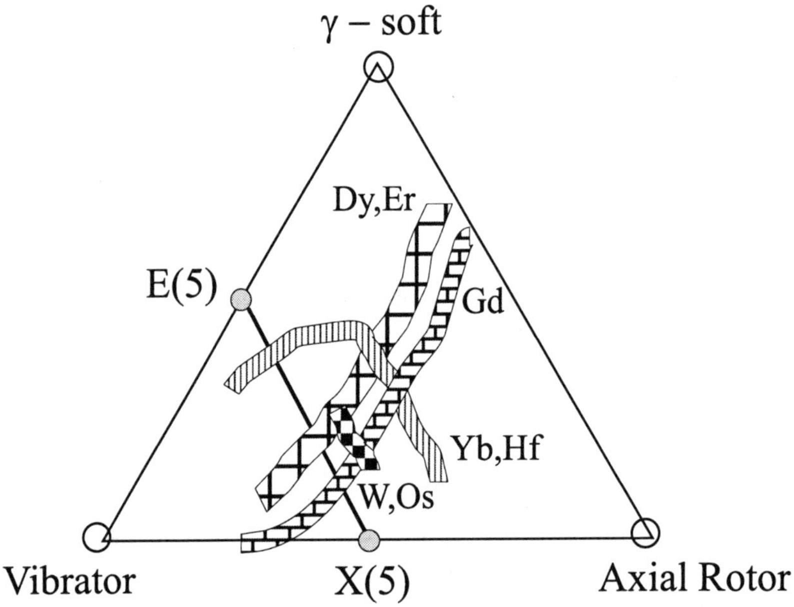

Figure 4. Schematic summary of the trajectories of structural evolution in the symmetry triangle of the IBA. Based on the detailed parameters in Ref. [5]. The regions of the triangle to the lower left and upper right of the slanting line of first order phase transitions represent spherical and deformed equilibrium configurations, respectively. The points in the triangle are defined by ϵ/κ, which is related to the distance of a point from the vibrator [U(5)] vertex, and χ which specifies the angle relative to the vibrator [U(5)] to axial rotor [SU(3)] leg.

4. Regularity vs. Chaos in Nuclear Spectra

The standard paradigms of nuclear structure, namely the vibrator, the symmetric rotor, and the γ–soft axially asymmetric rotor all display patterns of low lying energy levels that seem quite regular. This was borne out quantitatively about a decade ago in a study by Alhassid and Whelan [12,13] of the onset of chaos throughout the symmetry triangle. Their results indeed verified the regularity of spectra near the vertices, as well as along the U(5) to O(6) leg where the O(5) symmetry is conserved. Almost everywhere else chaotic behavior characterized the spectra. However, one surprising exception occurs on an unexpected arc of regularity [12,13] *within* the triangle but linking U(5) to SU(3). Here, rather suddenly as a function of position in the triangle, regular or near-regular behavior reappears. This arc is illustrated in Fig. 5. Presumably, this points to the emerging validity of some unidentified quantum number(s) and to the partial validity of some new, as yet unknown, symmetry.

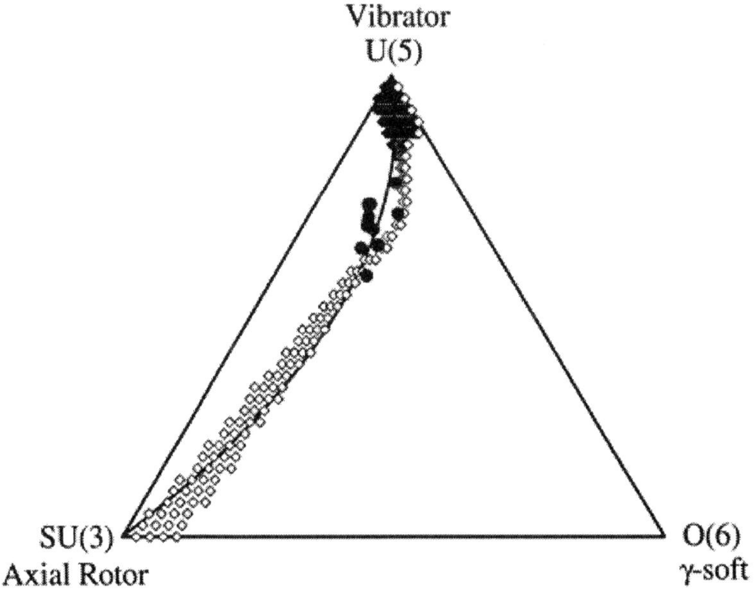

Figure 5. Symmetry triangle of the IBA (rotated relative to that in Fig. 4 for historical reasons) showing the arc of regularity. Based on Refs. [6,12,13]. The locations of the twelve nuclei along this arc of regularity are indicated (solid dots), based on Refs. [5,6]. The diamond hatching, starting at U(5) and descending along the arc towards SU(3), marks the locus of near degeneracy of the 0_2^+ and 2_2^+ states discussed in Ref. [6].

Unfortunately, at the time of Refs. [12,13], and subsequently, our best understanding of where nuclei lie in the triangle suggested no candidates for this arc of regularity. However, this has now changed. As a result of the new fits to rare earth nuclei in Ref. [5], discussed above, which were inspired in turn by the success of X(5) and the renewed interest in understanding 0_2^+ states, we have now discovered [6] twelve nuclei on or close to this arc. This set of nuclei comprises particular isotopes in Gd, Dy, Er, Yb, Hf, W, and Os ranging from A = 156 to 180. The locations of these nuclei are also indicated in Fig. 5. Note that, though these nuclei appear to lie close to each other along the arc, structure changes highly non–linearly in this region of the triangle (a point that is likewise illustrated by the very existence of the sharp deep valley of regularity itself). Thus, the twelve nuclei range in character from near vibrational, to transitional, to well deformed, and back towards less deformed character.

Moreover, we have discovered a robust empirical signature of this regularity, namely a near degeneracy of the 0_2^+ and 2_2^+ levels. The locus of this near degeneracy is shown in Fig. 5 by the hatching that tracks the arc. The twelve nuclei are themselves all characterized by close lying 0_2^+ and 2_2^+ states. It is important to stress the key role of this degeneracy condition. It is essentially synonymous with the arc of regularity. Moreover, it occurs *nowhere* else. Thus it is a unique signature that should be a necessary and sufficient condition to identify collective nuclei along the arc of regularity. Of course, while this degeneracy uniquely identifies nuclei in the regular region, it does not specify *where* they lie in this region. Other observables are needed to pinpoint their location and this has been done in placing these twelve nuclei in Fig. 5. The correlation between the 0_2^+–2_2^+ near degeneracy and the arc of regularity is so distinctive that it may provide clues to the conserved quantum numbers and symmetry underlying the regularity in this region of the triangle.

Needless to say it will be an important and fascinating program to search for additional examples of this regular behavior, especially in exotic nuclei. The unique signature of nearly degenerate 0_2^+ and 2_2^+ states is a sufficient indicator since no other region of the symmetry triangle displays this feature. The energies of these states should be rather simple to measure and, once again, β–decay may well provide the best technique.

Acknowledgments

We are grateful to our many collaborators in this work, especially to E.A. McCutchan, Victor Zamfir, Peter von Brentano, Jan Jolie, Volker Werner, Stefan Heinze, and Pavel Cejnar. Work supported by USDOE grant number DE-FG02-91ER-40609.

References

1. F. Iachello, Phys. Rev. Lett. **87**, 052502 (2001).
2. R.F. Casten and N.V. Zamfir, Phys. Rev. Lett. **87**, 052503 (2001).
3. R. Krücken *et al.*, Phys. Rev. Lett. **88**, 232501 (2002).
4. E.A. McCutchan *et al.*, Phys. Rev. **C 69**, 024308 (2004).
5. E.A. McCutchan, N.V. Zamfir and R.F. Casten, Phys. Rev. **C69**, 064306 (2004).
6. J. Jolie, R.F. Casten, P. Cejnar, S. Heinze, E.A. McCutchan and N.V. Zamfir, Phys. Rev. Lett. (to be published).
7. R.F. Casten, Phys. Rev. Lett. **54**, 1991 (1985).
8. R.F. Casten, D.S. Brenner and P.E. Haustein, Phys. Rev. Lett. **58**, 658 (1987).
9. E.A. McCutchan, N.V. Zamfir and R.F. Casten, *Symmetries in Nuclear Structure*, eds. A. Vitturi and R.F. Casten, World Scientific (2003).
10. R.F. Casten, *Interacting Bose–Fermi Systems in Nuclei*, ed. F. Iachello (Plenum, New York, 1981), p. 1.
11. W.-T. Chou, N.V. Zamfir and R.F. Casten, Phys. Rev. **C 56**, 829 (1997).
12. Y. Alhassid and N. Whelan, Phys. Rev. Lett. **67**, 816 (1991).
13. N. Whelan and Y. Alhassid, Nucl. Phys. **A 556**, 42 (1993).

KEY TOPICS IN NUCLEAR STRUCTURE
Proceedings of the 8th Int. Spring Seminar on Nuclear Physics
edited by Aldo Covello
© 2005 World Scientific Publishing Co.

FIRST MEASUREMENT OF A MAGNETIC MOMENT OF A SHORT-LIVED STATE WITH AN ACCELERATED RADIOACTIVE BEAM: ^{76}KR *

N. BENCZER-KOLLER, G. KUMBARTZKI, K. HILES
Department of Physics and Astronomy, Rutgers University, New Brunswick, NJ 08903, USA
E-mail: nkoller@physics.rutgers.edu

J. R. COOPER, L. BERNSTEIN, L. AHLE, A. SCHILLER
Lawrence Livermore National Laboratory, Livermore, CA 94550, USA

T. J. MERTZIMEKIS
NSCL, Michigan State University, East Lansing, MI 48824, USA

M. J. TAYLOR
School of Engineering, University of Brighton, Brighton BN2 4GJ, UK

M. A. MCMAHAN, L. PHAIR, J. POWELL, C. SILVER, D. WUTTE
Lawrence Berkeley National Laboratory, Berkeley, CA 94720, USA

P. MAIER-KOMOR
Technische Universität München, D-85748 Garching, Germany

K.-H. SPEIDEL
Helmholtz-Institut für Strahlen-und Kernphysik, Universität Bonn, D-53115 Bonn, Germany

*This work was partially supported by the U S National Science Foundation and the U S Department of Energy.

The understanding of the structure of nuclei very far-from-stability constitutes the next major challenge in the description of nucleon interactions resulting in low lying excited states. New techniques of measurement of magnetic moments of short-lived nuclear states combining Coulomb excitations of *beams* and the transient hyperfine magnetic interaction have led to the determination of magnetic moments of low-lying, short-lived, states with a precision that can distinguish between various theoretical calculations. The technique is particularly applicable to the study of radioactive beams and was used for the first time to measure the g factor of the 2_1^+ state of ^{76}Kr. The ^{76}Kr beam was produced and accelerated in batch mode at the Lawrence Berkeley National laboratory 88-Inch cyclotron. Peak rates of 10^8 particles/s were obtained yielding a g factor measurement of $g(^{76}; 2_1^+) = 0.37(11)$.

1. Introduction

Recent measurements of magnetic moments of short lived excited states in nuclei across the periodic table have set stringent constraints on the nuclear models proposed to explain the structure of these states. Recent g factor measurements of excited states of *stable* even A Kr isotopes stimulated this study toward lighter radioactive Kr nuclei. The IBA-II description provided a reasonable good picture of the structure far away from the closed neutron shell at $N = 50$ and deviates only for the very lightest nuclei ^{72,74,76}Kr. The neutrons and protons in these nuclei occupy the same orbitals and hence may provide evidence for neutron proton pairing correlations. Much more important is what happens far away from the valley of stability. It is expected that the spin-orbit coupling might be much reduced, the usual shell structure would be very different from the one that applies to the stable nuclei. In addition, it is not clear what the effective g factors would be under such conditions. In view of the anticipated novel physics that may be encountered far from stability, developing new approaches to the measurements of magnetic moments might become a very promising area of work.

Experimentally, the use of projectile excitation in inverse kinematics makes it possible to carry out fairly precise measurements which can probe the microscopic details of the wave functions of the nuclear states involved. This technique is particularly applicable to nuclei far-from-stability which are accessible only as radioactive beams. The first radioactive beam that has been produced that is amenable to such a measurement is a beam of ^{76}Kr ($T_{1/2} = 14.8$ h) which was produced and accelerated at the 88-Inch cyclotron at Lawrence Berkeley National Laboratory. The elements of the procedure are outlined below. The details are written up in two papers, Refs. [1,2]

1.1. *Production of ^{76}Kr*

The ^{76}Kr radioactive ion beam was produced using a batch mode method. The Lawrence Berkeley National Laboratory 88-Inch Cyclotron was used to produce the krypton isotope via the reaction ^{74}Se(α,2n)^{76}Kr. Approximately 10^{14} ^{76}Kr nuclei were produced during a 17-hour production period using a 6 particle-μA ^{4}He beam of 38 MeV on a 165 mg/cm^2 thick metallic ^{74}Se target. After irradiation the selenium was melted to release the krypton, which was transferred via a He gas flow to a cryogenic trap located near the upgraded Berkeley Advanced Electron Cyclotron Resonance (AECR-U) ion source. After the transfer, the charcoal trap was heated and the krypton gas was released into the AECR-U ion source. The 88-Inch Cyclotron then accelerated ^{76}Kr^{+15} to 230 MeV producing currents as high as 3x10^8 particles per second and yielding an average current of 4x10^7 particles per second for two hours on target. The use of the same cyclotron for production and acceleration has led to the name re-cyclotron method.

Three batches were produced during the g factor experiment. The first batch, which lasted three days, suffered from low ion source ionization efficiency and only 1x10^{11} particles were delivered to the experimental area. The last two batches, each one day long, delivered approximately 3x10^{11} particles to the target giving an integrated beam current of 7(1)x10^{11} particles during the five day experiment. For comparison with radioactive beam facilities providing a continuous beam, this intensity is equivalent to a constant beam of 1.6x10^6 particles per second for five days.

1.2. *Experimental setup*

The target chamber used during the stable krypton experiments was modified to allow for a moving tape, as shown in Fig. 1. The tape was mounted 2mm behind the target and moved at 8 cm/min to remove the stopped beam from the target area. The four Eurysis Clover high purity Ge detectors, used for γ ray detection, were shielded from the radiation collected on the cartridge by lead bricks.

1.3. *The magnetic moment measurement*

The same multi layer target used in all previous g-factor experiments on Kr isotopes was used again [3]. This target consisted of 0.9 mg/cm^2 of enriched ^{26}Mg evaporated on a 4.0 mg/cm^2 gadolinium layer itself deposited on a 1.1 mg/cm^2 tantalum foil, backed by a 3.9 mg/cm^2 thick copper layer. The

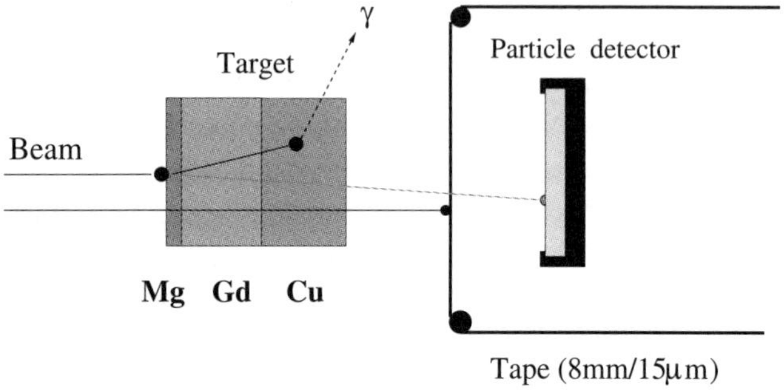

Figure 1. Modified setup for the radioactive beam experiment showing an enlarged view of the target, the particle detector and the moving tape which served as beam stopper.

gadolinium layer was magnetized by an external magnetic field of 0.06 T applied in a direction perpendicular to the γ–ray detection plane, either up or down. The target magnetization was measured before and checked after the experiment and found not to vary, with $\overrightarrow{M}(77\text{K}) = 0.1872$ T in the region $60 < \text{T} < 100$K. The target was kept at a nominal temperature of 77K by a liquid nitrogen reservoir. The kinematics applying to the run with ^{76}Kr and a calibration run carried out at the end of the experiment with a beam of stable ^{78}Kr are shown in Table 1.

Table 1. Summary of reaction kinematics characteristics: $< E >_{in}$, $< E >_{out}$, $< v/v_o >_{in}$, and $< v/v_o >_{out}$ are the average energies and velocities of the Kr isotopes entering into and exiting from the gadolinium foil. $v_o = e^2/\hbar$ is the Bohr velocity.

	E_{beam} [MeV]	$< E >_{in}$ [MeV]	$< E >_{out}$ [MeV]	$< v/v_o >_{in}$	$< v/v_o >_{out}$
^{76}Kr	230.0	56.1	9.8	5.3	2.3
^{78}Kr	224.0	60.0	11.9	5.6	2.5

The Kr ions that were Coulomb excited in the Mg layer were exposed to the transient magnetic field in the gadolinium layer before stopping in the copper backing. The beam itself traversed the copper layer and was stopped in the moving tape that was located between the target and the solar cell particle detector placed at 0° to the beam. The solar cell was of rectangular shape and subtended an angle of ±31° in the vertical directions and ±9° in the horizontal direction. The recoiling Mg ions traversed the whole target plus the tape and were detected in the solar cell. For the precession measurements deexcitation γ rays were detected in coincidence with the forward scattered Mg ions in four Clover Ge detectors placed at ± 67° and ± 113° with respect to the beam direction. Coincidence measurements are standard procedure for this type of experiments, but are particularly important for work with radioactive beams because they reduce the background caused by activity buildup accumulated in the target area. However, the activity contributes only randomly and can readily be subtracted from the coincidence data as shown in Fig. 2.

The precession of the spins aligned by the Coulomb excitation reaction in the transient magnetic field experienced by the moving ions in the gadolinium target, are extracted from the coincidence rates in the same manner as for experiments with stable beams. In short, the parameters of the angular correlation of the emitted γ-rays are determined, and the effect on the precession due to changing the polarization of the gadolinium with an external field are measured. The details can be found in Refs. [3,4,5].

The determination of this anisotropy requires an extra measurement and may not be feasible for a radioactive beam experiment. In principle, the angular correlation coefficients can be obtained from a COULEX code or a measurement under the "same" experimental conditions using a stable beam of the same element. In view of the similarity between the energy level structure of ^{76}Kr and ^{78}Kr, additional angular correlation and precession measurements were carried out with a ^{78}Kr beam. The energy of the beam, 224 MeV, was chosen so that the velocities of the ^{76}Kr and ^{78}Kr recoils in gadolinium were almost the same. The results of that run are also displayed in Table 2.

Table 2. Summary of the measured precession effects ϵ, the observed precession angles $\Delta\theta$ in mrad, and the resulting g factors.

	ϵ	$\Delta\theta$	g
^{76}Kr	-0.0475(136)	20.4(58)	+0.37(11)
^{78}Kr	-0.0552(15)	23.7(7)	+0.43(1)

2. Results

The accuracy of the g factor result depends on the measured photopeak intensity and the number of random coincidences and background in the spectra. In the six hours of ^{76}Kr beam on target each Clover counted about 800 events in the photopeak of the $2_1^+ \rightarrow 0_1^+$ transition for each field direction. Roughly 10% of the recorded rates were random, but no further

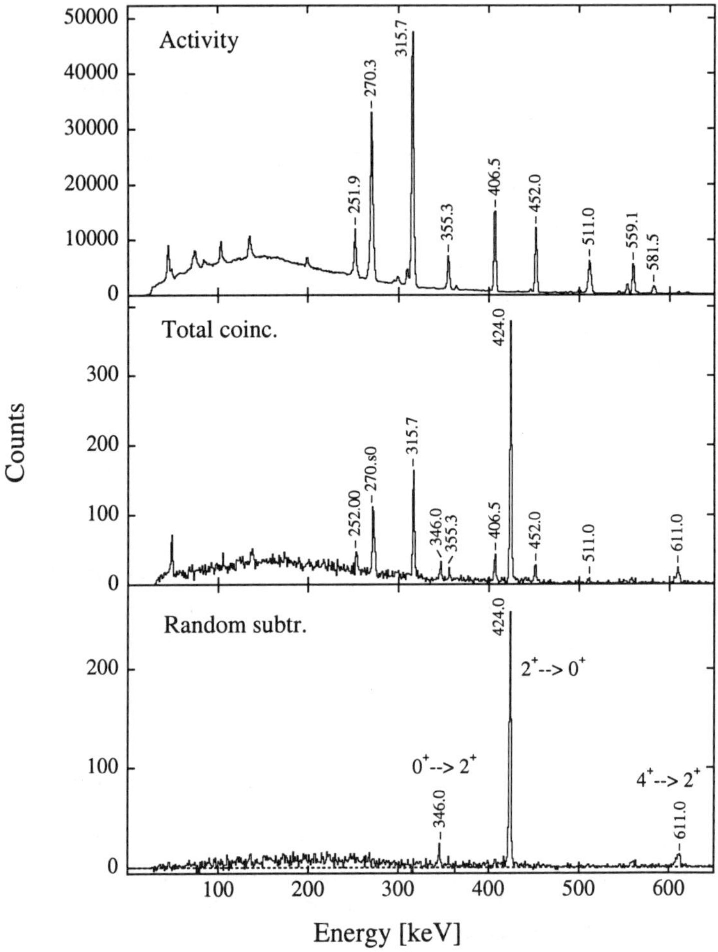

Figure 2. Top: a background spectrum taken after the end of a ^{76}Kr beam batch cycle (10 min singles rate in one Clover segment only). Middle: a γ-ray spectrum taken in coincidence with particles in the solar cell (total coincidence width = 500 ns during one production cycle for a full Clover). Bottom: the same Clover spectrum as shown in the middle panel with randoms subtracted. Only the ^{76}Kr γ-ray lines remain.

background had to be subtracted. For comparison, with the stable ^{78}Kr isotope beam 7x10^4 counts/Clover and field direction were recorded in the 455 keV peak in only 2.5 hours.

The data for ^{76}Kr can be analyzed now with the angular correlation data taken from the ^{78}Kr measurements. However, in view of the similarity of the level scheme and kinematic conditions for the two isotopes, a much more direct procedure that does not require knowledge of the transient magnetic field can be carried out. The g factor of the 2_1^+ state in ^{76}Kr can be directly written in terms of the known g factor of the 2_1^+ state in ^{78}Kr,

$$g(^{76}Kr; 2_1^+) = g(^{78}Kr; 2_1^+) \times \frac{\epsilon(^{76}Kr)}{\epsilon(^{78}Kr)} \tag{1}$$

The result of this approach yields:

$$g(^{76}Kr; 2_1^+) = +0.37(11) \tag{2}$$

3. Discussion

The g factors of the 2_1^+ states in the Kr isotopes have been measured across the region from the semi-magic ^{86}Kr to the lightest, radioactive ^{76}Kr and are summarized in Fig. 3.

^{86}Kr has a closed shell with 50 neutrons, and its large positive g factor of +1.12(14) (off scale in Fig. 3) is a clear indication of proton excitations. ^{84}Kr has a much smaller g factor, which is understood as arising from the additional contribution from the two neutron holes in the $g_{9/2}$ orbit. However, as more neutrons are removed to form the lighter isotopes, the g factors of the 2_1^+ states increase progressively toward the collective value of Z/A. At the same time, the g factors of the 4_1^+ and 2_2^+ states also tend to be equal to the nominal Z/A value [3]. Calculations based on the interacting boson model IBA-II, a "pairing-corrected" collective model and the shell model are presented in a recent paper [6] which interprets some of these results and examines the general systematics in the $A = 80$ mass region.

The N dependence of the IBA-II fits of the 2_1^+ g factors in Se isotopes suggested a possible subshell closure at $N = 38$ [4]. In the neighboring Kr isotopes similar fits proved inconclusive (Fig. 3). A distinction between the Z/A value and predictions of the IBA-II calculations can first be noted for ^{76}Kr. However, the accuracy of the g factor is not sufficient to unambiguously decide on this criterion. Furthermore, the experimental B(E2) values for the 2_1^+ states are in total disagreement with the assumption of a closed shell at $N = 38$ and reinforce the notion of shell closure at $N = 28$.

Shell model calculations were performed using the OXBASH code with a very truncated basis space and with two different interactions [6]. While

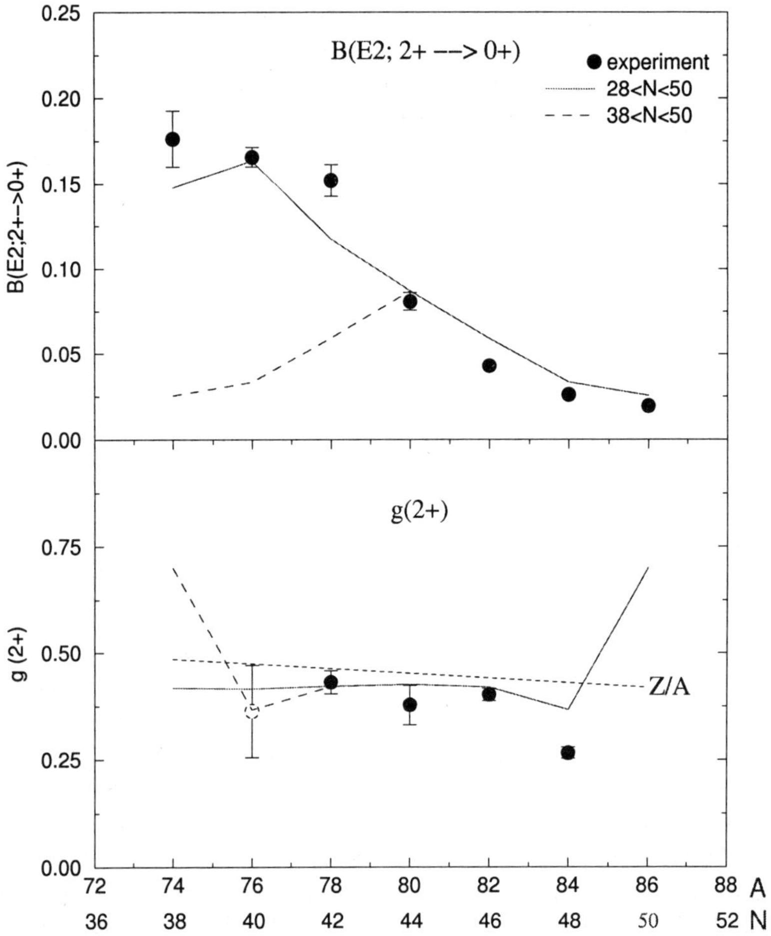

Figure 3. B(E2) values in e^2b^2 and g factors for even Kr isotopes. The new ^{76}Kr point is shown as open circle. The curves are IBA-II calculations as described in Ref.3 assuming shell closure for $N = 28$ or 38.

exact values cannot be predicted by these calculations, the trends in g factors with mass were accurately described. The similarity between the g factors of the 2_1^+ states in the light Kr and the Se isotopes further suggests that these nuclei tend to reflect mainly collective properties. Detailed shell calculations are prohibitive due to the large number of particles outside closed shells.

In summary, this experiment provided the first measurement of a g

factor carried out by the Coulomb excitation/transient field technique on a radioactive beam. The experiment was not hindered by the fact that the beam was radioactive. The result confirms the collective nature of the structure of the 2_1^+ state of ^{76}Kr. Future experiments on lighter Kr isotopes, with higher statistical accuracy, should help map out the nature of the collective excitations in this region.

4. Future

In the future the Rare Isotopes Accelerator will be built. However, in the mean time the method described in this paper as well as other techniques being developped at Argonne National Laboratory and Michigan State University, among others, need to be pursued. The recyclotron technique has limited scope. But at Oak Ridge National Laboratory, where the isotope can be produced in one accelerator and injected into a second one, a variety of beams have been produced. In particular, relatively intense beams of ^{132,134,136}Te have been accelerated and the B(E2;$0_1^+ \rightarrow 2_1^+$) values have been measured. A measurement of the g factor of the 2_1^+ state in ^{132}Te has been planned for the near future, and hopefully with planned improvements in the ion source, the other two Te isotopes will also become amenable to measurement. Unfortunately these measurements require fairly intense beams and improvements on the beam delivery and the experimental efficiency need to be made.

References

1. J. Cooper *et al.* Nucl. Instrum. Methods Phys. Res.(2004)
2. G. Kumbartzki *et al.* Phys. Lett. B (2004).
3. T. J. Mertzimekis *et al.* Phys. Rev. C**64**,024314 (2001).
4. K.-H. Speidel *et al.* Phys. Rev. C**57**,2181 (1998).
5. J. Holden *et al.* Phys. Rev. C**63**, 024315 (2001).
6. T. J. Mertzimekis *et al.* Phys. Rev. C**68**,054304 (2003).

KEY TOPICS IN NUCLEAR STRUCTURE
Proceedings of the 8th Int. Spring Seminar on Nuclear Physics
edited by Aldo Covello
© 2005 World Scientific Publishing Co.

NEW INTERACTIONS, EXOTIC PHENOMENA AND SPIN SYMMETRY FOR ANTI-NUCLEON SPECTRUM IN RELATIVISTIC APPROACH

J. MENG* S. F. BAN L. S. GENG J. Y. GUO J. LI W. H. LONG
H. F. LÜ J. PENG G. SHEN S. Q. ZHANG S. S. ZHANG
W. ZHANG S. G. ZHOU

School of Physics, Peking University, Beijing 100871, China

New effective interactions, PK1, PK1r and PKDD, in the relativistic mean field (RMF) theory, are proposed with the center-of-mass correction included in a microscopic way. They are able to provide an excellent description not only for the properties of nuclear matter and neutron stars, but also for the nuclei near and far from the β-stability line, including halos and giant halos at the neutron drip line in nuclei and hypernuclei. Based on the solution of the RMF equations, a good spin symmetry is found in anti-nucleon spectra. The magic proton and neutron numbers are searched in the superheavy region by the relativistic continuum Hartree-Bogoliubov (RCHB) theory. Spherical doubly magic superheavy nuclei are systematically investigated and their stabilities against deformation are discussed.

1. Introduction

We review our recent works on exotic nuclei, new effective interactions, new symmetry and mass limits in atomic nuclei. With the self-consistent microscopic relativistic continuum Hartree-Bogoliubov (RCHB) theory [1,2], the ground state properties including radii, density distributions and one neutron separation energies for C, N, O and F isotopes up to the neutron drip line have been systematically studied and the charge-changing cross sections for C, N, O, F and Na isotopes are calculated using the Glauber model. Good agreement with the data has been achieved [3,4]. The proton magic O, Ca, Ni, Zr, Sn, and Pb isotopes from the proton drip line to the neutron drip line are investigated and halos and giant halos in Ca isotopes with $A > 60$ and Ne-Na-Mg drip line nuclei are investigated [5,6]. Based on the systematic investigation of the data available for nuclei with $A \geq 40$, a

*e-mail: mengj@pku.edu.cn

new isospin dependent formula for the nuclear charge radii is proposed [7]. In order to investigate the halo nuclei with deformation, the Woods-Saxon (WS) basis is suggested to replace the widely used harmonic oscillator basis for solving the RMF equations [8,9]. Along this line, the spectra in the Dirac sea, or equivalently the spectra in the anti-particle channel, are studied in detail and a new symmetry, i.e., the spin symmetry in the anti-particle spectra is found [10]. The origin of the spin symmetry in the anti-particle spectra are found to be similar as that of the pseudo-spin symmetry in the particle spectra [11,12].

We also investigated other interesting topics in nuclear physics, including the structure and synthesis for superheavy nuclei [13,14,15], the hyperon-nucleon interaction, properties of neutron star [16], the existence of hyperon halos and neutron halos in hyper-nuclei [17,18], the solution of the Dirac equation with special Hulthen potentials[19], the energies and widths for single-particle resonant states in square-well, harmonic-oscillator and Woods-Saxon potentials by analytic continuation in the coupling constant method [20], the magnetic rotation and chiral doublets for nuclei in $A \sim 100$ and $A \sim 130$ regions with cranking RMF theory and the particle-rotor model [21,22], the properties of light nuclei, deformed proton emitters and alpha-decay properties of lately synthesized superheavy elements with Z=115 and 113 [23], etc..

Here in this paper, a brief introduction on the new interactions in RMF theory, PK1, PK1r and PKDD, the spin symmetry in anti-nucleon spectra, the exotic phenomena and mass limits in atomic nuclei will be presented.

2. New effective interactions in RMF theory

In the relativistic mean field theory, the effective interactions are adjusted to reproduce various properties of nuclear matter and finite nuclei. A number of effective interactions of meson-baryon couplings based on the RMF theory have been developed, including nonlinear self-couplings for the σ-meson or ω-meson, such as NL1, NL2 [24], NL3 [25], NLSH [26], TM1 and TM2 [27]. These nonlinear interactions have problems of stability at high densities, as well as the question of the physical foundation [24]. A more natural alternative is to introduce a density dependence in the couplings. Based on the results of the Dirac-Brueckner theory, Typel and Wolter proposed the density-dependent effective interaction TW-99 and expected that the model could be extrapolated to nuclear matter at extreme conditions of isospin and/or density [28]. Along this line, Nikšić et al. developed another effective interaction DD-ME1 [29].

Using the nonlinear effective interactions, NL1, NL2, NL3, NLSH, TM1, TM2, GL-97 [30] and the density-dependent effective interactions TW-99, DD-ME1, the density dependence of the effective interaction strengths and the influence on nuclear matter and neutron stars are studied and carefully compared. The formulism and numerical details can be found in Ref. [31], which gives us strong information that the density dependency for all interactions needs to be treated carefully.

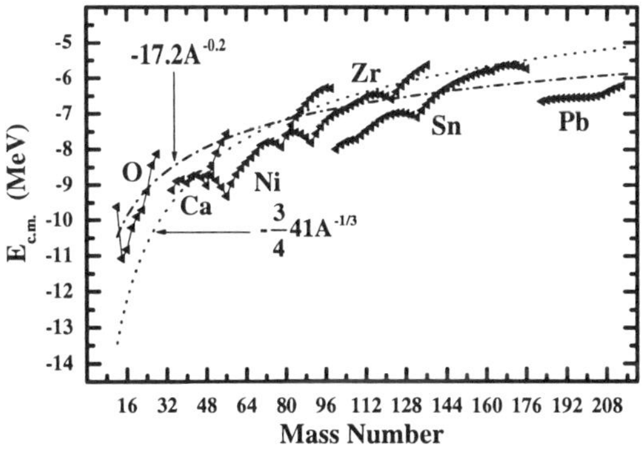

Figure 1. The microscopic center-of-mass correction in comparison with phenomenological ones.

Besides the density dependence, the isospin dependence has to be properly taken into account. Along the β-stability line, NL1 gives excellent results for binding energies and charge radii. Far away from the stability line the results are less satisfactory due to the large asymmetry energy $a \simeq 44$ MeV. In addition the calculated neutron skin thickness shows systematic deviations from the data. NLSH produces an asymmetry energy $a \simeq 36$ MeV while giving a slight over-binding along the line of β-stability. It also fails to reproduce the super-deformed minima in Hg-isotopes. For the nuclear matter incompressibility, K_{NL1}=212 MeV while K_{NLSH}=355 MeV. Both fail to reproduce the isoscalar giant monopole resonances for Pb and Zr nuclei. As an improvement, the effective interactions NL3 and TM1, provide reasonable incompressibility ($K_{\mathrm{NL3}} = 271.7$ MeV and $K_{\mathrm{TM1}} = 281.16$ MeV), and asymmetry energy ($a_{\mathrm{NL3}} = 37.4$ MeV and $a_{\mathrm{TM1}} = 36.89$ MeV), but they give fairly small saturation density ($\rho_{\mathrm{NL3}} = 0.148$ fm^{-3} and $\rho_{\mathrm{TM1}} = 0.145$

fm^{-3}). One should note that in all above parameterizations (NL1, NLSH, NL3 and TM1), the center-of-mass corrections are treated in a phenomenological way. In Fig. 1, the microscopic center-of-mass correction from the RCHB calculation [1,2] for proton magic isotopes is shown in comparison with usual phenomenological estimations. For the binding energies, this center-of-mass correction is sizable in light nuclei (about 9% in ^{16}O) but much less important in medium and heavy nuclei (about 0.4% in ^{208}Pb) as shown in Fig. 1.

As the effective interaction without nonlinear ω terms leads to strong repulsive potential for nuclear matter at high density, new effective interactions with nonlinear ω self-coupling (PK1) and also nonlinear ρ self-coupling (PK1r) have been developed. Following the density dependent interactions TW-99 and DD-ME1, PKDD is also proposed. These three new interactions can be found in Ref.[32] compared with other interactions. The density dependency of the new interaction strengths has been considered and studied in Ref.[32]. Their isospin dependence and the microscopic estimation for center-of-mass corrections have been improved by the fitting of new effective interactions.

We also compared the masses and charge radii of finite nuclei obtained with other effective interactions, our newly obtained ones reproduce well the experimental masses [34]. For these new effective interactions, only $4 \sim 5$ nuclear masses deviate by more than 1 MeV, the new effective interactions PK1, PK1r and PKDD also well describe the charge radii for these nuclei, especially for Pb isotopes. One can get a clear idea about the improvement of the new parameter sets on the description of bulk properties for finite nuclei from the root of relative square (rrs) deviation δ. The rrs deviation for the total binding energy, the rrs deviations from the new parameter sets are much smaller than those from old ones. For the charge radius, the rrs deviations from the new interactions are comparable with those from NL3 and DD-ME1, but a bit smaller than those from TM1 and TW-99 [32].

3. Halos and giant halos

From the binding energies we can extract systematics in the two-neutron separation energies $S_{2n} = E_B(N,Z) - E_B(N-2,Z)$. In Fig. 2, the systematic behavior of two-neutron separation energies against the neutron number for the proton magic nuclei O, Ca, Ni, Zr, Sn, and Pb, predicted by the new effective interactions PK1 and PKDD, is shown. Seen from these figures, the newly obtained interactions give a good description for

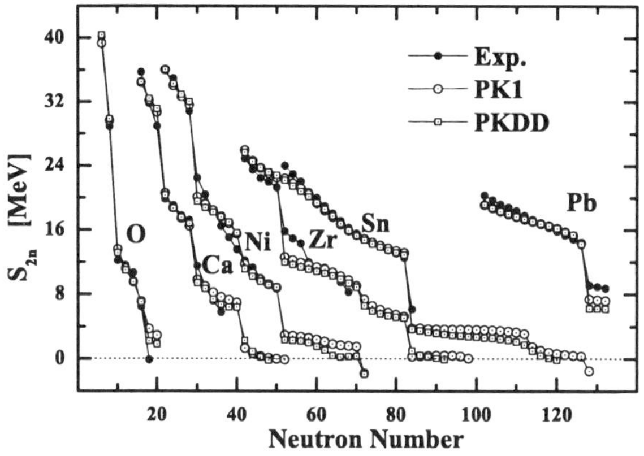

Figure 2. The two neutron separation energies calculated with PK1 and PKDD for proton magic isotopes.

two-neutron separation energies. Here all the theoretical results are from RCHB theory [2]. The interesting phenomena, the so-called giant halo predicted in Ca and Zr isotopes near the neutron drip line [2,5,6,17,18], exist also for these new interactions.

Unfortunately the predicted giant halo Ca or Zr nuclei are too neutron rich to be produced in nowadays accelerator. So it is an interesting question whether there exist giant halos in more wider mass regions. In ref.[6], the two-neutron separation energies S_{2n} in neutron drip-line region for Ne, Na, Mg, Al and Ar, as well as K, Ca, Sc, and Ti even-neutron nuclei are studied respectively. It is found that the two-neutron separation energies S_{2n} for all these isotope chains are almost parallel, and more than one lines lie within 2 MeV in the drip line region. Therefore there are quite a large mass region where giant halos may exist. Recently, it has been reported that both ^{37}Na and ^{34}Ne are bound, while ^{33}Ne and ^{36}Na are missing in experiment [36]. As ^{37}Na and ^{34}Ne nuclei lie in the drip line region and approach to the predicted giant halo nuclei, we suggest that much more efforts be devoted to the experimental study in this area.

4. Spin symmetry in anti-nucleon spectra

It's well known that the harmonic oscillator (HO) basis is not suitable for studying exotic nuclei. In order to give a proper description for exotic nuclear phenomena, such as halo, one should adopt the coordinate space (r space). However, for deformed nuclei, working in r space becomes much more difficult and numerically very sophisticated [8]. A reconciler between the HO basis and r space may be the WS basis. As a first attempt, the spherical relativistic Hartree theory in the WS basis (SRHWS) is proposed in Ref. [9] where the WS basis is obtained by solving either the Schrödinger equation (SRHSWS) or the Dirac equation (SRHDWS). The WS basis in the SRHDWS theory is much smaller than that in the SRHSWS theory, which will largely facilitate the deformed problem.

For stable nuclei, all approaches give identical results for properties such as total binding energies and the neutron, proton charge rms radii as well as neutron density distributions. For neutron drip line nuclei, e.g. ^{72}Ca, which has a very small neutron Fermi energy, $\lambda_n \sim 0.2$ MeV, both SRHR and SRHWS easily approach convergence by increasing the box size, while SRHHO does not. Furthermore, SRHWS can satisfactorily reproduce the neutron density distribution from SRHR, but SRHHO fails with similar cutoff's, as shown in ref. [9].

In SRHDWS calculations, negative energy states in the Dirac sea must be included in the basis in terms of which nucleon wave functions are expanded. We studied in detail the effects and contributions of negative energy states. Without the inclusion of negative energy levels, the calculated nuclear properties depend very much on the initial potentials. It is found that a small component from negative states in the wave functions, about $10^{-4\sim-5}$, contributes to the physical observances such as E/A and $r_{\rm rms}$ by the magnitude of 1%~10%. Again we notice that when the initial potentials differ more from the converged ones, the contribution from negative energy levels becomes more important. It can be concluded that the WS basis provides a compromise between the harmonic oscillator basis and the coordinate space which may be used to describe exotic nuclei both properly and efficiently [9].

When we studied in detail the effects and contributions of negative energy states, we found a very well developed spin symmetry in anti-nucleon spectra. In Fig. 3, we present the spin-orbit splitting in anti-neutron spectra of ^{16}O and ^{208}Pb. For ^{16}O, the spin-orbit splitting is around 0.2-0.5 MeV for p states. With increasing particle number A the spin symmetry in the anti-particle spectra becomes even more exact. For ^{208}Pb, the spin-orbit splitting is about 0.1 MeV for p states and less than 0.2 MeV even for h states. The dominant components of the wave functions of the spin doublet are almost identical. This spin symmetry in anti-particle spectra and the pseudo-spin symmetry in particle spectra have the same origin[10~12]. However the spin symmetry in anti-nucleon spectra is much better developed than the pseudo-spin symmetry in normal nuclear single particle spectra [10].

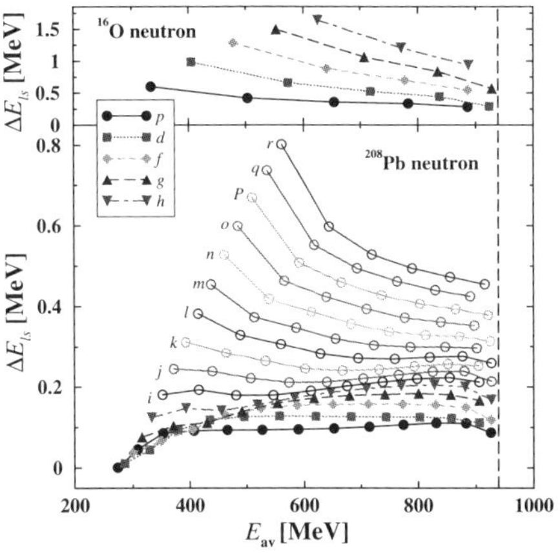

Figure 3. Spin-orbit splitting $\epsilon_t A(nl_{l-1/2}) - \epsilon_t A(nl_{l+1/2})$ in anti-neutron spectra of ^{16}O and ^{208}Pb versus the average energy of a pair of spin doublets. The vertical dashed line shows the continuum limit.

5. Magic numbers in superheavy nuclei

The magic numbers in superheavy region have been searched by the RCHB calculation with NL1, NL3, NLSH, TM1, TW-99, DD-ME1, PK1 and PK1r interactions for more than 1200 even-even nuclei, with proton number Z ranging from 100 to 140 and neutron number $N=(Z+30)$ - $(2Z+32)$. Based on the two nucleon separation energies S_{2n} and S_{2p}, the two nucleon gaps δ_{2n} and δ_{2p}, the shell correction energies E^n_{shell} and E^p_{shell}, the pairing energies E^n_{pair} and E^p_{pair}, and the pairing gaps Δ_n and Δ_p calculated with various effective interactions, Z = 120, 132, and 138, and N = 172, 184, 198, 228, 238 and 258 are inferred to be magic numbers [37], as shown in Table 1. In order to see the validity of spherical configuration assumption in the RCHB calculation, the deformation-constrained RMF calculations have been done. It is found that both the spherical and large-deformed minima for $^{292}120$ are well developed. The two minima with different deformations and almost same energies indicate that the so-called "shape coexistence" may exist. The shell effects stabilizing the superheavy nuclei are emphasized by extracting the shell correction energies.

Table 1. The possible magic proton number suggested by two-proton separation energies S_{2p} , two-proton gaps δ_{2p}, shell correction energies E^p_{shell}, pairing energies E^p_{pair} and effective pairing gaps Δ_p for proton with interactions NL1, NL3, NL-SH, TM1, TW-99, DD-ME1, PK1 and PK1r, respectively.

Effective interactions	Quantity	Magic proton number candidates					
		106	114	120	126	132	138
NL1	S_{2p}			■		■	■
	δ_{2p}			■	■	■	■
	E^p_{shell}	■		■	■	■	■
	E^p_{pair}	■		■	■	■	■
	Δ_p	■		■	■	■	■
NL3	S_{2p}	■		■		■	■
	δ_{2p}	■		■		■	■
	E^p_{shell}	■		■		■	■
	E^p_{pair}	■	■	■		■	■
	Δ_p	■	■	■		■	■
NLSH	S_{2p}	■		■		■	■
	δ_{2p}	■	■	■		■	■
	E^p_{shell}	■	■	■		■	■
	E^p_{pair}	■	■	■		■	■
	Δ_p	■	■	■		■	■
TM1	S_{2p}			■		■	■
	δ_{2p}			■		■	■
	E^p_{shell}			■		■	■
	E^p_{pair}			■		■	■
	Δ_p			■		■	■
TW99	S_{2p}	■		■		■	■
	δ_{2p}	■	■	■		■	■
	E^p_{shell}	■	■	■		■	■
	E^p_{pair}	■	■	■	■	■	■
	Δ_p	■	■	■	■	■	■
DD-ME1	S_{2p}			■		■	■
	δ_{2p}	■		■		■	■
	E^p_{shell}	■		■		■	■
	E^p_{pair}	■	■	■	■	■	■
	Δ_p	■	■	■	■	■	■
PK1	S_{2p}	■		■		■	■
	δ_{2p}	■		■		■	■
	E^p_{shell}	■	■	■		■	■
	E^p_{pair}	■	■	■		■	■
	Δ_p	■	■	■		■	■
PK1r	S_{2p}	■		■		■	■
	δ_{2p}	■		■		■	■
	E^p_{shell}	■	■	■		■	■
	E^p_{pair}	■	■	■		■	■
	Δ_p	■	■	■		■	■

6. Summary

The new effective interactions PK1, PK1r, and PKDD are proposed. They are able to provide an excellent description not only for the properties of nuclear matter and neutron stars, but also for nuclei in the nuclear chart from the proton to the neutron drip line, including halos and giant halos at the neutron drip line in nuclei and hyper-nuclei. To describe the deformed nuclei close to the drip-line, the Woods-Saxon basis has been suggested to replace the widely used harmonic oscillator basis for solving the RMF theory. We find that the spin symmetry is well developed in the single anti-nucleon spectra in nuclei. The origin of the spin symmetry in anti-nucleon spectra and that of the pseudo-spin symmetry in nucleon spectra are the same but the former is much more conserved in real nuclei. Based on the systematic calculation from RCHB theory, new magic numbers Z=120, 132 and 138, and N=172, 184, 198, 228, 238 and 258 for superheavy nuclei are suggested.

Acknowledgments

This work is partly supported by the Major State Basic Research Development Program Under Contract Number G2000077407 and the National Natural Science Foundation of China under Grant No. 10025522 and 10221003.

References

1. J. Meng and P. Ring, Phys. Rev. Lett.**77**, 3963 (1996); J. Meng and P. Ring, Phys. Rev. Lett. **80**, 460 (1998).
2. J. Meng, Nucl. Phys. A **635**, 3 (1998).
3. J. Meng, I. Tanihata and S. Yamaji, Phys. Lett. B **419**, 1 (1998).
4. J. Meng, S. G. Zhou and I. Tanihata, Phys. Lett. B **532**, 209 (2002).
5. J. Meng, H. Toki, J. Y. Zeng, S. Q. Zhang and S. G. Zhou, Phys. Rev. C **65**, 041302 (2002).
6. S. Q. Zhang, J. Meng and S. G. Zhou, Science of China G **46**, 632(2003).
7. S. Q. Zhang, J. Meng, S. G. Zhou and J. Y. Zeng, Euro. Phys. J. A **13**, 285 (2002).
8. S. G. Zhou, J. Meng, S. Yamaji and S. C. Yang, Chin. Phys. Lett. **17**, 717 (2000).
9. S. G. Zhou, J. Meng and P. Ring, Phys. Rev. C **68**, 034323 (2003).
10. S. G. Zhou, J. Meng and P. Ring, Phys. Rev. Lett. **91**, 262501 (2003).
11. J. Meng, K. Sugawara-Tanabe, S. Yamaji, P. Ring and A. Arima, Phys. Rev. C **58**, R628 (1998); J. Meng, K. Sugawara-Tanabe, S. Yamaji and A. Arima, Phys. Rev. C **59**, 154 (1999).

12. T. S. Chen, H. F. Lü, J. Meng and S. G. Zhou, Chin. Phys. Lett. **20**, 358 (2003).
13. J. Meng and N. Takigawa, Phys. Rev. C **61**, 064319 (2000).
14. W. H. Long, J. Meng and S. G. Zhou, Phys. Rev. C **65**, 047306 (2002).
15. W. Zhang, J. Meng and S. Q. Zhang, HEP&NP, **28**, 61 (2004) (in Chinese).
16. J. Li, S. F. Ban, H. Y. Jia, et al., HEP & NP **28**, 140 (2004) (in chinese); H. Y. Jia, B. X. Sun, J. Meng and E. G. Zhao, Chin. Phys. Lett. **18**, 1517 (2001); H. Y. Jia, H. F. Lü and J. Meng, HEP & NP **26**, 1050 (2002) (in chinese); B. X. Sun, H. Y. Jia, J. Meng and E. G. Zhao, Commum. Theor. Phys. **36**, 446 (2001).
17. J. Meng, H. F. Lü, S. Q. Zhang and S. G. Zhou, Nucl. Phys. A **722**, 366c (2003).
18. H. F. Lü, J. Meng, S. Q. Zhang and S. G. Zhou, Euro. Phys. J. A **17**, 19 (2003). H. F. Lü and J. Meng, Chin. Phys. Lett. **19**, 1775 (2002).
19. J. Y. Guo , J. Meng, F. X. Xu, Chin. Phys. Lett. **20(5)** 602 (2003).
20. S. S. Zhang, J. Y. Guo, S. Q. Zhang and J. Meng, Chin. Phys. Lett. **21**, 632 (2004); S. S. Zhang, J. Meng and J. Y. Guo, HEP & NP, **27**, 1095 (2003) (in chinese).
21. H. Madokoro, J. Meng, M. Matsuzaki and S. Yamaji, Phys. Rev. C **62**, 061301 (2000).
22. J. Peng, J. Meng, S. Q. Zhang, Phys. Rev. C **68**, 044324 (2003); J. Peng, J. Meng, S. Q. Zhang, Chin. Phys. Lett. **20**, 1223 (2003).
23. L. S. Geng, H. Toki, S. Sugimoto and J. Meng, Prog. Theor. Phys., **110**, 921 (2003); L. S. Geng, H. Toki, A. Ozawa and J. Meng, Nucl. Phys. A **730**, 80 (2004); L. S. Geng, H. Toki and J. Meng, nucl-th/0309016; L. S. Geng, H. Toki and J. Meng, Phys. Rew. C **68** 061303 (2003).
24. Lee Suk-Joon et. al, Phys. Rev. Lett. **57**, 2916 (1986).
25. G. A. Lalazissis, J. Konig and P. Ring, Phys. Rev. C **55**, 540 (1997).
26. M. M. Sharma, M. A. Nagarajan and P. Ring, Phys. Lett. B **312**, 377 (1993).
27. Y. Sugahara, H. Toki and P. Ring, Theor. Phys. **92**, 803 (1994).
28. S. Typel and H. H. Wolter, Nucl. Phys. A **656**, 331 (1999).
29. T. Nikšić, D. Vretenar, P.Finelli and P. Ring, Phys. Rev. C **66**, 024306 (2002).
30. N. K. Glendenning, Compact Stars, (Springer-Verlag, New York, 1997), 232 (1997).
31. S. F. Ban, J. Li, S. Q. Zhang, H. Y. Jia, J. P. Sang and J. Meng, Phys. Rev. C 69, 058405 (2004).
32. W. H. Long, J. Meng, N. V. Giai and S. G. Zhou, Phys. Rev. C **69**, 034319 (2004).
33. J. Meng, S. F. Ban, J. Li, et. al., Yadernaya Fizika, **67(9)** (2004).
34. G. Audi, A. H. Wapstra, Nucl. Phys. A **595**, 409 (1995).
35. W. Koepf and P. Ring, Z. Phys. A 339, 81 (1991).
36. M. Notani et. al., Phys. Lett. B **542**, 49 (2002); S. M. Lukyanov, Yu E. Penionzhkevich, R. Astabatyan et. al., J. Phys. G **28**, L41 (2002).
37. W. Zhang, J. Meng, S. Q. Zhang, L. S. Geng and H. Toki, submitted to Phys. Rev. C. See also http://xxx.itp.ac.cn/abs/nucl-th/0403021

KEY TOPICS IN NUCLEAR STRUCTURE
Proceedings of the 8th Int. Spring Seminar on Nuclear Physics
edited by Aldo Covello
© 2005 World Scientific Publishing Co.

NUCLEAR STRUCTURE BEYOND THE PROTON DRIP-LINE*

LíDIA S. FERREIRA
Centro de Física das Interacções Fundamentais, and Departamento de Física, Instituto Superior Técnico, Av. Rovisco Pais, P1049-001 Lisbon, Portugal.
E-mail:flidia@ist.utl.pt

The non–adiabatic quasi-particle approach for proton radioactivity from deformed drip–line nuclei is discussed. All experimental data available on decay from ground and isomeric states, and fine structure of odd-even and odd–odd nuclei, are consistently interpreted without free parameters, by this theoretical approach.

1. Introduction

Recent studies with exotic nuclei lead to the discovery of a new form of radioactivity[1,2] in nuclei lying beyond the proton drip-line with $50 < Z < 83$, where the Coulomb barrier is very high and protons can be trapped in a resonance state. Spherical[3] as well as deformed proton radioactive nuclei, even with large deformations[4] were observed, decaying mainly from the ground state, but decay from isomeric excited states of the parent nucleus and fine structure[5] have also been observed. The escape energy of the emitted proton is very small, therefore resonances lie very low in the continuum, and correspond essentially to single particle excitations, in contrast with what happens in stable nuclei.

From the experimental point of view, proton radioactivity provides measurements of the separation energy of the proton, which are important to test mass formulae, and to map the proton drip-line. Using tagging techniques with the proton, it is possible to measure the decay spectrum of the nucleus, which is very difficult to obtain otherwise. On the theoretical side, it is a quite important tool to study nuclear structure on exotic nuclei far away from the stability valley.

---

*The present work was done in collaboration with Enrico Maglione.

Due to the single particle character of this decays, and the large potential energy barrier, if the emitter is spherical a simple WKB calculations can already give a good estimate of the experimental data[3]. The purpose of this work is to show how decay from deformed nuclei can be studied as decay from a Nilsson resonance in a deformed system[6,7]. The parent nucleus can be treated in the strong coupling limit of the particle-rotor model[8], as a first approach. The inclusion[9] of Coriolis mixing, requires a proper treatment of the pairing residual interaction, leading to a unified interpretation of all available data on these emitters.

2. Decay widths for deformed proton emitters: Adiabatic approach

Considering proton radioactivity from deformed nuclei with an even number of neutrons and odd Z, the partial decay width can be determined from the overlap between the initial and final states. Since nuclei on the drip–line have a Fermi level very close or even immersed in the continuum, decay of odd–Z even–N nuclei has been interpreted, as decay from a single particle Nilsson resonance of the unbound core–proton system. The states close to the Fermi surface are the most probable ones for decay to occur and the corresponding wave function of the decaying proton, can than be obtained from the exact solution of the Schrödinger equation with a deformed mean field with deformed spin–orbit, imposing outgoing wave boundary conditions, as discussed in Ref[10,11]. The simplest approach to determine the wave function of the parent nucleus is to impose the strong coupling limit[8], where the nucleus behaves as a particle plus rotor with infinite moment of inertia.

Within these assumptions, if decay occurs to the ground state, only the component of the s.p. wave function with the same angular momentum as the ground state contributes, and the decay width becomes,

$$\Gamma_{l_p j_p}(r) = \frac{\hbar^2 k}{\mu(j_p + 1/2)} \frac{|u_{l_p j_p}(r)|^2}{|G_{l_p}(kr) + iF_{l_p}(kr)|^2} u_{K_i}^2, \tag{1}$$

where F and G are the regular and irregular Coulomb functions, respectively, and $u_{l_p j_p}$ the component of the wave function with momentum j_p, equal to the spin of the decaying nucleus. The quantity $u_{K_i}^2$ is the probability that the single particle level in the daughter nucleus is empty, evaluated in the BCS approach.

In case of decay to the ground state of the daughter nucleus, angular momentum conservation leads to $j_p = J_i = K_i$, only one component of

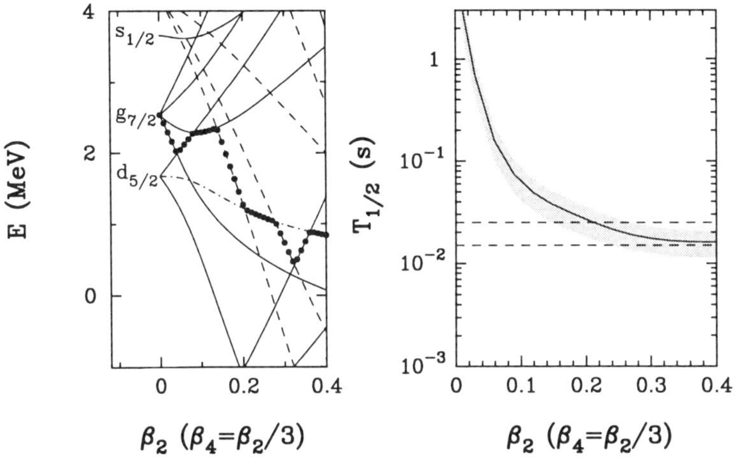

Figure 1. Proton Nilsson levels in ^{117}La (left part). The full circles represent the Fermi surface, the dashed lines the negative parity states, and the dashed–dotted line, the decaying state. A hexadecapole deformation $\beta_4 = \beta_2/3$ was included. Half–life for decay from the ground state of ^{117}La as a function of deformation (right side). The experimental value[12] is within the dashed lines.

the wavefunction is tested, and could even be very small. Decay to excited states, allow few combinations for $l_p j_p$ according to angular momentum coupling rules, and consequently different components of the parent wave function are then tested. Similar calculations were done within the coupled-channel Green's function model[4].

The decay width obtained from Eq. 1 depends on deformation, and is very sensitive to the wave function of the decaying state. Therefore, if it is able to reproduce the experimental value, will give clear information on the deformation and properties of the decaying state. The method is illustrated in Figure 1 for the proton emitter ^{117}La. The state $K = 3/2^+$ reproduces the experimental half–life for a deformation $\beta_2 \approx .2 - .3$, with a small hexadecapole contribution $\beta_4 = .1$, in close agreement with the theoretical predictions of Ref. [13].

Emission from deformed systems with an odd number of protons and neutrons can be discussed in a similar fashion[7]. However, in contrast with decay to ground state of odd-even nuclei where the proton is forced to escape with a specific angular momentum, many channels will be open due to the angular momentum coupling of the proton and daughter nucleus, $\vec{J}_d + \vec{j}_p$, giving the total width for decay as a sum of partial widths allowed by parity and momentum conservation. This dependence on the quantum numbers of

Table 1. Total angular momentum and deformation that reproduce the experimental half-lives for the measured deformed odd–even proton emitters compared with the predictions of Ref. [15]. The theoretical results are from Refs. [4,18,19]. The label m refers to decays from isomeric states.

	Proton decay		Möller–Nix	
	J	β	J	β
^{109}I	1/2+	0.14	1/2+	0.16
^{113}Cs	3/2+	0.15 ÷ 0.20	3/2+	0.21
^{117}La	3/2+	0.20 ÷ 0.30	3/2+	0.29
^{117m}La	9/2+	0.25 ÷ 0.35		
^{131}Eu	3/2+	0.27 ÷ 0.34	3/2+	0.33
^{141}Ho	7/2–	0.30 ÷ 0.40	7/2–	0.29
^{141m}Ho	1/2+	0.30 ÷ 0.40		
^{151}Lu	5/2–	-0.18 ÷-0.14	5/2–	-0.16
^{151m}Lu	3/2+	-0.18 ÷-0.14		

the unpaired neutron, gives to the neutron the important role of "influential spectator" contributing significantly with its angular momentum to the decay, thus allowing a determination of the neutron s.p. level.

We have applied[6,7,14,15] our model to all measured deformed proton emitters including isomeric decays. The results are shown in Table 1. The experimental half-lives are perfectly reproduced by a specific state, with defined quantum numbers and deformation, thus leading to unambiguous assignments of the angular momentum of the decaying states. Extra experimental information provided by isomeric decay observed in ^{117}La, ^{141}Ho and ^{151}Lu, and fine structure in ^{131}Eu can also be successfully accounted by the model. The experimental half–lives for decay from the excited states were reproduced in a consistent way with the same deformation that describes ground state emission.

Emission from deformed systems with an odd number of protons and neutrons can be discussed in a similar fashion[7]. The decaying nucleus is described by a wave function of two particles–plus–rotor in the strong coupling limit represented in terms of the single particle functions of the odd nucleons. However, in contrast with decay from ground state of odd-even nuclei where the proton is forced to escape with a specific angular momentum, many channels will be open due to the angular momentum coupling of the proton and daughter nucleus, $\vec{J}_d + \vec{j}_p$, giving the total width for decay as a sum of partial widths allowed by parity and momentum conservation,

As in the case of odd-even nuclei, Table 2 shows a perfect description of experimental decay rates for deformations in agreement with predictions

Table 2. As in Table 1 for the measured odd–odd proton emitters. Results from the calculation of Ref. [5]. The quantities K_p and K_n are the magnetic quantum numbers of the proton and neutron Nilsson wave functions, J the total angular momentum of the parent nucleus, and β and β_M the deformation coming from the proton decay calculation and the prediction of Ref. [15].

	K_p	K_n	J	β	β_M
^{112}Cs	3/2+	3/2+	0+, 3+	0.12 ÷ 0.22	0.21
^{140}Ho	7/2–	9/2–, 7/2+	8+, 0–	0.26 ÷ 0.34	0.30
^{150}Lu	5/2–	1/2–	2+	–0.15 ÷–0.17	–0.16
^{150m}Lu	3/2+	1/2–	1–, 2–	–0.22 ÷ 0.00	

made by other models[13]. The same Nilsson state of the odd proton is used in the calculation of odd–odd and neighbour odd–even nuclei, as can be seen comparing Tables 1 and 2. Similar deformations were also found for the odd-odd and nearby odd-even nuclei. This represents a further consistency check of the model. The largest contribution of the residual interaction between the odd-neutron and odd-proton, i.e. the diagonal part, was taken into account exactly.

The important conclusion to be drawn for odd-odd emitters is that the total decay width depends on the quantum numbers of the unpaired neutron which cannot be considered only a spectator, but influences significantly with its angular momentum, the decay.

3. The non–adiabatic quasi-particle approach: contributions from Coriolis mixing and pairing residual interaction

As we have discussed in the previous section, calculations within the strong coupling limit were able to reproduce the experimental results. According to this model, the daughter nucleus has an infinite moment of inertia, and the rotational spectrum collapses into the ground state. Considering a finite moment of inertia, the Hamiltonian of the decaying nucleus can be decomposed into a term acting on the degrees of freedom of the rotor, a recoil term acting on the coordinates of the valence proton, and a term representing a purely kinematic coupling between the degrees of freedom of both, known as the Coriolis coupling. Therefore, the wave functions of the rotor are modified with respect to the adiabatic approach, since its rotational spectrum is included, and another interaction is acting on the nucleus, the Coriolis force.

The effect of a finite moment of inertia of the daughter nucleus on proton decay, was studied within the non-adiabatic coupled channel[16], and

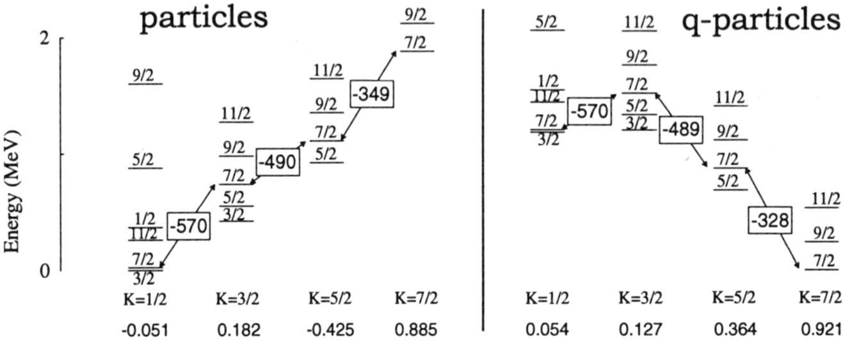

Figure 2. The left section of the figure represents the level scheme for particle states at energies equal to the diagonal matrix elements of the nuclear Hamiltonian. The numbers in boxes are the off-diagonal matrix elements of the Coriolis force (in keV). States with same angular momentum are connected. The right section as in the left for quasi-particles. The numbers in the bottom row are the components of the wave function of the $7/2^-$ decaying state of ^{141}Ho.

coupled-channel Green's function[17] methods, but the excellent agreement with experiment found in the adiabatic context was lost. The results differ by factors of three or four from the experiment, and even the branching ratio for fine structure decay is not reproduced[16,17,18]. This result is surprising, since calculations that include the Coriolis mixing should undoubtedly be better. The use in the calculations of Ref.[16,18] of a spherical spin–orbit mean field is a strong handicap of their model, and is responsible for the large deviations observed in ^{131}Eu and ^{117}La, nuclei with low angular momentum where the Coriolis coupling should be small. However, the strange behaviour found also in the calculation of Ref. [17] which includes deformed spin–orbit, for the decay of ^{141}Ho needs an explanation.

Decay rates in deformed nuclei, are extremely sensitive to small components of the wave function. The Coriolis interaction mixes different Nilsson wave functions, and can be responsible for strong changes in the decay widths. However, the residual pairing interaction can modify this mixing of states, an effect not considered in the calculations of Ref.[16,17,18]. We have included[9] beside the Coriolis mixing, the pairing residual interaction in the BCS approach. The mixing of states is modified by the residual interaction and transformed trough a Bogoliubov transformation into a mixing between quasi-particle states instead of particle ones, as was used in Refs.[16,17,18].

Let us consider for example the decay of the $7/2^-$ ground state of ^{141}Ho to the ground and first 2^+ excited states in ^{140}Dy. This was the most strange case in the calculation of Ref.[17]. Since it involves the spherical $h11/2$ state, which has a very high angular momentum, a strong Coriolis force is expected. Including the Coriolis interaction, the decay width to the ground state decreases drastically, leading to an increase of the branching ratio. With the residual pairing interaction the decay width to the ground state increases, leaving the width for decay to the excited state unchanged, and reducing the branching ratio.

This can be understood from the analysis of the level scheme corresponding to the basis states displayed in Fig. 2, at energies equal to the diagonal matrix elements of the Hamiltonian of the nucleus with the Coriolis interaction for particles, and after the transformation to quasi-particles. The difference between both representations, is an inversion of the level ordering, while the off-diagonal matrix elements are practically unchanged. After diagonalization, the wavefunction that describes the decaying nucleus corresponds, in the particle case, to the highest state in energy, while in the quasi-particles to the lowest one, as it should be. This inversion implies a change of sign of the wave function components, leading to an interference between these components, in the calculation of the decay width. For particles the interference is destructive and the width decreases, while it turns out to be constructive for quasi-particles. The width is enhanced, and the adiabatic results are recovered.

Such non-adiabatic treatment of the Coriolis coupling, brings back the perfect agreement with data observed in the strong coupling limit. Therefore, the previous disagreement between the calculation with Coriolis[16,17,18] and the experimental data or between calculations with Coriolis and the ones in the strong coupling limit[6], were simply due to an inadequate treatment of the residual pairing interaction.

4. Conclusions

We have presented a unified model to describe proton radioactivity from deformed nuclei. Decay is understood as decay from single particle Nilsson resonances that are evaluated exactly for single particle potentials that fit large sets of data on nuclear properties. The rotational spectra of the daughter nucleus, and the pairing residual interaction in the BCS approach, are taken into account, leading to a treatment of the Coriolis coupling in terms of quasi-particles. All available experimental data on even–odd and odd–odd deformed proton emitters from the ground and isomeric states and fine structure, are accurately and consistently reproduced by the model.

The calculation provides valuable nuclear structure information on deformation and angular momentum J of the decaying nucleus, and also on the state of the unpaired neutron in decay from odd–odd nuclei, thus giving unambiguous assignments to the decaying states.

Proton radioactivity provides a unique tool to access nuclear structure properties of nuclei far away from the stability domain.

Acknowledgements

Support from Fundação de Ciência e Tecnologia (Portugal), Grant N. 36575/99 and FEDER are acknowledge.

References

1. P. J. Woods and C. N. Davids, Annu. Rev. Nucl. Part. Sci. **47** (1997) 541.
2. A. A. Sonzogni, Nuclear Data Sheets **95** (2002) 1.
3. S. Åberg, P. B. Semmes and W. Nazarewicz, Phys. Rev. **C56** (1997) 1762.
4. C. N. Davids, *et al.*, Phys. Rev. Lett. **80** (1998) 1849.
5. A. A. Sonzogni, *et al.*, Phys. Rev. Lett. **83** (1999) 1116.
6. E. Maglione, L. S. Ferreira and R. J. Liotta, Phys. Rev. Lett. **81** (1998) 538; Phys. Rev. **C59** (1999) R589.
7. L. S. Ferreira and E. Maglione, Phys. Rev. Lett. **86** (2001) 1721.
8. D. D. Bogdanov, V. P. Bugrov and S. G. Kadmenskii, Sov. J. Nucl. Phys. **52** (1990) 229.
9. G. Fiorin, E. Maglione and L. S. Ferreira, Phys. Rev. **C67** (2003) 054302.
10. L. S. Ferreira, E. Maglione and R. J. Liotta, Phys. Rev. Lett. **78** (1997) 1640.
11. L. S. Ferreira, E. Maglione, and D. Fernandes, Phys. Rev. **C65** (2002) 024323.
12. F. Soramel, *et al.*, Phys. Rev. **C63** (2001) 031304(R).
13. P. Möller, *et al.*, At. Data Nucl. Data Tab. **59** (1995) 185; ibd. **66** (1997) 131.
14. L. S. Ferreira and E. Maglione, Phys. Rev. **C61** (2000) 021304(R).
15. E. Maglione and L.S. Ferreira, Phys. Rev. **C61** (2000) 47307.
16. A. T. Kruppa, B. Barmore, W. Nazarewicz and T. Vertse, Phys. Rev. Lett. **84** (2000) 4549; B. Barmore, A. T. Kruppa, W. Nazarewicz and T. Vertse, Phys. Rev. **C62** (2000) 054315.
17. H. Esbensen and C. N. Davids, Phys. Rev. **C63** (2001) 014315.
18. W. Królas, *et al.*, Phys. Rev. **C65** (2002) 031303(R).

KEY TOPICS IN NUCLEAR STRUCTURE
Proceedings of the 8th Int. Spring Seminar on Nuclear Physics
edited by Aldo Covello
© 2005 World Scientific Publishing Co.

COMPLEX SHELL MODEL WITH ANTIBOUND STATES

R. ID BETAN[1,2], R. J. LIOTTA[1], N. SANDULESCU[1,3], AND T. VERTSE[1,4]

[1] *Royal Institute of Technology, AlbaNova University Center, SE-10691, Stockholm, Sweden*

[2] *Departamento de Física, FCEIA, UNR, Avenida Pellegrini 250, 2000 Rosario, Argentina*

[3] *Institute of Physics and Nuclear Engineering, P.O.Box MG-6, Bucharest-Magurele, Romania*

[4] *Institute of Nuclear Research of the Hungarian Academy of Sciences, H-4001 Debrecen, Pf. 51, Hungary*

An unified shell model scheme to evaluate simultaneously the contributions of bound single-particle states, Gamow resonances, antibound (virtual) states and continuum complex scattering states is presented. The formalism could be very suitable to study processes occurring in the continuum part of the nuclear spectra.

1. Introduction

Beams of exotic nuclei offere the opportunity for investigate nuclear structure in the drip line region. The construction of different types of radioactive beam facilities confirm the scientific interest in this field of physics.

Theory takes its inspiration from experiment in guiding the structure of the models while nuclear experiment takes its inspiration from theory in helping to choose which experiments are most important to prove or reject theoretical models and theoretical methods[1]. Thus nowadays it is well established that halos in nuclei are produced by particles moving in single-particle states which extend far in space. In the neutron halo case this implies that the neutron configurations are form mainly from loosely bound or unbound *s*- and/or *p-waves* since in this case no barrier large enough to trap the nucleons inside the nuclear core will be present.

In the Borromean nuclei case it is the pairing interaction that holds the escaping neutron together [2]. In this kind of process the continuum plays an important role and therefore one can not use bound representations. Instead one must use a representation which includes explicitly the continuum part of the particle spectrum. One solution to this problem is given by the *Complex Shell Model*, where the continuum is given by a contour in the complex energy plane plus the resonance states enclosed between the positive energy real axis and the complex contour. In this proceeding we show how to generalize the Complex Shell Model in such a way as to include antibound (virtual) states in the single particle representation. In this way we give an unified shell model scheme to evaluate two particle bound and resonance states. In section 2 we describe the formalism. In section 3 we applied it to the ^{11}Li nuclei and in section 4 we draw some conclusions.

2. Formalism

The shell model in the complex energy plane is based on the Berggren representation[3]. It is a complete basis formed by a discrete set of wave functions corresponding to the poles of the S-matrix, plus a set of scattering states with energies belonging to a continuous path in the complex energy plane. In this representation the completeness relation can be written as[3,4,5]

$$\delta(r-r') = \sum_{n_b,n_a,n_d} u_{nlj}(r)\, u_{nlj}(r') + \int_L dk\; u_{lj}(k,r)\, u_{lj}(k,r') \qquad (1)$$

The summation in the expression above runs over all the bound states and over those poles of the S-matrix which are enclosed between the positive real energy axis and the contour L. This poles could be resonance and antibound (virtual) states. The integrand are form by the complex scattering states lying on the contour. One may choose the phase for the complex scattering states in such a way to write the completeness relation using the wave function instead its complex conjugate.

The contour in the complex energy plane can have in principle any form. Nevertheless, when we are interested in two particle resonance states, it is convenient to choose class of rectangular contours in order to isolate the pole, otherwise it could be embeded in the two-particle energy continuum[6].

In order to treat numerically the scattering states one must discretizes the complex contour integral. In order to do that one parametrizes the contour integral and uses some approximation method to discretize the integral. After that we are able to define the complex single particle representation form by the following set of discrete states [7],

$$\phi_{nlj}(r) = \begin{cases} u_{nlj}(r) & \text{Bound, antibound, reson. stat.} \\ \sqrt{\Delta_n \dot{L}_n}\; u_{lj}(k_n, r) & \text{Scattering states} \end{cases}$$

where k_n and Δ_n are defined by the procedure one uses to approximate the integral. $\dot{L}_n$ is the derivative of the complex contour respect to the parametrization variable. In the Gaussian method k_n are the Gaussian points and Δ_n the corresponding weights. Because the discrete scattering wave function are weight by the derivative of the contour, it must be continuous and its derivative must be continuous too, but because we are using Gaussian approximation an it never take the extreme of the contour we are able in practice to use a kind of contour for which the derivative is not continue.

After we have completed the radial part of the wave function with the angular part we can write the following completeness relation in a single particle representation,

$$\sum_{nljm} \left[\frac{\phi_{nlj}(r')}{r'} Y^*_{ljm}(\hat{r}') \right] \left[\frac{\phi_{nlj}(r)}{r} Y_{ljm}(\hat{r}) \right] = \delta(\bar{r} - \bar{r}') \tag{2}$$

We observe that the conjugation appears only in the angular part. The same is true for whatever matrix elements in this representation. This is the meaning of the Berggren metric. This set of discrete states defines the Berggren representation use in the Complex Shell Model (CXSM)calculations.

The new elements include in this paper in the CXSM are the antibound (virtual) states. From the mathematical point of view they are the outgoing solutions of the Schroedinger equation with negative imaginary wave numbers, i.e., $k = -i|k|$. Thus the energy corresponding to an antibound state is real and negative, as for the bound states, but the tail of the corresponding wave function diverges exponentially at large distances. From the physical point of view one can thing an antibound (virtual) state as an state for which the mean field strength is not enough to keep it bound. An antibound state near the threshold has an observable effect on the cross section.

On positive real energy axis an antibound state close to threshold manifests itself through the localization properties of the low-lying scattering states. This can be shown by considering a mean field that has an antibound s-state with energy E_0 ($k_0 = -i|k_0|$) lying near threshold. One thus finds that the radial scattering wave function with energies $E = \hbar^2 k^2/2\mu$ (k real and positive) close to zero can be approximated inside the mean field region by [8]

$$u_{lj}(k,r) \approx \sqrt{\frac{2k|k_0|a}{k^2+|k_0|^2}}\, u_{lj}(|k_0|,r), \tag{3}$$

where a is a constant depending on the normalization chosen for the scattering wave function $u_{lj}(|k_0|,r)$. This expression shows that close to threshold the radial dependence of the scattering wave functions inside the mean field region depends upon the energy only through the square root factor. This factor is maximum at $k = |k_0|$. Therefore in an energy interval located around $|E_0|$ the scattering states have an increased localization.

The same effect can be seen in Fig. 1, where we draw the localization of the scattering states, $L(E)$, defined as

$$L(E) = \int_0^b u^2_{s_{1/2}}(k,r)dr \tag{4}$$

where $b = 3.1$ fm and we take as an example the ^{10}Li nuclei.

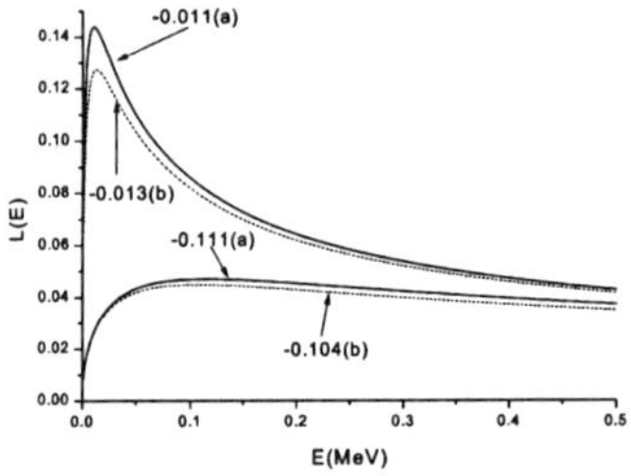

Figure 1. Localization $L(E)$, Eq. 4, in the presence of low-lying antibound (a) and bound (b) s-states. The numbers labeling the curves are the energies of the poles in MeV.

In the figure we show the localization function for different values of the strength in the mean field. We can see that when the strength is such that there is an antibound or a loosely bound state near the threshold the localization function has a sharp bump in the low lying scattering region.

These scattering states will represent indirectly the effect of the antibound state in any type of continuum shell model calculations based on real energy representations. Within the CXSM formalism one is able to study the direct effects of the antibound states straightforward as will do in the nex section.

3. Applications

To show the convenience of the formalism presented above we will apply it to one of the cases where an antibound states is known to be important. This is particularly the case of halo type nuclei ^{11}Li.

The existence of a low-lying virtual s-state in ^{10}Li has important consequences for the correlations developed in ^{11}Li [9]. As discussed above, an antibound state close to the continuum threshold enhances the localization of the low-lying scattering states. Therefore the s-wave content of the ground state of ^{11}Li is also increased, reaching the corresponding (large) experimental value. Moreover, the antibound state in ^{10}Li can affect the excited spectrum of ^{11}Li as well as the ground state. These effects of the antibound states will be studied here from the viewpoint of the CXSM.

It is by now well-known that in the description of ^{11}Li the two relevant single-particle states, as specified by the experimental spectrum of ^{10}Li, consist of a low-lying antibound (or virtual) $s_{1/2}$ state at about -25 keV , and a $p_{1/2}$-resonance at about 240 keV [10,11].

The two-body correlations induce a bound ground state in ^{11}Li at about -295 keV. Where we are using the shell-model language, where the core (^{9}Li) is considered as inert, the single-particle states are given by ^{10}Li and the two-body nucleus is ^{11}Li [12].

In the first step of the CXSM calculation one evaluates the single-particle states of the unbound nucleus ^{10}Li. As in Refs.[13,14], for the central field we choose a Woods-Saxon potential with different depths for even and odd orbital angular momenta l. One thus simulates the effect of particle-vibration upon single-particle states [15,16].

The Woods-Saxon mean-field potential is given by a = 0.67 fm, r_0 = 1.27 fm, V_0 = 50.55 (39.97) MeV and V_{so} = 19.31 MeV for l even (odd). With these parameters we found the single-particle bound states $0s_{1/2}$ at -23.689 MeV and $0p_{3/2}$ at -4.500 MeV forming the ^{9}Li core. The valence poles are the low lying resonances $0p_{1/2}$ at (0.240,-0.064) MeV and $0d_{5/2}$ at (2.281,-0.362) MeV and the wide resonance $0d_{3/2}$ at (6.613,-5.582) MeV. Besides, the state $1s_{1/2}$ appears as an antiboud state [a] at -0.025 MeV. We also found other resonances at high energies. However we include in the basis single-particle states lying up to 10 MeV of excitation energy only. We found that expanding the basis from this limit does not produce any

[a]The principal quantum number n labelling the single-particle states indicates that the corresponding wave functions are localized in a region inside the nucleus and that its real part has in that region n nodes, excluding the origin.

effect upon the calculation.

For the partial waves p and d we choose as complex scattering states a set of discrete state belonging to the contour as shown if fig. 2,

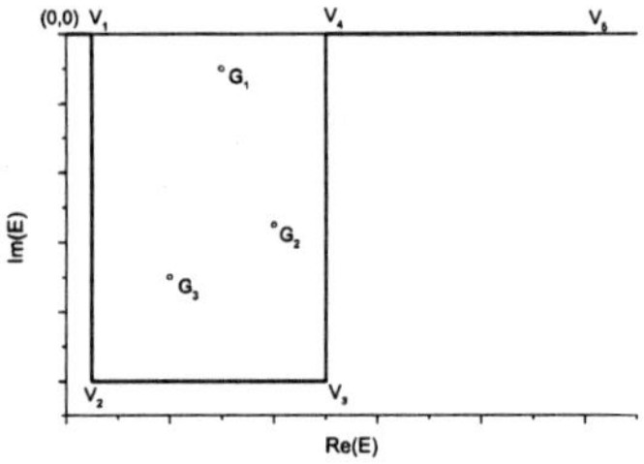

Figure 2. One-particle complex energy plane for p and d partial waves. The broad line indicates the contour. The points V_i are the vertices defining the contour. The open circles labelled by G_i indicate the complex energy of the Gamow resonances enclosed by the contour.

where the points were the following $V_1 = (0,0)Mev$, $V_2 = (0,-0.7)Mev$, $V_3 = (5,-0.7)Mev$, $V_4 = (5,0)Mev$ and $V_5 = (10,0)Mev$.

In order to include the antibound state in the model space we must choose a different kind of contour for the complex scattering partial waves s as is shown in fig. 3

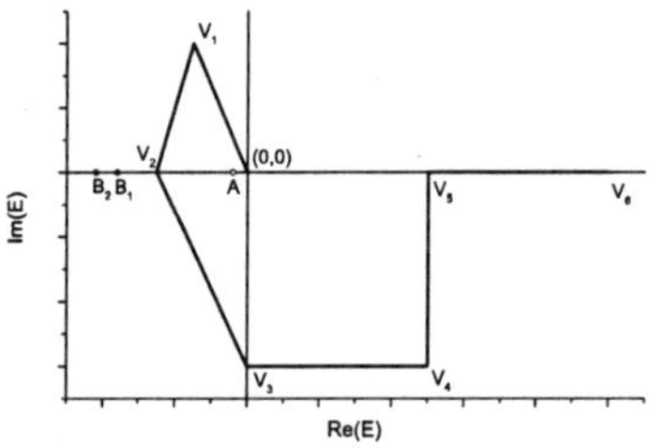

Figure 3. One-particle complex energy plane for the partial s wave state. The broad line indicates the contour embracing the antibound state A, indicated by an open circle. The points V_i are the vertices defining the contour. The points B_i indicate bound states.

Where the points were the following $V_1 = (-0.025, 0.1)Mev$, $V_2 = (-0.05, 0)Mev$, $V_3 = (0, -0.7)Mev$, $V_4 = (5, -0.7)Mev$, $V_5 = (5, 0)Mev$ and $V_6 = (10, 0)Mev$. The number of discretize complex scattering states was the following: 26 for the $p_{1/2}$, 40 for the $d_{5/2}$, 52 for the $s_{1/2}$, 16 for the $p_{3/2}$ and 18 for the $d_{3/2}$.

As in any standard shell model, in CXSM the multi-particle basis states are formed by the tensorial product of the ordered single-particle states belonging to the chosen Berggren representation [17,18,19]. The matrix elements of the residual interaction are calculated within this representation by using the Berggren metric. Thus for a separable interaction the matrix elements have the form $<\widetilde{kl};\alpha|V|ij;\alpha>=-G_\alpha f_\alpha(kl)f_\alpha(ij)$, where G_α is the strength of the force. One can see that due to the Berggren metric on the r.h.s. appears the form factor $f_\alpha(kl)$ and not $f_\alpha(kl)^*$ as in the standard Hilbert metric. Consequently, the standard dispersion relation for a two-particle system corresponding to a separable force becomes

$$-\frac{1}{G_\alpha}=\sum_{i\leq j}\frac{f^2_\alpha(ij)}{\omega_\alpha-\epsilon_i-\epsilon_j} \tag{5}$$

where ω_α are the correlated energies. For the field in the separable interaction we use the derivative of the Woods-Saxon, with R'=4.5 fm and a'=1.5 fm.

To evaluate the ground state of ^{11}Li we adjust the strength G_0 of the separable interaction to reproduce the corresponding energy, i. e. -295 keV. We thus obtained $G_0 = 15.3\ MeV$.

With the mean field and the two-body interaction thus established we evaluated the ground state wave function. First we performed the calculations by choosing the real energy as a contour. In this case the wave function is spread over many components. The largest of these components corresponds to configurations $p_{1/2}\otimes p_{1/2}$ lying close to 480 keV (i. e. about twice the energy of the $0p_{1/2}$ resonance) and $s_{1/2}\otimes s_{1/2}$ lying close to threshold (i. e. close to twice the energy of the antibound state). The wave function consists of 47 % s-states, 46 % p-states and 7 % d-states, as expected [20,21].

We will analyze the effects of the antibound and the Gamow poles upon the ground state of ^{11}Li by using the contours of Figs. 2 and 3. We therefore present in Table 1 the contribution of different configurations to that ground state.

Table 1.

	$(s_{1/2})^2$	$(p_{1/2})^2$	$(d_{5/2})^2$
pole-pole	(2.960, -0.001)	(0.583, -0.195)	(0.080, 0.015)
pole-scat.	(-7.825, 0.003)	(-0.145, 0.210)	(-0.017, -0.016)
scat.-scat.	(5.335, -0.002)	(0.002, -0.015)	(-0.001, 0.002)
total	**(0.470, 0.000)**	**(0.440, 0.000)**	**(0.062, 0.000)**

The corresponding complex amplitudes depend on the chosen contours and have no direct physical meaning. But the total content of a given partial wave in the bound ground state wave function, which is a physical quantity, does not depend upon the chosen contour and they are real quantities. From Table 1 we can see that for the p and d waves the configurations are built mainly on the corresponding Gamow resonances. The situation is different for the s-wave since apart from the configurations built upon the antibound state there is also an important contribution coming from the complex scattering states. This contribution is given mainly by those s scattering states located on the segments $(0,0) - V_1$ and $V_1 - V_2$ of Fig. 3, which are the closest to the antibound state.

Another thing we can learn from the Table 1 is that we are not able to leave out the contour for the complex scattering s partial wave in the model space. Is we do that the $l = 0$ content of the wave function increases to 77%. If however we readjust the value of G in order to get the state at the correct position at -295 keV then the $l = 0$ content of the wave function increases further up to 98%. This shows that the antibound pole and the scattering states along the $l = 0$ complex path are adding up with very strong destructive interference and this reduces the $l = 0$ content of the wave function somewhat below the $l = 1$ content.

Using the CXSM formalism we can see how the correlated poles move as a function of the strength G. We will show this for the ground state and for the first excited state 0^+ in ^{11}Li system.

In fig. 4 we shown the position of the ground state as a function of the strength. Because there is a cut in this region in the two particle energy plane one can not observe the correlated pole before the strength is strong enough to put the pole beyond the cut. This happen for a G value around 12.4 MeV and an energy $E_0 = -72\ KeV$ [b].The wave function is built by 60% of s, 37% of p and 3% of d. From here on, when we increase the strength the s contribution decreases while the p contribution increases up to the experimental value ($G = 15.3 MeV$) shown in table 1.

In fig. 5 we show how the 0_1^+ excited state in ^{11}Li is built up by the two-body interaction starting from the zeroth-order configuration $(0p_{1/2})^2$. This state is generated almost 100 % of $p_{1/2}$-states for all value of the strength. The main contributions came from the configuration where both particles are in the resonance state or one is in the resonance state an the other in the continuum.

[b]In order to get this pole one has to choose $V_2 = (-0.030, 0) MeV$ in fig. 3

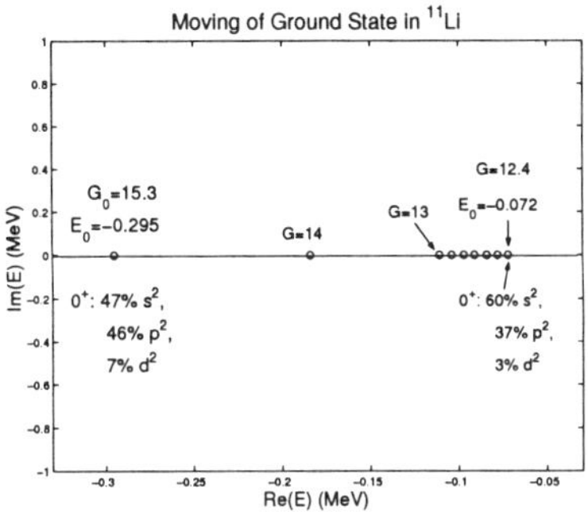

Figure 4.

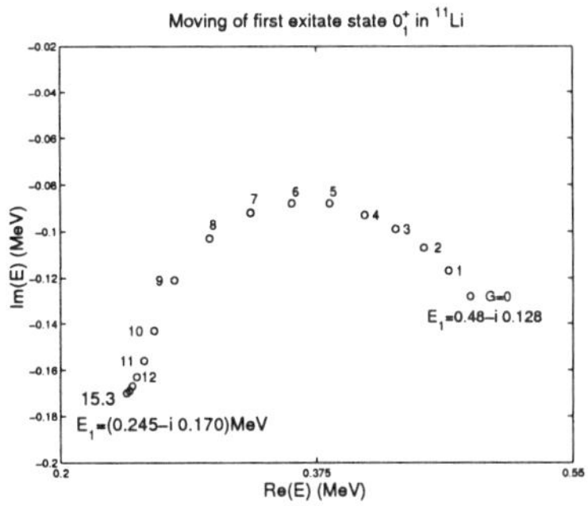

Figure 5.

As the attractive interaction increases the resonance becomes narrower and approaches threshold. However, a point is reached where continuum configurations become important and the resonance widens. This happens around G=6 MeV. As we can see in table 2, up to this point the main configuration is which one where the two particle are in the resonance state and then the state is localized inside the nucleus. That is, it is a physically meaningful resonance. But from here on the configuration where one particle is in the resonance state and the other one in the continuum become important. The contribution to the norm from configurations where both particles are in the continuum are less than 10^{-3} for all values of G.

The energy for this first $^+0_1$ state for the 'physical' strength $G_0 = 15.3MeV$ is $E_1 = (0.245 - i0.170)MeV$ and it is built mainly by configurations where one particle is the resonance state and the other in continuum states.

Table 2.

G	Pole-Pole	Pole-Scat.
3	(1.01,0.00)	(0.00,-0.01)
7	(1.07,0.07)	(0.02,-0.05)
10	(0.71,0.23)	(0.36,-0.07)
11	(0.49,0.04)	(0.57,0.08)
13	(0.18,-0.03)	(0.85,0.08)
G_0	(0.09,-0.01)	(0.93,0.04)

4. Conclusions

In conclusion, we have presented in this paper a new formalism to treat antibound (virtual) states exactly on the same footing as bound states and Gamow resonances. The antibound states and the Gamow resonances are selected by appropriate contours in the complex energy plane. Due to the complex single-particle representation used in the present shell model formalism, the contribution of the pole-pole, pole-continuum and continuum-continuum configurations in the two-particle systems can be easily analyzed. The effects induced by antibound states and the continuum encircling the poles can be studied separately. The advantage of the formalism was illustrated for the halo type nuclei ^{11}Li. We confirm that antibound states lying close to the continuum threshold are of a fundamental importance to build up the halo. But we found that in the ground state of the ^{11}Li the large contribution of the antibound pole is partly cancelled by the complex continuum. We also found that an excited low-lying two-particle resonance may exist in these nuclei which is strongly mixed with the continuum background.

This work has been supported by FOMEC and Fundación Antorchas (Argentina), by the Hungarian OTKA fund Nos. T37991 and T29003 and by the Swedish Foundation for International Cooperation in Research and Higher Education (STINT).

References

1. B. Jonson, Physics Reprts **389**, 1 (2004).
2. M. Zhukov, B. Danilin, D. Fedorov, J. Bang, I. Thompson, Physics Reports **231**, 151 (1993).
3. T. Berggren, Nucl. Phys. **A 109**, 265 (1968).
4. T. Vertse, P. Curutchet, R.J. Liotta, J. Bang, Acta Physica Hungaria **65**, 305 (1989).
5. P. Lind, Phys. Rev. **C 47**, 1903 (1993).
6. R. Id Betan, R. J. Liotta, N. Sandulescu and T. Vertse, Phys. Rev. Lett. **89**, 042501 (2002).

7. R. J. Liotta, E. Maglione, N. Sandulescu and T. Vertse, Phys. Lett. **B 367**, 1 (1996).
8. A. B. Migdal, A. M. Perelomov, and V. S. Popov, Yad. Fiz. **14**, 874 (1971) [Sov. J. Nucl. Phys **14** 488 (1972)].
9. I. J. Thompson and M. V. Zhukov, Phys. Rev. C **49**, 1904 (1994).
10. H. G. Bohlen et al, Nucl. Phys. **A616** 254c (1997).
11. www.tunl.duke.edu, Preliminary version on 'Energy Levels of Light Nuclei A=10'.
12. G. F. Bertsch and H. Esbensen, Ann. of Phys. **209**, 327 (1991).
13. H. Esbensen, G. F. Bertsch and K. Hencken, Phys. Rev. C **56**, 3054 (1997).
14. J. C. Pacheco, N. Vinh Mau, Phys. Rev. C **65** 044004 (2002).
15. N. Vinh Mau, Nuclear Physics **A 592**, 33 (1995).
16. F. Barranco et al, Eur. Phys. J. A **11** 385 (2001).
17. R. Id Betan, R. J. Liotta, N. Sandulescu and T. Vertse, Phys. Rev. C **67**, 014322 (2003).
18. N. Michel, W. Nazarewicz, M. Ploszajczak and K. Bennaceur, Phys. Rev. Lett. **89**, 042502 (2002).
19. N. Michel, W. Nazarewicz, M. Ploszajczak and J. Okolowicz, Phys. Rev. C **67**, 054311 (2003).
20. Structure and Reactions of Light Exotic Nuclei, Y. Suzuki, R. G. Lovas, K. Yabana and K. Varga, Taylor and Francis, London, 2003.
21. K. Varga, Y. Suzuki, R. G. Lovas, Phys. Rev. C **66**, 041302(R) (2002).

SECTION II

NUCLEAR FORCES AND NUCLEAR STRUCTURE

KEY TOPICS IN NUCLEAR STRUCTURE
Proceedings of the 8th Int. Spring Seminar on Nuclear Physics
edited by Aldo Covello
© 2005 World Scientific Publishing Co.

STUDIES OF PHASE-SHIFT EQUIVALENT LOW-MOMENTUM NUCLEON-NUCLEON POTENTIALS

T. T. S. KUO AND JASON D. HOLT

Department of Physics and Astronomy
State University of New York at Stony Brook
Stony Brook, NY 11794, USA
E-mail: kuo@tonic.physics.sunysb.edu

Motivated by the renormalization group (RG) and effective field theory (EFT) approach, a low-momentum NN interaction V_{low-k} has been derived by integrating out the high momentum components of modern NN potential models V_{NN}. Although the various V_{NN} models are significantly different, the V_{low-k}'s extracted from them are nearly identical to each other when a decimation momentum $\Lambda \approx 2fm^{-1}$ or smaller is employed. Starting from the Lee-Suzuki (or folded diagram) non-Hermitian low momentum NN interactions, a family of phase-shift equivalent V_{low-k}'s are obtained by way of a Schmidt orthogonalization method. We have found that V_{low-k} can be very accurately represented by a counter term expressed as a low order momentum expansion of the form $\sum C_n k^n$.

1. Introduction

First I (TTSK) would like to thank Prof. Aldo Covello and his collaborators at Naples for inviting me to this beautiful conference. I have been to their International Spring Seminars in Nuclear Physics almost everytime since it was started about twenty years ago. Together with the Naples group, we [1,2,3,4,5,6,7] have been working on the low-momentum nucleon nucleon (NN) interaction V_{low-k} in the past few years. In this report, I would like to discuss some recent works [9,10] we did in this area, namely the counter terms and a comparison of several phase-shift equivalent Hermitian low-momentum NN interactions.

A basic topic in nuclear physics is the nucleon-nucleon (NN) interaction, commonly denoted as V_{NN}. What is V_{NN}? It is a strong interaction and is a difficult problem. But a lot of progress has been made. There are now a number of realistic NN potential models. Among the early ones were the Paris [11] and Bonn [12] potentials. With the availability of more accurate NN scattering data, several high precision NN potential models were constructed, such as the CD-Bonn [13], Argonne-V18 [15], Nijmegen [14] and Idaho chiral [16] potentials.

These high precision V_{NN} models all fit the deuteron binding energy and NN scatterings up to $E_{lab} \approx 350$ MeV equally perfectly, all with χ^2 per datum close to 1.1. However, these models themselves are significantly different. As illustrated in Fig.1, the k-space matrix elements of them are quite different. Should one have a unique NN potential?

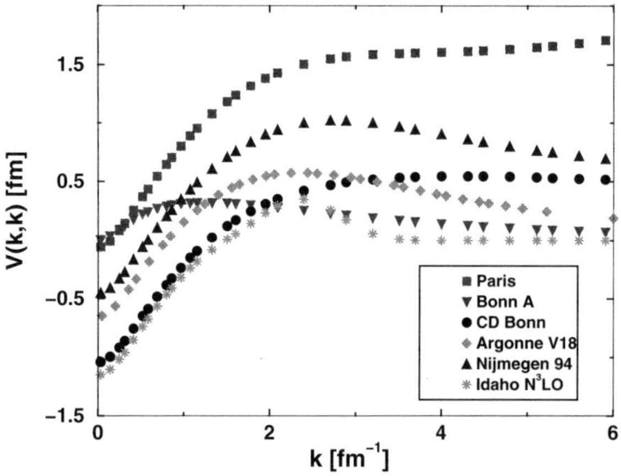

Figure 1. Comparison of k-space matrix elements $V_{NN}(k,k)$ of realistic NN potential models.

Since the pioneering work of Weinberg [17], there has been much progress and interest in treating low-energy nuclear physics using the renormalization group (RG) and effective field theory (EFT) approach [18,19,20,21,22,23]. A central idea here is that physics in the infra red region must be insensitive to the details of the short range (high momentum) dynamics. In low energy nuclear physics, we are probing nuclear systems with low energy probes of wave length λ; such probes certainly cannot reveal the short range details at distances much smaller than λ. Furthermore, our understanding about the short range dynamics is still preliminary and model dependent. Because of these considerations, a central step in the RG-EFT approach is to divide the fields into two categories: slow fields and fast fields, separated by a chiral symmetry breaking scale $\Lambda_\chi \sim 1$ GeV. Then by integrating out the fast fields, one obtains an effective field theory for the slow fields only.

This RG-EFT approach may be helpful in understanding the above differences among the various NN potential models. In order to have an effective interaction appropriate for complex nuclei in which typical nucleon momenta are $< k_F$, several authors [1,2,3,4,5,6,7,8] have employed a similar RG-EFT idea in studying a low-momentum NN potential V_{low-k} by integrating out the high-momentum components of the various modern models for V_{NN}. Here they separate fast and slow modes by a much smaller scale, namely $\Lambda \sim 2\text{fm}^{-1}$. In fact nucleon-nucleon experiments give a unique effective interaction only up to this scale (which is the pion production threshold).

A T-matrix equivalence approach has been employed by them to obtain the low-momentum NN interaction. We start from the T-matrix equation for the V_{NN} potential

$$T(k', k, k^2) = V_{NN}(k', k) + \frac{2}{\pi} P \int_0^\infty q^2 dq V_{NN}(k', q) \frac{1}{k^2 - q^2} T(q, k, k^2), \tag{1}$$

where the intermediate state momentum q is integrated from 0 to ∞. We then define an effective low-momentum T-matrix by

$$\begin{aligned} T_{low-k}(p', p, p^2) &= V_{low-k}(p', p) \\ &+ \frac{2}{\pi} P \int_0^\Lambda q^2 dq \frac{V_{low-k}(p', q) T_{low-k}(q, p, p^2)}{p^2 - q^2} \end{aligned} \tag{2}$$

where Λ denotes a momentum space cut-off (decimation momentum) and $(p', p) \leq \Lambda$. We choose $\Lambda \sim 2fm^{-1}$, essentially the momentum up to which the experiments give us information in the phase shift analysis. (This point will be further discussed later.) Note that here the intermediate state momentum is integrated from 0 to Λ. We require the above T-matrices satisfying the condition

$$T(p', p, p^2) = T_{low-k}(p', p, p^2); \ (p', p) \leq \Lambda. \tag{3}$$

The above equations define the effective low momentum interaction V_{low-k}, and are satisfied by the solution

$$V_{low-k} = \hat{Q} - \hat{Q}' \int \hat{Q} + \hat{Q}' \int \hat{Q} \int \hat{Q} - \hat{Q}' \int \hat{Q} \int \hat{Q} \int \hat{Q} + \ \dots \ , \tag{4}$$

which is just the Kuo-Lee-Ratcliff (KLR) folded-diagram effective interaction [24,25]. This V_{low-k} preserves the deuteron binding energy, since eigenvalues are preserved by the KLR effective interaction. It also preserves phase shifts, as phase shifts are given by the fully on-shell T-matrix $T(p, p, p^2)$.

For any decimation momentum Λ, the above V_{low-k} can be calculated highly accurately (essentially exactly) using either the Andreozzi-Lee-Suzuki [26,27] (ALS) or the Krenciglowa-Kuo [28] iteration methods. A main result of these authors is the following: Although the various NN potential models are quite different, the V_{low-k}'s derived from them are quite close to each other, leading to a nearly unique low-momentum NN potential. As shown Fig. 2, the V_{low-k}'s obtained from several modern potential models, for a decimation momentum $\Lambda=2fm^{-1}$, are indeed quite close to each other.

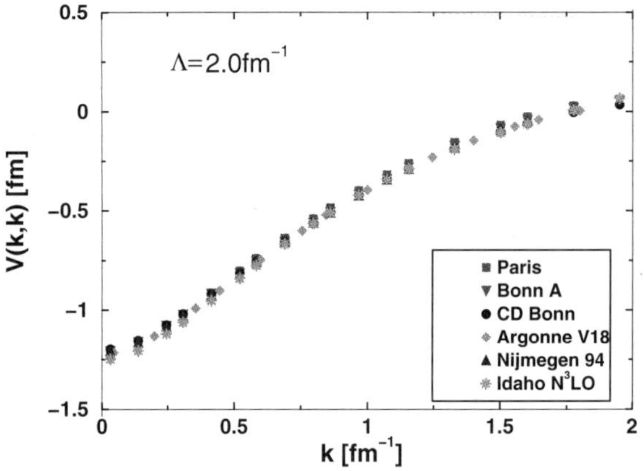

Figure 2. Comparison of k-space matrix elements $V_{low-k}(k,k)$ of realistic NN potential models.

In the following, let me discuss two specific topics recently studied by Holt, Kuo, Brown and Bogner [9] and by Holt, Kuo and Brown [10]. Namely the counter terms for V_{low-k} and a comparison of several phase-shift equivalent Hermitian low-momentum NN potentials. In addition, the choice of the decimation momentum Λ will also be addressed.

2. Counter terms

A central result of modern renormalization theory is that a general RG decimation generates an infinite series of counter terms [18] consistent with the input interaction. When we derive our low momentum interaction, the high momentum modes of the input interaction are integrated out. Does this decimation also generate a series of counter terms? If so, what are the counter terms so generated?

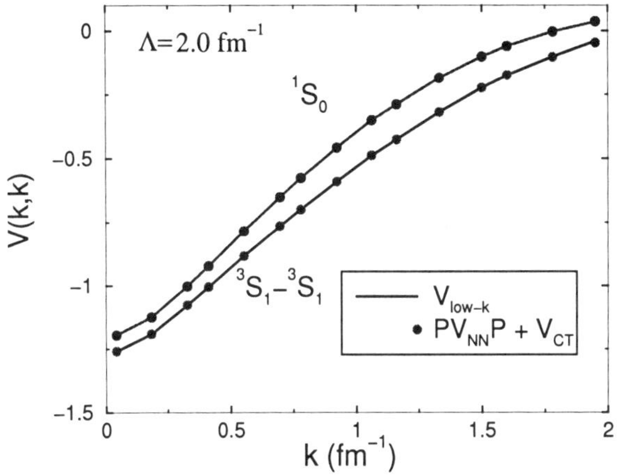

Figure 3. Comparison of V_{low-k} with $PV_{NN}P$ plus counter terms, for $^1S_0$ and $^3S_1$ channels.

We study here if the low-momentum interaction V_{low-k} can be well represented by the low-momentum part of the original NN interaction supplemented by some counter terms. Specifically, we consider

$$V_{low-k}(q,q') \simeq V_{bare}(q,q') + V_{counter}(q,q'); \ (q,q') \leq \Lambda, \tag{5}$$

where V_{bare} is the bare NN potential from which V_{low-k} is derived and the counter potential is given as a power series

$$\begin{aligned} V_{counter}(q,q') = C_0 + C_2q^2 + C_2'q'^2 + C_4(q^4+q'^4) + C_4'q^2q'^2 \\ +C_6(q^6+q'^6) + C_6'q^4q'^2 + C_6''q^2q'^4 + \ldots. \end{aligned} \tag{6}$$

The counter term coefficients are determined using standard fitting techniques so that the right hand side of Eq.(5) provides a best fit to the left hand side of the same equation. We perform this fitting over all partial wave channels, and find consistently good agreement. In Fig. 3 we compare some $^1S_0$ and $^3S_1$ matrix elements of $(PV_{bare}P + V_{CT})$ with those of V_{low-k} for momenta below the cutoff Λ. Here P denotes the projection operator for states with momentum less than Λ. The agreements for other channels are also very good.

Now let us examine the counter terms themselves. In Table 1, we list some of the counter term coeffieients, using CD-Bonn as our bare potential. In the table we list only the counter terms for the $^1S_0$ and $^3S_1 - ^3D_1$ partial waves; we have found that the counter terms for all the other waves are much smaller. This tells us an interesting result, namely, except for the above two channels, V_{low-k} is very similar to V_{bare} alone. We also point out that the coefficients C_6 are found to be very small, indicating that the above power series expansion converges rapidly. In the last row of the table, we list the rms deviations between V_{low-k} and $PV_{bare}P + V_{counter}$; the fit is indeed very good. We note that the low-momentum NN potential given by Eqs.(1-4) is not Hermitian. Our numerical results are obtained using a Hermitian version of V_{low-k} calculated with the Okubo transformation method as to be discussed in Section 3. The counter terms obtained for the interaction of Eq.(4) and those for the Hermitian one are in fact quite similar to each other.

As shown in the table, the counter terms are all rather small except for C_0 and C_2 of the S waves. This is consistent with the RG-EFT approach where the counter term potential is given as a delta function plus its derivatives [18].

Comparing counter term coefficients for different V_{bare} potentials can illustrate key differences between those potentials. For example, we have found that the $^1S_0$ C_0 coefficients for the CD-Bonn [13], Nijmegen [14], Argonne [15] and Paris [11] NN potentialsare respectively -0.158, -0.570, -0.753 and -1.162. Similarly, the $^3S_1$ C_0 coefficients for these potentials are respectively -0.467, -1.082, -1.148 and -2.224. That the C_0 coefficients for these potentials are significantly different is a reflection that the short range repulsion built into these potentials are different. For instance, the Paris potential effectively has a very strong short-range repulsion and consequently its C_0 is much larger compared with the others.

Table 1. Coefficients of the counter terms for V_{low-k} obtained from the CD-Bonn potential using $\Lambda = 2fm^{-1}$. The unit for the combined quantity $C_n k^n$ is fm, with momentum k in units of fm^{-1}.

	$^1S_0$	$^3S_1$	$^3S_1 -^3 D_1$	$^3D_1$
C_0	-0.1580	-0.4646	0	0
C_2	-0.0131	0.0581	-0.0017	-0.0005
C_2'	-0.0131	0.0581	0.0301	-0.0005
C_4	0.0004	-0.0011	-0.0013	0.0006
C_4'	-0.0011	-0.0113	-0.0047	-0.0018
C_6	-0.0004	-0.0004	0.0006	-0.0001
C_6'	-0.0005	0.0005	-0.0001	-0.0001
C_6''	-0.0005	0.0005	-0.0003	-0.0001
Δ_{rms}	0.0002	0.0003	0.0028	0.0003

3. Hermitian low-momentum interactions

The V_{low-k} given by the T-matrix equivalence approach mentioned earlier is not Hermitian, and this is not a desirable feature. One would like to have a NN interaction which is Hermitian. There are however a family of phase shift equivalent Hermitian V_{low-k}'s [9]. Let us denote this non-Hermitian low-momentum NN interaction as V_{LS}. (Recall that we have used the Lee-Suzuki method in its derivation.) V_{LS} preserves the half-on-shell T-matrix $T(k', k, k^2)$ for $(k', k) \leq \Lambda$. If we relax this half-on-shell constraint, we can obtain low-momentum NN interactions which are Hermitian.

There are several methods for obtaining a Hermitian effective interaction, such as those of Okubo [29], Suzuki and Okamoto [30], and Andreozzi [27]. Which of these methods should one use? How different are the Hermitian V_{low-k}'s given by them? To investigate these questions, let us first review some basic formulations about the model space effective interaction.

We start from the full-space Schroedinger equation

$$(H_0 + V)\Psi_n = E_n\Psi_n, \tag{7}$$

where H_0 is the unperturbed Hamiltonian and V the interaction. This equation can be reduced to a model space equation

$$P(H_0 + V_{eff})P\chi_m = E_m\chi_m, \tag{8}$$

where $\{E_m\}$ is a subset of $\{E_n\}$ of Eq.(7) and $\chi_m = P\Psi_m$. Here P is the model space projection operator. In the present work, P represents all the momentum states with momentum less than the cut-off scale Λ. There are a number of ways to derive V_{eff}, but, as indicated by Eq.(4), our effective interaction is obtained by the folded diagram method [24,25] and can be calculated conveniently using the Lee-Suzuki-Andreozzi [27] or Krenciglowa-Kuo [28] iteration methods. We denote the effective interaction so obtained as V_{LS}.

It is convenient to rewrite the above effective interaction in terms of the wave operator ω, namely

$$PV_{LS}P = Pe^{-\omega}(H_0 + V)e^{\omega}P - PH_0P, \tag{9}$$

where ω possesses the usual properties: $\omega = Q\omega P; \chi_m = e^{-\omega}\Psi_m; \omega\chi_m = Q\Psi_m$. Here Q is the complement of P, $P + Q = 1$.

While the eigenvectors Ψ_n of Eq.(7) are orthogonal to each other, it is clear that the eigenvectors χ_m of Eq.(8) are not so and the effective interaction V_{LS} is not Hermitian. We now make a Z transformation such that

$$Z\chi_m = v_m;$$
$$\langle v_m \mid v_{m'} \rangle = \delta_{mm'}; \; m, m' = 1, d, \tag{10}$$

where d is the dimension of the model space. This transformation reorients the vectors χ_m such that they become orthonormal to each other. We assume that χ_m's (m=1,d) are linearly independent so that Z^{-1} exists, otherwise the above transformation is not possible. Since v_m and Z exist entirely within the model space, we can write $v_m = Pv_m$ and $Z = PZP$.

Using Eq.(10), we transform Eq.(8) into

$$Z(H_0 + V_{LS})Z^{-1}v_m = E_m v_m, \tag{11}$$

which implies

$$Z(H_0 + V_{LS})Z^{-1} = \sum_{m \epsilon P} E_m \mid v_m \rangle \langle v_m \mid . \tag{12}$$

Since E_m is real (it is an eigenvalue of Eq.(7)) and the vectors v_m are orthonormal to each other, $Z(H_0 + V_{LS})Z^{-1}$ must be Hermitian. The original problem is now reduced to a Hermitian model-space eigenvalue problem

$$P(H_0 + V_{herm})Pv_m = E_m v_m \tag{13}$$

with the Hermitian effective interaction

$$V_{herm} = Z(H_0 + V_{LS})Z^{-1} - PH_0P, \tag{14}$$

or equivalently

$$V_{herm} = Ze^{-\omega}(H_0 + V)e^{\omega}Z^{-1} - PH_0P. \tag{15}$$

To calculate V_{herm}, we must first have the Z transformation. Since there are certainly many ways to construct Z, this generates a family of Hermitian effective interactions, all originating from V_{LS}. For example, we can construct Z using the familiar Schmidt orthogonalization procedure, namely:

$$\begin{aligned} v_1 &= Z_{11}\chi_1 \\ v_2 &= Z_{21}\chi_1 + Z_{22}\chi_2 \\ v_3 &= Z_{31}\chi_1 + Z_{32}\chi_2 + Z_{33}\chi_3 \\ v_4 &= ......, \end{aligned} \tag{16}$$

with the matrix elements Z_{ij} determined from Eq.(10). We denote the Hermitian effective interaction using this Z transformation as V_{schm}. Clearly there are more than one such Schmidt procedures. For instance, we can use v_2 as the starting point, which gives $v_2 = Z_{22}\chi_2$, $v_3 = Z_{31}\chi_1 + Z_{32}\chi_2$, and so forth. This freedom in how the orthogonalization is actually achieved, gives us infinitely many ways to generate a Hermitian interaction, and this is our family of Hermitian interactions produced from V_{LS}.

We now show how some well-known Hermitization transformations relate to (and in fact, are special cases of) ours. We first look at the Okubo transformation [29]. From the properties of the wave operator ω, we have

$$\langle \chi_m \mid (1 + \omega^+ \omega) \mid \chi_{m'} \rangle = \delta_{mm'}. \tag{17}$$

It follows that an analytic choice for the Z transformation is

$$Z = P(1 + \omega^+\omega)^{1/2} P. \tag{18}$$

This leads to the Hermitian effective interaction

$$V_{okb-1} = P(1 + \omega^+\omega)^{1/2} P(H_0 + V_{LS}) P(1 + \omega^+\omega)^{-1/2} P - PH_0P. \tag{19}$$

It is easily seen that the above is equal to the Okubo Hermitian effective interaction

$$V_{okb} = P(1 + \omega^+\omega)^{-1/2}(1 + \omega^+)(H_0 + V)(1 + \omega)(1 + \omega^+\omega)^{-1/2} P - PH_0P, \tag{20}$$

giving us an alternate expression, Eq.(19), for the Okubo interaction.

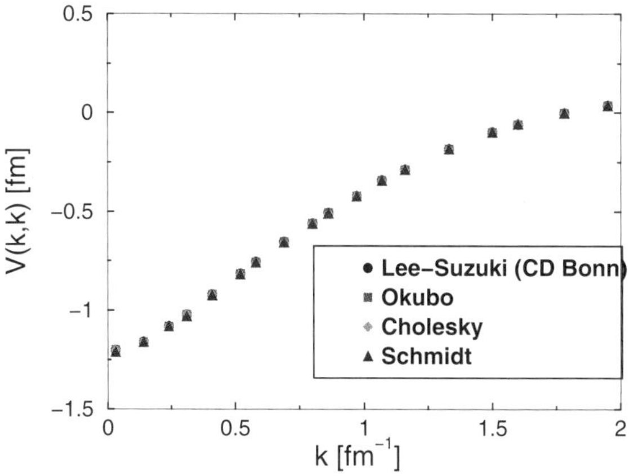

Figure 4. Comparison of non-Hermitian (Lee-Suzuki) and Hermitian (Okubo, Cholesky(Andreozzi), Schmidt) low-momentum NN interactions.

There is another interesting choice for the transformation Z. As pointed out by Andreozzi [27], the positive definite operator $P(1+\omega^{+}\omega)P$ can be decomposed into two Cholesky matrices, namely

$$P(1+\omega^{+}\omega)P = PLL^{T}P, \tag{21}$$

where L is a lower triangle Cholesky matrix, L^{T} being its transpose. Since L is real and it is within the P-space, we have

$$Z = L^{T} \tag{22}$$

and the corresponding Hermitian effective interaction from Eq.(15) is

$$V_{cho} = PL^{T}P(H_0 + V_{LS})P(L^{-1})^{T}P - PH_0P. \tag{23}$$

This is the Hermitian effective interaction of Andreozzi [27].

The Hermitian effective interaction of Suzuki and Okamota [30,31] is of the form

$$V_{suzu} = Pe^{-G}(H_0 + V)e^{G}P - PH_0P \tag{24}$$

with $G = tanh^{-1}(\omega - \omega^{\dagger})$ and $G^{\dagger} = -G$. It has been shown that this interaction is the same as the Okubo interaction [30]. In terms of the Z transformation, it is readily seen that the operator e^{-G} is equal to $Ze^{-\omega}$ with Z given by Eq. (18). Thus, three well-known and particularly useful Hermitian effective interactions indeed belong to our family.

So, starting from V_{LS} one can construct a family of Hermition effective interactions by way of a Z transformation. It has been shown [10] that all such interactions preserve the fully on shell T-matrix $T(k,k,k^2)$; $k \leq \Lambda$. Hence they are all phase shift equivalent. (Recall that V_{LS} satisfies the constraint of preserving the half-on-shell T-matrix $T(k',k,k^2)$; $(k',k) \leq \Lambda$. This constraint is relaxed for the low-momentum Hermitian interactions which preserve only the fully on shell T-matrix.)

Using a solvable matrix model, the above Hermitian effective interactions V_{schm}, V_{okb} and V_{cho} can be in general quite different [10] from each other and from V_{LS}, especially when V_{LS} is largely non-Hermitian. For the V_{NN} case, it is fortunate that the V_{low-k} coresponding to V_{LS} is only slightly non-Hermitian. As a result, the Hermitian low-momentum NN inteactions corresponding to V_{schm}, V_{okb} and V_{cho} are all quite similar to each other and to the one corresponding to V_{LS}, as illutrated in Fig.4 for the ${}^1S_0$ channel. Note that for the ${}^3S_1$ channel the differences among them are slightly larger than the ${}^1S_0$ case.

4. Summary and discussion

Motivated by RG-EFT ideas, a low momentum NN interaction V_{low-k} has been constructed via integrating out the high momentum, model dependent regions of different realistic NN potentials. The result appears to give an approximately unique representation of the NN potential. We have found that V_{low-k} is nearly identical to $V_{NN} + V_{CT}$ over all partial waves, where V_{CT} represents the counter terms. V_{CT} is mainly a $\delta(r)$ force and can be accurately represented by a low-order power series in momentum. The V_{low-k} given by the Andreozzi-Lee-Suzuki (ALS) method is not Hermitian. By performing a Schmidt orthogonality transformation, a family of phase-shift equivalent low-momentum NN potentials can be generated. Since the ALS V_{low-k} is only slightly non-Hermitian, the three Hermitian V_{low-k}'s (Okubo, Andreozzi and Schmidt) investigated in Ref. [10] are all numerically close to each other and to the ALS one.

In low-energy nuclear physics, we are probing the nuclear systems with low- energy probes of wave length λ. Such probes certainly can not reveal the details of the short-range (high momentum) details of the NN potential much shorter than λ. Furthermore, NN scattering phase shifts are available only up to $E_{lab} \approx 350$ MeV (pion production threshold). This E_{lab} corresponds to a decimation momentum $\Lambda \approx 2.1 fm^{-1}$. These considerations support the choice of Λ in the vicinity of 2.0 fm^{-1}. Beyond this momentum our knowledge about the NN interaction is indeed quite uncertain.

Acknowledgments

We thank G.E. Brown, S. Bogner and A. Schwenk for many helpful discussions. This work was supported in part by the U.S. DOE Grant No. DE-FG02-88ER40388.

References

1. S. Bogner, T.T.S. Kuo and L. Coraggio, Nucl. Phys. **A684**, 432c (2001).
2. S. Bogner, T.T.S. Kuo, L. Coraggio, A. Covello and N. Itaco, Phys. Rev. **C65** , 051301(R) (2002).
3. T.T.S. Kuo, S. Bogner and L. Coraggio, Nucl. Phys. **A704**, 107c (2002).
4. L. Coraggio, A. Covello, A. Gargano, N. Itako, T.T.S. Kuo, D.R. Entem and R. Machleidt, Phys. Rev. **C66** , 021303(R) (2002).
5. L. Coraggio, A. Covello, A. Gargano, N. Itako and T.T.S. Kuo, Phys. Rev. **C66** , 064311 (2002).
6. L. Coraggio, N. Itaco, A. Covello, A. Gargano and T.T.S.Kuo, Phys. Rev. **C48**, 034320 (2003).

7. S.K. Bogner, T.T.S. Kuo and A. Schwenk, Phys. Rep. **386** (2003) 1.
8. A. Schwenk, G.E. Brown and B. Friman, Nucl. Phys. **A 703** 745 (2002).
9. Jason D. Holt, T.T.S. Kuo, G.E. Brown and S.K. Bogner, Nuc. Phys. **A733**, 153 (2004).
10. Jason D. Holt, T.T.S. Kuo and G.E. Brown, Phys. Rev. C, **69** 034329 (2004).
11. M. Lacombe et al., Phys. Rev. **C21**, 861 (1980).
12. R. Machleidt, Adv. Nucl. Phys. **19**, 189 (1989).
13. R. Machleidt, Phys. Rev. **C63**, 024001 (2001).
14. V.G.J. Stoks and R. Klomp, C. Terheggen and J. de Schwart, Phys. Rev. **C49**, 2950 (1994).
15. R. B. Wiringa, V.G.J. Stoks and R. Schiavilla, Phys. Rev. **C51**, 38 (1995).
16. D.R. Entem, R. Machleidt and H. Witala, Phys. Rev. **C65**, 064005 (2002).
17. S. Weinberg, Phys. Lett. **B251**, 288 (1990); Nucl. Phys. **B363**, 3 (1991).
18. P. Lepage, "How to Renormalize the Schroedinger Equation" in Nuclear Physics (ed. by C.A. Bertulani et al.), World Scientific Press (1997); [nuc-th/9706029].
19. D.B. Kaplan, M.J. Savage and M.B. Wise, Phys. Lett. **B424**, 390 (1998); Nucl. Phys. **B534**, 329 (1998).
20. E. Epelbaum, W. Glöckle, A. Krüger and Ulf-G. Meissner, Nucl. Phys. **A645**, 413 (1999).
21. P. Bedaque et. al. (eds.), Nuclear Physics with Effective Field Theory II, (1999) World Scientific Press.
22. U. van Kolck, Prog. Part. Nucl. Phys. **43**, 409 (1999).
23. W. Haxton and C.L. Song, Phys. Rev. Lett. **84**, 5484 (2000).
24. T.T.S. Kuo, S.Y. Lee and K.F. Ratcliff, Nucl. Phys. **A176**, 65 (1971).
25. T.T.S. Kuo and E. Osnes, Springer Lecture Notes of Physics, **Vol. 364**, p.1 (1990).
26. K. Suzuki and S. Y. Lee, Prog. Theor. Phys. **64**, 2091 (1980).
27. F. Andreozzi, Phys. Rev. **C54**, 684 (1996).
28. E. M. Krenciglowa and T.T.S. Kuo, Nucl. Phys. **A235**, 171 (1974).
29. S. Okubo, Prog. Theor. Phys. **12**, 603 (1954).
30. K. Suzuki and R. Okamoto, Prog. Theo. Phys. **70**, 439 (1983)
31. K. Suzuki, R. Okamoto, P.J. Ellis and T.T.S. Kuo, Nucl. Phys. **A567** 576 (1994).

KEY TOPICS IN NUCLEAR STRUCTURE
Proceedings of the 8th Int. Spring Seminar on Nuclear Physics
edited by Aldo Covello
© 2005 World Scientific Publishing Co.

DEPENDENCE OF NUCLEAR BINDING ENERGIES ON THE CUTOFF MOMENTUM OF LOW-MOMENTUM NUCLEON-NUCLEON INTERACTION

S. FUJII

Department of Physics, University of Tokyo,
Tokyo 113-0033, Japan
E-mail: sfujii@nt.phys.s.u-tokyo.ac.jp

H. KAMADA, R. OKAMOTO AND K. SUZUKI

Department of Physics, Kyushu Institute of Technology,
Kitakyushu 804-8550, Japan

Binding energies of ^{3}H, ^{4}He, and ^{16}O are calculated, using low-momentum nucleon-nucleon interactions ($V_{\mathrm{low}-k}$) for a wide range of the cutoff momentum Λ. In addition, single-particle energies in nuclei around ^{16}O are computed. The dependence of the binding energies and the single-particle energies in these nuclei on the cutoff momentum Λ of the $V_{\mathrm{low}-k}$ is examined. Furthermore, the availability of the $V_{\mathrm{low}-k}$ in nuclear structure calculations is discussed.

1. Introduction

One of the fundamental objectives in nuclear structure calculations is to describe nuclear properties, starting with high-precision nucleon-nucleon interactions. Since this kind of interaction has a repulsive core at a short distance, one has been forced to derive an effective interaction or G matrix in a model space for each nucleus from the realistic interaction, except for precise few-nucleon structure calculations.

Recently, Bogner *et al.* have constructed low-momentum nucleon-nucleon interactions $V_{\mathrm{low}-k}$ from high-precision nucleon-nucleon interactions to use them as microscopic input to the nuclear many-body problem.[1] The $V_{\mathrm{low}-k}$ can be derived using techniques of conventional effective interaction theory or renormalization group method. They have shown that the $V_{\mathrm{low}-k}$ conserves the properties of the original interaction, such as the half-on-shell T matrix and the phase shift within a cutoff momentum Λ which specify the low-momentum region. The $V_{\mathrm{low}-k}$ for the typical cutoff

$\Lambda = 2.1$ fm^{-1} corresponding to $E_{\rm lab} \simeq 350$ MeV are almost the same and are not dependent on the realistic nucleon-nucleon interactions employed. Thus, as a *unique* low-momentum interaction, the $V_{{\rm low}-k}$ at approximately $\Lambda \sim 2$ fm^{-1} has been employed directly in nuclear structure calculations, such as the shell-model[2] and the Hartree-Fock calculations.[3] Especially, the calculated excitation spectra in the shell-model calculations show the good agreement with the experimental data and are even better than those using the sophisticated G matrix. Thus, the application of $V_{{\rm low}-k}$ to nuclear structure calculations has been growing. We should notice, however, that the $V_{{\rm low}-k}$ is derived introducing the cutoff momentum Λ, and thus the calculated results using the $V_{{\rm low}-k}$ have the Λ dependence to some extent.

One of the central aims of the present work is to examine the Λ dependence in structure calculations. First, we calculate binding energies for few-nucleon systems for which precise calculations can be performed, and confirm the validity of the $V_{{\rm low}-k}$ in the structure calculation by comparing the obtained results with the *exact* values. Second, we proceed to heavier systems such as ^{16}O and investigate not only the total binding energy itself but also the single-particle energy which is defined as the relative energy of neighboring two nuclei such as ^{16}O and ^{15}O. Through the obtained results, we discuss the applicability of the $V_{{\rm low}-k}$ to nuclear structure calculations.

2. Results and discussion

In the following structure calculations, we use the $V_{{\rm low}-k}$ which is derived from the CD-Bonn potential[4] by means of a unitary transformation theory.[5,6] The details of deriving the $V_{{\rm low}-k}$ and its numerical accuracy can be seen in Ref. 7.

2.1. ^{3}*H and* ^{4}*He*

In order to investigate the sensitivity of Λ to the binding energies of ^{3}H and ^{4}He precisely, we have performed the Faddeev and the Yakubovsky calculations, respectively.[a] For simplicity, only the neutron-proton interaction is used for all the channels.

Figure 1(a) exhibits the calculated ground-state energies of ^{3}H by a 34-channel Faddeev calculation as a function of the cutoff momentum Λ. The

[a]The collaboration with E. Epelbaum and W. Glöckle in this part of the present work which has already been done in Ref. 7 is highly appreciated.

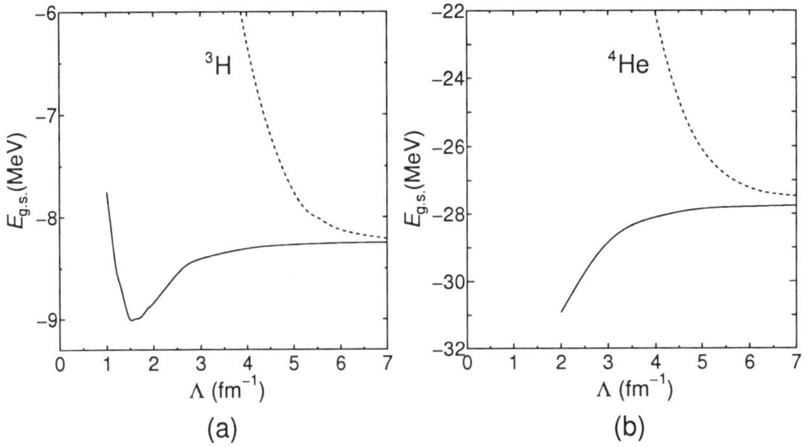

Figure 1. Calculated ground-state energies of ^{3}H (a) and ^{4}He (b) as a function of the cutoff momentum Λ. The solid lines represent the results using the $V_{\text{low}-k}$ for each Λ. The short-dashed lines are the results using the original CD-Bonn potential, where the high-momentum components beyond Λ are simply truncated in the structure calculation.

exact value using the original CD-Bonn potential on the above assumptions is −8.25 MeV. The solid line depicts the results using the $V_{\text{low}-k}$ from the CD-Bonn potential. The short-dashed line represents the results using the original CD-Bonn potential, where the high-momentum components beyond Λ are simply truncated in the structure calculation. For the case of the original CD-Bonn potential, we need $\Lambda \geq 8$ fm^{-1} to reach the *exact* value if the accuracy of 100 keV is required. This situation is largely improved if we use the $V_{\text{low}-k}$. Even if we require the accuracy of 1 keV, we do not need the high-momentum components beyond $\Lambda \sim 8$ fm^{-1}. However, it should be noted that the results using the $V_{\text{low}-k}$ for the values smaller than $\Lambda \sim 5$ fm^{-1} vary considerably, and there occurs the energy minimum at around $\Lambda = 1.5$ fm^{-1}. The magnitude of the difference between the *exact* value and the calculated result using the $V_{\text{low}-k}$ for the representative cutoff value $\Lambda = 2.0$ fm^{-1} is about 600 keV for ^{3}H.

A similar tendency can also be seen in the case of ^{4}He. We have performed the S-wave (5+5-channel) Yakubovsky calculation for ^{4}He without the Coulomb interaction. The *exact* ground-state energy using the original CD-Bonn potential on the above assumptions is −27.74 MeV. In Fig. 1(b), the calculated results for ^{4}He are shown. The shape of the energy curve is similar to that for ^{3}H within the region $\Lambda \geq 2$ fm^{-1}. The calculated results become more overbound as the value of Λ becomes smaller. The magnitude of the difference between the *exact* value and the calculated result using the $V_{\mathrm{low}-k}$ for $\Lambda = 2.0$ fm^{-1} is about 3 MeV for ^{4}He. This amounts to five times larger than the result of ^{3}H. In the case of ^{4}He, the results for $\Lambda < 2.0$ fm^{-1} are not shown due to the numerical instability in the structure calculation.

Concerning the investigation of the Λ dependence of the ground-state energies of the few-nucleon systems, a detailed study with three-nucleon forces has recently been reported by Nogga *et al.*[8]

2.2. ^{16}O

In order to examine the Λ dependence in heavier systems, we calculate the ground-state energy of ^{16}O within the framework of the unitary-model-operator approach (UMOA).[6] The details of recent calculated results for ^{16}O and its neighboring nuclei using modern nucleon-nucleon interactions can be seen in Ref. 9. In the present study, we follow the same calculation method in that work except for the determination method of the harmonic-oscillator energy $\hbar\Omega$ and the size of the model space. In Ref. 9, we have searched for the optimal value of $\hbar\Omega$ that leads to the energy minimum point by investigating the $\hbar\Omega$ dependence of the ground-state energy for each modern nucleon-nucleon interaction. Then, we have found that the optimal values are at around 14 MeV of which values are very close to the value determined by empirical formula such as $\hbar\Omega = 45A^{-1/3} - 25A^{-2/3}$ MeV. Since the optimal value was $\hbar\Omega = 15$ MeV for the CD-Bonn potential, we use this value in this work for each $V_{\mathrm{low}-k}$. Furthermore, we employ the same size of the optimal model space which is specified by the quantity ρ_1 as $\rho_1 = 2n_a + l_a + 2n_b + l_b = 12$, where $\{n_a, l_a\}$ and $\{n_b, l_b\}$ are the sets of harmonic-oscillator quantum numbers for two-body states. We note that these values of $\hbar\Omega$ and ρ_1 are not necessarily the optimal ones for each $V_{\mathrm{low}-k}$ in the present study.

In Fig. 2, the Λ dependence using the $V_{\mathrm{low}-k}$ from the CD-Bonn potential of the ground-state energy of ^{16}O is shown. We have used the neutron-neutron, neutron-proton, and proton-proton interaction of the CD-Bonn potential correctly for the corresponding channels, and included the Coulomb interaction. The partial waves up to $J = 6$ are taken into account in the calculation. The value of the ground-state energy of ^{16}O in the *full* calculation given in Ref. 9 using the original CD-Bonn potential is -115.61 MeV. Thus, the result for $\Lambda = 5.0$ fm^{-1} almost reproduces this

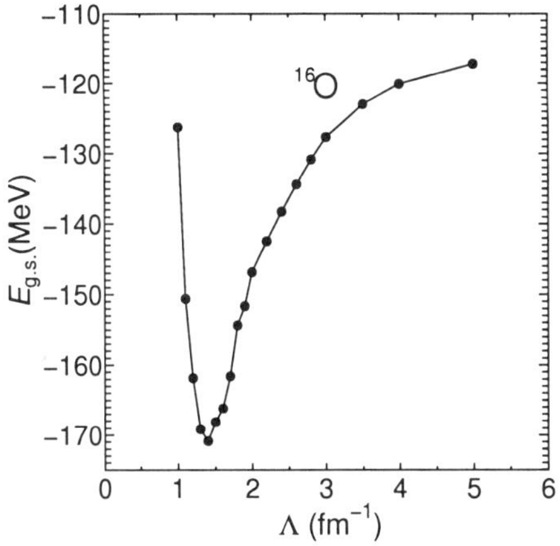

Figure 2. The Λ dependence of the ground-state energy of ^{16}O using the $V_{\text{low}-k}$ from the CD-Bonn potential.

value. The calculated energy curve shows a similar tendency to the results of ^{3}H and ^{4}He, but the magnitude of the difference between the result of the *full* calculation and the value at the energy minimum point is considerably larger than those for ^{3}H and ^{4}He due to the large difference of the mass number. The magnitude of the difference in ^{16}O amounts to 55 MeV. Even if we choose the typical cutoff $\Lambda = 2.0$ fm^{-1}, we still observe the significant overbinding of which magnitude is about 31 MeV. Thus, we may conclude from the results of ^{3}H, ^{4}He, and ^{16}O that the $V_{\text{low}-k}$ for the typical cutoff momentum $\Lambda \sim 2$ fm^{-1} cannot reproduce the *exact* values, showing the significant overbinding.

It should be noted, however, that this does not necessarily mean that the $V_{\text{low}-k}$ for $\Lambda \sim 2$ fm^{-1} is no longer valid in nuclear structure calculations. In fact, the shell-model calculations have shown that the $V_{\text{low}-k}$ for $\Lambda \sim 2$ fm^{-1} can work as well as the G matrix.[2]

2.3. ^{15}O and ^{17}O

In the previous sections, we have seen the results of the total binding energies. We here examine the Λ dependence of single-particle energies of the neutron for hole states in ^{16}O which correspond to the energy levels in ^{15}O and of neutron particle states in ^{17}O. The calculation procedure is essentially the same as in Ref. 9. In the present study, however, we do not search for the optimal values of $\hbar\Omega$ and ρ_1 for each single-hole or -particle

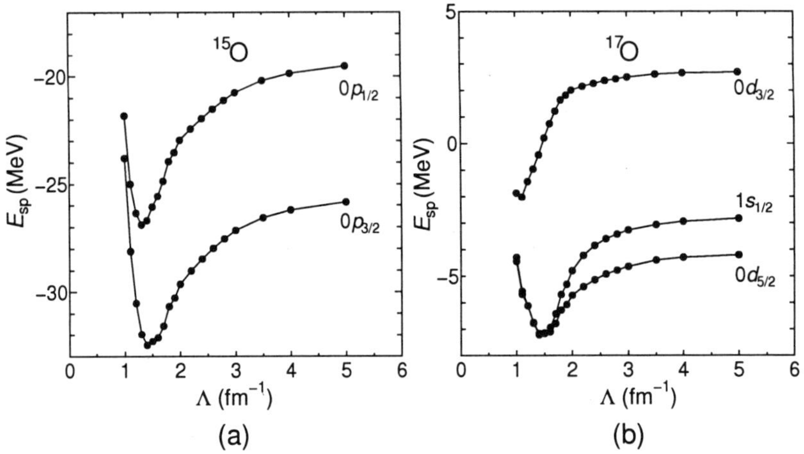

Figure 3. The Λ dependence of the single-particle energies of the neutron for the $0p$ hole states in ^{16}O which correspond to the energy levels in ^{15}O (a) and of the neutron particle states in ^{17}O (b) using the $V_{\mathrm{low}-k}$ from the CD-Bonn potential.

state for simplicity as in the case of ^{16}O. In the following calculations, we use the values of $\hbar\Omega = 15$ MeV and $\rho_1 = 12$ which are the same as in the calculation of ^{16}O in the previous section.

Figure 3(a) shows the Λ dependence of the calculated single-particle energies of the neutron for the $0p$ hole states in ^{16}O which correspond to the single-hole energy levels in ^{15}O. The values of the *full* calculation of the single-particle energy are -19.34 and -25.37 MeV for the $0p_{1/2}$ and $0p_{3/2}$ states, respectively. Though the present results for $\Lambda = 5.0$ fm^{-1} are fairly close to these values, there remain some discrepancies. These discrepancies may be due to the fact that we do not search for the optimal value of $\hbar\Omega$ for each state in the present study. The search for the optimal values of $\hbar\Omega$ and also ρ_1 in the structure calculation with the $V_{\mathrm{low}-k}$ should be done for completeness in future.

It is seen from Fig. 3(a) that the single-particle energies for the $0p$ states become more attractive as the Λ becomes smaller as in the results of the ground-state energies. However, what is interesting here is that the magnitudes of the spacing between the single-particle levels, namely the spin-orbit splitting, hold their values up to $\Lambda \sim 2$ fm^{-1}, although the structure of the single-particle levels is broken within the area $\Lambda < 2$ fm^{-1}.

A similar tendency can also be observed in the results of ^{17}O. In Fig. 3(b), the calculated results of the single-particle energies of the neutron for the $1s$ and $0d$ states in ^{17}O are shown. The values of the *full* calculation of the single-particle energy are 2.67, −2.76, and −4.11 MeV for the $0d_{3/2}$, $1s_{1/2}$, and $0d_{5/2}$ states, respectively. The present results for $\Lambda = 5.0$ fm^{-1} are not so different from these values. The tendency of the Λ dependence is essentially the same as in ^{15}O. It can be seen again that the magnitudes of the spacings between the single-particle levels do not vary very much within the region $\Lambda \geq 2$ fm^{-1}, while those are considerably broken within the area $\Lambda < 2$ fm^{-1}. These results may suggest that the $V_{\text{low}-k}$ for $\Lambda \sim 2$ fm^{-1} is valid as far as relative energies from a state such as the ground state are concerned.

3. Conclusions

We investigated the dependence of the ground-state energies of ^{3}H, ^{4}He, and ^{16}O on the cutoff momentum Λ of the low-momentum nucleon-nucleon interaction $V_{\text{low}-k}$. In all the cases, there appear the energy minima at around $\Lambda = 1.5$ fm^{-1}. We have found that the $V_{\text{low}-k}$ for the typical cutoff momentum $\Lambda \sim 2$ fm^{-1} cannot reproduce the *exact* values for the original interaction, showing the significant overbinding. If we try to reproduce the *exact* values, we need $\Lambda \geq 5$ fm^{-1}. On the other hand, the magnitudes of the spacings between the single-particle levels in nuclei around ^{16}O do not so vary within the region $\Lambda \geq 2$ fm^{-1}. This may suggest that the $V_{\text{low}-k}$ for the typical cutoff $\Lambda \sim 2$ fm^{-1} is valid in nuclear structure calculations as far as relative energies from a state such as the ground state are concerned as in the shell-model calculation.

Acknowledgments

This work was supported by a Grant-in-Aid for Scientific Research (C) (Grant No. 15540280) from Japan Society for the Promotion of Science and a Grant-in-Aid for Specially Promoted Research (Grant No. 13002001) from the Ministry of Education, Culture, Sports, Science and Technology in Japan.

References

1. S. K. Bogner, T. T. S. Kuo and A. Schwenk, *Phys. Rep.* **386**, 1 (2003).
2. Scott Bogner, T. T. S. Kuo, L. Coraggio, A. Covello and N. Itaco, *Phys. Rev.* **C65**, 051301(R) (2002).
3. L. Coraggio, N. Itaco, A. Covello, A. Gargano and T. T. S. Kuo, *Phys. Rev.* **C68**, 034320 (2003).

4. R. Machleidt, F. Sammarruca and Y. Song, *Phys. Rev.* **C53**, R1483 (1996).
5. S. Ōkubo, *Prog. Theor. Phys.* **12**, 603 (1954).
6. K. Suzuki and R. Okamoto, *Prog. Theor. Phys.* **92**, 1045 (1994).
7. S. Fujii, E. Epelbaum, H. Kamada, R. Okamoto, K. Suzuki and W. Glöckle, *nucl-th*/0404049.
8. A. Nogga, S. K. Bogner and A Schwenk, *nucl-th*/0405016.
9. S. Fujii, R. Okamoto and K. Suzuki, *Phys. Rev.* **C69**, 034328 (2004).

KEY TOPICS IN NUCLEAR STRUCTURE
Proceedings of the 8th Int. Spring Seminar on Nuclear Physics
edited by Aldo Covello
© 2005 World Scientific Publishing Co.

THE *AB INITIO* LARGE-BASIS NO-CORE SHELL MODEL

B. R. BARRETT*, P. NAVRÁTIL†, A. NOGGA‡, W. E. ORMAND†,
I. STETCU*, J. P. VARY§, and H. ZHAN*

**Department of Physics, P.O. Box 210081, University of Arizona, Tucson, AZ 85721, USA*

† University of California, Lawrence Livermore National Laboratory, Livermore, CA 94551, USA

‡ Institute for Nuclear Theory, University of Washington, Box 351550, Seattle, WA 98195, USA

§ Department of Physics and Astronomy, Iowa State University, Ames, IA 50011, USA

We describe the development and application of the *ab initio* No-Core shell Model (NCSM), in which the effective Hamiltonians are derived microscopically from realistic nucleon-nucleon (NN) plus theoretical three-nucleon (NNN) potentials, as a function of the finite harmonic-oscillator (HO) basis space. For presently feasible no-core model spaces, we evaluate the effective Hamiltonians in a cluster approach, which is guaranteed to provide exact results for sufficiently large model spaces and/or sufficiently large clusters. A number of recent applications of the NCSM are given.

1. INTRODUCTION

The major outstanding problem in nuclear-structure physics is to calculate the properties of finite nuclei starting from the basic interactions among the nucleons. Such calculations have been performed so far only for light nuclei up to $A = 10$.[1] We have developed a new *ab initio* technique for accurately computing nuclear properties, in which all A nucleons are taken to be active, interacting by realistic NN plus theoretical NNN interactions. We call this approach the *ab-initio* No-Core Shell Model, which we simply refer to as 'NCSM'.[2,3,4,5,6,7,8,9,10,11] This new method is very flexible with respect to the interactions used. As such, it allows us to test a wide range of NN and NNN interaction models using light nuclei as a laboratory.

2. NO-CORE SHELL-MODEL APPROACH

The NCSM is based on an effective Hamiltonian derived from realistic "bare" interactions and acting within a finite Hilbert space. All A-nucleons are treated on an equal footing. The approach is both computationally tractable and demonstrably convergent to the exact result of the full (infinite) Hilbert space.

Initial investigations used two-body interactions[2] based on a G-matrix approach. Later, we implemented the Lee-Suzuki procedure[12] to derive two-body[3] and three-body[5] effective interactions based on realistic NN and theoretical NNN interactions.

For pedagogical purposes, we outline the NCSM approach only with NN interactions and refer the reader to the literature on how to include NNN interactions. However, some results with NNN interactions will be given. We begin with the purely intrinsic Hamiltonian for the A-nucleon system, i.e.,

$$H_A = T_{\rm rel} + \mathcal{V} = \frac{1}{A}\sum_{i<j}^{A} \frac{(\vec{p}_i - \vec{p}_j)^2}{2m} + \sum_{i<j}^{A} V_{ij}^{\rm N} , \tag{1}$$

where m is the nucleon mass and $V_{ij}^{\rm N}$, the NN interaction in pair (ij), with both strong and electromagnetic components. Note the absence of a phenomenological single-particle potential. We may use either local potentials in coordinate-space, such as the Argonne potentials[1] or non-local ones, such as the CD-Bonn[13].

Next, we add the center-of-mass (CM) HO Hamiltonian, $H_{\rm CM} = T_{\rm CM} + U_{\rm CM}$, where $U_{\rm CM} = \frac{1}{2}Am\Omega^2\vec{R}^2$, $\vec{R} = \frac{1}{A}\sum_{i=1}^{A}\vec{r}_i$, to H_A. In the full Hilbert space the added $H_{\rm CM}$ term has no influence on the intrinsic properties. However, when we introduce our cluster approximation below, the added $H_{\rm CM}$ term will be very important for the determination of the effective interactions. The modified Hamiltonian, with a pseudo-dependence on the HO frequency Ω, can be cast into the form

$$H_A^{\Omega} = H_A + H_{\rm CM} = \sum_{i=1}^{A}\left[\frac{\vec{p}_i^{\,2}}{2m} + \frac{1}{2}m\Omega^2\vec{r}_i^{\,2}\right] + \sum_{i<j}^{A}\left[V_{ij}^{\rm N} - \frac{m\Omega^2}{2A}(\vec{r}_i - \vec{r}_j)^2\right] . \tag{2}$$

In the spirit of Da Providencia and Shakin[14] and Lee, Suzuki and Okamoto[12], we introduce a unitary transformation, which is able to accommodate the short-range two-body correlations in a nucleus, by choosing an antihermitian operator S, acting only on intrinsic coordinates, such that

$$\mathcal{H} = e^{-S} H_A^{\Omega} e^{S} \ . \tag{3}$$

In our approach, S is determined by the requirements that $\mathcal{H}$ and H_A^{Ω} have the same symmetries and eigenspectra over the subspace $\mathcal{K}$ of the full Hilbert space. In general, both S and the transformed Hamiltonian are A-body operators. Our simplest, non-trivial approximation to $\mathcal{H}$ is to develop a two-body ($a = 2$) effective Hamiltonian, where the upper bound of the summations "A" is replaced by "a", but the coefficients remain unchanged. A three-body effective Hamiltonian, ($a = 3$), is obtained in a similar manner.

If the full Hilbert space is divided into a finite model space ("P-space") and a complementary infinite space ("Q-space"), using the projectors P and Q with $P + Q = 1$, it is possible to determine the transformation operator S_a from the decoupling condition

$$Q_a e^{-S^{(a)}} H_a^{\Omega} e^{S^{(a)}} P_a = 0 \ , \tag{4}$$

and the simultaneous restrictions $P_a S^{(a)} P_a = Q_a S^{(a)} Q_a = 0$. Note that a-nucleon-state projectors (P_a, Q_a) appear in Eq. (4). P_a is chosen to project onto the set of all a body states, which are included in P. The unitary transformation and decoupling condition, introduced by Suzuki and Okamoto and referred to as the unitary-model-operator approach[15] (UMOA), has a solution that can be expressed in the following form

$$S^{(a)} = \operatorname{arctanh}(\omega - \omega^{\dagger}) \ , \tag{5}$$

with the operator ω satisfying $\omega = Q_a \omega P_a$, and solving its own decoupling equation,

$$Q_a e^{-\omega} H_a^{\Omega} e^{\omega} P_a = 0 \ . \tag{6}$$

Given the eigensolutions, $H_a^{\Omega}|k\rangle = E_k|k\rangle$, then the operator ω can be determined from

$$\langle \alpha_Q|\omega|\alpha_P\rangle = \sum_{k\in\mathcal{K}} \langle \alpha_Q|k\rangle\langle \tilde{k}|\alpha_P\rangle \ , \tag{7}$$

where we denote by tilde the inverted matrix of $\langle \alpha_P|k\rangle$, i.e., $\sum_{\alpha_P} \langle \tilde{k}|\alpha_P\rangle\langle \alpha_P|k'\rangle = \delta_{k,k'}$ and $\sum_k \langle \alpha'_P|\tilde{k}\rangle\langle k|\alpha_P\rangle = \delta_{\alpha'_P,\alpha_P}$, for $k, k' \in \mathcal{K}$. In the relation (7), $|\alpha_P\rangle$ and $|\alpha_Q\rangle$ are the model-space and the Q-space basis states, respectively, and $\mathcal{K}$ denotes a set of d_P eigenstates, whose properties are reproduced in the model space. Necessarily, d_P is equal to the dimension of the model space.

With the help of the solution for ω (7) we obtain a simple expression for the matrix elements of the hermitian effective Hamiltonian

$$\langle\alpha_P|\bar{H}_{\mathrm{a-eff}}|\alpha'_P\rangle = \sum_{k\in\mathcal{K}}\sum_{\alpha''_P}\sum_{\alpha'''_P}\langle\alpha_P|(P_a+\omega^\dagger\omega)^{-1/2}|\alpha''_P\rangle\langle\alpha''_P|\tilde{k}\rangle E_k\langle\tilde{k}|\alpha'''_P\rangle \times\langle\alpha'''_P|(P_a+\omega^\dagger\omega)^{-1/2}|\alpha'_P\rangle\ . \qquad (8)$$

We note that in the limit $a \to A$, we obtain the exact solutions for d_P states of the full problem for any finite basis space.

On account of our cluster approximation, a dependence of our results on the model-space size and on the HO frequency Ω arises. For a fixed cluster size, the smaller the basis space, the larger the dependence on Ω.

In order to construct the operator ω (7) we need to select the set of eigenvectors $\mathcal{K}$. Because of the added CM Hamiltonian, the a-body clusters are confined, which ensures that all eigenvectors are bound states. We keep the lowest states obtained in each two-body channel. It turns out that these states also have the largest overlap with the model space for the range of $\hbar\Omega$ and the P-spaces we have investigated.

We input the effective Hamiltonian, now consisting of a relative 2-body operator and the pure H_{CM} term introduced earlier, into an m-scheme Lanczos diagonalization process to obtain the P-space eigenvalues and eigenvectors. At this stage we also add the term H_{CM} again with a large positive coefficient to separate the physically interesting states with $0s$ CM motion from those with excited CM motion. We retain only the states with pure $0s$ CM motion when evaluating observables. All observables that are expressible as functions of relative coordinates, such as the rms radius and radial densities, are then evaluated free of CM motion effects.

We close our presentation on the theoretical framework with the observation that all observables require the same transformation as implemented on the Hamiltonian. To date, we have found rather small effects on the rms radius operator when we transformed it to a P-space effective rms operator at the a=2 cluster level.[6] On the other hand, substantial renormalization was observed for the kinetic energy operator when using the a=2 transformation to evaluate its expectation value.[16]

3. RESULTS

Our $A = 3$ results indicate the feasibility of our approach, showing that accurate values of physics properties can be obtained in sufficiently large model spaces, which we define by the maximal HO excitation above the

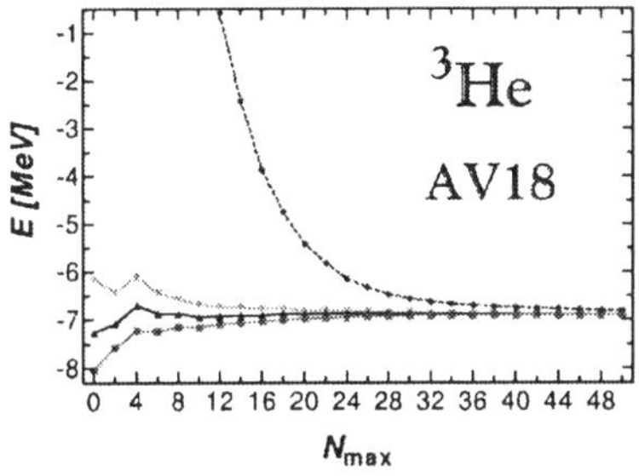

(a)

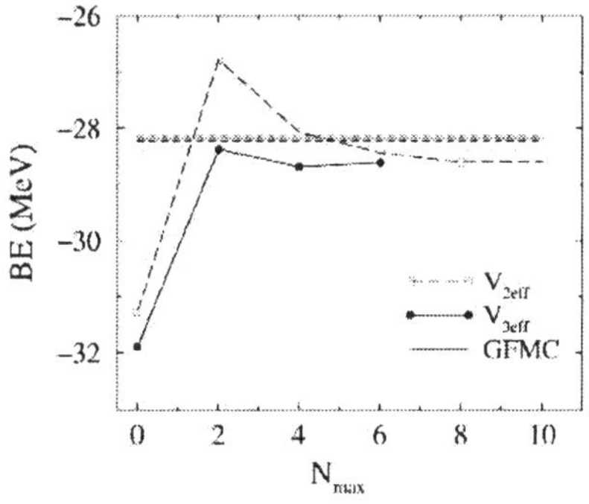

(b)

Figure 1. (a) Ground-state energy dependence on the model-space size for ^{3}He interacting by the AV18 NN potential The dashed line shows the result based on the bare interaction. The solid lines with down-pointing triangles, up-pointing triangles and squares are results based on the effective interaction for $\hbar\Omega = 32$, 28 and 24 MeV, respectively. (b) Same for ^{6}Li using the AV8' NN potential with Coulomb. The plotted energies occur at the HO frequency minima for the given value of N_{max} The results using both the two-body and the three-body effective interaction are compared with the GFMC results from Ref. 1. The figure is from Ref. 8.

minimal A-body configuration N_{max} included in the basis. Even for the AV18 NN potential[1], which has a strongly repulsive core, the $N_{max} = 50$ model space is sufficient for obtaining a converged result with an error less than 10 keV, as shown in Fig. 1(a). It is also seen that the utilization of the effective NN interaction speeds up the convergence significantly compared with the bare interaction.

A bigger challenge for the NCSM is the p-shell, where model spaces increase rapidly in size with increasing N_{max}. Consequently, model spaces larger than $N_{max} = 8$ are not presently feasible for most p-shell nuclei. However, besides increasing N_{max} to improve convergence, one can also increase the cluster size of the effective interaction. This has been investigated by Navrátil and Ormand[8] for several p-shell nuclei. E.g. for ^{6}Li, it was demonstrated that three-body effective interactions accelerate convergence. This is is shown in Fig. 1(b).

Our ability to calculate the effective Hamiltonian at the three-body cluster level as well as for two-body cluster makes it possible for us to investigate the nature of different NNN interaction models. The spectra of the light nuclei are well suited for analyzing NNN forces, because they are especially sensitive to their spin/isospin structure. In addition, the NCSM results for the spectra typically converge faster than the binding en-

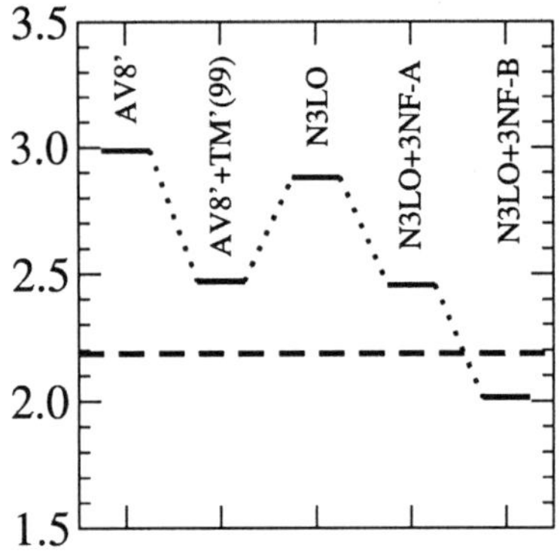

Figure 2. Excitation energy of the 3^+ state of ^{6}Li for NN and NNN interactions. The dashed line marks experiment. All results are for $N_{max} = 6$.

ergies (see, *e.g.*, Ref. 6) and are generally more accurately predicted than the binding energies. To exemplify the sensitivity of the spectra to the three-nucleon interaction, we show in Fig. 2 our results for the excitation energy of the 3^+ state in ^{6}Li. NN forces alone overpredict this observable. Note that the combinations AV8' with the 2π exchange Tucson-Melbourne (TM'(99)) NNN force[17] are already close to the experimental number. This becomes even more pronounced for ^{10}B, where TM'(99) corrects the wrong ordering of ground and excited state predicted by NN interactions only or in combination with the Urbana-IX 3N force[1]. We also studied the chiral NN interaction Idaho-N3LO[18] in combination with the consistently defined leading chiral 3N interaction[19]. Here we identified two sets of parameters, which describe the ^{3}H and ^{4}He binding energies equally well. The excitation energy is different for both sets of parameters, clearly showing the sensitivity of this quantity to the NNN force structure. Calculations for other nuclei are in progress.

To improve the accuracy of our predictions, it is desirable to further increase the model space sizes of these calculations. We recently developed an extrapolation method for estimating the binding energies of NCSM calculations without diagonalizing the complete Hamiltonian in the extremely large basis space.[20] It is motivated by the observation that the binding energy $E_0 = \langle H \rangle$ evaluated in an approximate ground state must approach the exact binding energy $\mathcal{E}_0$ as the energy variance $\Delta E^2 = \langle H^2 \rangle - \langle H \rangle^2$ vanishes.[21] From second-order perturbation theory, we expect that there exists an approximate linear correlation between E_0 and ΔE^2, if approximate ground states are calculated from Hamiltonians truncated from the

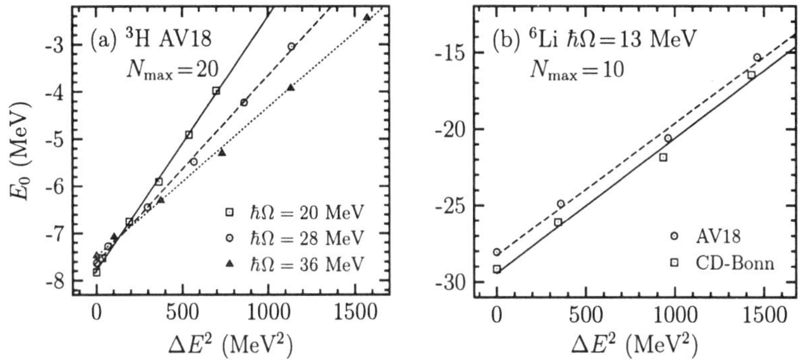

Figure 3. (a) The linear relation between E_0 and ΔE^2, for ^{3}H with the AV18 potential. Symbols from right to left correspond to $\tilde{N}_{max} = 10, 12, \ldots, 20$ for each model. The lines fit the results from $\tilde{N}_{max} = 10$ to 18. (b) The same as (a), but for the ^{6}Li nucleus. Symbols from right to left correspond to $\tilde{N}_{max} = 4, 6, 8$, and 10 for each model. The lines fit the results from $\tilde{N}_{max} = 4, 6$, and 8.

large-space Hamiltonian by HO quantum numbers $\tilde{N}_{max}$. This linear scaling is used to extrapolate large-space results from smaller-space calculations in context of the NCSM.

We obtained the results that the converged binding energy scales with the energy variance, ΔE^2, as shown in Figs. 3(a) and 3(b) for ^{3}H and ^{6}Li nuclei, respectively. Compared to direct diagonalization, the extrapolation has an error of a few tens of keV for ^{3}H and several hundred keV for ^{6}Li. A higher value of N_{max} will further reduce the extrapolation error. Since the CD-Bonn and AV18 predictions for the binding energies deviate from each other by more than 1 MeV, the extrapolation errors are acceptably small. We are presently extending this procedure to heavier mass nuclei.

4. CONCLUSIONS

In this contribution we described the *ab initio* NCSM approach and demonstrated its usefulness by applications to the $A = 3$ system, for which we obtain well-converged results. For the accurate description of p-shell nuclei not only NN, but also NNN interactions are important. In order to include the latter, we need three-body effective interactions. These also improve the rate of convergence in smaller model spaces, although more work remains to be done regarding the role and properties of NNN and perhaps NNNN effective interactions. We have shown how the NCSM approach at the NNN cluster level can be used to analyze the nature of different theoretical models for the NNN forces. It should be noted that our calculations contain no adjustable parameters. The favorable comparison with available data that we obtain is a consequence of the underlying NN and NNN interaction. We have also established an extrapolation method to extend our calculations to even larger model spaces. It can serve as a way to estimate the uncertainties of NCSM results arising from $\hbar\Omega$ dependence and choices of interactions. This will be an important application of our method in future investigations of nuclei within the NCSM.

Acknowledgments

B.R.B., I.S. and H.Z. acknowledge partial support by NSF grants PHY0070858 and PHY0244389. J.P.V. acknowledges partial support by USDOE grant No. DE-FG-02-87ER-40371. The work was performed in part under the auspices of the U. S. Department of Energy by the University of California, Lawrence Livermore National Laboratory under contract No. W-7405-Eng-48. P.N. and W.E.O. received support from LDRD contract 00-ERD-028. A.N. acknowledges support by USDOE grants DE-F02-01ER41187 and DE-FG02-00ER41132.

References

1. R. B. Wiringa, V. G. J. Stoks and R. Schiavilla, Phys. Rev. C **51**, 38 (1995); B. S. Pudliner, V. R. Pandharipande, J. Carlson, S. C. Pieper and R. B. Wiringa, Phys. Rev. C **56** 1720, (1997); R. B. Wiringa, Nucl. Phys. **A 631**, 70c (1998); S. Pieper and R. B. Wiringa, Annu. Rev. Nucl.Part.Sci. 51, 53 (2001); S. Pieper, K. Varga and R. B. Wiringa, Phys. Rev. C 66, 044310 (2002).
2. D. C. Zheng, B. R. Barrett, L. Jaqua, J. P. Vary, and R. L. McCarthy, Phys. Rev. C **48**, 1083 (1993); D. C. Zheng, J. P. Vary, and B. R. Barrett, Phys. Rev. C **50**, 2841 (1994); D. C. Zheng, B. R. Barrett, J. P. Vary, W. C. Haxton, and C. L. Song, Phys. Rev. C **52**, 2488 (1995).
3. P. Navrátil and B. R. Barrett, Phys. Rev. C **54**, 2986 (1996); Phys. Rev. C **57**, 3119 (1998).
4. P. Navrátil and B. R. Barrett, Phys. Rev. C **57**, 562 (1998), Phys. Rev. C **59**, 1906 (1999).
5. P. Navrátil, G. P. Kamuntavičius and B. R. Barrett, Phys. Rev. C **61**, 044001 (2000). E-print archive No. nucl-th/9907054.
6. P. Navrátil, J. P. Vary and B. R. Barrett, Phys. Rev. Lett. **84**, 5728 (2000); Phys. Rev. C **62**, 054311 (2000).
7. P. Navrátil, J. P. Vary, W. E. Ormand, and B. R. Barrett, Phys. Rev. Lett. **87**, 172502 (2001).
8. P. Navrátil and W. E. Ormand, Phys. Rev. Lett. **88**, 152502 (2002).

9. C.P. Viazminsky and J.P. Vary, J. Math. Phys., **42**, 2055(2001).
10. P. Navrátil and W. E. Ormand, Phys. Rev. C 68, 034305 (2003).
11. P. Navrátil and E. Caurier, Phys. Rev. C 69, 014311 (2004).
12. K. Suzuki and S.Y. Lee, Prog. Theor. Phys. **64**, 2091 (1980); K. Suzuki, Prog. Theor. Phys. **68**, 246 (1982). K. Suzuki and R. Okamoto, Prog. Theor. Phys. **70**, 439 (1983).
13. R. Machleidt, F. Sammarruca and Y. Song, Phys. Rev. C **53**, 1483 (1996); R. Machleidt, Phys. Rev. C **63**, 024001 (2001).
14. J. Da Providencia and C. M. Shakin, Ann. of Phys. **30**, 95 (1964).
15. K. Suzuki, Prog. Theor. Phys. **68**, 1999 (1982); K. Suzuki and R. Okamoto, Prog. Theor. Phys. **92**, 1045 (1994).
16. H. Kamada, et. al, Phys. Rev. C **64**, 044001. (2001).
17. S. A. Coon and H. K. Han, Few Body Systems **30**, 131 (2001).
18. D. R. Entem and R. Machleidt, Phys. Rev. C 68, 041001(R) (2003).
19. U. van Kolck, Phys. Rev. C 49, 2932(1994); E. Epelbaum et. al., Phys. Rev. C 66, 064001 (2002).
20. H. Zhan, A. Nogga, B. R. Barrett, J. P. Vary and P. Navrátil, Phys. Rev. C **69**, 034302 (2004).
21. T. Mizusaki and M. Imada, Phys. Rev. C **65**, 064319 (2002); Phys. Rev. C **67**, 041301(R) (2003).

KEY TOPICS IN NUCLEAR STRUCTURE
Proceedings of the 8th Int. Spring Seminar on Nuclear Physics
edited by Aldo Covello
© 2005 World Scientific Publishing Co.

A MONOPOLE PRIMER

ANDRÉS P. ZUKER
IReS, Bât27, IN2P3-CNRS/Université Louis Pasteur BP 28, F-67037 Strasbourg Cedex 2, France

An elementary introduction to the monopole Hamiltonian is proposed. Emphasis is put on comparison with experimental data, especially those indicating the need of three-body interactions.

1. Getting acquainted with $\mathcal{H}_m$

In the mid-1960's many people were worried by the first two excited states in ^{16}O: a 3^- of one particle-one hole nature (1p1h) was nearly degeneate, at about 6 MeV, with a 4p4h 0^+. According to conventional wisdom, the Hamiltonian was assumed to contain a central field H_0 that accounted for the "unperturbed energy", and a residual part V responsible for "correlations" *i.e.*, configuration mixing. Hence, if the unperturbed 3^- was at, say, 8 Mev, the unperturbed 0^+ should be at 32 MeV and it was very difficult to imagine how correlations could bring it down to 6 MeV. The puzzle was solved by a small shell model calculation in which ^{12}C was taken as a core and ^{16}O was described by four particles moving in the $p_{1/2} \equiv h$ and $s_{1/2}\, d_{5/2} \equiv p$ orbits some 3 MeV above (1) (ZBM). The key to the puzzle amounts to note that the unperturbed position of a state is given by the monopole Hamiltonian $\mathcal{H}_m$ which contains not only the single particle term but *everything* that is diagonal in the basis. In Sec. 2 we shall present it in some detail. Here we deal with its modest form for two shells

$$H_m = \varepsilon_p m_p + \varepsilon_h m_h + \tag{1}$$
$$-\frac{1}{2}|V_{pp}|m_p(m_p-1) - \frac{1}{2}|V_{hh}|m_h(m_h-1) - |V_{ph}|m_p m_h,$$

where m_r is the number operator for shell r. If we calculate the excitation energy of kpkh states with respect to the closed shell in which orbit h is full and contains D_h particles, 4 in our case)

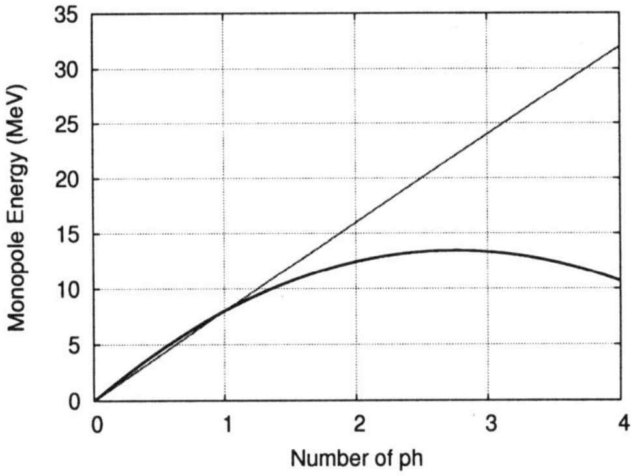

Figure 1. Illustrating Eq. (2)

$$\epsilon(k) = \left[(\varepsilon_p - \varepsilon_h) - \frac{1}{2}(|V_{pp}| - |V_{hh}|)(D_h - 1)\right] k$$
$$+ (\frac{1}{2}|V_{pp} + V_{hh}| - |V_{ph}|)k(D_h - k). \qquad (2)$$

everything becomes clear in Fig. 1: the unperturbed 4p4h state is no longer at 32 MeV but at 11 MeV. Furthermore the 1p1h state is at 8 Mev. Had we kept only the $(\varepsilon_p - \varepsilon_h)k$ part it would have been at 3 Mev as in ^{13}C. The two-body (2b) terms in Eq. (1) are seen to be responsible for transforming the single-particle spectrum in ^{13}C into the single-particle spectrum in ^{17}O (and single-hole spectrum in ^{15}O, in this case reduced to the ground state). We call this phenomenon **monopole drift**.

It is also worth noting that the ZBM calculation was done with two sets of matrix elements: one empirical and one realistic. The results for ^{16}O were comparably good but the price to pay for the realistic set was a very unrealistic $\varepsilon_p - \varepsilon_h = 0$, indicating that something was wrong with the centroids V_{rs}.

1.1. *The isospin puzzle*

The ZBM calculations contained a further lesson unearthed in (2). Let us consult another simple part of $\mathcal{H}_m$, the famous Bansal French formula(3)

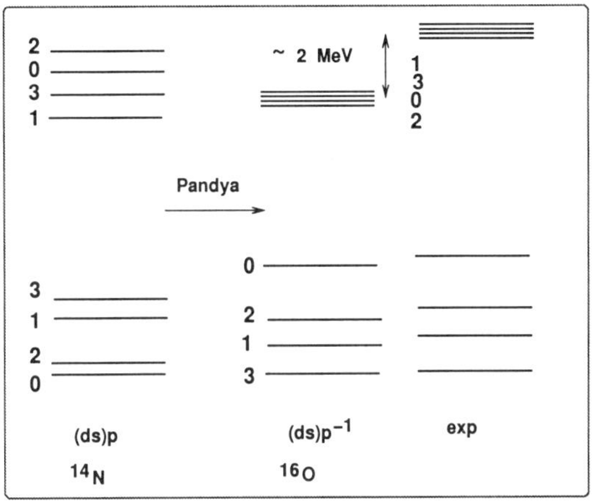

Figure 2. The isospin puzzle: b_{ph} in Eq. (3) is not the same for ^{16}O and ^{16}N. The lower levels have $T = 0$ and the higher ones $T = 1$. $p_{1/2} \equiv p$ and $s_{1/2}\, d_{5/2} \equiv sd$. See text.

(T_r is the isospin operator for shell r, T is the total isospin)

$$H_{BF} = a_{ph} m_p m_h + b_{ph} T_p \cdot T_h, \tag{3}$$

$$T_p \cdot T_h = \frac{1}{2}[T(T+1) - T_p(T_p+1) - T_h(T_h+1)] \tag{4}$$

In Fig. 2 the two particle states of negative parity in ^{14}N (taken from experiment) became particle-hole states in ^{16}O under a Pandya transformation. The distance between their centroids is given by the b_{ph} parameter in H_{BF}. While the ordering for the lower ($T = 0$) and upper ($T = 1$) multiplets is nearly perfect, ^{16}O is seen to demand a b_{ph} parameter larger than the one in ^{14}N by some 2 MeV. [Same effect in ^{40}Ca with $h \equiv d_{3/2}$, $p \equiv f_{7/2}$].

Now: For a 2b force b_{ph}, must be constant. It is not, therefore it seems to demand a three-body (3b) additive.

Conclusions

- Quadratic *i.e.*, 2b effects are crucial in producing monopole drift.
- Three body terms seem necessary.
- There is a monopole problem with the realistic interactions

2. General form of the monopole Hamiltonian $\mathcal{H}_m$

A purely 2b Hamiltonian can be written as

$$\mathcal{H} = \mathcal{K} + \sum_{r\leq s,\, t\leq u,\, \Gamma} \mathcal{V}^{\Gamma}_{rstu} Z^{+}_{rs\Gamma} \cdot Z_{tu\Gamma}$$

where $\mathcal{K}$ is the kinetic energy, $Z^{+}_{rs\Gamma}$ creates a pair with $\Gamma \equiv JT$ in orbits rs, $Z_{rs\Gamma}$ is its Hermitean conjugate and $\cdot$ stands for scalar product. The plan is to proceed with 2b forces so as to make clear where 3b ones are needed.

$\mathcal{H}$ can be separated as

$$\mathcal{H} = \mathcal{H}_m + \mathcal{H}_M, \tag{5}$$

The monopole $\mathcal{H}_m$ contains $\mathcal{K}$ plus 2b quadratics in $a^{+}_{r_x} \cdot a_{s_y}$; $j_r = j_s$, $\pi_r = \pi_s$ x, y =neutron or proton. The multipole $\mathcal{H}_M$ contains all the rest (quadrupole, pairing, etc.) The first proof of the separation was given in (4). See (5) for a comprehensive study of $\mathcal{H}_M$ for realistic forces (5), and (6) for an updated review. Here, only basic facts are itemized

- $\mathcal{H}_m$ is unique, but its diagonal part $\mathcal{H}^d_m$, which contains only m and T operators, can take two forms that reproduce average energies of configurations either at fixed m and T ($\mathcal{H}^d_{mT}$ in "T" formalism) or at fixed number of protons and neutrons ($\mathcal{H}^d_{np}$ in "np" formalism).
- Closed shells and the single-particle and single-hole states built on them (the $cs \pm 1$ set) are configurations with only one state *i.e.*, Slater determinants.
- The full $\mathcal{H}_m$ is closed under unitary transformations of the underlying fermion operators.
- Hence spherical Hartre Fock (HF) variation involves only $\mathcal{H}_m$.
- However $\mathcal{H}_m$ goes well beyond HF. An open field.
- The realistic 2b $\mathcal{H}_m$ has many problems. $\mathcal{H}_M$ on the contrary is quite good.

2.1. *Forms of* $\mathcal{H}^d_m$

The full form of $\mathcal{H}_m$ is given in Appendix B of (6), from which we borrow a compact set of formulas that give all possible variants of $\mathcal{H}^d_m$ relying on the power of Bruce French's product notations (7):

$D_r = 2j_r + 1$, $(-1)^{2r} = -1$, $\Gamma \equiv J,$, $[J] = 2J + 1$ in np formalism

$D_r = 2(2j_r + 1)$, $(-1)^{2r} = +1$, $\Gamma \equiv JT$, $[\Gamma] = (2J + 1)(2T + 1)$ in T formalism

$$\mathcal{V}_{rs} = \frac{\sum_\Gamma \mathcal{V}^\Gamma_{rsrs}[\Gamma](1-(-1)^\Gamma\delta_{rs})}{D_r(D_s-\delta_{rs})} \tag{6a}$$

$$\mathcal{V}^T_{rs} = \frac{4\sum_J \mathcal{V}^{JT}_{rsrs}[J](1-(-1)^{J+T}\delta_{rs})}{D_r(D_s+2\delta_{rs}(-1)^T)} \tag{6b}$$

$$a_{rs} = \frac{1}{4}(3\mathcal{V}^1_{rs}+\mathcal{V}^0_{rs}) = \mathcal{V}_{rs} + \frac{3}{4}\frac{\delta_{rs}}{D_r-1}b_{rs}$$
$$b_{rs} = \mathcal{V}^1_{rs} - \mathcal{V}^0_{rs} \tag{6c}$$

$$\mathcal{H}^d_m = \mathcal{K} + \sum \frac{1}{(1+\delta_{rs})}\left[\mathcal{V}_{rs}\, n_r(n_s-\delta_{rs}) + \right.$$
$$\left. + b_{rs}\left(T_r\cdot T_s - \frac{3n_r\bar{n}_r}{4(D_r-1)}\delta_{rs}\right)\right] \tag{7}$$

This expression correspons to $\mathcal{H}^d_{mT}$, but by dropping the b_{rs} terms, one obtains $\mathcal{H}_m$ 0 (averages at fixed m only) *or* $\mathcal{H}^d_{np}$.

In the np scheme each orbit r goes into two r_n and r_p and the centroids can be obtained through ($x, y = n$ or p, $x \neq y$)

$$\mathcal{V}_{r_x s_y} = \frac{1}{2}\left[\mathcal{V}^1_{rs}\left(1-\frac{2\delta_{rs}}{D_r}\right) + \mathcal{V}^0_{rs}\left(1+\frac{2\delta_{rs}}{D_r}\right)\right]$$
$$\mathcal{V}_{r_x s_x} = \mathcal{V}^1_{rs}. \tag{8}$$

3. Inventing $\mathcal{H}^d_m$

As the realistic 2b potentials pose serious monopole problems we are forced to "(re)invent" $\mathcal{H}^d_m$. We shall start by investigating what it does, and then try to reproduce it with few parameters.

As it happens $\mathcal{H}^d_m$ does many things. In particular it must contain the liquid drop (LD) energy (Bethe Weizsäcker). We use (V^d_{Cm} is the Coulomb energy.)

$$E(LD) = 15.5A - 17.8A^{2/3} - 28.6\frac{4T(T+1)}{A} + 40.2\frac{4T(T+1)}{A^{4/3}} - V^d_{Cm}, \tag{9}$$

not a fit, but designed to have mostly positive "shell effects".

Once $E(LD)$ is subtracted from measured binding energies, the shell effects shown in Fig. 3 emerge. Magicity shows as spikes, overwhelmingly in the "intruder-extruder" (EI) closures, very little in the the harmonic oscillator (HO) ones . Note that binding energies are taken to be positive.

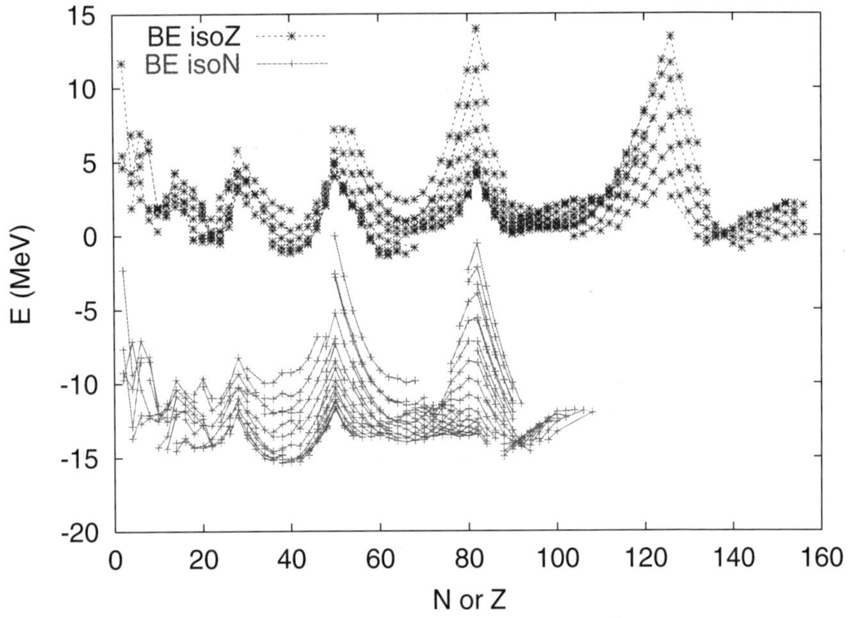

Figure 3. Experimental shell effects (BE(exp)-$E(LD)$) along isotope and isotone lines. Both plots contain the same information. The isotone lines are displaced by -14 MeV.

Again, most of these shell-effects come from $\mathcal{H}_m^d$. Furthermore it is also responsible for the spectra of $cs \pm 1$ states.

To invent $\mathcal{H}_m^d$ we start by following the strategy adopted in (8). It assumes that the realistic 2b monopole properties are simply wrong and have to be changed. Then

I) Separate LD using a hint from realistic 2b potentials that produce a collective monopole term W that collects the bulk effects of LD, that can be "eliminated" by using a combination that cancels exactly the A and $A^{2/3}$ contributions ($4K$ is the kinetic energy, p the principal HO quantum number):

$$W - 4K = \left(\sum_p \frac{m_p}{\sqrt{D_p}}\right)^2 - 2\sum_p m_p(p + 3/2), \qquad (10)$$

II) Assume $cs + 1$ spectra on HO closures is well given by 1b $l \cdot s + l \cdot l$.

III) Add 2b monopole drift to reproduce $cs \pm 1$ in extruder-intruder (EI) closures.

Then fit the resulting six-parameter H_m^s (s for shell effects) to all known $cs \pm 1$ spectra (some 90 data) with a rmsd=220 keV (8).

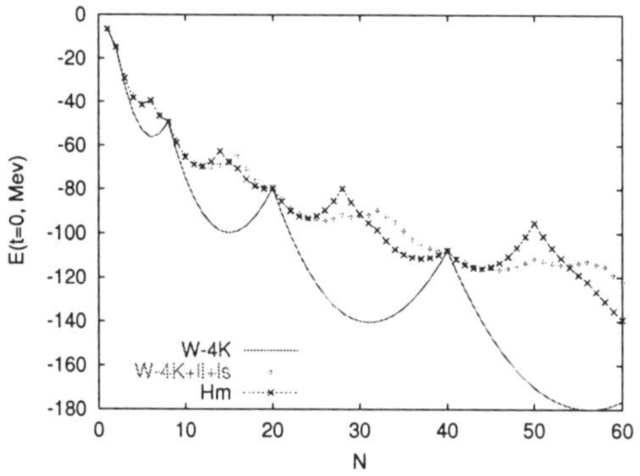

Figure 4. The different contributions to H_m^s for N=Z nuclei

Fig 4 shows the energies produced by the model H_m^s by filling orbits as prescribed by the $l \cdot s + l \cdot l$ term: $W - 4k$ produces enormous HO closures. They are very much erased by the 1b $l \cdot s + l \cdot l$ term, and then the EI closures emerge through the 2b monopole drift. The pronounced decrease in energy is unphysical: it goes as $-A^{1/3}$ and should be corrected.

3.1. *Predicted monopole shell effects*

In Fig. 5 we compare the shell effects predicted by H_m^s—corrected as we explain later—for $N - Z = 8$ nuclei with the observed ones. The agreement is typical of other isospin chains and definitely satisfactory. It illustrates two general trends:

- If a closure exists, it is there, but;
- If it is there, it does not necessarily exist.

Proceeding from left to right we find first the $N = 16, Z = 8$ closure (the object of much attention recently, for which reliable mass data are missing). At $N = 22$, $Z = 14$ the closure is erased by correlations. Then at $N = 28$, $Z = 20$ and $N = 36$, $Z = 28$ two good closures, followed by one erased at $N = 40, Z = 32$, and finally two good closures at $N = 50$, $Z = 42$

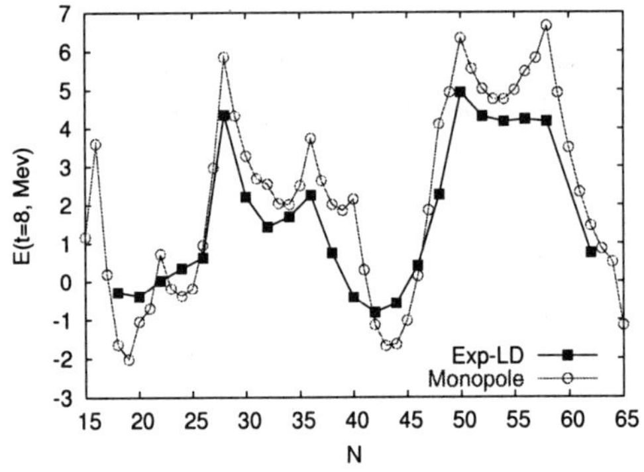

Figure 5. H_m^s for $N-Z=8$ nuclei compared to experimental shell effects (only even-even cases shown for the latter).

and $N = 58$, $Z = 50$. Note that by following nuclei at constant isospin, both neutron and proton magicity is detected.

As noted in Fig. 4 a $A^{1/3}$ correction is needed, and a similar one associated to $T(T+1)A^{-5/3}$, which amounts to a constant in our case. At this point we are still within a HF context. Correlations (*i.e.*, configuration mixing) will always *reduce* the shell effects and the expedient way to introduce them is through an overall multiplicative factor. To eliminate the unwanted closure effects, configuration mixing must be done a bit more carefully. See (4) for a detailed discussion, and (9) for a simple way to account for well deformed states. The mass fit in this reference is a precursor of the ideas developed here.

Conclusion: the model H_m^s in (8) is based on a fit of to the $cs \pm 1$ spectra that proves sufficient to account for shell effects in a very goodd first approximation.

4. Problems

There are two kinds of problems: (a) those arising from the data, and (b) those arising from the 2b interaction. They are examined in turn.

4.1. *Comparing H_m^s with data*

The golden rule of monopole drift (4) may be stated as follows:

- Filling the largest j (intruder) orbit favors large l orbits.

This applies whether the intruder acts on the same HO or other shells, and on the same or the other fluid. For example: in Sb ($Z = 51$) isotopes the $d_{5/2}$ proton orbit is below the $g_{7/2}$ one (by some 700 keV) for low neutron number due to the $l \cdot s$ field. As the $h_{11/2}$ starts filling around $N = 64$ the drift brings down (almost) linearly the $g_{7/2}$ orbit and at $N = 78$ it is 700 keV below the $d_{5/2}$ orbit.

The rule suffers two exceptions illustrated in the top panel of Fig. 6: in ^{12}C and ^{28}Si, the filling of the $p_{3/2}$ and $d_{5/2}$ orbits favour the low l orbits. In ^{56}Ni the general trend is restored. H^s_m deals very well with the general case but not with the exceptions, its main detectable defect. An *ad-hoc* 2b cure may exist, but a 3b one is likely to be better.

Another remarkable exception to systematics occurs above the HO closures: the lower panel indicates that the $l \cdot l$ 1b term must change sign at the pf shell. H^s_m accomodates the situation through an *ad-hoc* factor $l \cdot l \Longrightarrow (p-3)/(p+3/2) l \cdot l$.

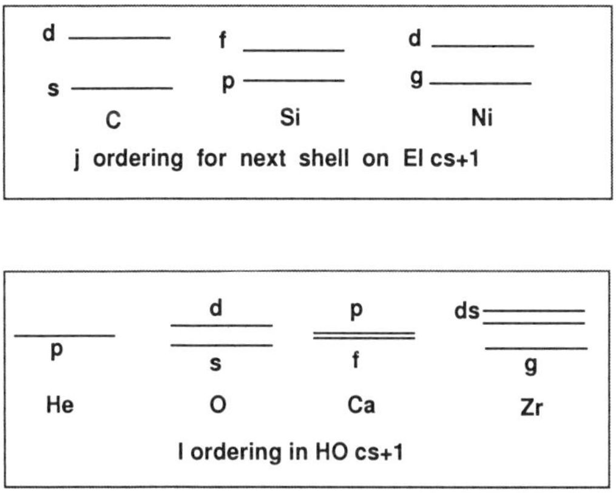

Figure 6. Single particle spectra above double magic nuclei. In the HO case l centroids are given. In the EI case only $l + 1/2$ states are shown, *e.g.*, $g \equiv g_{9/2}$, $f \equiv f_{7/2}$, etc. The scales are arbitrary.

4.2. *Comparing H^s_m with realistic forces*

H^s_m was invented under the assumption that the realistic 2b forces could not be trusted. However, nowadays it seems plausible that the 2b forces are right but that 3b forces are needed; and in Sections 1.1 and 4.1 it was shown that the data demand them. Therefore, since H^s_m is 2b, it is possible

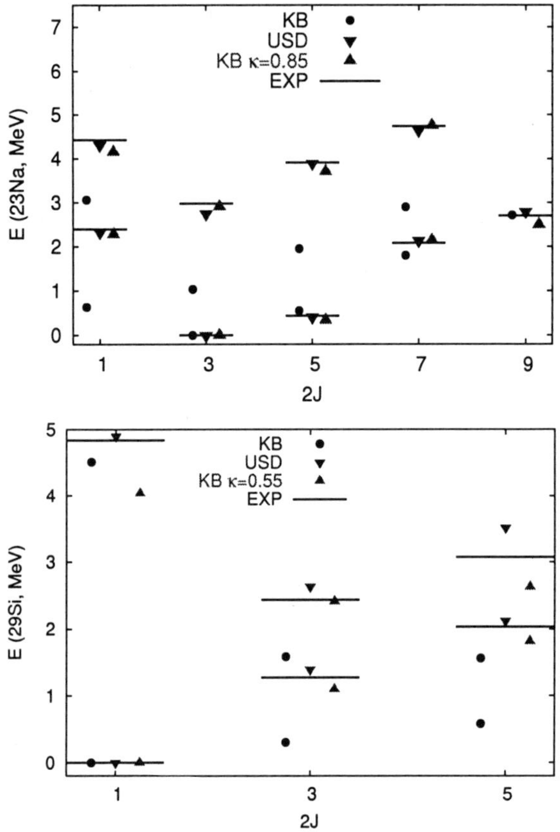

Figure 7. The excitation specta of ^{23}Na and ^{29}Si for different interactions.

that it was forced to do something that should be 3b. Now: there are many monopole things the realistic 2b forces do not do well, but there is one that is catastrophic: they fail to produce the EI closures, the very heart of nuclear structure. The problem was solved some 30 years ago (10) by the following recipe

$$\begin{aligned} V_{fr}^{T}(\mathrm{R}) &\Longrightarrow \mathrm{V}_{\mathrm{fr}}^{\mathrm{T}}(\mathrm{R}) - (-)^{\mathrm{T}}\,\kappa \\ V_{ff}^{T}(\mathrm{R}) &\Longrightarrow \mathrm{V}_{\mathrm{ff}}^{\mathrm{T}}(\mathrm{R}) - 1.5\,\kappa\delta_{\mathrm{T0}}, \end{aligned} \tag{11}$$

where R stands for realistic, and the notations stand for $f \equiv (p_{3/2}, d_{5/2}, f_{7/2})$ generically in the (p, sd, pf) shells respectively, and $r =$

$p_{1/2}$ and $r \equiv d_{3/2}, s_{1/2}$ for the p and sd shells and κ is a constant for each HO shell. The recipe works very well in pf shell (for which it was designed) but not so well in the sd shell.

The solution was found recently (11). It amounts to use $\kappa = \kappa_0 + (m - m_0)\,\kappa_1$ *i.e.,* to introduce a 3b effect. In Fig. 7 the new recipe ($m_0 = 6$, $\kappa_0 = .5$, $\kappa_1 = .05$) is seen to transform the very bad spectra produced by the classic KB interaction (12) into good ones that can compete with the USD fit to all matrix matrix elements (13).

It appears the strategy to produce better models than H^s_m demands keeping the realistic 2b monopoles—instead of inventing them—and invent whatever 3b terms are necessary properly to describe the $cs \pm 1$ spectra.

References

1. A. P. Zuker, B. Buck, and J. B. McGrory, Phys. Rev. Lett. **21**, 39 (1968).
2. A. P. Zuker, Phys. Rev. Lett **23**, 983 (1969).
3. R. K. Bansal and J. B. French, Phys. Lett. **11**, 145 (1964).
4. A. P. Zuker, Nucl. Phys. **A 576**, 65 (1994).
5. M. Dufour and A. P. Zuker, Phys. Rev. C **54**, 1641 (1996).
6. E. Caurier *et al.*, The Shell Model as Unified View of Nuclear Structure, 2004, nucl-th/0402046, to be published in RMP (some day).
7. J. B. French, in *Enrico Fermi Lectures XXXVI*, edited by C. Bloch (Academic Press, New York, 1966).
8. J. Duflo and A. P. Zuker, Phys. Rev. C **59**, R2347 (1999).
9. J. Duflo and A. P. Zuker, Phys. Rev. C **52**, R23 (1995).
10. E. Pasquini, Ph.D. thesis, Université Louis Pasteur, Strasbourg, 1976, report CRN/PT 76-14.
11. A. P. Zuker, Phys. Rev. Lett. **90**, 042502 (2003).
12. T. T. S. Kuo and G. E. Brown, Nucl. Phys. **A 85**, 40 (1966).
13. B. H. Wildenthal, Prog. Part. Nucl. Phys. **11**, 5 (1984).

KEY TOPICS IN NUCLEAR STRUCTURE
Proceedings of the 8th Int. Spring Seminar on Nuclear Physics
edited by Aldo Covello
© 2005 World Scientific Publishing Co.

COUPLED CLUSTER APPROACHES TO NUCLEI, GROUND STATES AND EXCITED STATES

D. J. DEAN[1,2], M. HJORTH-JENSEN[2,3,4,5], K. KOWALSKI[6], T. PAPENBROCK[1,7], M. WLOCH[6], AND P. PIECUCH[5,6]

[1] *Physics Division, Oak Ridge National Laboratory, P.O. Box 2008, Oak Ridge, TN 37831, USA*
[2] *Center of Mathematics for Applications, University of Oslo, N-0316 Oslo, Norway*
[3] *Department of Physics, University of Oslo, N-0316 Oslo, Norway*
[4] *PH Division, CERN, CH-1211 Geneve 23, Switzerland*
[5] *Department of Physics and Astronomy, Michigan State University, East Lansing, MI 48824, USA*
[6] *Department of Chemistry, Michigan State University, East Lansing, MI 48824, USA*
[7] *Department of Physics and Astronomy, University of Tennessee, Knoxville, TN 37996, USA*

We present recent coupled-cluster studies of nuclei, with an emphasis on ground state and excited states of closed shell nuclei. Perspectives for future studies are delineated.

1. Introduction

Physical properties, such as masses and life-times, of very short-lived, and hence very rare, nuclei are important ingredients that determine element production mechanisms in the universe. Given that present nuclear structure research facilities and the proposed Rare Isotope Accelerator will open significant territory into regions of medium-mass and heavier nuclei, it becomes important to investigate theoretical methods that will allow for a description of medium-mass systems that are involved in such element production. Such systems pose significant challenges to existing nuclear structure models, especially since many of these nuclei will be unstable and short-lived. How to deal with weakly bound systems and coupling to resonant states is an unsettled problem in nuclear spectroscopy.

The *ab initio* coupled-cluster theory is a particularly promising candidate for such endeavors due to its enormous success in quantum chemistry.

Here we describe applications of coupled cluster techniques to nuclear structure. The coupled-cluster methods are very promising, since they allow one to study ground- and excited-state properties of nuclei with dimensionalities beyond the capability of present shell-model approaches, with a much smaller numerical effort when compared to the more traditional shell-model methods aimed at similar accuracies. Even though the shell-model combined with appropriate effective interactions offers in general a very good description of several stable and even weakly bound nuclei, the increasing single-particle level density of weakly bound systems makes it imperative to identify and investigate methods that will extend to unstable systems, systems whose dimensionality is beyond reach for present shell-model studies, typically limited today to systems with at most $\sim 10^9$ basis states.

The coupled-cluster approach offers several advantages. It is fully microscopic and allows one to start with the free nucleon-nucleon interaction, or eventually three-body interaction models. It contains only linked diagrams, it is size extensive. and can be improved upon systematically, for example by the inclusion of three-body interactions and more complicated correlations. One can study both closed-shell systems and valence systems and it is possible to derive effective two and three-body interactions for open shell systems, with the inclusion of complex interactions, of great relevance for the study of weakly bound systems. Finally, it is amenable to parallel computing.

Here we present several coupled-cluster results from recent calculations with singles, doubles, and noniterative triples and their generalizations to excited states applied to the ^{4}He and ^{16}O nuclei. A comparison of coupled cluster results with the results of the exact diagonalization of the Hamiltonian in the same model space shows that the quantum chemistry inspired coupled cluster approximations provide an excellent description of ground and excited states of nuclei. The results presented here are a summary of recent works listed in Refs. [1,2].

2. Coupled Cluster approach to Nuclei

Nuclear many-body theory often begins with a G-matrix interaction which is derived from an underlying bare nucleon-nucleon interaction. This G-matrix can in turn be used in perturbative many-body approaches in order to derive effective interactions for the nuclear shell model, see for example Refs. [3,4] for recent reviews. These approaches have shown be to rather successful in shell-model studies of several nuclear systems. However, to derive effective interactions within the framework of many-body perturbation theory is hard to expand upon in a systematic manner by including for example three-body diagrams. In addition, there are no clear signs of convergence, even in terms of a weak interaction such as the G-matrix. Even in atomic and molecular physics, many-body perturbative methods are not much favoured any longer, see for example Ref. [5] for a critical discussion. The lessons from atomic and molecular many-body systems clearly point to the need of non-perturbative resummation techniques of large classes of diagrams.

An alternative to such resummation techniques is however offered by the so-called no-core approach. There one typically defines a two-body or three-body effective interaction within a large, but limited model space. This is parallel to our own approach below, where we limit the discussion to the no-core G-matrix so that all particles are active within our chosen model space. Using a given basis expansion of the many-body wave function we could then solve the nuclear problem by diagonalization as has been pursued by the no-core shell model collaboration [6,7,8,9]. In fact, the current and most advanced no-core techniques have approached ^{12}C, with nearly converged solutions [10].

It should be evident, however, that diagonalization procedures scale almost combinatorially with the number of particles in a given number of single-particle orbitals. Because of this scaling, diagonalization simply becomes untenable at some point. The efforts to expand diagonalization into p-shell nuclei with all nucleons active, an effort that spans over ten years, illustrates the problem. The computational complexity of the nucleus grows dramatically as the size of the nucleus increases. As a simple example consider oscillator single-particle states, and single-particle spaces consisting of 4 and 7 major oscillator shells, and compare the number of uncoupled many-body basis states there are for 4,8,12, and 16 particles. From table 1 we see an enormous growth of the standard shell-model diagonalization problem within the space. We calculated the number of $M = 0$ states for He and B within the model space consisting of 4 major shells and estimated the number of basis states for C and O. Also indicated are similar estimates for seven major oscillator shells. The important lesson to learn from these numbers is that the model-space expansion becomes astronomical quite quickly.

Yet, because of the advent of radioactive nuclear beam accelerators, such as the proposed Rare Isotope Accelerator (RIA) in the U.S., we face the daunting task of moving beyond p-shell nuclei in *ab initio* calculations.

Table 1. Dimensions of the shell-model problem in four major oscillator shells and 7 major oscillator shells with $M = 0$.

System	4 shells	7 shells
^{4}He	4E4	9E6
^{8}B	4E8	5E13
^{12}C	6E11	4E19
^{16}O	3E14	9E24

We should therefore investigate several ways of approaching the nuclear many-body problem in order to successfully make the move into the RIA era. Here we will discuss the coupled-cluster technique which can be used to pursue nuclear many-body calculations to heavier systems beyond the p-shell.

Coupled cluster theory originated in nuclear physics [11,12] around 1960. Early studies in the seventies [13] probed ground-state properties in limited spaces with free nucleon-nucleon interactions available at the time. The subject was revisited only recently by Bishop *et al.* [14], for further theoretical development, and by Mihaila and Heisenberg [15], for coupled cluster calculations using realistic two- and three-nucleon bare interactions and expansions in the inverse particle-hole energy spacings. However, much of the impressive development in coupled cluster theory made in quantum chemistry in the last 20-30 years [16,17,18,19,20], after the introduction of coupled-cluster theory to quantum chemistry by Čížek in the 1960's [21,22], still awaits applications to the nuclear many-body problem.

Many solid theoretical reasons exist that motivate a pursuit of coupled-cluster methods. First of all, the method is fully microscopic and is capable of systematic and hierarchical improvements. Indeed, when one expands the cluster operator in coupled-cluster theory to all A particles in the system, one exactly produces the fully-correlated many-body wave function of the system. The only input that the method requires is the nucleon-nucleon interaction. The method may also be extended to higher-order interactions such as the three-nucleon interaction. Second, the method is size extensive which means that only linked diagrams appear in the computation of the energy (the expectation value of the Hamiltonian) and amplitude equations. As discussed, for example, in Refs. [16,18] all shell model calculations that use particle-hole truncation schemes actually suffer from the inclusion of disconnected diagrams in computations of the energy. Third, coupled-cluster theory is also size consistent which means that the energy of two non-interacting fragments computed separately is the same as that computed for both fragments simultaneously. In chemistry, where the study of reactions is quite important, this is a crucial property not available in the interacting shell model (named configuration interaction in chemistry). Fourth, while the theory is not variational, the energy behaves as a variational quantity in most instances. Finally, from a computational point of view, the practical implementation of coupled cluster theory is amenable to parallel computing.

We are in the process of applying quantum chemistry inspired coupled cluster methods [16,17,18,19,20,21,22,23,24,25,26] to finite nuclei [1,2]. We show one result from our current studies, namely the convergence of ^{16}O as a function of the model space in which we perform the calculations.

The basic idea of coupled-cluster theory is that the correlated many-body wave function $\mid \Psi\rangle$ may be obtained by application of a cluster operator, T, such that

$$\mid \Psi\rangle = \exp(T) \mid \Phi\rangle \ , \tag{1}$$

where Φ is a reference Slater determinant chosen as a convenient starting point. For example, we use the filled $0s$ state as the reference determinant for ^{4}He.

The cluster operator T is given by

$$T = T_1 + T_2 + \cdots T_A \ , \tag{2}$$

and represent various n-particle-n-hole (np-nh) excitation amplitudes such as

$$T_1 = \sum_{a\langle \varepsilon_f, i\rangle \varepsilon_f} t_a^i a_a^\dagger a_i \ , \tag{3}$$

$$T_2 = \frac{1}{4} \sum_{i,j\langle \varepsilon_f; ab\rangle \varepsilon_f} t_{ab}^{ij} a_a^\dagger a_b^\dagger a_j a_i \ , \tag{4}$$

and higher-order terms from T_3 to T_A. The basic approximation is obtained by truncating the many-body expansion of T at the $2p - 2h$ cluster component T_2. This is commonly referred to in the literature as the coupled-cluster singles and doubles approach (CCSD).

We compute the ground-state energy from

$$E_{\text{g.s.}} = \langle \Phi \mid \exp(-T)\, H \exp(T) \mid \Phi\rangle \ . \tag{5}$$

The Campbell-Hausdorff-Baker relation may be used to rewrite the similarity transformation as an expansion in terms of nested commutators.

The expansion terminates exactly at four nested commutators when the Hamiltonian contains, at most, two-body terms, and at six-nested commutators when three-body potentials are present. This can also be seen diagrammatically, since $e^{-T}He^{T}$ is equivalent to the connected product of the Hamiltonian and e^{T}, which has to terminate at the quartic terms in T when interactions are pairwise (the Hamiltonian has at most four lines that can be connected with the T vertices) and at the T^6 terms when interactions are three-body (the Hamiltonian has at most six lines that can be connected with the T vertices) [18,21,22]. The equations for amplitudes are found by left projection of excited Slater determinants so that

$$0 = \langle \Phi_i^a \mid \exp(-T)\, H \exp(T) \mid \Phi\rangle \,,$$
$$0 = \langle \Phi_{ij}^{ab} \mid \exp(-T)\, H \exp(T) \mid \Phi\rangle \,. \tag{6}$$

The commutators also generate nonlinear terms within these expressions. To derive these equations, we use the diagrammatic approach. In order to obtain the computationally efficient algorithms, which lead to the lowest operation count and memory requirements, we use the idea of recursively generated intermediates and diagram factorization [17]. The resulting equations can be cast into a computationally efficient form, where diagrams representing intermediates multiply diagrams representing cluster operators. The resulting equations can be solved using efficient iterative algorithms, see for example Refs. [1,17].

In our coupled-cluster study of Ref. [1], we performed calculations of the ^{16}O ground state for up to seven major oscillator shells as a function of $\hbar\omega$. Fig. 1 indicates the level of convergence of the energy per particle for $N = 4, 5, 6, 7$ shells. The experimental value resides at 7.98 MeV per particle. This calculation is practically converged. By seven oscillator shells, the $\hbar\omega$ dependence becomes rather minimal and we find a ground-state binding energy of 7.52 MeV per particle in oxygen using the Idaho-A potential. Since the Coulomb interaction should give approximately 0.7 MeV/A of repulsion, and is not included in this calculation, we actually obtain approximately 6.90 MeV of nuclear binding in the 7 major shell calculation which is somewhat above the experimental value (most likely, due to the neglect of three-body interactions in the calculations). We note that the entire procedure (G-matrix plus CCSD) tends to approach from below converged solutions. We have recently performed calculations with eight major shells, and the results are practically converged.

We also considered chemistry inspired noniterative corrections to the CCSD energy due to three-body clusters T_3 (labelled triples in quantum

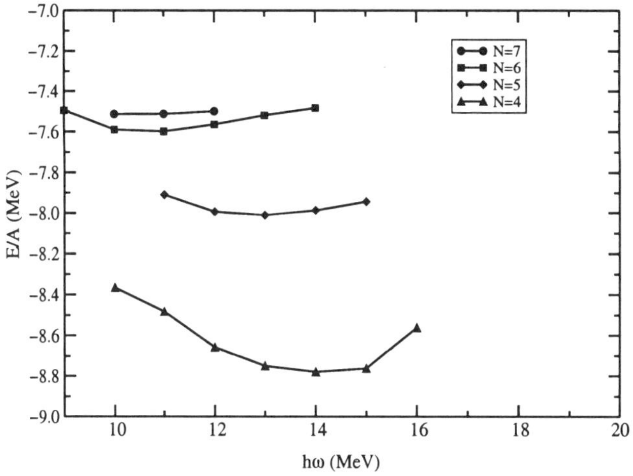

Figure 1. Dependence of the ground-state energy of ^{16}O on ω as a function of increasing model space.

chemistry). We performed this study in the model space consisting of four major oscillator shells, since we can perform exact shell-model calculations for nuclei such as ^4He. Table 2 shows the total ground-state energy values obtained with the CCSD and one of the triples-correction approaches (labeled CR-CCSD(T) [19,20,25,26] in the table). Slightly differing triples-corrections yield similar corrections to the CCSD energy. The coupled cluster methods recover the bulk of the correlation effects, producing the results of the SM-SDTQ, or better, quality. SM-SDTQ stands for the expensive shell-model (SM) diagonalization in a huge space spanned by the reference and all singly (S), doubly (D), triply (T), and quadruply (Q) excited determinants. To understand this result, we note that the CCSD T_1 and T_2 amplitudes are similar in order of magnitude. (For an oscillator basis, both T_1 and T_2 contribute to the first-order MBPT wave function.) Thus, the T_1T_2 *disconnected* triples are large, much larger than the T_3 *connected* triples, and the difference between the SM-SDT (SM singles, doubles, and triples) and SM-SD energies is mostly due to T_1T_2.The small T_3 effects, as estimated by CR-CCSD(T), are consistent with the SM diagonalization calculations. If the T_3 corrections were large, we would observe a significant lowering of the CCSD energy, far below the SM-SDTQ result. Moreover, the CCSD and CR-CCSD(T) methods bring the nonnegligible higher-than-quadruple excitations, such as $T_1^3T_2$, $T_1T_2^2$, and T_2^3, which are not present in SM-SDTQ. It is, therefore, quite likely that the CR-CCSD(T) results are very close to the results of the exact diagonalization, which cannot be performed.

Table 2. The ground-state energy of ^{16}O calculated using various coupled cluster methods and oscillator basis states.

Method	Energy
CCSD	-139.310
CR-CCSD(T)	-139.467
SM-SD	-131.887
SM-SDT	-135.489
SM-SDTQ	-138.387

These results indicate that the bulk of the correlation energy within a nucleus can be obtained by solving the CCSD equations. This gives us confidence that we should pursue this method in open shell systems and to excited states. We have recently [2] performed excited state calculations on ^{4}He using the EOMCCSD (equation of motion CCSD) method. For the excited states $|\Psi_K\rangle$ and energies $E_K^{(\text{CCSD})}$ $(K > 0)$, we apply the EOMCCSD ("equation of motion CCSD") approximation [23,24] (equivalent to the response CCSD method [27]), in which

$$|\Psi_K\rangle = R_K^{(\text{CCSD})} \exp(T^{(\text{CCSD})})|\Phi\rangle. \tag{7}$$

Here $R_K^{(\text{CCSD})} = R_0 + R_1 + R_2$ is a sum of the reference (R_0), one-body (R_1), and two-body (R_2) components obtained by diagonalizing $\bar{H}^{(\text{CCSD})}$ in the same space of singly and doubly excited determinants $|\Phi_i^a\rangle$ and $|\Phi_{ij}^{ab}\rangle$ as used in the ground-state CCSD calculations. These calculations may also be corrected in a non-iterative fashion using the completely renormalized theory for excited states [19,20,25,26,28]. The low-lying $J = 1$ state most likely results from the center-of-mass contamination which we have removed only from the ground state. The $J = 0$ and $J = 2$ states calculated using EOMCCSD and CR-CCSD(T) are in excellent agreement with the exact results. We have recently also computed excited states in ^{16}O, with a particular emphasis on the first 3_1^- state, which is known to be of a 1p-1h nature. Our results based on the EOMCCSD method yields 13.57 MeV for five shells and 12.98 MeV for six shells, to be compared with the experimental value of 6.13 MeV. We expect that with seven shells and the inclusion of triples to get closer to the experimental value. For states like this and for two-body

Table 3. The excitation energies of ^{4}He calculated using the oscillator basis states (in MeV).

State	EOMCCSD	CR-CCSD(T)	CISD	Exact
J=1	11.791	12.044	17.515	11.465
J=0	21.203	21.489	24.969	21.569
J=2	22.435	22.650	24.966	22.697

interactions it is well known in quantum chemistry that EOMCCSD is a very accurate approach, producing excitation energies within 10 % of the exact values. Thus, we will be able to predict the result corresponding to an Idaho-A potential that we used in these calculations once we complete our work for the seven shells and extrapolate the energies to the complete basis set limit. These results will be presented elsewhere, see Ref. [29].

Our experience thus far with the quantum chemistry inspired coupled cluster approximations to calculate the ground and excited states of the ^{4}He and ^{16}O nuclei indicates that this will be a promising method for nuclear physics. By comparing coupled cluster results with the exact results obtained by diagonalizing the Hamiltonian in the same model space, we demonstrated that relatively inexpensive coupled cluster approximations recover the bulk of the nucleon correlation effects in ground- and excited-state nuclei. These results are a strong motivation to further develop coupled cluster methods for the nuclear many-body problem, so that accurate *ab initio* calculations for small- and medium-size nuclei become as routine as molecular electronic structure calculations.

3. Perspectives and Future Plans

The study of exotic nuclei opens new challenges to nuclear physics. The challenges and the excitement arise because exotic nuclei will present new and radically different manifestations of nucleonic matter that occur near the bounds of nuclear existence, where the special features of weakly bound, quantal systems come into prominence. Furthermore, many of these nuclei are key to understanding matter production in the universe. Given that present and future nuclear structure research facilities will open significant territory into regions of medium-mass and heavier nuclei, it becomes important to investigate theoretical methods that will allow for a description of medium-mass nuclear systems. Such systems pose significant challenges to existing nuclear structure models, especially since many of these nuclei will be unstable and short-lived. How to deal with weakly bound systems and coupling to resonant states is an unsettled problem in nuclear spectroscopy.

Many-body methods like the coupled cluster theory offer possibilities for extending microscopic *ab initio* calculations to nuclei of the size of ^{40}Ca. Especially the coupled-cluster methods are very promising, since they allow one to study ground- and excited-state properties of nuclei with dimensionalities beyond the capability of present shell-model approaches. As demonstrated here and in Ref. [2] we show for the first time how to calculate excited states for a nucleus using coupled cluster methods from quantum chemistry. For the weakly bound nuclei to be produced by future low-energy nuclear structure facilities it is almost imperative to increase the degrees of freedom under study in order to reproduce basic properties of these systems. We are presently working on deriving complex effective interactions, see for example Ref. [30], for weakly bound systems to be used in coupled cluster studies of these weakly bound nuclei.

We have based most of our analysis using two-body nucleon-nucleon interactions only. We feel this is important since techniques like the coupled cluster methods allow one to include a much larger class of many-body terms than done earlier. Eventual discrepancies with experiment such as the missing reproduction of e.g., the first excited 2^+ state in a $1p0f$ calculation of ^{48}Ca, can then be ascribed to eventual missing three-body forces, as indicated by the studies in Refs. [9,31,32,33,34,35] for light nuclei. The inclusion of real three-body interactions belongs to our future plans.

Acknowledgments

Supported by the U.S. Department of Energy under Contract Nos. DE-FG02-96ER40963 (University of Tennessee), DE-AC05-00OR22725 with UT-Battelle, LLC (Oak Ridge National Laboratory), and DE-FG02-01ER15228 (Michigan State University), the National Science Foundation (Grant No. CHE-0309517; Michigan State University), the Research Council of Norway, and the Alfred P. Sloan Foundation.

References

1. D.J. Dean, and M. Hjorth-Jensen, Phys. Rev. **C69** (2004) 054320.
2. K. Kowalski, D.J. Dean, M. Hjorth-Jensen, T. Papenbrock, and P. Piecuch, Phys. Rev. Lett. **92** (2004) 132501.
3. M. Hjorth-Jensen, T.T.S. Kuo and E. Osnes, Phys. Rep. **261** (1995) 125.
4. D.J. Dean, T. Engeland, M. Hjorth-Jensen, M.P. Kartamyshev, and E. Osnes, Prog. Part. Nucl. Phys. **53** (2004) 419.
5. T. Helgaker, P. Jørgensen, and J. Olsen, *Molecular Electronic Structure Theory. Energy and Wave Functions*, (Wiley, Chichester, 2000).
6. P. Navrátil and B.R. Barrett, Phys. Rev. **C57** (1998) 562.

7. P. Navrátil, J.P. Vary, and B.R. Barrett Phys. Rev. Lett. **84** (2000) 5728.
8. P. Navrátil, J.P. Vary, and B.R. Barrett, Phys. Rev. **C62** (2000) 054311.
9. P. Navrátil and W.E. Ormand, Phys. Rev. Lett. **88** (2002) 152502.
10. A.C. Hayes, P. Navrátil, and J.P. Vary, Phys. Rev. Lett. **91** (2003) 012502.
11. F. Coester, Nucl. Phys. **7** (1958) 421.
12. F. Coester and H. Kümmel, Nucl. Phys. **17** (1960) 477.
13. H. Kümmel, K.H. Lührmann and J.G. Zabolitzky, Phys. Rep. **36** (1977) 1.
14. R.F. Bishop, E. Buendia, M.F. Flynn and R. Guardiola, J. Phys. G: Nucl. Part. Phys. **17** (1991) 857; ibid. **18** (1992) 1157; ibid. **19** (1993) 1663; R. Guardiola, P.I. Moliner, J. Navarro, R.F. Bishop, A. Puente and N.R. Walet, Nucl. Phys. **A609** (1996) 218; R.F. Bishop and R. Guardiola.
15. J.H. Heisenberg, and B. Mihaila, Phys. Rev. **C59** (1999) 1440
16. T.D. Crawford and H.F. Schaefer III, Rev. Comp. Chem. **14** (2000) 33..
17. S.A. Kucharski, R.J. Bartlett, Theor. Chim. Acta **80** (1991) 387; P. Piecuch, S.A. Kucharski, K. Kowalski, and M. Musiał, Comp. Phys. Comm, **149** (2002) 72, and references therein.
18. J. Paldus and X. Li, Adv. Chem. Phys. **110** (1999) 1.
19. P. Piecuch and K. Kowalski and I.S.O. Pimienta and M.J. McGuire, Int. Rev. Phys. Chem. **21** (2002) 527.
20. P. Piecuch and K. Kowalski and P.-D. Fan and I.S.O. Pimienta, eds. J. Maruani, R. Lefebvre and E. Brändas, *Topics in Theoretical Chemical Physics* vol. **12**, series Progress in Theoretical Chemistry and Physics, (Kluwer, Dordrecht, 2004) 119.
21. J. Čížek, J. Chem. Phys. **45** (1966) 4256.
22. J. Čížek, Adv. Chem. Phys. **14** (1969) 35.
23. J. F. Stanton and R. J. Bartlett, J. Chem. Phys. **98** (1993) 7029.
24. P. Piecuch and R. J. Bartlett, Adv. Quantum Chem. **34** (1999) 295.
25. K. Kowalski and P. Piecuch, J. Chem. Phys. **113** (2000) 18.
26. K. Kowalski and P. Piecuch, J. Chem. Phys. **120** (2004) 1715.
27. H. Monkhorst, Int. J. Quantum Chem. Symp. **11** (1977) 421.
28. K. Kowalski and P. Piecuch, J. Chem. Phys. **115** (2001) 2966.
29. M. Wloch, D.J. Dean, J.R. Gour, M. Hjorth-Jensen, K. Kowalski, T. Papenbrock, and P. Piecuch, in preparation for Phys. Rev. Lett.
30. G. Hagen, J.S. Vaagen, and M. Hjorth-Jensen, J. Phys. A:Math. Gen. **37** (2004) 8991.
31. S.C. Pieper, V.R. Pandharipande, R.B. Wiringa, and J. Carlson, Phys. Rev. **C64** (2001) 014001
32. S.C. Pieper, K. Varga, and R.B. Wiringa, Phys. Rev. **C66** (2002) 0044310
33. R.B. Wiringa and S.C. Pieper, Phys. Rev. Lett. **89** (2002) 182501
34. A. Akmal, V.R. Pandharipande and D.G. Ravenhall, Phys. Rev. **C58** (1998) 1804.
35. P. Navrátil and W.E. Ormand, Phys. Rev. **C68** (2003) 034305.

KEY TOPICS IN NUCLEAR STRUCTURE
Proceedings of the 8th Int. Spring Seminar on Nuclear Physics
edited by Aldo Covello
© 2005 World Scientific Publishing Co.

MICROSCOPIC CORRELATIONS IN NUCLEAR STRUCTURE CALCULATIONS

M. TOMASELLI, T. KÜHL, AND D. URSESCU
GSI-Gesellschaft für Schwerionenforschung mbH, D64291 Darmstadt, Germany
E-mail: m.tomaselli@gsi.de

L.C. LIU
T-Division, Los Alamos National Laboratory, Los Alamos, NM 87545, USA

We present the structure of ^{6}Li and ^{18}O calculated with the use of a dynamic correlation model. In our model, the nucleon-nucleon correlations are built into a set of nonlinear equations of motion, which we solve self-consistently to obtain the eigenstates of the nucleons that are dressed by their interactions with the nuclear medium. We have found that the solution does not depend on the original choice of the two-body matrix elements. The inputs to solve the dynamical eigenvalue-equations are the n-body matrix elements. We have developed a cluster factorization method to relate the n-body matrix elements to the (n-1)-body matrix elements by recursion relations. Hence all the needed n-body matrix elements can be built up from the basic 2-body matrix elements.

1. Introduction

As early as 1963 Villars [1] proposed the unitary-model-operator (UMO) method (also known as the e^{iS} method) to construct effective operators. The method was firstly implemented with the use of short-range repulsive two-body potential by Shakin et al. [2] for the study of nuclear structure. Many variants of the e^{iS} method have since been proposed and solved in a perturbative framework. In recent years, the nonlinear theory of the *Dynamic Correlation Model* (DCM) has been developed [3]. The theory does not use effective operators, instead it solves self-consistently the dynamical equations to obtain nonperturbative eigenenergy solutions of nuclei. The method has had great success in describing the complex nuclear structures of odd-even nuclei and in treating mesonic and polarization corrections. The observation that N even nucleons can be considered as $N/2$ bosons has further led to the development of the *Boson-Dynamic Correlation Model*

(BDCM) for nuclei having an even number of valence nucleons [4]. In the BDCM, we also include the full complexity of nuclear dynamics arising from the valence as well as the core excitations. This inclusion of core excitation allows us to address the situation where valence clusters and core clusters are almost energetically degenerated and may, therefore, coexist. Many studies of core excitations have been presented in the literature, where the core excitations were treated perturbatively. In the present work, we show that the perturbative approach does not give a fully satisfactory description of medium nuclei, in particular, the ordering of the nuclear levels. On the other hand, when G-matrix and core excitations are calculated nonperturbatively at the same time, a very good description of the nuclear energy spectra, nuclear charge distributions, and various other nuclear properties is obtained.

2. The Boson Dynamic Correlation Model

In the BDCM [4] we introduce valence bosons as either neutron-neutron or neutron-proton pairs. The one valence-boson state (one valence particle pair) is defined as

$$|\Phi_J^{(1)}\rangle = \sum_{\alpha_1} A_1^\dagger(\alpha_1; J)|0\rangle. \tag{1}$$

The creation operator is defined by

$$A_1^\dagger(\alpha_1; J) \equiv N_{\alpha_1;J}\,(a_{j_1}^\dagger \otimes a_{j_2}^\dagger)^J\ ,$$

where J is the total spin and $N_{\alpha_1;J}$ is the normalization constant. For a simpler notation, the third components of the j's, the isospin and the radial quantum numbers are not written explicitly but are denoted collectively by α_1. In the BDCM the core excitation is included through coupling the valence bosonic states, Eq. (1), to intrinsic bosonic states corresponding to particle-hole excitations of the nuclear core. We have considered the following mixed-mode bosonic states: a) valence bosonic states coupled to the dynamic particle-hole states of normal parity; b) valence bosonic states coupled to the dynamic particle-hole states of nonnormal parity. The couplings are implemented through the strong two-body force V, which has a central component and a tensor component. The tensor component of V causes simultaneously the excitation of the valence particles to higher shell-model states and the deformation of the nuclear core. The mixed-mode state of n valence bosons and n' core bosons (or particle-hole pairs) can be written as

$$|\Phi_J^{(n,n')}\rangle = \Bigg[\sum_{\alpha_n(J_1J_2\ldots J_n)} X_{\alpha_n(J_1J_2\ldots J_n);J} A_n^\dagger(\alpha_n(J_1J_2\ldots J_n);J) + \sum_{\alpha_{n+1'}(J_1J_2\ldots J_{n+1'})} X_{\alpha_{n+1'}(J_1J_2\ldots J_{n+1'});J} A_{n+1'}^\dagger(\alpha_{n+1'}(J_1J_2\ldots J_{n+1'});J) + \ldots \sum_{\alpha_{n+n'}(J_1J_2\ldots J_{n+n'})} X_{\alpha_{n+n'}(J_1J_2\ldots J_{n+n'});J} A_{n+n'}^\dagger(\alpha_{n+n'}(J_1J_2\ldots J_{n+n'});J) \Bigg] |0\rangle, \tag{2}$$

where X's are the mixing coefficients. The operators $A_{n+m}^\dagger$, $(m = 1, 2, ...n)$ contain consequently hole creators $b_j^\dagger$. For the two nuclei we discuss here, ^{18}O and ^{6}Li, $n = 1$ in the last equation. In summary, in the BDCM one starts with the valence bosonic states of Eq. (1) and constructs subsequently mixed-mode nuclear states by including components having additional bosons formed by the particle-hole pairs of the core excitation, Eq. (2). The resulting nuclear states are then classified in terms of configuration mixing wave functions (CMWFs) of increasing degrees of complexity (number of particle-hole pairs). Of course, the effectiveness of this procedure depends crucially on the ability to perform easy and exact calculations for the complex matrix elements of the many-body operators, and this ability is achieved by the use of the cluster factorization method introduced in Ref. [4]. The basic dynamic equations of the BDCM are the commutator equations derived from the Equations of Motion (EOM) of the creation operators [4]. We obtain after a lengthy but elementary creation/annihilation operator algebra the following results for the commutator. Commutator for the one-boson state:

$$[H, A_1^\dagger(\alpha_1 J_1; J)] = \sum_{\beta_1 J_1'} \langle A_1(\alpha_1 J_1; J) \| H \| A_1^\dagger(\beta_1 J_1'; J)\rangle \, A_1^\dagger(\beta_1 J_1'; J) + \sum_{\beta_2(J_1'J_2')} \langle A_1(\alpha_1 J_1; J) \| H \| A_2^\dagger(\beta_2(J_1'J_2'); J)\rangle \, A_2^\dagger(\beta_2(J_1'J_2'); J), \tag{3}$$

where

$$\langle A_1(\alpha_1 J_1; J) \| H \| A_1^\dagger(\beta_1 J_1'; J)\rangle = \epsilon_{j_1} + \epsilon_{j_2} + \langle j_1 j_2 | v | j_1' j_2' \rangle_a^J$$

and

$$\langle A_1(\alpha_1 J_1; J) \| H \| A_2^\dagger(\beta_2(J_1'J_2'); J)\rangle$$

are the diagonal and the off-diagonal matrix elements and ϵ_{j_1}, ϵ_{j_2} are the single-particle energies. ($|A^\dagger\rangle$ and $\langle A|$ denote $A^\dagger|0\rangle$ and $\langle 0|A$, respectively.) Similar commutator equation has been obtained for the operators $A_2^\dagger$ and $A_3^\dagger$. We have linearized the commutator equation for the $A_3^\dagger$, Eq. (3), to convert the set EOM to a finite-dimensional eigenvalue problem. This linearization at the level of $A_3^\dagger$ [4] as the consequence of limiting the model space to containing states built up with 2 valence particles and 3p1h excitations. Our studies have shown that at the 3p1h level we can already give a very good decription for ^{18}O and ^{6}Li.

3. Results

In this section we present and discuss the results of the BDCM for the low-lying positive- and negative-parity levels of ^{18}O and for the positive- parity levels of ^{6}Li. Comparisons between the BDCM calculations, shell-model calculations, V_{low-k} method, and experimental data are given in Figs. 1 to 3. The experimental levels are those of Ref. [5] for ^{18}O and of Ref. [6] for ^{6}Li. These last references also contain an extensive literature on shell-model and cluster-model calculations of ^{18}O and ^{6}Li energy levels. We refer the interested readers to the references contained therein. For the particle-particle interaction, we use the matrix elements calculated in Ref. [7]. For the particle-hole interaction, we use the phenomenological potential of Ref. [8]. For ^{18}O the oscillator parameter of 2.76 fm was used. For the ^{6}Li, the value 2.56 fm was used. The antisymmetrization between the valence and the core particles increases the saturation properties of the system through lowering the higher-energy part of the nuclear spectrum of ^{18}O and assuring the correct behavior of the binding-energy per nucleon of ^{18}O and ^{6}Li.

3.1. *The calculated spectrum of ^{18}O*

In Table 1 we give the most relevant components of the 0^+ ground state of ^{18}O. The matrix diagonalized is of the order 300. The additional 3p1h components (not given in the table), though small, are important for reproducing the correct energy levels. In Fig. 1 we compare the BDCM results with those given in Fig. 3 of Ref. [9]. It is worth noting that the BDCM gives the correct ordering of the low-lying states, namely, $0_1^+, 2_1^+, 4_1^+, 0_2^+, 2_2^+, \ldots$. Except the empirical fit of Wildenthal (Wild) [10,11] and Bonn A, shell-model calculations with the Reid [12], Paris [13], Hml [14], Bonn B , and C [9] potentials all gave a spectrum in which the 2_2^+ and 0_2^+ states have their energy ordering reversed. However, in Wild [10] the matrix elements were adjusted to fit the data and were not calculated microscopically. Furthermore, both the Wild and Bonn A gave high-energy spectra that compare badly with the data.

Table 1: List of the most significant components of the ground state wave function 0^+ of ^{18}O

	$1d_{\frac{5}{2}}1d_{\frac{5}{2}}$	$2s_{\frac{1}{2}}2s_{\frac{1}{2}}$	$1d_{\frac{3}{2}}1d_{\frac{3}{2}}$
nn	.8279	.2879	.3930
	$2p_{\frac{3}{2}}2p_{\frac{3}{2}}$	$2p_{\frac{1}{2}}2p_{\frac{1}{2}}$	
	-.09151	-.2239	
3p1h	$(1d_{\frac{5}{2}}2p_{\frac{3}{2}})^{21}\ (1d_{\frac{5}{2}}1p_{\frac{1}{2}}^{-1})^{20}$	$(1d_{\frac{3}{2}}2p_{\frac{3}{2}})^{01}\ (1d_{\frac{3}{2}}1p_{\frac{3}{2}}^{-1})^{01}$	$(1d_{\frac{3}{2}}2p_{\frac{1}{2}})^{20}\ (1d_{\frac{3}{2}}1p_{\frac{1}{2}}^{-1})^{21}$
	.02981	-.03726	-.01983
	$(2s_{\frac{1}{2}}1d_{\frac{3}{2}})^{21}\ (1d_{\frac{5}{2}}1s_{\frac{1}{2}}^{-1})^{21}$		
	-.0089		

Although Bonn A differed from Bonn B and C (the latter two were not shown in Fig. 1) in the level ordering, they all overestimated the $0_2^+ - 4_1^+$ splitting with respect to the data and to the BDCM. One should also note from Fig. 1 that the 3rd experimental 2^+ and the 3rd experimental 0^+ levels are only present in the BDCM spectrum. Therefore, we may ascribe the presence of 2_3^+ and 0_3^+ to the use of nonperturbative calculation in BDCM. An overall good agreement between the nonperturbative BDCM

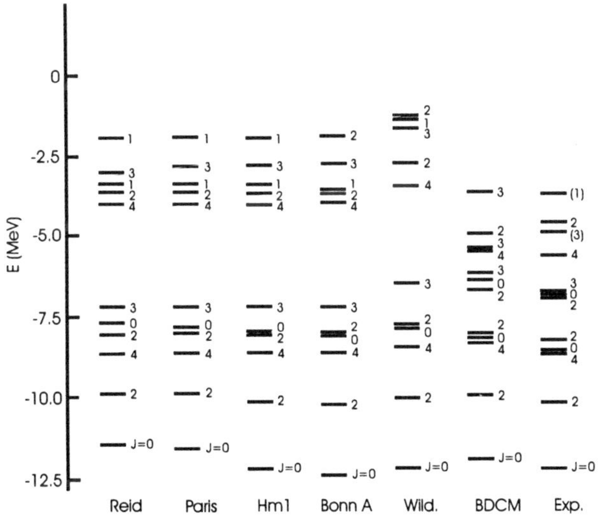

Figure 1. Calculated low-lying positive- and negative-parity states of ^{18}O

and the experimental ^{18}O levels is obtained for the low- as well as for the high-energy sectors while shell-model calculations give rise to high-lying levels (4_2^+ and up) that are situated too high in energy. We have noted that the success of the BDCM can be ascribed to two principal reasons: (a) it has a very large model space; (b) the nonperturbative calculation in a large model space generates important collective effects that move down the energies of the spectrum and give a more complete spectrum. Recently, a V_{low-k} method has been applied to nuclear structure calculations [15,16]. The five low-lying positive-parity states of ^{18}O calculated with V_{low-k} are compared with the corresponding BDCM levels in Fig. 2. (For comparison, the energy scale used in Refs. [15,16] was converted to the energy scale used in Fig. 1.) A dependence of the V_{low-k} results on the momentum cutoff parameter Λ is noticed. While in Fig. 1 the BONN-A potential gives a too big energy splitting between the 0_2^+ and 4_1^+ states, the $V_{Bonn-CD,low-k}$ [17] gives a reduced splitting which is quite comparable with the BDCM result. So far, there is no information on the high-energy sector of the ^{18}O levels calculated with V_{low-k}. We note that the V_{low-k} method has not changed the underlying nuclear structure theory (the folded-diagram expansions). While details of these modifications have not been given in the literature, it

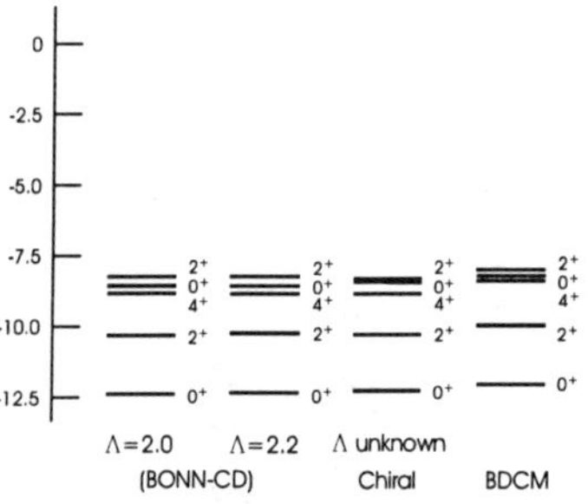

Figure 2. Calculated spectrum of ^{18}O: BDCM versus V_{low-k}

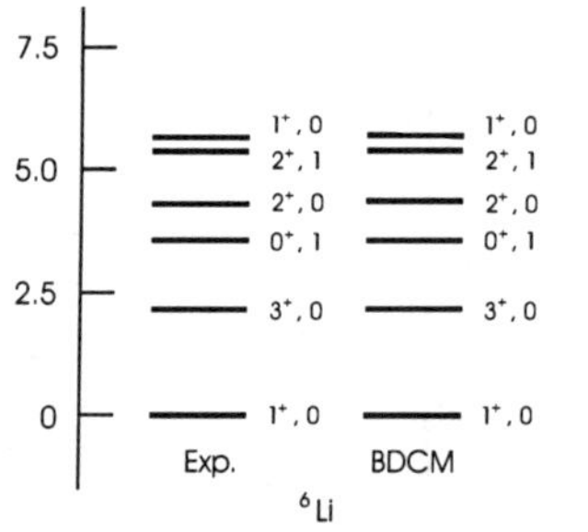

Figure 3. Calculated positive parity-states of ^{6}Li

was mentioned [15] that V_{low-k} "is a smooth potential". It is very likely that this smoothness has improved the quality of perturbative calculations. On the other hand, the BDCM contains folded diagrams in addition to other nuclear structure dynamics [4] and it uses self-consistent, nonperturbative calculations to guide the n-body system to the best solution of the dynamics. Future information on high-lying levels given by the V_{low-k} method and their comparison with the corresponding BDCM levels are of great interest. Hence, the differences between the low-lying energy levels obtained with $V_{Bonn(low-k)}$ and V_{Bonn} can be ascribed to modifications of dynamics at the nucleon-nucleon level. We may infer from the procedure employed in the V_{low-k} method, namely, using only the low-momentum part of the NN matrix elements to reproduce the NN phase shifts at on-shell momenta $\leq \Lambda$, that the shape and the strength of the original configuration-space potential has been inevitably modified, especially in the small-r region.

3.2. *Spectrum and charge radius of ^6Li*

In Fig. 3 the calculated spectrum of the low lying positive parity states of ^{6}Li is compared with the experimental levels of Ref. [6]. The 1^+, $T = 0$ ground state and the excited states of ^{6}Li are defined by exciting two valence particles to high single-particles states ($2\hbar\omega$) and by including the $\{3p-1h\}$ excitations up to an energy of 80 MeV. We solve the Schrödinger equation for the Wood-Saxon potential to obtain the single-particle energies to be used in the calculation of eigenvalue equation.

Table 2: Two particle components of the ground state wave function 1^+ of ^6Li

	$1p_{\frac{3}{2}}1p_{\frac{3}{2}}$	$1p_{\frac{3}{2}}1p_{\frac{1}{2}}$	$1p_{\frac{3}{2}}2p_{\frac{3}{2}}$	$1p_{\frac{3}{2}}1f_{\frac{5}{2}}$	$1p_{\frac{3}{2}}2p_{\frac{1}{2}}$	$1p_{\frac{1}{2}}1p_{\frac{1}{2}}$	$1p_{\frac{1}{2}}2p_{\frac{3}{2}}$	$1p_{\frac{1}{2}}2p_{\frac{1}{2}}$
pn	.8159	.5576	.0133	.0091	.0154	-.1133	-.0179	-.0106
pn+3p1h	.8102	.5598	.0142	.0091	.0158	-.1165	-.0175	-.0110
	$1d_{\frac{5}{2}}1d_{\frac{5}{2}}$	$1d_{\frac{5}{2}}1d_{\frac{3}{2}}$	$2s_{\frac{1}{2}}2s_{\frac{1}{2}}$	$2s_{\frac{1}{2}}1d_{\frac{3}{2}}$	$1d_{\frac{3}{2}}1d_{\frac{3}{2}}$	$1f_{\frac{7}{2}}1f_{\frac{7}{2}}$	$1g_{\frac{9}{2}}1g_{\frac{9}{2}}$	-
pn	-.0459	-.0775	-.0069	.0001	.0347	.0114	-.0017	-
pn+3p1h	-.0498	-.0849	-.0070	-.0008	.0409	.0116	-.0019	-

The final single-particle radial wave functions are then reproduced by using harmonic oscillator wave functions having a state-dependent oscillator parameter [18]. In Table 2 we give the most significant two-particle components calculated without and with the 3p1h CMWFs.

Table 3: Spectroscopic factor of the ground state wave function 1^+ of ^6Li

pn in $p_{\frac{3}{2}}$ shell	pn in (psd) shell	pn in (psd) shell and 3p1h
1.0	0.6657	0.6564

In Table 3 the ground-state spectoscopic factor calculated by using p-shell, psd-shell, and (psd+3p1h) configurations is given. The BDCM charge radii of ^{6}Li are given in Table 4 as functions of the different CMWFs. The experimental result of 2.55 fm [19] is well reproduced by the radius calculated using the full CMWFs space.

Table 4: Charge Radii of ^{6}Li calculated using different CMWFs

three protons in p-shell model states	Point radius	Folded radius
	2.15fm	2.31fm
three protons in psd-shell model states	Point radius	Folded radius
	2.32fm	2.47fm
three protons in psd+3p1h-shell model states	Point radius	Folded radius
	2.40 fm	2.55 fm

In Table 5 we compare the calculated charge radius of ^{6}Li with the results of Refs. [20,21]. As we can see, the radius calculated by [20] is smaller than the experimental value,

Table 5: Point charge radii of ^{6}Li calculated in the BDCM compared with the results of Navrátil-Barrett [20] and Pieper-Wiringa [21]

BDCM	Ref. [20]	Ref. [21]
2.40 fm	2.045 fm	2.39 fm

while the radius of Ref. [21] reproduces the experimental and the BDCM results.

References

1. F. Villars, Proceedings of the Enrico Fermi International School of Physics, XXIII, (1961); Academic Press, N.Y. (1963).
2. C.M. Shakin and Y.R. Waghmare, Phys. Rev. Lett. 16, 403 (1966); M.H. Hull and C.M. Shakin, Phys. Lett. 19, 506 (1965).
3. M. Tomaselli, Phys. Rev. C37, 349 (1988); Ann. Phys. (N.Y.) 205, 362 (1991); Phys. Rev. C48, 2290 (1993).
4. M. Tomaselli, L.C. Liu, S. Fritzsche, and T. Kühl, to be published in J. Phys. G.
5. D.R. Tilley *et al.*, Nucl. Phys. A595, 1 (1995).
6. D.R. Tilley *et al.*, Nucl. Phys. A708, 3 (202).
7. C.M. Shakin, Y.R. Waghmare, M. Tomaselli, and M.H. Hull, Phys. Rev. 161, 1015 (1967).
8. D.J. Millener and D. Kurath, Nucl. Phys. A255, 315 (1975).
9. M.F. Jiang, R. Machleidt, D.B. Stout, and T.T.S Kuo, Phys. Rev. C46, 910 (1992).
10. B.H. Wildenthal, Prog. Part. Nucl. Phys. 11, 5 (1984).
11. B.A. Brown, W.A. Richter, R.E. Julius, and B.H. Wildenthal, Ann. Phys. 182, 191 (1988).
12. R.V. Reid, Ann. Phys. (N.Y.) 50, 411 (1968).
13. M. Lacombe *et al.*, Phys. Rev. C21, 861, (1980).
14. K. Hollinde and R. Machleidt, Nucl. Phys. A247, 495 (1975).
15. Scott Bogner, T.T.S. Kuo, L. Coraggio, A. Covello, and N. Itaco, Phys. Rev. C65, 051301(R) (2002).
16. L. Coraggio, A. Covello, A. Gargano, N. Itaco, T.T.S. Kuo, D.R. Entem, and R. Machleidt, Phys. Rev. C66, 021303(R) (2002).
17. R. Machleidt, Phys. Rev. C63, 024001 (2001).
18. V. Gillet, B. Giraud, and M. Rho, Nucl. Phys. A103,257 (1967).
19. G.C. Li *et al.*, Nucl. Phys. A152, 583 (1971).
20. P. Navrátil and B.R. Barrett, Phys. Rev. C57, 3119 (1998).
21. S.C. Pieper and R.B. Wiringa, Annu. Rev. Nucl. Part, Sci. 51, 53 (2001).

KEY TOPICS IN NUCLEAR STRUCTURE
Proceedings of the 8th Int. Spring Seminar on Nuclear Physics
edited by Aldo Covello
© 2005 World Scientific Publishing Co.

PARTICLE-NUMBER-PROJECTED HFB METHOD WITH SKYRME FORCES AND DELTA PAIRING

M.V. STOITSOV[1−4], J. DOBACZEWSKI[2−5], W. NAZAREWICZ[2−5],
P.-G. REINHARD[6], J. TERASAKI[7]

[1] *Institute of Nuclear Research and Nuclear Energy, Bulgarian Academy of Sciences, Sofia-1784, Bulgaria*
[2] *Department of Physics & Astronomy, University of Tennessee, Knoxville, Tennessee 37996, USA*
[3] *Physics Division, Oak Ridge National Laboratory, P.O. Box 2008, Oak Ridge, Tennessee 37831, USA*
[4] *Joint Institute for Heavy-Ion Research, Oak Ridge, Tennessee 37831, USA*
[5] *Institute of Theoretical Physics, Warsaw University, ul. Hoża 69, 00-681 Warsaw, Poland*
[6] *Institut für Theoretische Physik, Universität Erlangen, Staudtstrasse 7, D-91058 Erlangen, Germany*
[7] *Department of Physics and Astronomy, University of North Carolina at Chapel Hill, Phillips Hall, Chapel Hill, NC 27599-3255, USA*

Particle-number restoration before variation is implemented in the HFB method employing the Skyrme force and zero-range delta pairing. Results are compared with those obtained within the Lipkin-Nogami method, with or without the particle-number projection after variation. Shift invariance property is proven to be valid also in the case of density functional calculations which allows the well known singularity ($u_n^2 = v_n^2$) in PNP HFB calculations to be safely avoided.

1. Introduction

Pairing correlations play a central role in describing properties of atomic nuclei. In mean-field approaches, they are best treated in the Hartree-Fock-Bogoliubov (HFB) approximation [1]. The HFB ansatz for the nuclear wave function, however, breaks the particle-number symmetry. The symmetry needs to be restored, in principle, especially if one looks at observables that strongly vary as functions of particle number.

Recently, it has been shown [2] that the total energy in the particle-number-projected (PNP) HFB approach can be expressed as a functional of the unprojected HFB density matrix and pairing tensor. Its variation leads to a set of HFB-like equations with modified Hartree-Fock fields and pairing potentials. The method has been illustrated within schematic models [2], and also implemented in HFB calculations with the finite-range Gogny force [3]. In the present paper we adopt it for the Skyrme functional and delta pairing, where the method must rely on the spatial locality of densities and mean fields. The HFB results using the Lipkin-Nogami (LN) approximation followed by the particle-number projection *after variation* (PLN) are compared to the HFB results with projection *before variation* (PNP).

2. Particle-Number-Projected Skyrme-HFB Method

2.1. *Skyrme-HFB method*

Due to the zero-range character of the Skyrme force, the Skyrme-HFB energy is an energy functional,

$$\mathsf{E}[\rho,\tilde{\rho}] = \frac{\langle\Phi|H|\Phi\rangle}{\langle\Phi|\Phi\rangle} = \int d\mathbf{r}\;\left[H(\mathbf{r}) + \tilde{H}(\mathbf{r})\right], \tag{1}$$

of *local* particle and pairing densities, where $H(\mathbf{r})$ and $\tilde{H}(\mathbf{r})$ are normal and pairing energy densities, respectively. Their explicit expressions [4] are given in terms of particle (pairing) local densities and currents. All local densities and currents are completely determined by particle $\rho_{n'n}$ and pairing $\tilde{\rho}_{n'n}$ density-matrix elements in the configurational space, i.e.,

$$\begin{aligned}
\rho(\mathbf{r}\sigma,\mathbf{r}'\sigma') &= \sum_{nn'} \rho_{nn'}\;\psi^*_{n'}(\mathbf{r}',\sigma')\psi_n(\mathbf{r},\sigma),\\
\tilde{\rho}(\mathbf{r}\sigma,\mathbf{r}'\sigma') &= \sum_{nn'} \tilde{\rho}_{nn'}\;\psi^*_{n'}(\mathbf{r}',\sigma')\psi_n(\mathbf{r},\sigma).
\end{aligned} \tag{2}$$

The use of the pairing density matrix $\tilde{\rho}(\mathbf{r}\sigma,\mathbf{r}'\sigma') = -2\sigma'\kappa(\mathbf{r},\sigma,\mathbf{r}',-\sigma')$ instead of the pairing tensor κ is convenient when the time-reversal symmetry is assumed [4].

The derivatives of the energy (1) with respect to $\rho_{nn'}$ and $\tilde{\rho}_{nn'}$ define the particle-hole and particle-particle matrices

$$\begin{aligned} h_{nn'} &= \frac{\partial \mathsf{E}(\rho,\tilde{\rho})}{\partial \rho_{n'n}} = \sum_{\sigma\sigma'} \int d\mathbf{r}\ \psi_n^*(\mathbf{r},\sigma) h(\mathbf{r},\sigma,\sigma') \psi_{n'}(\mathbf{r},\sigma'), \\ \tilde{h}_{nn'} &= \frac{\partial \mathsf{E}(\rho,\tilde{\rho})}{\partial \tilde{\rho}_{n'n}} = \sum_{\sigma\sigma'} \int d\mathbf{r}\ \psi_{n'}^*(\mathbf{r},\sigma) \tilde{h}(\mathbf{r},\sigma,\sigma') \psi_n(\mathbf{r},\sigma'), \end{aligned} \tag{3}$$

respectively, which enter the Skyrme-HFB equations

$$\begin{pmatrix} h-\lambda & \tilde{h} \\ \tilde{h} & -h+\lambda \end{pmatrix} \begin{pmatrix} U \\ V \end{pmatrix} = E \begin{pmatrix} U \\ V \end{pmatrix}. \tag{4}$$

2.2. *Particle-number-projection*

Let us consider, in the context of HFB theory, the PNP state:

$$|\Psi\rangle \equiv P^N|\Phi\rangle = \frac{1}{2\pi}\int_0^{2\pi} d\phi\, e^{i\phi(\hat{N}-N)}|\Phi\rangle, \tag{5}$$

where $\hat{N}$ is the number operator, N is the particle number, and $|\Phi\rangle$ is the HFB wavefunction which does not have a well-defined particle number. As shown in Ref. [2], the PNP HFB energy

$$\mathsf{E}^N[\rho,\bar{\rho}] = \frac{\langle\Phi|HP^N|\Phi\rangle}{\langle\Phi|P^N|\Phi\rangle} = \frac{\int d\phi \langle\Phi|He^{i\phi(\hat{N}-N)}|\Phi\rangle}{\int d\phi \langle\Phi|e^{i\phi(\hat{N}-N)}|\Phi\rangle}, \tag{6}$$

is again an energy functional of the unprojected densities ρ and $\bar{\rho}$.

The situation, however, is not so simple in the case when the energy is deduced from an energy density functional. The source of the problem lies in the fact that the unprojected energy (1) is defined only in connection with one mean field state, i.e., one identifies formally $E[\rho,\bar{\rho}] \Longleftrightarrow \langle\Phi|\hat{H}|\Phi\rangle$ where $\rho(\mathbf{r}\sigma,\mathbf{r}'\sigma')$ and $\bar{\rho}(\mathbf{r}\sigma,\mathbf{r}'\sigma')$ are densities associated with $|\Phi\rangle$.

The question is how to use the energy-density functionals in projection techniques which require the knowledge of off-diagonal (or "transitional") expectation values:

$$\langle\Phi(0)|\hat{H}|\Phi(\phi)\rangle, \quad |\Phi(\phi)\rangle = e^{i\phi(\hat{N}-N)}|\Phi\rangle. \tag{7}$$

These are not automatically given by DFT. Extensions of the formalism are necessary and they are not unique. Various recipes are discussed in Ref. [3].

Let us consider in particular the "mixed density" recipe that treats all local densities as "mixed" (or "transitional") ones defined between the states $\Phi(0)$ and $\Phi(\phi)$. In the case of the Skyrme force, the projected energy (6) reads:

$$\mathsf{E}^N[\rho,\tilde{\rho}] = \frac{\langle\Phi|HP^N|\Phi\rangle}{\langle\Phi|P^N|\Phi\rangle} = \int d\phi\; y(\phi) \int d\mathbf{r}\left(H(\mathbf{r},\phi) + \tilde{H}(\mathbf{r},\phi)\right), \tag{8}$$

where

$$x(\phi) = \frac{1}{2\pi}\frac{e^{-i\phi N}\det(e^{i\phi}I)}{\sqrt{\det C(\phi)}},\quad y(\phi) = \frac{x(\phi)}{d\phi'\, x(\phi')}, \tag{9}$$

I is the unit matrix, and the gauge-angle dependent energy densities $H(\mathbf{r},\phi)$ and $\tilde{H}(\mathbf{r},\phi)$ are derived from the unprojected ones by simply replacing particle (pairing) local densities by their gauge-angle dependent counterparts. The latter ones are defined by the gauge-angle dependent "mixed" density matrices

$$\begin{aligned}\rho(\mathbf{r}\sigma,\mathbf{r}'\sigma',\phi) &= \sum_{nn'} \rho_{nn'}(\phi)\; \psi^*_{n'}(\mathbf{r}',\sigma')\psi_n(\mathbf{r},\sigma),\\ \tilde{\rho}(\mathbf{r}\sigma,\mathbf{r}'\sigma',\phi) &= \sum_{nn'} \tilde{\rho}_{nn'}(\phi)\; \psi^*_{n'}(\mathbf{r}',\sigma')\psi_n(\mathbf{r},\sigma),\end{aligned} \tag{10}$$

where

$$\begin{aligned}\rho_{n'n}(\phi) &= \frac{\langle\Phi|c^\dagger_n c_{n'} e^{i\phi\hat{N}}|\Phi\rangle}{\langle\Phi|e^{i\phi\hat{N}}|\Phi\rangle} = \sum_m C_{nm}(\phi)\rho_{mn'},\\ \tilde{\rho}_{n'n}(\phi) &= e^{-i\phi}\sum_m C_{nm}(\phi)\tilde{\rho}_{mn'},\\ C(\phi) &= e^{2i\phi}\left(1+\rho(e^{2i\phi}-1)\right)^{-1}.\end{aligned} \tag{11}$$

Obviously, the projected energy (8) is again a functional of the unprojected density matrices ρ and $\tilde{\rho}$. Its derivatives with respect to $\rho_{n'n}$ and $\tilde{\rho}_{n'n}$ lead to the PNP Skyrme-HFB equations

$$\begin{pmatrix} h^N & \tilde{h}^N \\ \tilde{h}^N & -h^N \end{pmatrix}\begin{pmatrix} U \\ V \end{pmatrix} = E^N \begin{pmatrix} U \\ V \end{pmatrix}, \tag{12}$$

where

$$\begin{aligned} h^N &= \int d\phi y(\phi) Y(\phi) E(\phi) + \int d\phi y(\phi) e^{-2i\phi}\, C(\phi) h(\phi) C(\phi) \\ &\quad - \tfrac{1}{2} \left[\int d\phi y(\phi) 2ie^{-i\phi} sin(\phi) \tilde{\rho}(\phi) \tilde{h}(\phi) C(\phi) + \text{h.c.} \right], \\ \tilde{h}^N &= \tfrac{1}{2} \int d\phi y(\phi) e^{-i\phi} \{ \tilde{h}(\phi) C(\phi) + (\tilde{h}(\phi) C(\phi))^T \} \\ Y(\phi) &= ie^{-i\phi} \sin\phi\, C(\phi) - i \int d\phi' y(\phi') e^{-i\phi'} \sin\phi'\, C(\phi'). \end{aligned} \tag{13}$$

The gauge-angle dependent field matrices $h(\phi)$ and $\tilde{h}(\phi)$ are obtained by simply replacing in the unprojected fields (3) the particle and pairing local densities with their gauge-angle dependent counterparts.

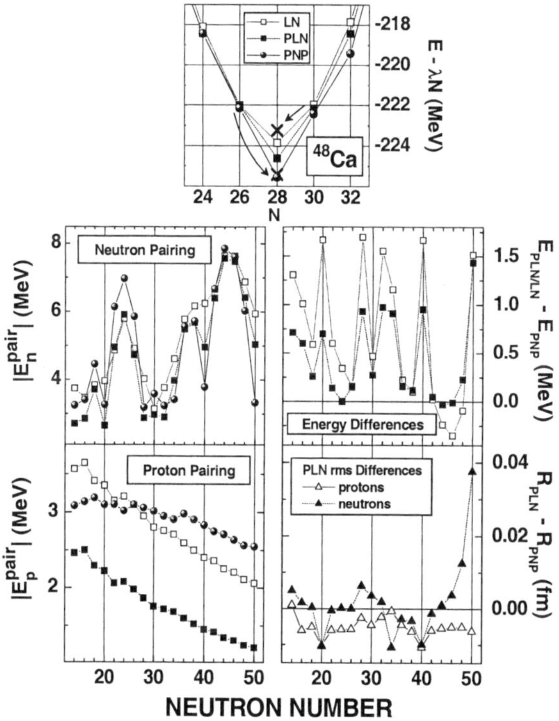

Figure 1. The LN and PLN (projection after variation), and PNP HFB (projection before variation) results obtained for the SLy4 force and mixed delta pairing. Arrows in the top panel indicate projection results from the neighboring nuclei.

2.3. *Delta pairing forces*

When using delta pairing forces, one has to restrict the quasiparticle space in order to avoid the divergences associated with the zero range. Within the unprojected HFB calculations, a pairing cut-off is introduced by using the so-called equivalent single-particle spectrum [4,5]. After each iteration performed with a given chemical potential λ, one calculates an equivalent spectrum $\bar{e}_n$ and pairing gaps $\bar{\Delta}_n$:

$$\bar{e}_n = (1 - 2P_n)E_n + \lambda, \quad \bar{\Delta}_n = 2E_n\sqrt{P_n(1 - P_n)}, \tag{14}$$

where E_n are the quasiparticle energies and P_n denotes the norms of the lower HFB wave function. Due to the similarity between $\bar{e}_n$ and the single-particle energies, one can take into account only those quasiparticle states for which $\bar{e}_n$ is less than the cut-off energy ϵ_{cut} (usually around 60 MeV).

One can see that this procedure cannot be directly applied to the PNP HFB calculations, because the Lagrange multiplier λ entering the unprojected HFB Eqs. (4) is no longer available in Eq. (12). This means that the local densities emerging after each HFB diagonalization (12) are not automatically normalized to the particle number N. As a result, all auxiliary quantities, as e.g. the analogues of the quasiparticle energies, E_n^N, and probabilities, P_n^N, do not have the usual meaning. However, one can always reintroduce the Lagrange multiplier λ into Eq. (12) without changing the results, and adjust it to give a correct *average* particle number in the *unprojected* state. In practice, it is enough to calculate for the solutions of Eq. (12) the average values E_n of the unprojected HFB matrix and use them in Eq. (14) together with $P_n \equiv P_n^N$. This allows for defining the Lagrange multiplier and implementing the cut-off procedure.

2.4. *Sample PNP HFB results*

Figure 1 gives the PNP HFB results for the complete chain of the calcium isotopes (proton-neutron drip to drip line), calculated for the Sly4 force [6] and mixed delta pairing [5]. Comparison is also made with the HFB Lipkin-Nogami (LN) results and projected (after variation) Lipkin-Nogami results (PLN). One can conclude that the PLN approximation works good for open-shell nuclei, where the total energy differences between various variants of calculations are less than 250 keV. For closed-shell nuclei [7], however, the energy differences increase to more than 1 MeV. In such cases, one can improve the PLN results by applying the projection to the LN solutions obtained for the neighboring nuclei [8], as illustrated in the top panel of

Fig. 1.

3. Shift Invariance Within DFT

Important consequences for the PNP HFB expectation values follow from the obvious shift invariance property [9]

$$\underbrace{e^{\psi(\hat{N}-N)}}_{\hat{S}_\psi}|\Psi\rangle = |\Psi\rangle \tag{15}$$

of the PNP wave function (5), where ψ is an arbitrary number. For example, the energy (6) is obviously shift–invariant under the transformation $\hat{S}_\psi$ since

$$E^N = \frac{\langle\Phi|\hat{H}|\Psi\rangle}{\langle\Phi|\Psi\rangle} = \frac{\langle\Phi|\hat{H}\hat{S}_\psi|\Psi\rangle}{\langle\Phi|\hat{S}_\psi|\Psi\rangle}. \tag{16}$$

Again, the situation becomes more complicated when the energy is deduced from an energy density functional.

3.1. *The shift invariance under the "mixed density" recipe*

In order to prove the shift invariance under the "mixed density" recipe, let us introduce the "mixed local densities" in their canonical representation

$$\begin{aligned}
\rho_\phi(\mathbf{r}) &= \sum_n \frac{v_n^* v_n\ e^{2\imath\phi}}{u_n^* u_n + v_n^* v_n\ e^{2\imath\phi}}\ \varphi_n^\dagger(\mathbf{r})\varphi_n(\mathbf{r}), \\
\kappa_\phi(\mathbf{r}) &= \sum_n \frac{u_n^* v_n\ e^{2\imath\phi}}{u_n^* u_n + v_n^* v_n\ e^{2\imath\phi}}\ \varphi_n^\dagger(\mathbf{r})\varphi_n(\mathbf{r}), \\
\bar{\kappa}_\phi(\mathbf{r}) &= \sum_n \frac{u_n v_n^*}{u_n^* u_n + v_n^* v_n\ e^{2\imath\phi}}\ \varphi_n^\dagger(\mathbf{r})\varphi_n(\mathbf{r}).
\end{aligned} \tag{17}$$

The energy (8) can be rewritten as

$$E^N = \frac{\int d\phi\ e^{-\imath\phi N}\mathbb{I}(\phi)\ \mathbb{E}\left(\kappa_\phi, \bar{\kappa}_\phi, \rho_\phi\right)}{\int d\phi\ e^{-\imath\phi N}\ \mathbb{I}(\phi)},\quad \mathbb{I}(\phi) = \langle\Phi(0)|\Phi(\phi)\rangle. \tag{18}$$

As a rule of thumb, we note that the phase factor $e^{2\imath\phi}$ is always linked to v_n, considering v_n^* as an independent variable.

In order to check whether the definition (17)-(18) does also guarantees the shift invariance (16), one needs to show that $E^N = E^{(\text{shift})}$ where

$$E^{(\text{shift})} = \frac{\int d\phi\ e^{-\imath\phi N-\psi N}\mathbb{I}(\phi,\psi)\ \mathbb{E}\left(\kappa_{\phi\psi}, \bar{\kappa}_{\phi\psi}, \rho_{\phi\psi}\right)}{\int d\phi\ e^{-\imath\phi N-\psi N}\ \mathbb{I}(\phi,\psi)}, \tag{19}$$

$$\mathbb{I}(\phi,\psi) = \langle\Phi(0)|e^{-\psi N}|\Phi(\phi)\rangle \tag{20}$$

and

$$\rho_{\phi\psi}(\mathbf{r}) = \sum_n \frac{v_n^* v_n \; e^{2\imath\phi+2\psi}}{u_n^* u_n + v_n^* v_n \; e^{2\imath\phi+2\psi}} \varphi_n^\dagger(\mathbf{r})\varphi_n(\mathbf{r}),$$

$$\kappa_{\phi\psi}(\mathbf{r}) = \sum_n \frac{u_n^* v_n \; e^{2\imath\phi+2\psi}}{u_n^* u_n + v_n^* v_n \; e^{2\imath\phi+2\psi}} \varphi_n^\dagger(\mathbf{r})\varphi_n(\mathbf{r}), \tag{21}$$

$$\bar{\kappa}_{\phi\psi}(\mathbf{r}) = \sum_n \frac{u_n v_n^*}{u_n^* u_n + v_n^* v_n \; e^{2\imath\phi+2\psi}} \varphi_n^\dagger(\mathbf{r})\varphi_n(\mathbf{r}).$$

The shift invariance is trivially maintained for the kinetic energy because this quantity is given by expectation value of an operator such that a reasoning as in Eq. (16) applies. To prove the invariance for other terms, we start from the observation that the shift is tightly linked to the r.h.s. occupation amplitude, i.e.

$$v_n \longleftrightarrow v_n e^{2\imath\phi+2\psi}. \tag{22}$$

Now we make the strong assumption that the energy expectation value can be expanded into a mixed power series with respect to v_n, v_n^*, u_n, and u_n^*. We collect terms having the same number of v_n. The numerator in the energy expression will then contain a kernel as

$$E^{(\mathrm{num})} = \int d\phi \, e^{-\imath\phi N-\psi N} \sum_{N_v} \underbrace{v_{n_1}..v_{n_{N_v}}}_{N_v \text{ terms}} e^{\imath\phi N_v+\psi N_v} \sum_{\text{anything}} v_m^*..u_k..u_l^*... \tag{23}$$

The ϕ integration filters the term $N = N_v$, yielding

$$\int d\phi \, e^{-\imath\phi N} e^{\imath\phi N_v} \Longrightarrow N = N_v \Longrightarrow e^{-\psi N} e^{\psi N_v} = 1.$$

The same reasoning applies to the denominator. Thus both the numerator and the denominator in the projected energy expression are separately shift-invariant. This holds for DFT with the extension recipe (17).

The above demonstration relies on a power series expansion in order v^N. But such an expansion will not converge just around the critical point $u_n = v_n$. However, as we discuss later, one can always avoid the case containing the critical point $u_n = v_n$, and one indeed does not need such a proof expansion.

3.2. *Other extensions of DFT*

There are alternatives to the recipe((17)-(18)) suggested in Ref. [10]. For example, one may use the projected densities as, e.g., $\rho^N(\mathbf{r})$ in the DFT energy expectation value. This reads, e.g., for the potential energy

$$E_{\text{pot}}^N = E_{\text{pot}}\left[\rho^N\right]. \tag{24}$$

With the same reasoning as above, one can show that $\rho^N(r)$ is shift invariant. The recipe (24) is then also shift-invariant. There are, however, objections for other reasons. For example, this recipe can be shown to be wrong in the simple case of a two-body point coupling force.

There is a proposal from [10] to extend the DFT definition just by adding two densities $\overline{\rho}_\phi(\mathbf{r})$ and $\overline{\rho}_0(\mathbf{r})$ associated with $\Phi(0)$ and $\Phi(\phi)$, respectively. The recipe consists in using an average value

$$E_{\text{pot}}(\rho) \longrightarrow \frac{1}{2}\left(E_{\text{pot}}\left[\overline{\rho}_0\right] + E_{\text{pot}}\left[\overline{\rho}_\phi\right]\right) \tag{25}$$

in the projection kernel. But note that $\overline{\rho}_\phi(r) = \overline{\rho}_0(r) = \rho(r)$ for that particular case of particle number projection. The phase factors $e^{2\imath\phi}$ just cancel out if used on bra and ket simultaneously. One then obtain

$$\frac{1}{2}\left(E_{\text{pot}}\left[\overline{\rho}_0\right] + E_{\text{pot}}\left[\overline{\rho}_\phi\right]\right) = E_{\text{pot}}[\rho], \tag{26}$$

i.e., one obtains the unprojected energy.

3.3. *The innocent singularity in DFT applications*

At first glance, the mixed density recipe (17)-(18) also has a problem. Looking at the denominator of the spatial densities (17), one notices a possible singularity at $u_n^2 = v_n^2 = 1/2$ for a gauge-angle $\phi \to \pi/2$. The shift invariance allows to show that this singularity is unimportant.

We assume that we have a discrete spectrum with a finite set of v_n and u_n. Now let it happen that $u_n^2 = v_n^2$. We apply a shift $v_n \longrightarrow v_n e^{2\psi}$ which guarantees that $u_n^2 \neq v_n^2$. In a discrete spectrum, one can always find a ψ such that all other amplitudes u_m and v_m stay different. We then can evaluate the projected energy without having dealt with singularity.

In numerical applications one can easily implement the shift $v_n \longrightarrow v_n e^{2\psi}$ by changing the normalization of the internal density

$$\int \bar{\rho}(\mathbf{r}) d\mathbf{r} = \bar{N}, \quad \bar{\rho}(\mathbf{r}) = \sum_n v_n^2 \varphi_n^\dagger(\mathbf{r}) \varphi_n(\mathbf{r}). \tag{27}$$

Different values of ψ correspond to different values of the normalization constant $\bar{N}$. Therefore, instead of ψ, one can keep the internal normalization constant $\bar{N}$ fixed during the PNP HFB iterations. This could be achieved by introducing a Lagrange multiplier μ by means of Eq. (27), and μ always goes to zero when the PNP HFB solution is achieved.

Indeed, all the numerical tests we have performed have shown that the PNP HFB results do not depend on the particular values of $\bar{N}$, and perfect shift invariance is always achieved. Changing the intrinsic normalization $\bar{N}$ in a wide range, occupation probability v^2, which is closer to the critical value 1/2, varies from 0.076 to 0.945, but all nuclear characteristics remain unchanged.

Acknowledgments

This work was supported in part by the U.S. Department of Energy under Contract Nos. DE-FG02-96ER40963 (University of Tennessee), DE-AC05-00OR22725 with UT-Battelle, LLC (Oak Ridge National Laboratory), and DE-FG05-87ER40361 (Joint Institute for Heavy Ion Research); by the National Nuclear Security Administration under the Stewardship Science Academic Alliances program through DOE Research Grant DE-FG03-03NA00083; by the Polish Committee for Scientific Research (KBN) under contract N0. 1 P03B 059 27 and by the Foundation for Polish Science (FNP).

References

1. P. Ring and P. Schuck, *The Nuclear Many-Body Problem* (Springer Verlag, New York, 1980).
2. J.A. Sheikh and P. Ring, Nucl. Phys. **A665**, 71 (2000).
3. M. Anguiano, J.L. Egido, and L.M. Robledo, Nucl. Phys. **A696**, 476 (2001).
4. J. Dobaczewski, H. Flocard, and J. Treiner, Nucl. Phys. **A422**, 103 (1984).
5. M.V. Stoitsov, W. Nazarewicz, and S. Pittel, Phys. Rev. **C58**, 2092 (1998); M.V. Stoitsov, J. Dobaczewski, P. Ring, and S. Pittel, Phys. Rev. C **61**, 034311 (2000); M.V. Stoitsov , J. Dobaczewski, W. Nazarewicz, S. Pittel, D.J. Dean, Phys. Rev. **C68**, 054312 (2003).
6. E. Chabanat, P. Bonche, P. Haensel, J. Meyer, and F. Schaeffer, Nucl. Phys. **A635**, 231 (1998).
7. J. Dobaczewski and W. Nazarewicz, Phys. Rev. **C47**, 2418 (1993).
8. P. Magierski, S. Ćwiok, J. Dobaczewski, and W. Nazarewicz, Phys.Rev. **C48**, 1686 (1993).
9. J.L. Egido, P. Ring, Nucl. Phys. **A383**, 189 (1982).
10. T. Duguet, *PhD thesis*; T. Duguet, P. Bonche, Phys.Rev. **C67**, 054308(2003).

KEY TOPICS IN NUCLEAR STRUCTURE
Proceedings of the 8th Int. Spring Seminar on Nuclear Physics
edited by Aldo Covello
© 2005 World Scientific Publishing Co.

RELATIVISTIC PSEUDOSPIN SYMMETRY AS A SUPERSYMMETRIC PATTERN IN NUCLEI

A. LEVIATAN
Racah Institute of Physics, The Hebrew University, Jerusalem 91904, Israel
E-mail: ami@phys.huji.ac.il

Shell-model states involving several pseudospin doublets and "intruder" levels in nuclei, are combined into larger multiplets. The corresponding single-particle spectrum exhibits a supersymmetric pattern whose origin can be traced to the relativistic pseudospin symmetry of a nuclear mean-field Dirac Hamiltonian with scalar and vector potentials.

1. Introduction

Pseudospin doublets [1] in nuclei refer to the empirical observation of quasi-degenerate pairs of certain shell-model orbitals with non-relativistic single-nucleon radial, orbital, and total angular momentum quantum numbers:

$$(n, \ell, j = \ell + 1/2) \quad \text{and} \quad (n-1, \ell+2, j' = \ell + 3/2)\ . \tag{1}$$

The doublet structure (for $n \geq 1$) is expressed in terms of a "pseudo" orbital angular momentum $\tilde{\ell} = \ell + 1$ and "pseudo" spin, $\tilde{s} = 1/2$, coupled to $j = \tilde{\ell} - 1/2$ and $j' = \tilde{\ell} + 1/2$. For example, $[ns_{1/2}, (n-1)\, d_{3/2}]$ will have $\tilde{\ell} = 1$, etc. The states in Eq. (1) involve only normal-parity shell-model orbitals. The states $(n = 0, \ell, j = \ell + 1/2)$, with aligned spin and no nodes, are not part of a doublet. This is empirically evident in heavy nuclei, where such states with large j, *i.e.*, $0g_{9/2}$, $0h_{11/2}$, $0i_{13/2}$, are the "intruder" abnormal-parity states, which are unique in the major shell.

Pseudospin symmetry is experimentally well corroborated and plays a central role in explaining features of nuclei [2] including superdeformation [3] and identical bands [4]. It has been recently shown to result from a relativistic symmetry of the Dirac Hamiltonian in which the sum of the scalar and vector nuclear mean field potentials cancel [5]. The symmetry generators [6] combined with known properties of Dirac bound states, provide a natural explanation [7] for the structure of pseudospin doublets and for the special status of "intruder" levels in nuclei.

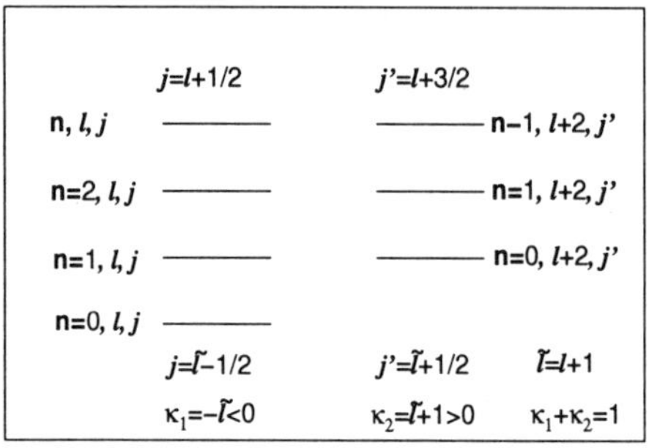

Figure 1. Nuclear single-particle spectrum composed of pseudospin doublets and an "intruder" level. All states share a common $\tilde{\ell}$ and $\tilde{s}=1/2$. The corresponding Dirac κ-quantum numbers are also indicated.

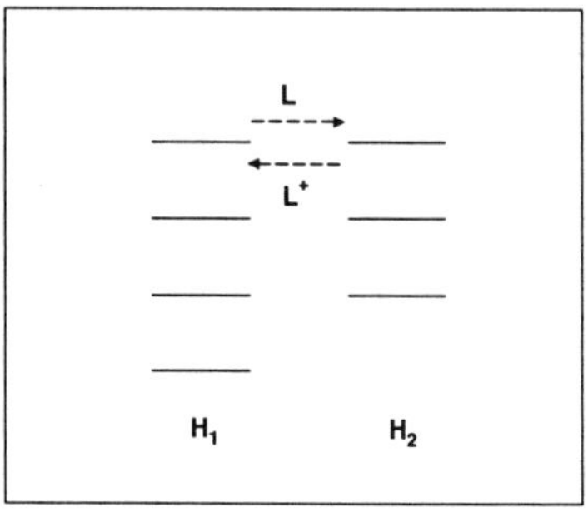

Figure 2. Typical supersymmetric pattern. The Hamiltonians H_1 and H_2 have identical spectra with an additional level for H_1 when SUSY is exact. The operators L and $L^\dagger$ connect the partner states.

Figure 1 portrays the level scheme of an ensemble of pseudospin doublets, Eq. (1), with fixed ℓ, j, j' and $n = 1,2,3,\ldots$ together with the "intruder" level ($n=0,\ell,j=\ell+1/2$). The single-particle spectrum exhibits towers of pair-wise degenerate states, sharing a common $\tilde{\ell}$, and an additional non-degenerate nodeless "intruder" state at the bottom of the spin-aligned tower. A comparison with Fig. 2 reveals a striking similarity with a supersymmetric pattern. In the present contribution we identify [8] the underlying supersymmetric structure associated with a Dirac Hamiltonian possessing a relativistic pseudospin symmetry.

2. Supersymmetric Quantum Mechanics

Supersymmetric quantum mechanics (SUSYQM), initially proposed as a model for supersymmetry (SUSY) breaking in field theory [9], has by now developed into a field in its own right, with applications in diverse areas of physics [10]. The essential ingredients of SUSYQM are the supersymmetric charges and Hamiltonian

$$Q_- = \begin{pmatrix} 0 & 0 \\ L & 0 \end{pmatrix} \ , \ Q_+ = \begin{pmatrix} 0 & L^\dagger \\ 0 & 0 \end{pmatrix} \ , \ \mathcal{H} = \begin{pmatrix} H_1 & 0 \\ 0 & H_2 \end{pmatrix} = \begin{pmatrix} L^\dagger L & 0 \\ 0 & LL^\dagger \end{pmatrix} \quad (2)$$

which generate the supersymmetric algebra

$$[\mathcal{H}, Q_\pm] = \{Q_\pm, Q_\pm\} = 0 \ , \ \{Q_-, Q_+\} = \mathcal{H} \ . \quad (3)$$

The partner Hamiltonians H_1 and H_2 satisfy an intertwining relation, $LH_1 = H_2 L$, where in one-dimension the transformation operator $L = \frac{d}{dx} + W(x)$ is a first-order Darboux transformation expressed in terms of a superpotential $W(x)$. The intertwining relation ensures that if Ψ_1 is an eigenstate of H_1, then also $\Psi_2 = L\Psi_1$ is an eigenstate of H_2 with the same energy, unless $L\Psi_1$ vanishes or produces an unphysical state (*e.g.* non-normalizable). Consequently, as shown in Fig. 2, the SUSY partner Hamiltonians H_1 and H_2 are isospectral in the sense that their spectra consist of pair-wise degenerate levels with a possible non-degenerate single state in one sector (when the supersymmetry is exact). The wave functions of the degenerate levels are simply related in terms of L and $L^\dagger$. Such characteristic features define a supersymmetric pattern. In what follows we show [8] that a Dirac Hamiltonian with pseudospin symmetry obeys an intertwining relation and consequently gives rise to a supersymmetric pattern.

3. Dirac Hamiltonian with Central Fields

A relativistic mean field description of nuclei employs a Dirac Hamiltonian, $H = \hat{\alpha}\cdot\boldsymbol{p} + \hat{\beta}(M + V_S) + V_V$, for a nucleon of mass M moving in external scalar, V_S, and vector, V_V, potentials. When the potentials are spherically symmetric: $V_S = V_S(r)$, $V_V = V_V(r)$, the operator $\hat{K} = -\hat{\beta}\,(\boldsymbol{\sigma}\cdot\boldsymbol{\ell} + 1)$, (with $\boldsymbol{\sigma}$ the Pauli matrices and $\boldsymbol{\ell} = -i\boldsymbol{r}\times\boldsymbol{\nabla}$), commutes with H and its non-zero integer eigenvalues $\kappa = \pm(j + 1/2)$ are used to label the Dirac wave functions

$$\Psi_{\kappa,\, m} = \frac{1}{r}\begin{pmatrix} G_\kappa [Y_\ell\, \chi]^{(j)}_m \\ iF_\kappa [Y_{\ell'}\, \chi]^{(j)}_m \end{pmatrix} . \tag{4}$$

Here $G_\kappa(r)$ and $F_\kappa(r)$ are the radial wave functions of the upper and lower components respectively, Y_ℓ and χ are the spherical harmonic and spin function which are coupled to angular momentum j with projection m. The labels $\kappa = -(j + 1/2) < 0$ and $\ell' = \ell + 1$ hold for aligned spin $j = \ell + 1/2$ ($s_{1/2}, p_{3/2}$, etc.), while $\kappa = (j + 1/2) > 0$ and $\ell' = \ell - 1$ hold for unaligned spin $j = \ell - 1/2$ ($p_{1/2}, d_{3/2}$, etc.). Denoting the pair of radial wave functions by

$$\Phi_\kappa = \begin{pmatrix} G_\kappa \\ F_\kappa \end{pmatrix} , \tag{5}$$

the radial Dirac equations can be cast in Hamiltonian form,

$$H_\kappa\, \Phi_\kappa = E\, \Phi_\kappa , \tag{6}$$

with

$$H_\kappa = \begin{pmatrix} M + \Delta & -\frac{d}{dr} + \frac{\kappa}{r} \\ \frac{d}{dr} + \frac{\kappa}{r} & -(M + \Sigma) \end{pmatrix} \tag{7a}$$

$$\Delta(r) = V_S + V_V \quad , \quad \Sigma(r) = V_S - V_V . \tag{7b}$$

The nuclear single-particle spectrum is obtained from the valence bound-state solutions of Eq. (6) with positive binding energy $(M-E) > 0$ and total energy $E > 0$. The non-relativistic shell-model wave functions are identified with the upper components of the Dirac wave functions (4). For relativistic mean fields relevant to nuclei, V_S is attractive and V_V is repulsive with typical values $V_S(0) \sim -400$, $V_V(0) \sim 350$, MeV. The potentials satisfy $rV_S,\ rV_V \to 0$ for $r \to 0$, and $V_S,\ V_V \to 0$ for $r \to \infty$. Under such circumstances one can prove [7] the following properties which are relevant for understanding the nodal structure of pseudospin doublets and intruder levels in nuclei.

(a) The radial nodes of F_κ (n_F) and G_κ (n_G) are related. Specifically,

$$\begin{aligned} n_F &= n_G & \text{for } \kappa < 0\ , \\ n_F &= n_G + 1 & \text{for } \kappa > 0\ . \end{aligned} \tag{8}$$

(b) Bound states with $n_F = n_G = 0$ can occur only for $\kappa < 0$.

4. Relativistic Pseudospin Symmetry in Nuclei

A relativistic pseudospin symmetry occurs when the sum of the scalar and vector potentials is a constant

$$\Delta(r) = V_S(r) + V_V(r) = \Delta_0\ . \tag{9}$$

A Dirac Hamiltonian satisfying (9) has an invariant $SU(2)$ algebra generated by [6]

$$\hat{\tilde{S}}_\mu = \begin{pmatrix} \hat{\tilde{s}}_\mu & 0 \\ 0 & \hat{s}_\mu \end{pmatrix}\ . \tag{10}$$

Here $\hat{s}_\mu = \sigma_\mu/2$ are the usual spin operators, $\hat{\tilde{s}}_\mu = U_p \hat{s}_\mu U_p$ and $U_p = \frac{\boldsymbol{\sigma}\cdot\boldsymbol{p}}{p}$ is a momentum-helicity unitary operator [11]. In the symmetry limit the Dirac eigenfunctions belong to the spinor representation of $SU(2)$. The relativistic pseudospin symmetry determines the form of the eigenfunctions in the doublet to be

$$\Psi_{\kappa_1<0,\,m} = \frac{1}{r}\begin{pmatrix} G_{\kappa_1}[\,Y_{\tilde{\ell}-1}\,\chi\,]^{(j)}_m \\ iF_{\kappa_1}[\,Y_{\tilde{\ell}}\,\chi\,]^{(j)}_m \end{pmatrix} \quad \kappa_1 = -\tilde{\ell} < 0\ ,\ j = \tilde{\ell} - 1/2\ , \tag{11a}$$

$$\Psi_{\kappa_2>0,\,m} = \frac{1}{r}\begin{pmatrix} G_{\kappa_2}[\,Y_{\tilde{\ell}+1}\,\chi\,]^{(j')}_m \\ iF_{\kappa_2}[\,Y_{\tilde{\ell}}\,\chi\,]^{(j')}_m \end{pmatrix} \quad \kappa_2 = \tilde{\ell} + 1 > 0\ ,\ j' = \tilde{\ell} + 1/2\ , \tag{11b}$$

and imposes the following conditions on their radial amplitudes:

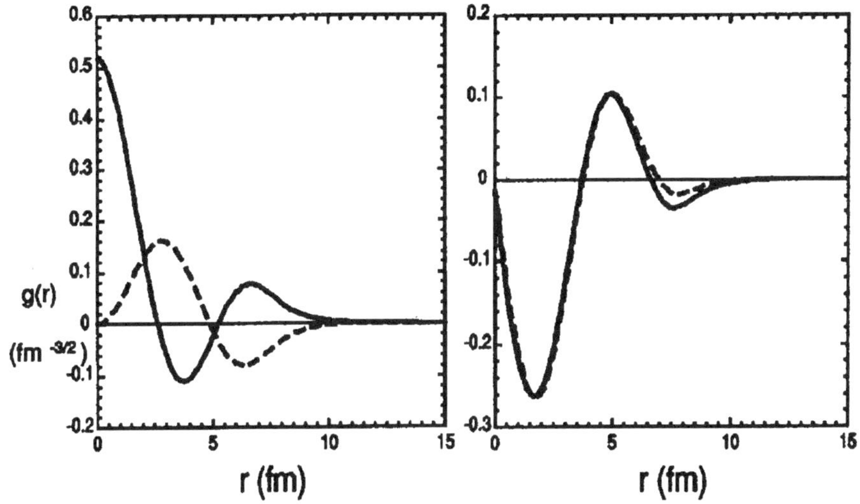

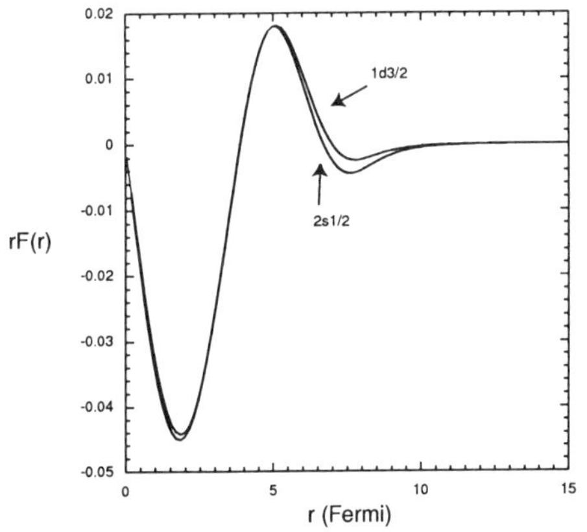

Figure 3. Top left panel: the upper components $g(r) = rG_\kappa(r)$ of the $2s_{1/2}$, (solid line) and $1d_{3/2}$ (dashed line) Dirac eigenfunctions in ^{208}Pb. Top right panel: testing the differential relation of Eq. (12b) for the upper components of $2s_{1/2}$ ($\kappa_1 = -1$) and $1d_{3/2}$ ($\kappa_2 = 2$). Bottom panel: the lower components $f(r) = rF_\kappa(r)$ of $2s_{1/2}$ and $1d_{3/2}$, testing relation (12a). Based on calculations in [12,13].

$$F_{\kappa_1} = F_{\kappa_2} \,, \tag{12a}$$

$$\frac{dG_{\kappa_1}}{dr} + \frac{\kappa_1}{r}G_{\kappa_1} = \frac{dG_{\kappa_2}}{dr} + \frac{\kappa_2}{r}G_{\kappa_2} \,. \tag{12b}$$

The two eigenstates in the doublet are connected by the pseudospin generators $\hat{\tilde{S}}_\mu$ (10). The lower components are connected by the usual spin operators and, therefore, have the same spatial wave functions. Consequently, the two states of the doublet share a common $\tilde{\ell}$ which is the orbital angular momentum of the lower component. The Dirac structure then ensures that the orbital angular momentum of the upper components in Eq. (11) is $\ell = \tilde{\ell} - 1$ for $j = \tilde{\ell} - 1/2 = \ell + 1/2$, and $\ell + 2 = \tilde{\ell} + 1$ for $j' = \tilde{\ell} + 1/2 = \ell + 3/2$.

This explains the particular angular momenta defining the pseudospin doublets in Eq. (1). The radial amplitudes of the lower components are equal (12a) and, in particular, have the same number of nodes $n_F = n$. Property (a) of the previous section then ensures that G_{κ_1} has n nodes and G_{κ_2} has $n-1$ nodes, in agreement with Eq. (1). Property (b) ensures that the Dirac state with $n_F = n_G = 0$, corresponding to the "intruder" shell-model state, has a wave function as in Eq. (11a) with $\kappa < 0$, and does not have a partner eigenstate (with $\kappa > 0$).

Realistic mean fields in nuclei approximately satisfy condition (9) with $\Delta_0 \approx 0$. The required breaking of pseudospin symmetry in nuclei is small. Quasi-degenerate doublets of normal-parity states and abnormal-parity levels without a partner eigenstate persist in the spectra. The relations (12) between wave functions have been tested in numerous realistic mean field calculations in a variety of nuclei, and were found to be obeyed to a good approximation, especially for doublets near the Fermi surface [12,13]. A representative example for neutrons in ^{208}Pb is shown in Fig. 3.

5. Relativistic Pseudospin Symmetry and SUSY

In the pseudospin limit, Eq. (9), the two Dirac states $\Psi_{\kappa_1<0,m}$ and $\Psi_{\kappa_2>0,m}$ of Eq. (11) with $\kappa_1 + \kappa_2 = 1$ are degenerate, unless both the upper and lower components have no nodes, in which case only $\Psi_{\kappa_1<0,m}$ is a bound state. Altogether, as shown in Fig. 4, the ensemble of Dirac states with $\kappa_1+\kappa_2 = 1$ exhibits a supersymmetric pattern of twin towers with pair-wise degenerate pseudospin doublets sharing a common $\tilde{\ell}$, and an additional non-degenerate nodeless state at the bottom of the $\kappa_1 < 0$ tower. An exception to this rule is the tower with $\kappa_2 = 1$ ($p_{1/2}$ states with $\tilde{\ell} = 0$), which is on its own, because states with $\kappa_1 = 0$ do not exist. The supersymmetric structure arises because in the pseudospin limit the respective radial Dirac Hamiltonians, H_{κ_1} and H_{κ_2}, satisfy an intertwining relation of the form

$$LH_{\kappa_1} = H_{\kappa_2}L \ , \tag{13}$$

with $\kappa_1 + \kappa_2 = 1$. The transformation operator is found to be

$$L = b\begin{pmatrix} 0 & \frac{d}{dr} - \frac{\kappa_2}{r} \\ -\frac{d}{dr} - \frac{\kappa_1}{r} & (2M + \Sigma + \Delta_0) \end{pmatrix} . \tag{14}$$

L connects the two doublet states

$$L\,\Phi_{\kappa_1} = b(M + \Delta_0 - E)\,\Phi_{\kappa_2} \qquad (\kappa_1 + \kappa_2 = 1) \ , \tag{15}$$

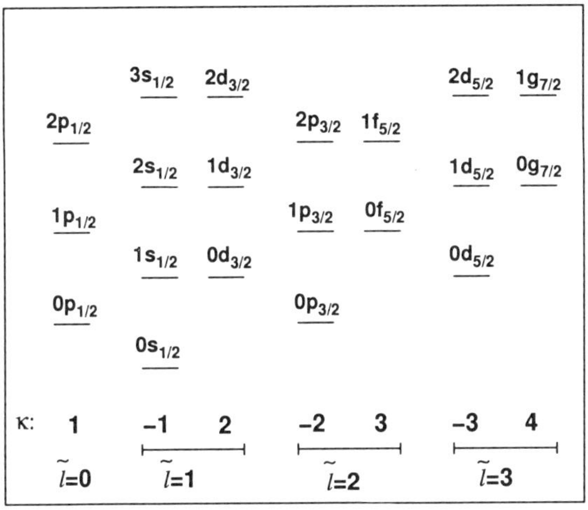

Figure 4. Schematic supersymmetric pattern in the pseudospin limit of the Dirac Hamiltonian [8].

and identifies the two states as supersymmetric partners. Eq. (15) relies on the input that the two states are eigenstates of the Dirac Hamiltonian with $E_{\kappa_1} = E_{\kappa_2} = E$ and their wave functions satisfy the relations in Eq. (12). The fact that nuclear wave functions, obtained in realistic mean-field calculations, obey these relations to a good approximation, confirms the relevance of supersymmetry to these nuclear states.

Constructing supersymmetric charges $Q_\pm$ and Hamiltonian $\mathcal{H}$ from L and H_{κ_1}, H_{κ_2} as in Eq. (2), ensures the fulfillment of Eq. (3), except for the last relation which now reads

$$\{Q_-, Q_+\} = b^2[\mathcal{H} - (M + \Delta_0)][\mathcal{H} - (M + \Delta_0)] \ . \tag{16}$$

Eq. (16) involves a polynomial of $\mathcal{H}$, indicating a quadratic deformation of the conventional supersymmetric algebra [14]. The latter arises because both the Dirac Hamiltonian, H_κ, and the transformation operator, L, are of first order. In real nuclei, the relativistic pseudospin symmetry is slightly broken, implying $\Delta(r) \neq \Delta_0$ in Eq. (9). Taking H_κ as in Eq. (7) and L as in Eq. (14) but with $\Delta_0 \to \Delta(r)$, we now find that

$$LH_{\kappa_1} - H_{\kappa_2}L = i\, b\, \frac{d\Delta}{dr}\, \sigma_2 \tag{17}$$

Furthermore, $\{Q_-, Q_+\}$ has the same formal form as in Eq. (16), but the appearance of $\Delta(r)$ instead of Δ_0 implies that the anticommutator is no longer just a polynomial of $\mathcal{H}$.

6. Summary

We discussed a possible grouping of shell-model single-particle states into larger multiplets, exhibiting a supersymmetric pattern. The multiplets involve several quasi-degenerate pseudospin doublets and intruder levels without a partner eigenstate. In contrast to previous studies of pseudospin in nuclei, the suggested grouping of nuclear states treats the intruder levels and pseudospin doublets on equal footing. The underlying supersymmetric structure is linked with an approximate relativistic pseudospin symmetry of the nuclear mean-field Dirac Hamiltonian. The relativistic pseudospin symmetry imposes relations between the upper and lower components of the two Dirac states forming the doublet. These relations, which are obeyed to a good approximation by realistic mean-field wave functions, imply that the Dirac Hamiltonian with pseudospin symmetry obeys an intertwining relation which gives rise to the indicated supersymmetric pattern.

Acknowledgments

This work was partly supported by the U.S.-Israel Binational Science Foundation, in collaboration with J.N. Ginocchio (LANL), and partly by the Israel Science Foundation.

References

1. K.T. Hecht and A. Adler, *Nucl. Phys.* **A137**, 129 (1969); A. Arima, M. Harvey and K. Shimizu, *Phys. Lett.* **B30**, 517 (1969).
2. A. Bohr, I. Hamamoto and B.R. Mottelson, *Phys. Scr.* **26**, 267 (1982).
3. J. Dudek *et al.*, *Phys. Rev. Lett.* **59**, 1405 (1987).
4. W. Nazarewicz *et al.*, *Phys. Rev. Lett.* **64**, 1654 (1990); F.S. Stephens at al., *Phys. Rev. Lett.* **65**, 301 (1990).
5. J.N. Ginocchio, *Phys. Rev. Lett.* **78**, 436 (1997).
6. J.N. Ginocchio and A. Leviatan, *Phys. Lett.* **B425**, 1 (1998).
7. A. Leviatan and J.N. Ginocchio, *Phys. Lett.* **B518**, 214 (2001).
8. A, Leviatan, *Phys. Rev. Lett.* **92**, 202501 (2004).
9. E. Witten, *Nucl. Phys.* **B188**, 513 (1981).
10. G. Junker, *Supersymmetric Methods in Quantum and Statistical Physics*, (Springer-Verlag, Berlin, Heidelberg 1996).
11. A. L. Blokhin, C. Bahri, and J.P. Draayer, *Phys. Rev. Lett.* **74**, 4149 (1995).
12. J.N. Ginocchio and D.G. Madland, Phys. Rev. C **57**, 1167 (1998); G.A. Lalazissis *et al.*, *ibid.* **58**, R45 (1998); J. Meng *et al.*, *ibid.* **59**, 154 (1999).
13. J.N. Ginocchio and A. Leviatan, Phys. Rev. Lett. **87**, 072502 (2001); J.N. Ginocchio, Phys. Rev. C **66**, 064312 (2002).
14. N. Debergh *et al.*, J. Phys. A **35**, 3279 (2002); L.M. Nieto *et al.*, Ann. Phys. **305**, 151 (2003).

KEY TOPICS IN NUCLEAR STRUCTURE
Proceedings of the 8th Int. Spring Seminar on Nuclear Physics
edited by Aldo Covello
© 2005 World Scientific Publishing Co.

PSEUDOSPIN SYMMETRY IN SPHERICAL AND DEFORMED NUCLEI *

JOSEPH N. GINOCCHIO
Theoretical Division,
MS B 283, Los Alamos National Laboratory
Los Alamos,NM, 87545, USA
E-mail: gino@lanl.gov

Pseudospin symmetry is a relativistic symmetry of the Dirac Hamiltonian. We show that the eigenfunctions of realistic relativistic nuclear mean fields approximately conserve pseudospin symmetry.

1. Introduction

Pseudospin doublets were introduced more than thirty years ago into nuclear physics to accommodate an observed near degeneracy of certain normal parity shell model orbitals with non-relativistic quantum numbers (n_r, ℓ, $j = \ell + 1/2$) and ($n_r - 1, \ell + 2$, $j = \ell + 3/2$) where n_r, ℓ, and j are the single-nucleon radial, orbital, and total angular momentum quantum numbers, respectively [1,2]. The doublet structure is expressed in terms of a "pseudo" orbital angular momentum, which is an average of the orbital angular momentum of the two orbits in the doublet, $\tilde{\ell} = \ell + 1$, coupled to a "pseudo" spin, $\tilde{s} = 1/2$ with $j = \tilde{\ell} \pm \tilde{s}$. For example, the shell model orbitals $(n_r s_{1/2}, (n_r - 1)d_{3/2})$ will have $\tilde{\ell} = 1$, $(n_r p_{3/2}, (n_r - 1)f_{5/2})$ will have $\tilde{\ell} = 2$, for the two states in the doublet. Then the single-particle energy is approximately independent of the orientation of the pseudospin leading to an approximate pseudospin symmetry. These doublets persist for deformed nuclei as well [3]. The axially-symmetric deformed single-particle orbits with non-relativistic asymptotic quantum numbers $[N, n_3, \Lambda]\Omega = \Lambda + 1/2$ and $[N, n_3, \Lambda' = \Lambda + 2]\Omega' = \Lambda + 3/2$ are quasi-degenerate. Here N is the total harmonic oscillator quantum number, n_3 is the number of quanta for oscil-

*This work is supported by the United States Department of Energy under contract W-7405-ENG-36 lations along the symmetry axis, taken to be in the z-direction, Λ and Ω are respectively the components of the orbital and total angular momentum projected along the symmetry axis [4]. In this case, the doublet structure is expressed in terms of a "pseudo" orbital angular momentum projection, $\tilde{\Lambda} = \Lambda + 1$, which is added to a "pseudo" spin projection, $\tilde{\mu} = \pm 1/2$ to yield the above mentioned doublet of states with $\Omega = \tilde{\Lambda} - 1/2$ and $\Omega' = \tilde{\Lambda} + 1/2$. This approximate pseudospin "symmetry" has been used to explain features of deformed nuclei, including superdeformation [5] and identical bands [6,7,8,9] as well.

Although there have been attempts to understand the origin of this "symmetry" [10,11], only recently has it been shown to arise from a relativistic symmetry of the Dirac Hamiltonian [12,13] which we review in Section 2. This relativistic symmetry implies conditions on the Dirac eigenfunctions [14] which we discuss in Section 3. These relationships have been studied extensively for spherical nuclei [14,15,16,17,18,19] and for deformed nuclei [20,21,22,23] and we shall review them in Section 4.

2. The Dirac Hamiltonian and Pseudospin Symmetry

The Dirac Hamiltonian, H, with an external scalar, $V_S(\vec{r})$, and vector, $V_V(\vec{r})$, potentials is given by:

$$H = \hat{\boldsymbol{\alpha}} \cdot \boldsymbol{p} + \hat{\beta}\,(\,M + V_S(\vec{r})\,) + V_V(\vec{r}) \ , \tag{1}$$

where $\hat{\boldsymbol{\alpha}}$, $\hat{\beta}$ are the usual Dirac matrices, M is the nucleon mass, and we have set $\hbar = c = 1$. The Dirac Hamiltonian is invariant under a SU(2) algebra for two limits: $V_S(\vec{r}) = V_V(\vec{r}) + C_s$ and $V_S(\vec{r}) = -V_V(\vec{r}) + C_{ps}$ where C_s, C_{ps} are constants [24]. The former limit has application to the spectrum of mesons for which the spin-orbit splitting is small [25] and for the spectrum of an antinucleon in the mean-field of nucleons [26,27]. The latter limit leads to pseudospin symmetry in nuclei [12]. This symmetry occurs independent of the shape of the nucleus: spherical, axial deformed, or triaxial.

3. Pseudospin Symmetry Generators

The generators for the pseudospin SU(2) algebra, $\tilde{S}_i$ $(i = x, y, z)$, which commute with the Dirac Hamiltonian, $[\,H\,,\,\tilde{S}_i\,] = 0$, for the pseudospin symmetry limit $V_S(\vec{r}) = -V_V(\vec{r}) + C_{ps}$, are given by [13]

$$\tilde{S}_i = \begin{pmatrix} \tilde{s}_i & 0 \\ 0 & s_i \end{pmatrix} = \begin{pmatrix} U_p\, s_i\, U_p & 0 \\ 0 & s_i \end{pmatrix} \tag{2}$$

where $s_i = \sigma_i/2$ are the usual spin generators, σ_i the Pauli matrices, and $U_p = \frac{\boldsymbol{\sigma} \cdot \boldsymbol{p}}{p}$ is the momentum-helicity unitary operator [11]. Thus the operators $\tilde{S}_i$ generate an SU(2) invariant symmetry of H_{ps}. Therefore, each eigenstate of the Dirac Hamiltonian has a partner with the same energy,

$$H_{ps}\, \Phi^{ps}_{\tilde{k},\tilde{\mu}}(\vec{r}) = E_{\tilde{k}}\, \Phi^{ps}_{\tilde{k},\tilde{\mu}}(\vec{r}) \tag{3}$$

where $\tilde{k}$ are the other quantum numbers and $\tilde{\mu} = \pm\frac{1}{2}$ is the eigenvalue of $\tilde{S}_z$,

$$\tilde{S}_z\, \Phi^{ps}_{\tilde{k},\tilde{\mu}}(\vec{r}) = \tilde{\mu}\, \Phi^{ps}_{\tilde{k},\tilde{\mu}}(\vec{r})\ . \tag{4}$$

The eigenstates in the doublet will be connected by the generators $\tilde{S}_{\pm} = \tilde{S}_x \pm i\tilde{S}_y$,

$$\tilde{S}_{\pm}\, \Phi^{ps}_{\tilde{k},\tilde{\mu}}(\vec{r}) = \sqrt{\left(\frac{1}{2} \mp \tilde{\mu}\right)\left(\frac{3}{2} \pm \tilde{\mu}\right)}\, \Phi^{ps}_{\tilde{k},\tilde{\mu}\pm 1}(\vec{r})\ . \tag{5}$$

The fact that Dirac eigenfunctions belong to the spinor representation of the pseudospin SU(2), as given in Eqs. (4)-(5), leads to the conditions on the corresponding Dirac amplitudes that are reviewed in the next Section.

4. Pseudospin Symmetry for Axially Deformed and Spherical Nuclei

The axial-symmetry of the potentials determines the ϕ-dependence of the Dirac wave functions, leading to the following form for the relativistic eigenstates [14]

$$\Phi_{\tilde{\eta},\tilde{\Lambda},\tilde{\mu},\Omega}(\vec{r}) = \begin{pmatrix} g^{+}_{\tilde{\eta},\tilde{\Lambda},\tilde{\mu}}(\rho, z)\, e^{i(\tilde{\Lambda}+\tilde{\mu}-\frac{1}{2})\phi} \\ g^{-}_{\tilde{\eta},\tilde{\Lambda},\tilde{\mu}}(\rho, z)\, e^{i(\tilde{\Lambda}+\tilde{\mu}+\frac{1}{2})\phi} \\ i f_{\tilde{\eta},\tilde{\Lambda}}(\rho, z)\, e^{i(\tilde{\Lambda}+\tilde{\mu}-\frac{1}{2})\phi} \\ i f_{\tilde{\eta},\tilde{\Lambda}}(\rho, z)\, e^{i(\tilde{\Lambda}+\tilde{\mu}+\frac{1}{2})\phi} \end{pmatrix}, \quad \Omega = \tilde{\Lambda} + \tilde{\mu}\ . \tag{6}$$

The first entry in the Dirac four vector is the spin up upper amplitude, the second entry is the spin down upper amplitude, the third entry is the spin up lower amplitude, and the fourth entry is the spin down lower amplitude. The generic label $\tilde{\eta}$ in $\Phi_{\tilde{\eta},\tilde{\Lambda},\tilde{\mu},\Omega}(\vec{r})$ replaces the harmonic oscillator labels N and n_3, which are not conserved for realistic axially-deformed potentials in nuclei.

Pseudospin symmetry as expressed in Eqs. (4)-(5) lead to the conditions

$$f^{+}_{\tilde{\eta},\tilde{\Lambda},-\frac{1}{2}}(\rho,z) = f^{-}_{\tilde{\eta},\tilde{\Lambda},\frac{1}{2}}(\rho,z) = 0 \ , \tag{7a}$$

$$f^{+}_{\tilde{\eta},\tilde{\Lambda},\frac{1}{2}}(\rho,z) = f^{-}_{\tilde{\eta},\tilde{\Lambda},-\frac{1}{2}}(\rho,z) \equiv f_{\tilde{\eta},\tilde{\Lambda}}(\rho,z) \ , \tag{7b}$$

$$g^{+}_{\tilde{\eta},\tilde{\Lambda},\frac{1}{2}}(\rho,z) = -g^{-}_{\tilde{\eta},\tilde{\Lambda},-\frac{1}{2}}(\rho,z) \equiv g_{\tilde{\eta},\tilde{\Lambda}}(\rho,z) \ , \tag{7c}$$

and the differential relations

$$\left(\frac{\partial}{\partial\rho} + \frac{\tilde{\Lambda}+1}{\rho}\right) g^{-}_{\tilde{\eta},\tilde{\Lambda},\frac{1}{2}}(\rho,z) = \left(\frac{\partial}{\partial\rho} - \frac{\tilde{\Lambda}-1}{\rho}\right) g^{+}_{\tilde{\eta},\tilde{\Lambda},-\frac{1}{2}}(\rho,z) \ , \tag{8a}$$

$$\frac{\partial}{\partial z}\, g^{\pm}_{\tilde{\eta},\tilde{\Lambda},\mp\frac{1}{2}}(\rho,z) = \pm\left(\frac{\partial}{\partial\rho} \pm \frac{\tilde{\Lambda}}{\rho}\right) g^{\pm}_{\tilde{\eta},\tilde{\Lambda},\pm\frac{1}{2}}(\rho,z) \ . \tag{8b}$$

In Fig. 1 we show an example how well these conditions are satisfied [23] from which we can draw a number of conclusions.

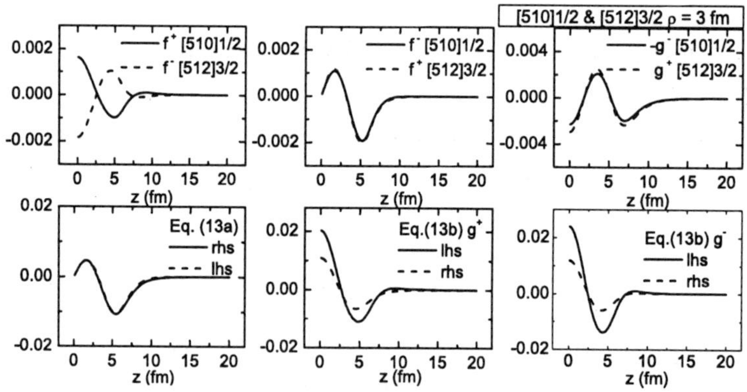

Figure 1. Eigenfunctions in $(\text{fm})^{-3/2}$ as a function of z for $\rho = 3$ fm for the neutron pseudospin doublet [510]1/2 and [512]3/2 ($\tilde{\Lambda} = 1$) in ^{168}Er. In each segment, the top row shows (from left to right) the relations in (i) Eq. (7a), involving $f^{+}_{\tilde{\eta},\tilde{\Lambda},-1/2}$ and $f^{-}_{\tilde{\eta},\tilde{\Lambda},1/2}$, (ii) Eq. (7b), involving $f^{-}_{\tilde{\eta},\tilde{\Lambda},-1/2}$ and $f^{+}_{\tilde{\eta},\tilde{\Lambda},1/2}$, (iii) Eq. (7c), involving $g^{+}_{\tilde{\eta},\tilde{\Lambda},1/2}$ and $-g^{-}_{\tilde{\eta},\tilde{\Lambda},-1/2}$. The bottom row shows (from left to right) the lhs and rhs of (i) Eq. (8a), involving $g^{-}_{\tilde{\eta},\tilde{\Lambda},\frac{1}{2}}$ and $g^{+}_{\tilde{\eta},\tilde{\Lambda},-\frac{1}{2}}$, (ii) Eq. (8b), involving $g^{+}_{\tilde{\eta},\tilde{\Lambda},-\frac{1}{2}}$ and $g^{+}_{\tilde{\eta},\tilde{\Lambda},+\frac{1}{2}}$, (iii) Eq. (8b), involving $g^{-}_{\tilde{\eta},\tilde{\Lambda},+\frac{1}{2}}$ and $g^{-}_{\tilde{\eta},\tilde{\Lambda},-\frac{1}{2}}$.

First, while the amplitudes $f^{+}_{\tilde\eta,\tilde\Lambda,-\frac{1}{2}}(\rho,z)$, $f^{-}_{\tilde\eta,\tilde\Lambda,\frac{1}{2}}(\rho,z)$ are not zero as predicted by Eq. (7a), they are much smaller than $f^{-}_{\tilde\eta,\tilde\Lambda,-\frac{1}{2}}(\rho,z)$, $f^{+}_{\tilde\eta,\tilde\Lambda,\frac{1}{2}}(\rho,z)$. Furthermore, $f^{+}_{\tilde\eta,\tilde\Lambda,\frac{1}{2}}(\rho,z)$ and $f^{-}_{\tilde\eta,\tilde\Lambda,-\frac{1}{2}}(\rho,z)$ have similar shapes as predicted by Eq. (7b). The amplitude $-g^{-}_{\tilde\eta,\tilde\Lambda,-\frac{1}{2}}(\rho,z)$ has the same shape as the amplitude $g^{+}_{\tilde\eta,\tilde\Lambda,\frac{1}{2}}(\rho,z)$, in line with the prediction of Eq. (7c), but they differ in magnitude. These amplitudes are much smaller than the other upper amplitudes, $g^{\pm}_{\tilde\eta,\tilde\Lambda,\mp\frac{1}{2}}(\rho,z)$.

The differential relation in Eq. (8a) between the dominant upper components, $g^{-}_{\tilde\eta,\tilde\Lambda,\frac{1}{2}}(\rho,z)$ and $g^{+}_{\tilde\eta,\tilde\Lambda,-\frac{1}{2}}(\rho,z)$, is well obeyed in all cases. The differential relations in Eq. (8b) relate the dominant upper components, $g^{\pm}_{\tilde\eta,\tilde\Lambda,\mp\frac{1}{2}}(\rho,z)$ to the small upper components $g^{\pm}_{\tilde\eta,\tilde\Lambda,\pm\frac{1}{2}}(\rho,z)$. The shapes of the left-hand-side and of the right-hand-side of Eq. (8b) are the same, but the corresponding amplitudes are quite different. Therefore, the differential relations in Eq. (8b) are less satisfied. These differences might partly originate from the differences in the magnitudes of the small upper components in Eq. (7c).

For spherical nuclei the Dirac eigenfunctions in the pseudospin limit can be written in the two row form

$$\Psi^{ps}_{\tilde n_r,\tilde\ell,j,M}(\vec r) = \begin{pmatrix} g_{\tilde n_r,\tilde\ell,j}(r)\ [Y^{(\tilde\ell_j)}(\theta,\phi)\ \chi]^{(j)}_M \\ i f_{\tilde n_r,\tilde\ell}(r)\ [Y^{(\tilde\ell)}(\theta,\phi)\ \chi]^{(j)}_M \end{pmatrix}. \tag{9}$$

where χ is the spin function and $\tilde\ell_j = \tilde\ell - 1$ for $j = \tilde\ell - \frac{1}{2}$, $\tilde\ell_j = \tilde\ell + 1$ for $j = \tilde\ell + \frac{1}{2}$ for the two states in the doublet. The lower amplitude is the same for the two states in the doublet [12] and indeed they have been shown to be very similar [15].

On the other hand the upper amplitudes are related by a first order differential equation [14]

$$\left(\frac{\partial}{\partial r} + \frac{\tilde\ell + 2}{r}\right) g_{\tilde n_r,\tilde\ell,\tilde\ell+\frac{1}{2}}(r) = \left(\frac{\partial}{\partial r} - \frac{\tilde\ell - 1}{r}\right) g_{\tilde n_r,\tilde\ell,\tilde\ell-\frac{1}{2}}(r). \tag{10}$$

In Fig. 2 we show one example of how well these conditions are satisfied. Although the upper amplitudes are very different in shape having different radial nodes the differential relations between the two upper amplitudes is well satisfied.

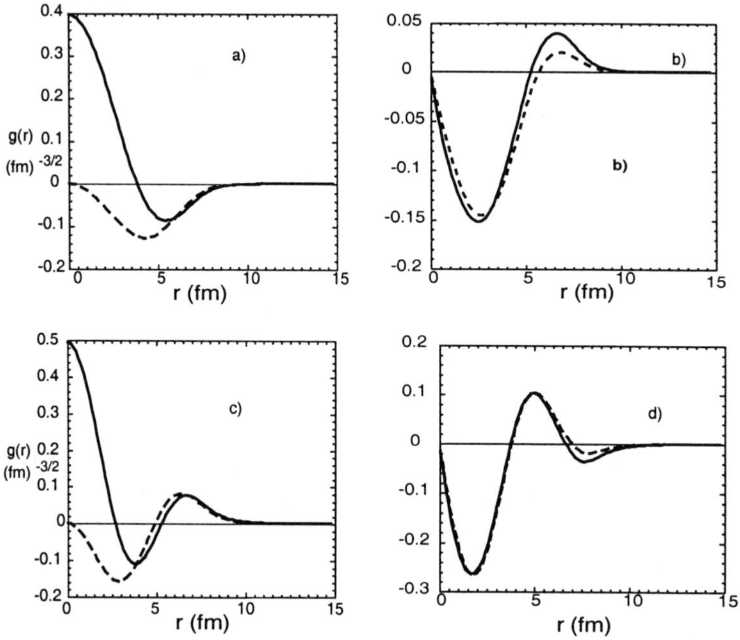

Figure 2. a) The upper amplitude $g(r)$ for the $1s_{\frac{1}{2}}$ (solid line) and $0d_{\frac{3}{2}}$ (dashed line) eigenfunctions, b) the differential equation on the right hand side (RHS) of Equation (10) with $\tilde{\ell} = 1$ for the $1s_{\frac{1}{2}}$ (solid line) eigenfunction and the differential equation on the left hand side (LHS) of Equation (10) with $\tilde{\ell} = 1$ for the $0d_{\frac{3}{2}}$ (dashed line) eigenfunction, c) the upper amplitude $g(r)$ for the $2s_{\frac{1}{2}}$ (solid line) and $1d_{\frac{3}{2}}$ (dashed line) eigenfunctions, and d) the differential equation on the RHS of Equation (10) with $\tilde{\ell} = 1$ for the $2s_{\frac{1}{2}}$ eigenfunction (solid line) and the differential equation on the LHS of Equation (10) with $\tilde{\ell} = 1$ for the $1d_{\frac{1}{2}}$ (dashed line) eigenfunctions.

5. Summary

We have reviewed the conditions that pseudospin symmetry places on the Dirac eigenfunctions and found that pseudospin symmetry is well conserved in these eigenfunctions. The pseudospin symmetry improves as the binding energy and pseudo-orbital angular momentum decrease for both spherical and deformed nuclei.

6. Future

Pseudospin symmetry and charge conjugation together predict that an anti-nucleon in nuclear matter has spin symmetry [26]. Anti-nucleon scattering from nuclei produces zero polarization and thus spin symmetry is confirmed for the limited data available [28]. More spin polarized anti-nucleon scattering data is needed to study anti-nucleon spin symmetry as a function of energy and scattering angle.

References

1. A. Arima, M. Harvey and K. Shimizu, Phys. Lett. B **30**, 517 (1969).
2. K.T. Hecht and A. Adler, Nucl. Phys. A **137**, 129 (1969).
3. A. Bohr, I. Hamamoto and B. R. Mottelson, Phys. Scr. **26**, 267 (1982).
4. A. Bohr and B. R. Mottelson, *Nuclear Structure*, Vol. **II** (W. A. Benjamin, Reading, Ma., 1975).
5. J. Dudek, W. Nazarewicz, Z. Szymanski and G. A. Leander, Phys. Rev. Lett. **59**, 1405 (1987).
6. W. Nazarewicz, P. J. Twin, P. Fallon and J.D. Garrett, Phys. Rev. Lett. **64**, 1654 (1990).
7. F. S. Stephens *et al*, Phys. Rev. Lett. **65**, 301 (1990); F. S. Stephens *et al*, Phys. Rev. C **57**, R1565 (1998).
8. J. Y. Zeng, J. Meng, C. S. Wu, E. G. Zhao, Z. Xing and X. Q. Chen, Phys. Rev. C **44**, R1745 (1991).
9. A.M. Bruce *et. al.*, Phys. Rev. C **56**, 1438 (1997).
10. C. Bahri, J. P. Draayer, and S. A. Moszkowski, Phys. Rev. Lett. **68**, 2133 (1992).
11. A. L. Blokhin, C. Bahri, and J. P. Draayer, Phys. Rev. Lett. **74**, 4149 (1995).
12. J.N. Ginocchio, Phys. Rev. Lett. **78**, 436 (1997).
13. J. N. Ginocchio and A. Leviatan, Phys. Lett. B **425**, 1 (1998).
14. J.N. Ginocchio, Phys. Rev. C **66**, 064312 (2002).
15. J. N. Ginocchio and D. G. Madland, Phys. Rev. C **57**, 1167 (1998).
16. G.A. Lalazissis, Y.K. Gambhir, J.P. Maharana, C.S. Warke and P. Ring, Phys. Rev. C **58**, R45 (1998).
17. J. Meng, K. Sugawara-Tanabe, S. Yamaji, P. Ring and A. Arima, Phys. Rev. C **58**, R628 (1998); J. Meng, K. Sugawara-Tanabe, S. Yamaji and A. Arima, Phys. Rev. C **59**, 154 (1999).
18. J.N. Ginocchio and A. Leviatan, Phys. Rev. Lett. **87**, 072502 (2001).
19. P.J. Borycki, J. Ginocchio, W. Nazarewicz, and M. Stoitsov, Phys. Rev. C **68**, 014304 (2003).
20. K. Sugawara-Tanabe and A. Arima, Phys. Rev. C **58**, R3065 (1998).
21. K. Sugawara-Tanabe, S. Yamaji, and A. Arima, Phys. Rev. C **62**, 054307 (2000).
22. K. Sugawara-Tanabe, S. Yamaji, and A. Arima, Phys. Rev. C **65**, 054313 (2002).

23. J. N. Ginocchio, A. Leviatan, J. Meng, and Shan-Gui Zhou, Phys. Rev. C **69**, 034303 (2004).
24. J. S. Bell and H. Ruegg, Nucl. Phys. B **98**, 151 (1975).
25. P.R. Page, T. Goldman, and J. N. Ginocchio, Phys. Rev. Lett. **86**, 204 (2001).
26. J. N. Ginocchio, Phys. Rep. **315**, 231 (1999).
27. J. N. Ginocchio, Phys.Rev. C **69**, 034318 (2004).
28. D. Garetta *et. al.*, Physics Letters B **151**, 473 (1985).

SECTION III

THE ROLE OF SHELL MODEL IN THE UNDERSTANDING OF NUCLEAR STRUCTURE

KEY TOPICS IN NUCLEAR STRUCTURE
Proceedings of the 8th Int. Spring Seminar on Nuclear Physics
edited by Aldo Covello
© 2005 World Scientific Publishing Co.

NUCLEAR STRUCTURE CALCULATIONS WITH MODERN NUCLEON-NUCLEON POTENTIALS

A. COVELLO, L. CORAGGIO, A. GARGANO AND N. ITACO
Dipartimento di Scienze Fisiche, Università di Napoli Federico II,
and Istituto Nazionale di Fisica Nucleare,
Complesso Universitario di Monte S. Angelo, Via Cintia, 80126 Napoli, Italy
E-mail: covello@na.infn.it

We have performed shell-model calculations starting from modern nucleon-nucleon potentials V_{NN}. We make use of a new approach to the renormalization of the short-range repulsion of V_{NN} in which a low-momentum potential $V_{\rm low-k}$ is derived by integrating out the high-momentum components of V_{NN} down to a cutoff momentum Λ. We present some results for nuclei around closed shells which have been obtained starting from the CD-Bonn potential. We have also performed calculations making use of different modern NN potentials. Comparison of the results obtained shows that they are only slightly dependent on the kind of potential used as input. The effects of changes in Λ are explored.

1. Introduction

A fundamental problem of nuclear physics is to understand the properties of nuclei starting from the forces among nucleons. Within the framework of the shell model, which is the basic approach to nuclear structure calculations in terms of nucleons, this problem implies the derivation of the two-body effective interaction $V_{\rm eff}$ from the free nucleon-nucleon (NN) potential V_{NN}.

A main difficulty encountered in the derivation of $V_{\rm eff}$ from any modern NN potential is the existence of a strong repulsive core which prevents its direct use in nuclear structure calculations. This difficulty is usually overcome by resorting to the well-known Brueckner G-matrix method. It should be recalled, however, that this method is somewhat involved, especially as regards the Pauli blocking dependence.

The idea of finding a more convenient way to handle this problem is not new. In fact, in the late 1960s a method was developed[1] for deriving directly from the observed nucleon-nucleon phase shifts a set of matrix elements of V_{NN} in oscillator wave functions. This resulted in the so called Sussex interaction which has been used in several nucler structure calculations. Recently, a completely new approach has been proposed[2,3] which consists in deriving from V_{NN} a low-momentum potential $V_{\rm low-k}$ defined within a given cutoff momentum Λ. This is a smooth potential which can be used directly to derive the shell-model effective interaction.

To assess the practical value of the $V_{\rm low-k}$ approach, we have performed shell-model calculations by using both this method and the traditional G-matrix one.[2,4,5] Comparison of the $V_{\rm low-k}$ and G-matrix results between them and with experimental data shows that the former are as good as or somewhat better than the latter.

The aim of this paper is twofold. Firstly, we report on our recent study[6,7] of nuclei in the regions of shell closures off stability. In this study, we have used as initial input the CD-Bonn potential[8] and derived $V_{\rm eff}$ within the framework of the $V_{\rm low-k}$ approach. Secondly, we discuss some preliminary results of a study performed with three different phase-shift equivalent potentials, Nijmegen II,[9] Argonne v_{18} (AV18)[10] and CD-Bonn.

Our presentation is organized as follows. In Sec. 2 we give an outline of the theoretical framework in which our shell-model calculations have been performed. In Sec. 3 we present some results for nuclei in the close vicinity to ^{100}Sn and ^{132}Sn, focusing attention on particle-hole multiplets. In Sec. 4 we compare the results obtained for the nucleus ^{130}Sn with the various potentials considered in this study. Sec. 5 presents some concluding remarks.

2. Outline of theoretical framework

As mentioned in the Introduction, in our shell-model calculations we have made use of realistic effective interactions derived from modern high-quality NN potentials which all fit very accurately (χ^2/datum ≈ 1) the NN scattering data below 350 MeV.[11]

The shell-model effective interaction $V_{\rm eff}$ is defined, as usual, in the following way. In principle, one should solve a nuclear many-body Schrödinger equation of the form

$$H\Psi_i = E_i\Psi_i, \tag{1}$$

with $H = T + V_{NN}$, where T denotes the kinetic energy. This full-space many-body problem is reduced to a smaller model-space problem of the form

$$PH_{\rm eff}P\Psi_i = P(H_0 + V_{\rm eff})P\Psi_i = E_iP\Psi_i. \tag{2}$$

Here $H_0 = T + U$ is the unperturbed Hamiltonian, U being an auxiliary potential introduced to define a convenient single-particle basis, and P denotes the projection operator onto the chosen model space,

$$P = \sum_{i=1}^{d} |\psi_i\rangle\langle\psi_i|, \quad (3)$$

d being the dimension of the model space and $|\psi_i\rangle$ the eigenfunctions of H_0. The effective interaction $V_{\rm eff}$ operates only within the model space P. In operator form it can be schematically written[12] as

$$V_{\rm eff} = \hat{Q} - \hat{Q}' \int \hat{Q} + \hat{Q}' \int \hat{Q} \int \hat{Q} - \hat{Q}' \int \hat{Q} \int \hat{Q} \int \hat{Q} + \ \dots \ , \quad (4)$$

where $\hat{Q}$, usually referred to as the $\hat{Q}$-box, is a vertex function composed of irreducible linked diagrams, and the integral sign represents a generalized folding operation. $\hat{Q}'$ is obtained from $\hat{Q}$ by removing terms of first order in the interaction. Once the $\hat{Q}$-box is calculated, the folded-diagram series of Eq. (4) can be summed up to all orders by iteration methods.

As pointed out in the Introduction, we renormalize the bare NN interaction by making use of a new approach which has proved to be an advantageous alternative[2,4,5] to the traditional G-matrix method. The basic idea underlying this approach is to construct a low-momentum NN potential, $V_{\rm low-k}$, that preserves the physics of the original potential V_{NN} up to a certain cutoff momentum Λ. In particular, the scattering phase shifts and deuteron binding energy calculated by V_{NN} are reproduced by $V_{\rm low-k}$.[2,13] This is achieved by integrating out the high-momentum components of V_{NN} by means of an iterative method. The resulting V_{low-k} is a smooth potential that can be used directly as input for the calculation of shell-model effective interactions.

A detailed decription of the derivation of $V_{\rm low-k}$ as well as a discussion of its main features can be found in Refs. 2 and 3, where a criterion for the choice of the cutoff parameter Λ is also given. The results presented in section 3 have been obtained using for Λ the value 2.1 fm^{-1}. Once the $V_{\rm low-k}$ is obtained, the calculation of the matrix elements of the effective interaction is carried out within the framework of the folded-diagram method outlined above. In summary, we first calculate the $\hat{Q}$-box including diagrams up to second order in V_{low-k} and then obtain $V_{\rm eff}$ by summing up the folded-diagram series using the Lee-Suzuki iteration method.[14,15]

3. Selected results of realistic shell-model calculations

The study of nuclei in the close vicinity to doubly magic ^{132}Sn and ^{100}Sn is a subject of great interest. From the experimental point of view, it is very difficult to obtain information on these nuclei. In recent years, however, substantial progress has been made to access the limits of nuclear stability, which has paved the way to spectroscopic studies of nuclei in the neighborhood of both ^{132}Sn and ^{100}Sn. This offers the opportunity for testing the basic ingredients of shell-model calculations, especially the matrix elements of the effective interaction, well away from the valley of stability.

Motivated by these experimental achievements, we have studied several nuclei[6,7,16–18] around ^{132}Sn and ^{100}Sn in terms of the shell model employing realistic effective interactions derived from the CD-Bonn NN potential via the $V_{\rm low-k}$ approach. As already mentioned in Sec. 2, we have used for Λ the value 2.1 fm^{-1}. We report here some selected results of the studies of Refs. 6, 7 and 17, to which we refer for details. In particular, we consider the four odd-odd nuclei ^{132}Sb, ^{134}I, ^{98}Ag and ^{102}In , which are a direct source of information on the proton-neutron effective interaction in the ^{132}Sn and ^{100}Sn regions. More precisely, we focus attention on proton-neutron hole and neutron-proton hole multiplets in the former and latter region, respectively.

Let us start with the ^{132}Sn region. We assume that ^{132}Sn is a closed core and let the valence proton and neutron holes occupy the five single-particle levels $0g_{7/2}$, $1d_{5/2}$, $1d_{3/2}$, $2s_{1/2}$, and $0h_{11/2}$ of the 50-82 shell. The single-proton and single neutron-hole energies have been taken from the experimental spectra of ^{133}Sb and ^{131}Sn, respectively (see Ref. 6 for details).

In Figs. 1 and 2 some calculated multiplets for ^{132}Sb and ^{134}I are reported and compared with the existing experimental data. We see that the calculated energies are in very good agreement with the observed ones. In fact, the discrepancies are all in the order of tens of keV, except for the 1^+ state of the $\pi d_{5/2}\nu d_{3/2}^{-1}$ multiplet in ^{132}Sb, which lies 300 keV above the experimental counterpart. For ^{132}Sb , some other calculated multiplets having the neutron hole in the $h_{11/2}$ level are reported in Ref. 6, where the structure of the wave functions is also discussed.

A main feature of the calculated multiplets shown in Fig. 1 is that the states with minimum and maximum J have the highest excitation energy, while the state with next to highest J is the lowest one. This pattern is in agreement with the experimental one for the $\pi g_{7/2}\nu d_{3/2}^{-1}$ multiplet and

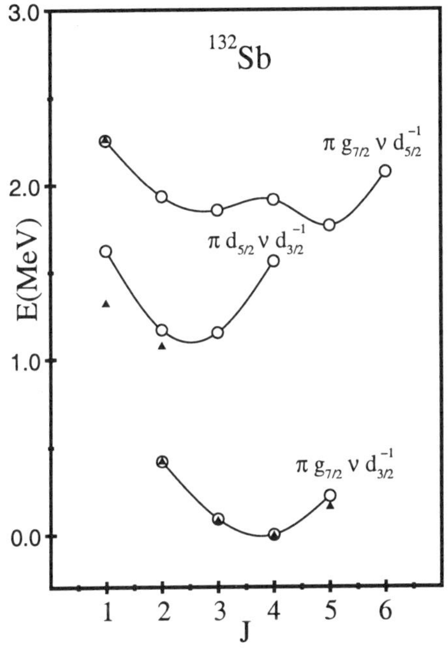

Figure 1. Proton particle-neutron hole multiplets in ^{132}Sb. The calculated results are represented by open circles while the experimental data by solid triangles. The lines are drawn to connect the points.

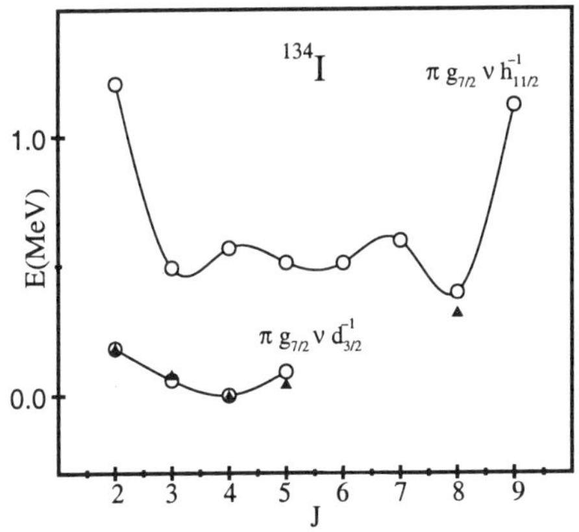

Figure 2. Same as Fig. 1, but for ^{134}I.

the experimental data available for the other multiplets (including those reported in Ref. 6) also go in the same direction. The nucleus ^{134}I has two additional protons with respect to ^{132}Sb. We see, however, that the behavior of the two multiplets shown in Fig. 2 is quite similar to that found for ^{132}Sb.

We turn now to the ^{100}Sn region. We assume that ^{100}Sn is a closed core and let the valence neutrons occupy the five levels $0g_{7/2}$, $1d_{5/2}$, $1d_{3/2}$, $2s_{1/2}$, and $0h_{11/2}$ of the 50-82 shell, while for the proton holes the model space includes the four levels $0g_{9/2}$, $1p_{1/2}$, $1p_{3/2}$, and $0f_{5/2}$ of the 28-50 shell. As regards the neutron single-particle and the proton single-hole energies, they cannot be taken from experiment since no spectroscopic data are yet available for ^{101}Sn and ^{99}In. We have therefore determined them by an analysis of the low-energy spectra of the odd Sn isotopes with $A \leq 111$ for the former and of the $N = 50$ isotones with $A \geq 89$ for the latter. More details about our choice and the adopted values are given in Ref. 7.

In the ^{100}Sn region the counterpart of ^{132}Sb is ^{100}In, for which studies of excited states are at present out of reach. We consider therefore the two neighboring odd-odd isotopes ^{98}Ag and ^{102}In, for which some experimental information is available, focusing attention on the $\pi g_{9/2}^{-1} \nu d_{5/2}$ multiplet.

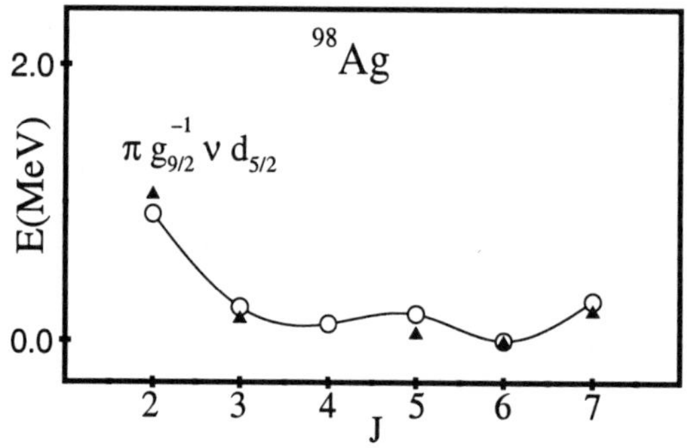

Figure 3. Proton hole-neutron particle $\pi g_{9/2}^{-1} \nu d_{5/2}$ multiplet in ^{98}Ag. The conventions of the presentation are the same as those used in Fig. 1.

In Figs. 3 and 4 we report the results of our calculations for ^{98}Ag and ^{102}In, respectively, and compare them with the experimental data. We see that the agreement between experiment and theory is of the same quality as that obtained in the ^{132}Sn region, the largest discrepancy being about 200 keV for the 7^+ state in ^{102}In. The pattern of the calculated multiplets turns out to be similar to that of the multiplets in the ^{132}Sn region, but with the states with minimum and maximum J less separate from the other ones. Actually, for ^{102}In we find that the energy of the 7^+ state is about the same as that of the 3^+ state. In this connection, it should be mentioned that for the former state the percentage of configurations other than those having a $g_{9/2}$ proton hole and a $d_{5/2}$ neutron is about 50%.

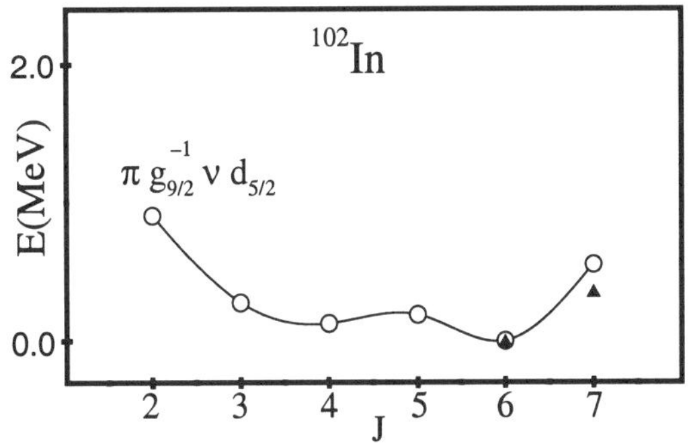

Figure 4. Same as Fig. 3, but for ^{102}In.

A detailed discussion of the structure of the calculated states can be found in Ref. 7, where the $\pi g_{9/2}^{-1} \nu g_{7/2}$ multiplet is also considered and our predictions for the hitherto unknown ^{100}In are reported. Before closing this section, it is worth noting that in all of our calculated multiplets the state of spin $(j_\pi + j_\nu - 1)$ is the lowest, in agreement with the early predictions of the Brennan-Bernstein coupling rule.[19]

4. Phase-shift equivalent nucleon-nucleon potentials and nuclear structure calculations

The results presented in the previous section have all been obtained with a $V_{\text{low-k}}$ derived from the CD-Bonn potential and confined within a cutoff momentum $\Lambda = 2.1\,\text{fm}^{-1}$. This value has been chosen according to the criterion discussed in Ref. 2.

A few years ago we performed a study[20] aimed at investigating the dependence of nuclear structure results on the NN potential used to derive the shell-model effective interaction through the G-matrix approach. Within this framework, it turned out that different NN potentials (we considered the Paris, Nijmegen93, CD-Bonn and Bonn A potentials) produce somewhat different nuclear structure results. This makes it very interesting to perform a similar study within the framework of the $V_{\text{low-k}}$ approach, which we are currently carrying out. Here we present only some preliminary results obtained for the nucleus with two valence neutron holes ^{130}Sn, which provides a good testing ground for this investigation.

As already mentioned in the Introduction, we consider the three phase-shift equivalent NN potentials NijmegenII, AV18 and CD-Bonn. As regards the cutoff Λ, we let it vary from 1.7 to 2.5 fm^{-1}. In all cases we compare the calculated spectrum of ^{130}Sn with the experimental one up to about 2.5 MeV excitation energy (this includes nine excited states) and calculate the *rms* deviation σ.[21]

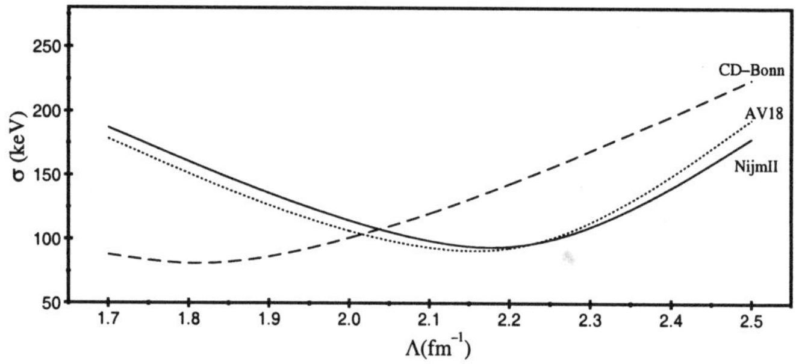

Figure 5. Behavior of the *rms* deviation σ relative to the spectrum of ^{130}Sn as a function of Λ for different NN potentials. See text for details.

In Fig. 5 we show the behavior of σ as a function of Λ for the three potentials. We see that the curves relative to NijmegenII and AV18 practically overlap each other while that for the CD-Bonn potential has a rather different pattern. In particular, the minimum of σ for CD-Bonn is at $\Lambda \sim 1.8\,\text{fm}^{-1}$ while for the other two potentials it lies at $\Lambda \sim 2.2\,\text{fm}^{-1}$. The minimum value of σ for the three potentials is however almost equal and also the energies of the various states are practically the same. By way of illustration, we report in Table 1 the ground-state energy of ^{130}Sn (relative to doubly magic ^{132}Sn) and the excitation energies of the first three positive-parity yrast states.

Table 1. Energy levels (in MeV) of ^{130}Sn. Predictions by different NN potentials are compared with experiment.

	NijmII	AV18	CD-Bonn	Expt
E_{gs}	12.410	12.427	12.406	12.474
$E(2^+)$	1.462	1.448	1.433	1.221
$E(4^+)$	2.057	2.055	2.057	1.966
$E(6^+)$	2.227	2.214	2.240	2.257

From Fig. 5 it also appears that all three curves are rather flat around the minimum. As a consequence, the quality of agreement between theory and experiment does not change significantly for moderate changes in the value of Λ. More precisely, for the CD-Bonn potential σ remains below 100 keV for values of Λ between 1.7 and 2.0 fm^{-1} while for the other two potentials this occurs for Λ between 2.0 and 2.3 fm^{-1}.

The main conclusion of this preliminary study is that, allowing for limited changes in the value of Λ, nuclear structure results obtained from $V_{\text{low-k}}$'s extracted from different NN potentials are practically independent of the input potential. It therefore appears that the low-momentum interaction $V_{\text{low-k}}$ gives an approximately unique representation of the NN potential. This is quite in line with the conclusions of Ref. 22.

At this point, a comment on the results presented in this section is in order. As mentioned in Sec. 2, we include in the $\hat{Q}$-box all diagrams up to second order in V_{low-k}. In the calculation of these diagrams we have inserted intermediate states composed of particle and hole states restricted to the two major shells above and below the Fermi surface. However, as a development of our study, we are currently investigating the effect of increasing the number of intermediate states. The first results indicate that the minimum value of σ occurs at values of Λ which are somewhat larger than those reported here. Our final findings will be the subject of a forthcoming publication.

5. Concluding remarks

The main conclusions of this paper may be summarized in the following remarks.

(i) Effective interactions derived from modern NN potentials are able to describe with quantitative accuracy the spectroscopic properties of nuclei in the regions of shell closures off stabilty. This gives confidence in their predictive power and may stimulate, and be helpful to, future experiments.

(ii) The $V_{\text{low-k}}$ approach to the renormalization of the bare NN potential is a valuable tool for nucler structure calculations. This potential may be used directly in shell-model calculations without first calculating the G matrix.

(iii) The $V_{\text{low-k}}$'s extracted from various NN potentials give practically the same results in shell-model calculations, suggesting the realization of a unique low-momentum NN potential.

Acknowledgments

This work was supported in part by the Italian Ministero dell'Istruzione, dell'Università e della Ricerca (MIUR).

References

1. J. P. Elliott, A. D. Jackson, H. A. Mavromatis, E. A. Sanderson and B. Singh, *Nucl. Phys.* **A121**, 241 (1968).
2. S. Bogner, T. T. S. Kuo, L. Coraggio, A. Covello and N. Itaco *Phys. Rev.* **C65**, 051301(R) (2002).
3. T. T. S. Kuo, S. Bogner, L. Coraggio, A. Covello and N. Itaco, in *Challenges of Nuclear Structure*, ed. A. Covello (World Scientific, Singapore, 2002), p. 129.
4. A. Covello, L. Coraggio, A. Gargano, N. Itaco and T. T. S. Kuo, in *Challenges of Nuclear Structure*, ed. A. Covello (World Scientific, Singapore, 2002), p. 139.
5. A. Covello, in *Proceedings of the International School of Physics "E. Fermi", Course CLIII*, edited by A. Molinari, L. Riccati, W. M. Alberico and M. Morando (IOS Press, Amsterdam, 2003), p. 79.
6. L. Coraggio, A. Covello, A. Gargano, N. Itaco and T. T. S. Kuo, *Phys. Rev.* **C66**, 064311 (2002).
7. L. Coraggio, A. Covello, A. Gargano and N. Itaco *Phys. Rev.* **C70**, 034310 (2004).
8. R. Machleidt, *Phys. Rev.* **C63**, 024001 (2001).
9. V. G. J. Stoks, R. A. M. Klomp, C. P. F. Terheggen and J. J. de Swart, *Phys. Rev.* **C49**, 2950 (1994).
10. R. B. Wiringa, V. G. J. Stoks and R. Schiavilla, *Phys. Rev.* **C51**, 38 (1995).
11. R. Machleidt and I. Slaus, *J. Phys. G* **27**, R69 (2001).
12. T. T. S. Kuo and E. M. Krenciglowa, *Nucl. Phys.* **A342**, 454 (1980).
13. S. K. Bogner, T. T. S. Kuo and A. Schwenk, *Phys. Rep.* **386**, 1 (2003).
14. S. Y. Lee and K. Suzuki, *Phys. Lett.* **B91**, 173 (1980).
15. K. Suzuki and S. Y. Lee, *Prog. Theor. Phys.* **64**, 2091 (1980).
16. J. Genevey, J. A. Pinston, H. R. Faust, R. Orlandi, A. Scherillo, G. S. Simpson, I. S. Tsekhanovic, A. Covello, A. Gargano and W. Urban, *Phys. Rev.* **C67**, 054312 (2003).
17. A. Gargano, L.Coraggio, A. Covello and N. Itaco, in *The Labirinth in Nuclear Structure*, edited by A. Bracco and C. A. Kalfas, AIP Conf. Proc. No. 701 (AIP, Melville, N.Y., 2004), p. 149.
18. A. Covello, L. Coraggio, A. Gargano, and N. Itaco, *Yad. Fiz.* **67**, 1637 (2004).
19. M. H. Brennan and A. M. Bernstein, *Phys. Rev.* **120**, 927 (1960).
20. A. Covello, L. Coraggio, A. Gargano and N. Itaco, in *Nuclear Structure 98*, ed. C. Baktash, AIP Conf. Proc. **481** (1999), p. 56.
21. We define $\sigma = \{(1/N_d)\, \Sigma_i\, [E_{exp}(i) - E_{calc}(i)]^2\}^{1/2}$, where N_d is the number of data.
22. T. T. S. Kuo and J. D. Holt, contribution to these Proceedings.

KEY TOPICS IN NUCLEAR STRUCTURE
Proceedings of the 8th Int. Spring Seminar on Nuclear Physics
edited by Aldo Covello
© 2005 World Scientific Publishing Co.

TESTING SHELL MODEL ON EXOTIC NUCLEI AT ^{135}Sb

H. MACH[1], A. KORGUL[2], B.A. BROWN[3], A. COVELLO[4], A. GARGANO[4], B. FOGELBERG[1], R. SCHUBER[1,5], W. KURCEWICZ[2], E.WERNER-MALENTO[2], R. ORLANDI[6,7], AND M. SAWICKA[2]

[1] *Department of Radiation Sciences, Uppsala University, Sweden*
[2] *Institute of Experimental Physics, Warsaw University, Poland*
[3] *Department of Physics and Astronomy and NSCL, Michigan State University, USA*
[4] *Dipartimento di Scienze Fisiche, Università di Napoli Federico II and Istituto Nazionale di Fisica Nucleare, Napoli, Italy*
[5] *Department of Physics, University of Konstanz, Germany*
[6] *Institut Lave-Langevin, Grenoble, France*
[7] *Schuster Laboratory, University of Manchester, UK*
E-mail: Henryk.Mach@studsvik.uu.se

Recently the first excited state in ^{135}Sb has been observed at the excitation energy of only 282 keV and interpreted as mainly $d_{5/2}$ proton coupled to the ^{134}Sn core. It was suggested that its low-excitation energy is related to a relative shift of the proton $d_{5/2}$ and $g_{7/2}$ orbits induced by the neutron excess. With the aim to provide more spectroscopic information on this anomalously low-lying $5/2^+$ state, we have measured its lifetime by the Advanced Time-Delayed $\beta\gamma\gamma$(t) method at the OSIRIS fission product mass separator at Studsvik. The $M1$ and $E2$ transition rates from the 282 keV state are strongly hindered, similarly to what occurs in ^{211}Bi for the transition de-populating the first excited state at 405 keV. However, more data are needed above ^{132}Sn especially on the transition matrix elements. Thus our investigation was extended to include lifetime measurement of the $5/2^+$ 243 keV state in ^{137}I, which has an extra pair of protons above ^{135}Sb. Results of shell model calculations are presented.

1. Introduction

A number of theoretical studies predict that very neutron-rich medium-heavy nuclei are governed by a shell structure different to that established along the line of stability. Although the 'neutron skin effects' are expected to occur at a very heavy neutron excess, thus closer to the neutron drip line, yet some limited effects related to specific orbits could perhaps be observed much earlier. This study is focussed on ^{135}Sb, for which new experimental results have been puzzling.

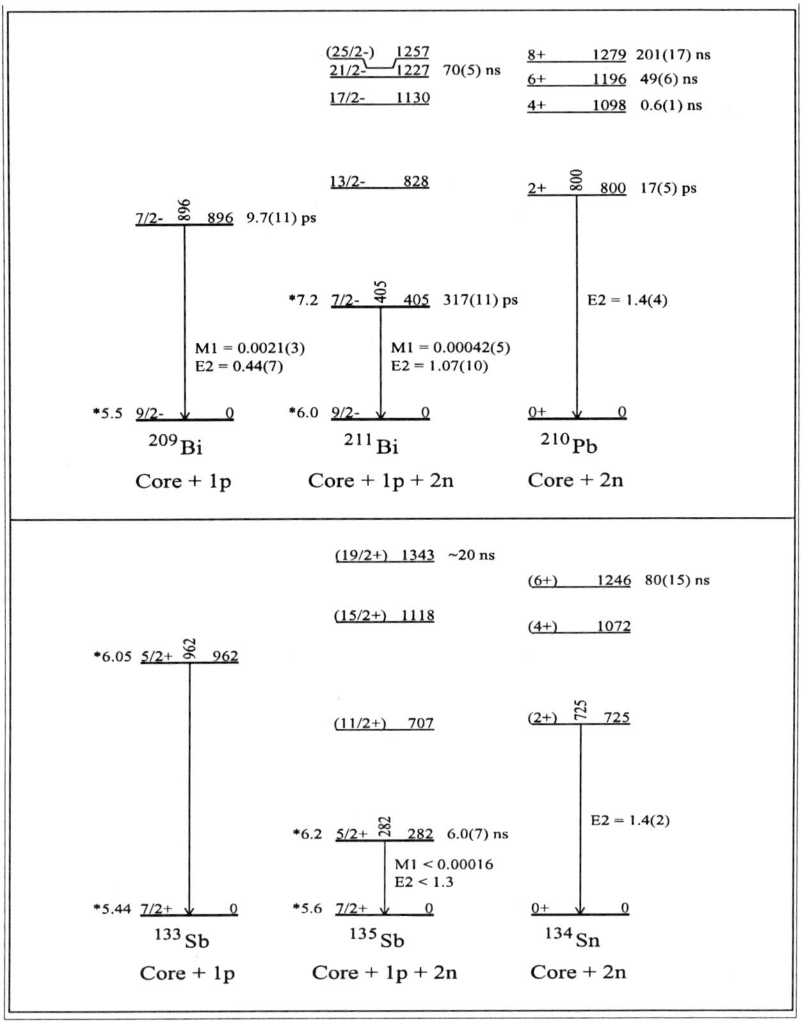

Figure 1. A partial summary of the experimentally known properties of simple nuclear systems above ^{132}Sn (BOTTOM) and ^{208}Pb (TOP): 'Core + 1 proton', 'Core + 2 neutrons' and 'Core + 1 proton + 2 neutrons'. Known level lifetimes are indicated on the right of the level, while logft values from β-decay of the parent are on the left hand side marked by an asterisk (*). $M1$ and $E2$ represent experimental $B(M1)$ and $B(E2)$ values expressed in W.u. Transition rates for ^{135}Sb are from this work; see text for discussion.

The nucleus ^{135}Sb has two neutrons and one proton above ^{132}Sn and is the most exotic nucleus beyond ^{132}Sn for which information exists on excited states. The first information on levels in ^{135}Sb came from the prompt fission study by Bhattacharyya *et al.*,[1] who have identified three core excited states originating from the $\pi g_{7/2}\nu f_{7/2}^2$ configuration at 707, 1118 and 1343 keV as $J^\pi = 11/2^+$, $15/2^+$ and $19/2^+$. The excitation energies of these states almost coincide with the 2^+, 4^+ and 6^+ states of mainly $\nu f_{7/2}^2$ configuration[2] in ^{134}Sn, see Figure 1. However, only recently in the study by Korgul *et al.*,[3] it was found that the first excited state in ^{135}Sb is located at an exceptionally low energy of only 282 keV. Subsequent study by Shergur *et al.*[4] at the ISOLDE facility has confirmed this result and extended the information to include the exceptionally low logft values to the ground and the 282 keV states. In the same study shell model calculations were performed and the systematics of the lowest-lying $5/2^+$ states in the odd-proton nuclei near ^{132}Sn were examined in order to understand the origin of the low-lying $5/2^+$ state in ^{135}Sb. It was concluded[4] that its low position provides support for the idea that nuclei with an N/Z ratio that exceeds 1.6 have a more diffuse nuclear surface that changes the relative binding energies of low-spin orbitals when compared to higher spin orbitals. It was also suggested[4] that lowering of the single-particle proton $d_{5/2}$ state by 300 keV does provide a better fit for that level without disturbing the otherwise excellent agreement between theory and experiment.

The idea[4] that the location of the $5/2^+$ state in ^{135}Sb is related to a strong relative shift between the proton $d_{5/2}$ and $g_{7/2}$ orbits just above ^{132}Sn, possibly due to the neutron diffuseness, can be examined via combined experimental and theoretical studies. Yet, little experimental data exist on nuclei with a few valence nucleons just above ^{132}Sn. In this exotic region ^{135}Sb provides at present the best case where there are some data on excited states from a few independent experimental probes. At the same time, which is a critical requirement, this simple nucleus having only three valence particle, can be modeled theoretically with high precision.

The aim of the present study was to measure the lifetime of this anomalously low-lying $5/2^+$ state in ^{135}Sb. Figure 1 illustrates the experimental situation near ^{135}Sb, which is one proton and two neutrons above the core of ^{132}Sn, and of ^{211}Bi, which is an equivalent nucleus above the core of ^{208}Pb. The data are from Refs. 1-8. In principle, since the $M1$ transition is forbidden between the $d_{5/2}$ and $g_{7/2}$ single particle states and the $E2$ collectivity is small in a weakly deformed nucleus, one would expect for the 282-keV state in ^{135}Sb a very slow $B(M1)$ rate if there is shift of the orbits, and a faster one if the lowering of the state is due to collective effects.

2. The Measurement and Discussion

The measurements were performed at the OSIRIS fission product mass separator at Studsvik in Sweden by using the Advanced Time-Delayed $\beta\gamma\gamma(t)$ method.[9] The activity was produced via thermal neutron induced fission of ^{235}U. The mass separated beam of A=135 isobars was implanted into an aluminized mylar tape at the experimental station, where fast timing β and BaF_2 γ detectors, as well as two Ge spectrometers were positioned in a close geometry. By selecting in Ge the 732 and 923 keV γ-rays feeding the 282 keV state from above[3] and selecting a very strong and pure 282-keV peak in the coincident BaF_2 spectrum, see Figure 2, one obtains the time-delayed $\beta\gamma(t)$ spectrum due to the lifetime of the 282 keV state in ^{135}Sb. It was verified that the feeding γ transitions do not carry any time-delayed components, which could affect fitting of the slope.

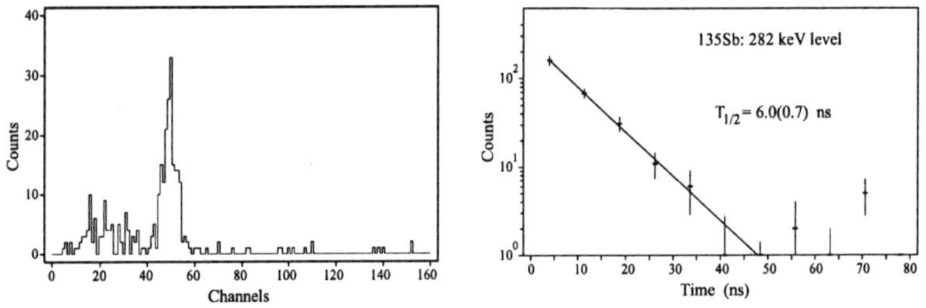

Figure 2. LEFT: the BaF_2 energy spectrum in coincidence with the 732 and 923 keV γ-rays recorded in Ge. By selecting the strong full energy peak at 282 keV (channel ~48) one obtains the time-delayed $\beta\gamma(t)$ spectrum due to the 282-keV level in ^{135}Sb shown on the RIGHT.

The half-life of the level is measured as $T_{1/2}$=6.0(7) ns (preliminary value). Since the $M1/E2$ mixing ratio for the transition is not known, we can only deduce the upper limits for the $B(M1)$ and $B(E2)$ rates, assuming in this evaluation either a 100% pure $M1$ or 100% pure $E2$ transition. In any case the results show strongly hindered $M1$ and $E2$ transition rates from the 282 keV state, which are almost identical to the equivalent case in ^{211}Bi, as seen in Figure 1, although the $B(M1)$ in ^{135}Sb is even lower.

Table 1 provides a comparison of the experimental $B(M1)$ values to the shell-model calculations by A. Covello and A. Gargano (CG) and by A. Brown (AB). CG use a two-body effective interaction derived from the CD-Bonn nucleon-nucleon potential[10] and single particle energies taken from the experimental spectra of ^{133}Sb and ^{133}Sn. In the derivation of the ef-

Table 1. Comparison of the experimental $B(M1)$ values and shell model calculations by A. Covello and A. Gargano (labelled CG) and A. Brown (labelled AB) for the lowest $5/2^+$ states in ^{135}Sb and ^{137}I, in the units of 10^{-3} μ_N^2, see text for discussion.

Nucleus	$J_i \rightarrow J_f$	$B(M1)^{exp}$	$B(M1)^{AB}_{free}$	$B(M1)^{AB}_{eff}$	$B(M1)^{CG}$
^{135}Sb	$5/2_1 \rightarrow 7/2_1$	<0.29	4.8	2.2	44
^{137}I	$5/2_1 \rightarrow 7/2_1$	~4.5(10)	0.30	20	
	$5/2_2 \rightarrow 7/2_1$		2.0	9.6	
	$5/2_2 \rightarrow 5/2_1$		5500	3700	

fective interaction the bare NN potential is renormalized by constructing a low-momentum potential $V_{\text{low-k}}$.[11] This is then employed to calculate the effective interaction using the $\hat{Q}$-box plus folded diagram method, the $\hat{Q}$-box being composed of $V_{\text{low-k}}$ diagrams through second order.[12] This calculation predicts a relatively pure $7/2^+$ ground state as predominantly $g_{7/2}$ proton coupled to the ^{134}Sn core with 78% $\pi g_{7/2} \nu f^2_{7/2}$, while the $5/2^+$ state is predicted at 560 keV and significantly admixed, with the first two leading terms 44% $\pi d_{5/2} \nu f^2_{7/2}$ and 27% $\pi g_{7/2} \nu f^2_{7/2}$. Making use of the free g-factors a $B(M1; 5/2^+ \rightarrow 7/2^+)$ value of 44 in units of 10^{-3} μ_N^2 is obtained.

A main point of this calculation is that there is no adjustable parameter. However, if one lowers the proton $d_{5/2}$ energy by 400 keV, then the predicted excitation energy of the $5/2^+$ state becomes 282 keV, matching the experiment, and the wavefunction of the state is more pure, with 65% $\pi d_{5/2} \nu f^2_{7/2}$ and 6% $\pi g_{7/2} \nu f^2_{7/2}$. Also, the $B(M1)$ value would go down to $9 \times 10^{-3} \mu_N^2$, which is still significantly larger than the experimental value. It is worth mentioning, however, that the $B(M1)$ may be reduced by taking an effective g_l factor for neutrons different from zero. Actually, to reproduce the experimental $B(M1)$ a value of 0.85 is needed, which becomes 0.65 when the proton $d_{5/2}$ energy is lowered. It remains to be seen if this is substantiated by a calculation of the effective magnetic operator consistent with the derivation of the effective interaction.

The AB calculation makes use of the wave functions obtained in Ref. 4. These were calculated with the CD-Bonn G matrix evaluated in an oscillator potential with a renormalization based on the $\hat{Q}$-box method that includes nonfolded diagrams to third order and folded diagrams to infinite order. The single-particle energies are from the ^{133}Sb and ^{133}Sn, except the $d_{5/2}$ energy which was shifted down by 300 keV. As discussed in

Ref. 4 part of this shift may be attributed to the difference between the G matrix monopole interaction for the $g_{7/2} - d_{5/2}$ splitting and that obtained in a Hartree-Fock (finite-well) potential. With free-nucleon g-factors $B(M1) = (0.172 - 0.102)^2\ \mu_N^2 = 4.8 \times 10^{-3}\ \mu_N^2$, where the terms inside the brackets are from the orbital and spin contributions, respectively. With the effective $M1$ operator from Table 1 in Ref. 13 one obtains $B(M1) = (0.160 - 0.045 - 0.163)^2\ \mu_N^2 = 2.2 \times 10^{-3}\ \mu_N^2$, where the terms inside the brackets are from the effective orbital, spin and tensor (proportional to $[Y_2 \otimes \mathbf{s}]^{(1)}$) operators, respectively. It was noted in Ref. 14 that the tensor contribution to the $M1$ effective operator was especially important for the magnetic moment of the $g_{7/2}$ ground state in ^{133}Sb. Owing to the cancellation between the orbital and tensor terms in the effective $M1$ matrix element, it is possible to obtain an arbitrary small $B(M1)$ with only a 20% change in the spin or tensor terms. This may be the origin of the unusually small $B(M1)$ for this transition. For other $M1$ transitions the relative contribution of the orbital, spin and tensor contributions are quite different from those in ^{135}Sb. For example, the $5/2^+$ to $7/2^+$ transition in ^{133}Sb is ℓ-forbidden and only depends upon the effective tensor $M1$ operator. Thus a systematic study of $M1$ transitions in this mass region just above ^{132}Sn is important.

We have extended our experimental study to include also ^{137}I. From the measured half-life of the $5/2^+$ 243 keV state in ^{137}I, $T_{1/2} = 470(20)$ ps (preliminary value), and by assuming that the $B(E2)$ rate is lower than 10 W.u. (note, that $B(E2; 2^+ \rightarrow 0^+)$ in the core nucleus ^{136}Te is only 5 W.u.[15]), one can deduce the $B(M1)$ value of $4.5(10) \times 10^{-3}\ \mu_N^2$. It is interesting that the AB-'eff' calculations correctly reproduce a strong jump in the $B(M1)$ rates by about one order of magnitude between transition rates in ^{135}Sb and ^{137}I, while the AB-'free' predict also one order of magnitude jump but in the wrong direction. We also note that the predicted $B(M1)$ values for the transition from the second excited $5/2^+$ state in ^{137}I is by three orders of magnitude higher. It has to be seen whether this interesting prediction is confirmed experimentally. Very large changes in the $B(M1)$ values for close-lying states in the same nucleus, that are predicted theoretically, offer an opportunity to verify the structure of the $M1$ operator.

The present comparison of the experimental and theoretical results must be viewed with caution. The 282-keV level in ^{135}Sb does represent a challenge to shell-model calculations but clearly more data are needed in this region before any firm conclusion can be drawn.

References

1. P. Bhattacharyya et al., *Eur. Phys. J.* **A3**, 109 (1998).
2. C.T. Zhang et al., *Z. Phys.* **A358**, 9 (1997).
3. A. Korgul et al., *Phys. Rev.* **C4**, 021302(R) (2001).
4. J. Shergur et al., *Phys. Rev.* **C65**, 034313 (2002).
5. M. Sanchez-Vega et al., *Phys. Rev.* **C60**, 024303 (1999).
6. M.J. Martin, *Nucl. Data Sheets* **63**, 723 (1991).
7. E. Browne, *Nucl. Data Sheets* **99**, 483 (2003).
8. A. Artna-Cohen, *Nucl. Data Sheets* **63**, 79 (1991).
9. H. Mach et al., *Nucl. Phys.* **A523**, 197 (1991), and references therein.
10. R. Machleidt, *Phys. Rev.* **C63**, 024001 (2001).
11. S. Bogner, T. T. S. Kuo, L. Coraggio, A. Covello, and N. Itaco, *Phys. Rev.* **C65**, 051301(R) (2002).
12. T. T. S. Kuo and E. M. Krenciglowa, *Nucl. Phys.* **A342**, 454 (1980).
13. G.N. White et al., *Nucl. Phys.* **A644**, 277 (1998).
14. N.J. Stone et al., *Phys. Rev. Lett.* **78**, 820 (1997).
15. D.C. Radford et al., *Phys. Rev. Lett.* **88**, 222501 (2002).

KEY TOPICS IN NUCLEAR STRUCTURE
Proceedings of the 8th Int. Spring Seminar on Nuclear Physics
edited by Aldo Covello
© 2005 World Scientific Publishing Co.

NEUTRON-RICH In AND Cd ISOTOPES IN THE ^{132}Sn REGION

J. GENEVEY,[1] J. A. PINSTON,[1*] A. SCHERILLO,[2,3†] A. COVELLO,[4‡] H. FAUST,[2] A. GARGANO,[4‡] R. ORLANDI,[2,5] G. S. SIMPSON,[2] AND I. TSEKHANOVICH[2]

[1] *LPSC IN2P3-CNRS/Université Joseph Fourier, F-38026 Grenoble Cedex, France*
[2] *Institut Laue-Langevin, F-38042 Grenoble Cedex 9, France*
[3] *Insitut für Kernphysik, Universität zu Köln, D-50937 Köln, Germany*
[4] *Dipartimento di Scienze Fisiche, Università di Napoli Federico II, and Istituto Nazionale di Fisica Nucleare, Complesso Universitario di Monte S. Angelo, via Cintia, I-80126 Napoli, Italy*
[5] *The University of Manchester, Brunswick Street, M13 9PL, United Kingdom*

In this work, μs isomers in In and Cd isotopes with $A = 123$ to 130 were investigated. These experiments were conducted at the ILL (Grenoble) using the LOHENGRIN mass spectrometer. The isomers were produced by thermal-neutron induced fission of Pu targets. The level schemes of the odd-mass $^{123-129}$In and new measurements of the μs half-lives of the odd-odd $^{126-130}$In are reported. In contrast, the expected 8^+ isomers in the even-Cd isotopes were not observed. A shell-model study of the heaviest In and Cd nuclei was also performed using a realistic effective interaction derived from the CD-Bonn nucleon-nucleon potential. Comparison shows that the calculated levels of ^{130}In and ^{129}In are in good agreement with experimental values while some discrepancies appear for the lighter In isotopes. The collectivity of ^{126,128}Cd is discussed in the framework of the shell model and in comparison with ^{204}Hg.

1. Introduction

Experimental progress has been made recently in the region surrounding the doubly-magic ^{132}Sn.[1] However, nuclear structure information is more complete for nuclei above the $Z = 50$ shell closure (for instance, Sb and Te

*e-mail: pinston@lpsc.in2p3.fr
†work partially supported by grant O6K-167 of BMBF
‡work partially supported by Ministero dell'Istruzione, dell' Università e della Ricerca (MIUR)

isotopes) than for the In and Cd isotopes, which are much more difficult to produce. Only recently, some spectroscopic information on ^{126,128}Cd, which have few proton and neutron holes inside the ^{132}Sn core, was obtained by Kautzsch *et al.*.[2] Although this information is rather scarce, it seems to indicate that these nuclei possess some degree of collectivity. This is one of the main reasons for the present study on neutron-rich nuclei in the ^{132}Sn region.

In the present work we have searched for and studied the decay of μs isomers in the neutron-rich mass $A = 123$ to 130 nuclei with the LOHENGRIN spectrometer at the ILL in Grenoble. The aim was to complete the previous data on the heavy Cd and In isotopes and to study the yrast lines of these nuclei. Apart from the above quoted study on the Cd isotopes, the low-spin levels up to $13/2^+$ in $^{123-127}$In were previously investigated from the β-decay of Cd isotopes.[3,4] Very recently, high spin ms isomers in $^{125-129}$In have been observed.[5,6] Preliminary reports were also presented by M. Hellström *et al.*[7,8] on the search for μs isomers in the heavy Cd and In isotopes at the FRS spectrometer at GSI, but no level schemes were proposed.

Motivated by the new data from the present experiment, we have performed calculations to test the ability of the shell model to describe the heavy Cd and In isotopes, with proton and neutron holes inside the ^{132}Sn core. In these calculations, a realistic effective interaction derived from the CD-Bonn nucleon-nucleon potential[9] is used. Similar calculations were performed for nuclei with proton particles and neutron holes around ^{132}Sn, and for ^{129}In in Ref. [10,11], the latter work containing details on the present calculations. In both cases good agreement with the experimental data was found.

2. Experimental procedure

In the present work we have searched for and studied the decay of μs isomers of very neutron-rich In and Cd isotopes. The experiments were performed at the ILL (Grenoble) using the LOHENGRIN spectrometer. The nuclei of mass $A = 123$ to 130 were produced by thermal-neutron induced fission of ^{239}Pu and ^{241}Pu. The spectrometer has been used to separate the fission fragments (FF) recoiling from thin targets according to their mass to ionic charge ratios (A/Q). The FFs were detected by an ionization chamber. The γ rays de-exciting the isomeric states were detected by two large volume Ge detectors, and conversion electrons and X rays were detected by two adjacent Si(Li) diodes. The details of the experimental set up are given in [11]. As an example, Ge and Si(Li) spectra of the μs isomer decay in ^{127}In are presented in Fig. 1. This nucleus decays by a cascade consisting of a strongly converted 47 keV $E2$ transition, and two γ rays of 221 and 233 keV.

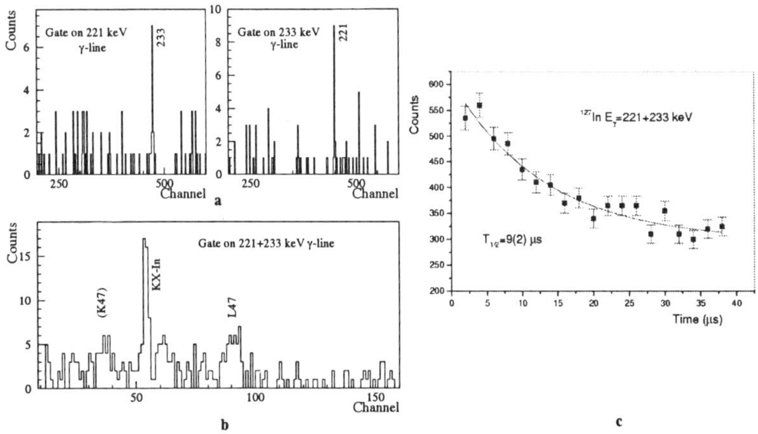

Figure 1. ^{127}In spectra. (a) coincidences between the 221 keV and 233 keV γ lines, (b) Si(Li) in coincidences with 221 or 233 keV γ lines, (c) time spectrum of the 221 and 233 keV γ lines.

In this work, new data have been obtained for ^{123,127,129}In. Moreover, the half-lives of the first excited state in the odd-odd ^{126,128,130}In have been re-measured, thus obtaining the values of 22(2) μs, 23(3) μs and 3.1(3) μs, respectively. In contrast the expected 8^+ isomers in the even-Cd nuclei were not observed.

3. Level scheme of the odd-In isotopes

The heavy In and Cd nuclei, with neutron and proton holes inside the ^{132}Sn core, are characterized by the presence of two high-spin levels, $\pi g_{9/2}^{-1}$ and $\nu h_{11/2}^{-1}$, at low excitation energy. The $p-n$ interaction in the $(\pi g_{9/2}^{-1} \nu h_{11/2}^{-1})_{10^-}$ state is very strong and is expected to produce very perturbed yrast lines. These features greatly complicate the construction of the level schemes and therefore different experimental techniques are needed to study these nuclei. The level schemes of $^{123-129}$In shown in Fig. 2 are the result of the synthesis of different works: the ms isomer experiments performed at the OSIRIS mass separator,[5,6] and the μs isomer experiments performed with the FRS at GSI[7,8] and the LOHENGRIN spectrometer.[11] All the reported levels are in the vicinity of the yrast line. The low-spin levels fed in the previous works by β-decay experiments are not shown in Fig. 2.

In ^{129}In, ms and μs isomers are present. Fogelberg *et al.*[5] have shown evidence of a high-spin $29/2^+$ yrast trap which decays by an $E3$ transition (110 ms half-life) to another yrast trap of spin $23/2^-$ (half-life 700 ms), which β decays to ^{129}Sn. An excitation energy of 1630(56) keV was deduced for the $23/2^-$ state from the β-decay spectra.[6] Genevey *et al.*[11] have shown evidence of a $17/2^-$ isomer of 8.5 μs, which decays by a γ cascade to the $9/2^+$ ground state.

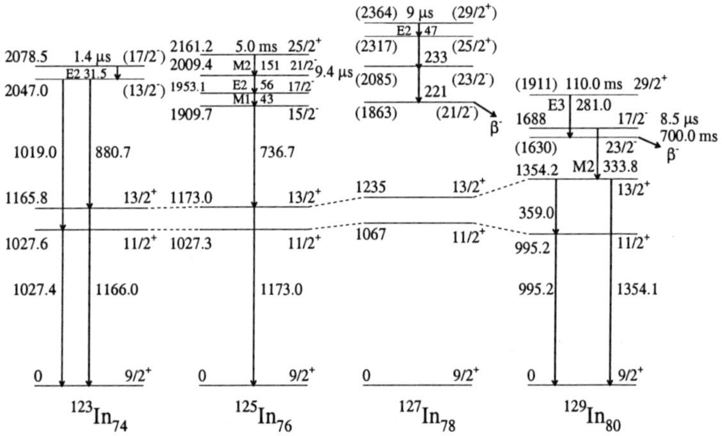

Figure 2. Level schemes of ^{123,125,127,129}In.

In ^{127}In, we have observed a 9 μs isomer which decays by a cascade consisting of a strongly converted $E2$ transition of 47 keV, and two γ-rays of 221 and 233 keV. This cascade has no overlap with the previously known γ-rays which feed directly or indirectly the $1/2^-$ isomer of 420(65) keV energy.[6] Moreover, the comparison with the neighboring ^{125}In and ^{129}In shows that the excited states above the $9/2^+$ ground state have energies higher than about 1 MeV, which excludes the possibility of the observed cascade ending at this level. Consequently, this cascade can only de-excite a $29/2^+$ isomer to a $21/2^-$ state decaying itself by β emission. This finding agrees with the measurement of Gausemel *et al.*[6] who have found a $21/2^-$

isomer in ^{127}In, of 1.0 s half-life and have deduced its energy, 1863(58) keV, from β-decay spectra.

Two other states at energies of 1067 keV and 1235 keV, feeding directly the $9/2^+$ ground state are added to the level scheme. They were reported by Hoff *et al.*[3] but without spin and parity assignments. By analogy with ^{125}In and ^{129}In, spin and parity assignments of $11/2^+$ and $13/2^+$, respectively, are proposed for these two states.

In ^{125}In, Fogelberg *et al.*[5] have found a 5 ms isomer which decays by an $M2$ transition to a 9.4 μs isomer which itself decays by a γ-ray cascade to the $9/2^+$ ground state. We have also observed the decay of the μs isomer and agree with the previous level scheme and the $E2$ and $M1$ multipolarities measured for the 56 and 43 keV transitions, respectively. Fogelberg *et al.* have suggested a spin and parity assignment of $23/2^-$ for the ms isomer, and positive parities for the states below the isomer. However, this hypothesis seems to be inconsistent with the non-observation of the cross over between the 1953.1 keV and 1173.0 keV levels and between the 1909.7 and 1027.3 keV levels, respectively. A $25/2^+$ spin and parity assignment for the ms isomer and a change of parity for the three successive levels seem necessary to explain the non-observation of the cross over in the experimental data.

We have observed in the present work a 1.4 μs isomer which decays by an $E2$ transition and a γ-ray cascade to the $9/2^+$ ground state of ^{123}In. The simultaneous feeding of the $11/2^+$ and $13/2^+$ levels at 1027.6 and 1165.8 keV, respectively, suggests a spin and parity assignment $13/2^-$ or $15/2^+$ for the 2047.0 keV level. The negative parity assignment is preferred by analogy with the heavier In isotopes and is reported in Fig. 2, but a positive parity assignment cannot be completely ruled out.

4. Shell-model calculations in In isotopes

4.1. *Odd In nuclei*

In Fig. 3a experimental levels of ^{129}In and ^{130}Sn are shown together with the calculated ones. For ^{129}In all the experimental levels, except the $1/2^-$ level at 369 keV, are reported, while only some selected yrast levels of ^{130}Sn are shown. The dominant configurations of all these levels are also indicated. The excitation energies of ^{130}Sn are rather well reproduced by the shell-model calculations. However, it is interesting to note that the first 2^+ state is overestimated by 162 keV. This is a common feature in this region, and may be traced to the model-space truncation.

The low-energy levels of ^{129}In are expected to result from the coupling of

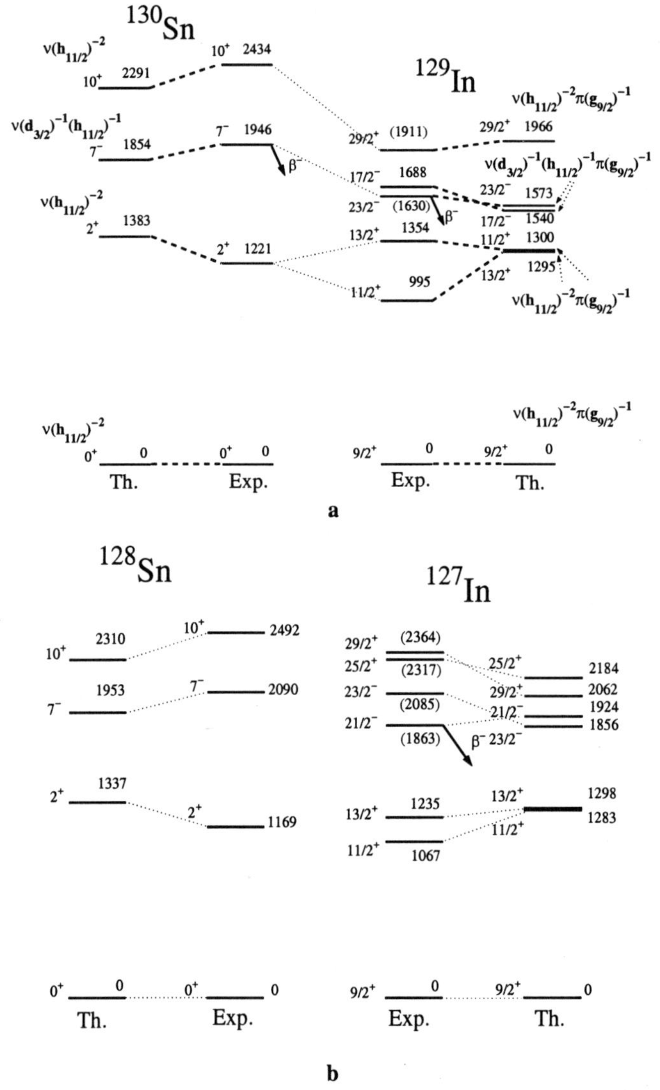

Figure 3. Experimental and calculated energies in ^{129}In and ^{130}Sn (a), and in ^{127}In and ^{128}Sn (b).

a $\pi g_{9/2}$ hole to the two-neutron hole states of ^{130}Sn. The observed decrease in energy of the $29/2^+$ aligned state with respect to the 10^+ in ^{130}Sn is explained by the strong $p-n$ interaction in the $(\nu h_{11/2}^{-1} \pi g_{9/2}^{-1})_{10^-}$ state. An analogous effect is observed for the other aligned state, $23/2^-$, of dominant configuration $\pi g_{9/2}^{-1} \nu (h_{11/2}^{-1} d_{3/2}^{-1})$. However, this effect is weaker because the $p-n$ interaction in the $(\pi g_{9/2}^{-1} \nu d_{3/2}^{-1})_{6^+}$ state is less attractive than in the $(\pi g_{9/2}^{-1} \nu h_{11/2}^{-1})_{10^-}$ state. The decrease in energy of the $23/2^-$ and $29/2^+$ levels is responsible for these two states to be long-lived isomers. This is very well reproduced by the shell-model calculation which also predicts the μs isomerism for the $17/2^-$ state, which decays by an $E2$ transition. The main discrepancy between experiment and theory concerns the $11/2^+$ level which is overestimated by 300 keV. This is not surprising because its dominant configuration results from the coupling of the 2^+ state in ^{130}Sn to the $g_{9/2}$ proton hole, the 2^+ energy, as we have already seen, being overestimated by the theory.

From Fig. 3b it appears that the observed $29/2^+$ and $23/2^-$ states in ^{127}In are closer to the 10^+ and 7^- states in ^{128}Sn, respectively, as compared to what is shown in Fig. 3a. This could be explained by a decrease in the effects of the $p-n$ interaction from ^{129}In to ^{127}In. Such a decrease is underestimated by theory, indicating that the distribution of the two extra neutron holes may not be properly described. Another feature, possibly connected to the effects of the $p-n$ interaction, is the inversion of the $29/2^+$ and $25/2^+$, and $23/2^-$ and $21/2^-$ levels respectively, in the calculated spectrum of ^{127}In. As a consequence, a $29/2^+$ ms isomer decaying by an $E3$ transition is predicted, whilst a μs one decaying by an $E2$ transition is measured. Moreover, the theory predicts a $23/2^-$ β-decaying isomer, whilst a $21/2^-$ one is measured.[6] Unfortunately, the number of configurations for ^{125}In and ^{123}In isotopes is too large to allow calculations with the OXBASH code.

4.2. *Even In nuclei*

The nuclear structure information is very scarce for the heavy odd-odd In nuclei. However, the 1^-, 3^+ and 1^+ states [13] are experimentally known in $^{126-130}$In. They result from the coupling of a proton hole $g_{9/2}$ to the neutron holes $h_{11/2}$, $d_{3/2}$ and $g_{9/2}$, respectively. In these three states the neutron and proton are in coplanar orbits and the $p-n$ interaction is expected to become strongly attractive, in particular for the 1^+ state. A weaker interaction is expected for the 3^+ state where a $d_{3/2}$ neutron is involved, and this level is used to normalize the level schemes of Fig. 4. The shell-model calculations reproduce rather well the relative energies of the three levels of ^{130}In as well as their evolution when going from ^{130}In to ^{126}In. It is interesting to note the strong variation of the position of the

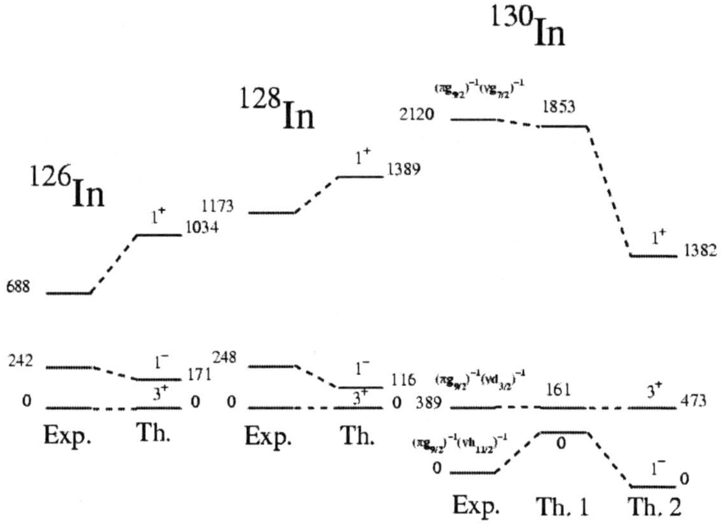

Figure 4. Experimental and calculated energies in the odd-odd In.

1^- state from ^{130}In to ^{128}In. Qualitatively, this effect could be explained by a decrease in the effects of the $p-n$ interaction when increasing the number of neutron holes. However, as for the aligned $29/2^+$ state in ^{127}In, the decrease is underestimated by the calculation.

The very recent results of the shell-model calculation for ^{130}In by Dillmann *et al.* [13] are also reported in Fig. 4 (Th2). Although in this work the two-body matrix elements are also deduced from the CD-Bonn potential, the energy of the 1^+ state is underestimated by 738 keV energy. This result is in strong disagreement with the outcome of our calculation.

In conclusion, the shell model reproduces rather well the levels of ^{129}In and ^{130}In. However, this model underestimates the decrease in the effects of the $p-n$ interaction when two neutrons are removed from ^{130}In or ^{129}In. These results show the limits of the predictive power of the model in the vicinity of ^{132}Sn.

5. Even Cd isotopes in the framework of the shell model

The μs isomers are very abundant in the vicinity of the two magic shells of ^{132}Sn and disappear rapidly far from them. However, below $Z = 50$ they disappear suddenly for Cd isotopes, no μs isomers have been identified up to now in the even-mass ones.

Neutron-rich ^{126}Cd and ^{128}Cd isotopes were recently produced[2] at ISOLDE from the β-decay of Ag isotopes. However, only the first 2^+ and 4^+ states were identified and are shown in Fig. 5. The value of $E(4^+)/E(2^+) \sim 2.2$ found experimentally suggests that some degree of collectivity is present in these two nuclei. and the authors of Ref. [2] have taken it as a possible evidence for a weakening of the spherical N=82 neutron shell below ^{132}Sn.

Our shell-model predictions for these two nuclei are reported in Fig. 5. The comparison between experiment and theory shows that the energies of both the 2^+ and 4^+ states are overestimated while the $4^+ - -2^+$ energy difference is correctly reproduced. The $E2$ transition rates are also reported. Unfortunately, the experimental data to compare with are not available, but the $B(E2)$ values and the energy of the $8^+ \rightarrow 6^+$ transitions allow us to predict a half-life of about 10 ns for the 8^+ state in both Cd isotopes. This value is much shorter than the time of flight of 2 μs of the FFs through the LOHENGRIN spectrometer and could explain the non-observation of these isomers in our work.

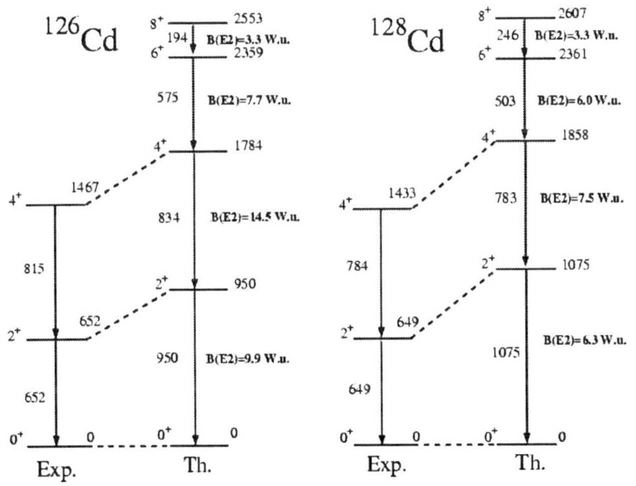

Figure 5. Experimental and calculated energies in ^{126,128}Cd.

It is interesting to note that the experimental spectrum of ^{204}Hg, which has also two neutron and proton holes inside doubly-magic ^{208}Pb, shows features similar to those observed in ^{128}Cd. For this nucleus, which is more easy to study because it is on the line of stability, the energy levels and the $E2$ transition rates have been reported in the literature. This nucleus presents a nice collective band based on the ground state and characterized by an $E(4^+)/E(2^+) \sim 2.6$ ratio, which is larger than the one measured in ^{128}Cd. Rydström *et al.* [14] have shown that the energy levels and the $E2$ transition rates up to the 6^+ level are well reproduced by the shell model. the largest discrepancy concerning the 2^+ state, whose energy is overestimated by about 250 keV. The authors have also predicted that the collective band ends at 8^+ owing to the limited number of active nucleons in this nucleus.

In conclusion, both ^{204}Hg and ^{128}Cd, with two neutron holes and two proton holes inside doubly-magic ^{208}Pb and ^{132}Sn, respectively, possess some degree of collectivity. This collectivity is well reproduced by the shell-model calculations for ^{204}Hg and is a consequence of a strong $p-n$ interaction. The nucleus ^{128}Cd shows a weaker collectivity, but the present information is too scarce to conclude that the collectivity is a consequence of the shell structure and more experimental data are needed to clarify the situation.

References

1. J. A. Pinston and J. Genevey, *J. Phys.* **G30**, R57 (2004).
2. T. Kautzsch et al., *Eur. Phys. J.* **A9**, 201 (2000).
3. P. Hoff et al., *Nucl. Phys.* **A459**, 35 (1986).
4. H. Huck et al., *Phys. Rev.* **C39**, 997 (1989).
5. B. Fogelberg et al. *Proc. 2nd Intern. Workshop on Fission and Fission-Product Spectroscopy (Seyssins, france) (AIP Conf. Proc. 447)* ed. G. Fioni, H. Faust, S. Oberstedt and F.-J. Hambsch p. 191.
6. H. Gausemel et al., *Phys. Rev.* **C69**, 054307-1 (2004).
7. M. Hellström et al., *Proc. 3rd Conf. on Fission and Properties of Neutron-Rich Nuclei (Sanibel Island,FL)* ed. J. H. Hamilton, A V. Ramayya, H. K. Carter (Singapore: World Scentific) p. 23.
8. M. Hellström et al., *Proc. of Intern. Workshop XXXI on Gross Properties of Nuclei and Nuclear Excitations (Hirscheg, Austria)* ed. H. Feldmeier, J. Knoll, W. Nörenberg, J. Wambach, p. 72.
9. R. Machleidt, *Phys. Rev.* **C63**, 024001 (2001).
10. L. Coraggio et al., *Phys. Rev.* **C66**, 064311 (2002).
11. J. Genevey et al., *Phys. Rev.* **C67**, 054312 (2003).
12. J. Genevey et al., *Eur. Phys. J.* **A9**, 201 (2000).
13. I. Dillmann et al., *Phys. Rev. Lett.* **91**, 162503 (2003).
14. L. Rydström et al., *Nucl. Phys.* **A512**, 287 (1990).

KEY TOPICS IN NUCLEAR STRUCTURE
Proceedings of the 8th Int. Spring Seminar on Nuclear Physics
edited by Aldo Covello
© 2005 World Scientific Publishing Co.

PAIR BREAKING IN A SHEARS BAND OF ^{104}IN*

O. YORDANOV, K. P. LIEB, E. GALINDO, M. HAUSMANN, A. JUNGCLAUS, G. A. MÜLLER

II. Physikalisches Institut, Universität Göttingen,
Friedrich-Hund-Platz 1
D-37077 Göttingen, Germany
E-mail: plieb@gwdg.de

F. BRANDOLINI

Dipartimento di Fisica, Universita di Padova,
I-35131 Padova, Italy
E-mail: brandolini@padova.infn.it

A. ALGORA, A. GADEA, D. NAPOLI, T. MARTINEZ

INFN-LNL,
I-35020 Legnaro (Padova), Italy
E-mail: daniel.r.napoli@lnl.infn.it

DSA lifetime measurements in the negative-parity yrast band of the odd-odd ^{104}In have given evidence for the interplay of the $[\pi g_{9/2} \otimes \nu h_{11/2}]$ proton-neutron shears band with the remaining $\nu(d_{5/2}, g_{7/2})$ valence neutrons. The recoupling of the $\nu^{2,4}(d_{5/2}, g_{7/2})$ part of the wave functions to either seniority 2 or 4 produces a pronounced up-bend of B(M1) which is reproduced by large-scale shell model calculations with effective single-particle energies and two-body matrix elements.

1. Introduction

The study of stretched magnetic dipole or shears bands in near-spherical nuclei as a new type of elementary mode of nuclear excitation has attracted great interest. Such bands feature very large reduced M1 strengths of several μ_N^2 and point to particular couplings of a few protons and neutrons (or holes) in well-defined single-particle orbits of high angular momenta. The shears mechanism involves the spin alignment of these particular nucleons and leads to a large transverse component $\mu_\perp$ of the total magnetic moment μ, relative to the total angular momentum I. Since the first observation of

$\Delta I = 1$ shears bands in ^{199}Pb [1], such structures have been located in many medium-mass and heavy nuclei and their nature has been confirmed by the measurement of regular energy spacings, large M1 strengths and/or small cross-over E2-to-M1 branching ratios.

As pointed out by Frauendorf and Reif [2], the situation is somewhat more complex in nuclei of the $A = 100$ region which have $Z < 50$ protons and $N > 50$ neutrons. While the high-j orbit for the proton hole(s) clearly is $g_{9/2}$, the valence neutrons outside ^{100}Sn are distributed over single-particle orbits with spins ranging from $j = 1/2$ to $j = 11/2$. The possible shears mechanism due to the simple $[\pi(g_{9/2}^{-1} \otimes \nu h_{11/2}]$ or $[\pi g_{9/2}^{-1} \otimes \nu g_{7/2}]$ coupling therefore may be strongly modified or even destroyed by the presence of neutrons in lower-j orbits. On the other hand, having few valence nucleons, these nuclei are accessible to shell model calculations, which may help us in understanding the role of particles in the various orbits and their rearrangement along the $\Delta I = 1$ sequences.

In the present paper, we report on a detailed lifetime study on the negative-parity yrast sequence in ^{104}In . Since this nucleus has 49 protons and 55 neutrons, it appears to be a very favorable case for studying the shears mechanism: the (only) high-j proton-hole is in the g $_{9/2}$ orbit and the (only) accessible high-j neutron orbit is $h_{11/2}$ to reach negative parity. In previous investigations, a regular $\Delta I = 1$ structure ranging from spin 8^- to (18^-) was established [3,4], extending up to 6152 keV, but interrupted by a close-lying 12^- doublet at 3560 and 3565 keV excitation (see the left -hand side of Fig. 2). In their lifetime measurements by means of the recoil distance Doppler-shift and DDCM methods[3], Kast and collaborators reported lifetime values of $\tau(13^-) < 5ps$ and $\tau(14^-) = 0.86(7)ps$, which are consistent with large M1 strengths. This work motivated the present Doppler Shift attenuation experiment for the higher-lying members of this band.

2. Experiment

The high-spin states in ^{104}In were populated in the reaction ^{58}Ni($^{50}Cr, 3pn$), using the 5 pnA ^{50}Cr beam at 200 MeV provided by the XTU tandem accelerator of the INFN laboratory at Legnaro. The target consisted of a 1.0 mg/cm^2 ^{58}Ni layer (enriched to 99.8 %) that was evaporated onto a 15 mg/cm^2 Au backing. Doppler-broadened lineshapes were measured in the GASP array that consisted of 40 Compton-suppressed germanium detectors positioned at the angles $\theta = 35°, 59°, 72°, 90°, 108°, 121°$

and 145° relative to the beam axis. Twofold $\gamma\gamma$-coincidence spectra were accumulated and sorted in seven matrices. In most cases, gates were set on all detectors of the array and the spectra were summed over all Ge detectors of each ring. The spectrum at $\theta = 90°$ was not analyzed for Doppler-broadenings, but was used to distinguish lines in close-lying multiplets. The analysis of the Doppler-broadened lineshapes was carried out by using the conventional DSA method [5] or the newly developed NGTB method[6] and shell-corrected Northcliffe-Schilling stopping powers [7]. In the conventional DSA analysis, gates are set onto a lower-lying transition and the appropriate feeding pattern has to be inserted including the average side-feeding time τSF, which is treated as an adjustable parameter. On the other hand, as discussed by Brandolini and Ribas[6], the NGTB method avoids the problem of delayed side-feedings by setting appropriate coincidence gates onto certain fractions of the subsequent Doppler-broadened transitions. An example of measured and fitted lineshapes of the 422 keV transition (DSA and NGTB) is presented in Fig. 1.

3. Discussion

The lifetime values deduced from both methods as well as those obtained previously are listed in Table 1. Within the typical errors of 5-15 %, the agreement between DSA and NGTB is good. As the NGTB results do not depend on the sidefeeding times, we have used these values to derive the M1 transition strengths listed in the last column of Table 1 and displayed in Fig. 2. We found a pronounced increase in M1 strength to occur between spins (16^-) and (17^-), which was superimposed over the general decrease of B(M1) for rising spin I, typical of magnetic rotation.

Table 1. Measured lifetimes and deduced M1 transition strengths in ^{104}In.

E_x (keV)	I^π	E_γ (keV)	NGTB	τ(ps)	DSA	B(M1) (μ_N^2)
3816	(13^-)	256		$<5^a$		>0.13
4102	(14^-)	286		$0.86(7)^a$		2.8(2)
4652	(15^-)	540	0.18(3)		0.18(2)	$2.0(2)^b$
5202	(16^-)	561	0.31(2)		0.33(1)	$1.04(7)^b$
5625	(17^-)	422	0.39(5)		0.45(3)	$1.9(3)^b$
6152	(18^-)	527			<0.82	>0.47

Note: a)Measured by DDCM[3]; b) Deduced from τ(NGTB)

In their comprehensive high-spin study of ^{104}In, Kast et al.[3] performed shell model calculations with the proton-hole in the $g_{9/2}$ or $p_{1/2}$ orbit, and the five neutrons distributed over the $s_{1/2}$, $d_{3/2}$, $d_{5/2}$, $g_{7/2}$ and $h_{11/2}$ orbits. At negative parity, configurations of the structure $[\pi p_{1/2} \otimes \nu^{3,5}(d_{5/2}g_{7/2})]$ and $[\pi g_{9/2} \otimes \nu^{2,4}(d_{5/2}g_{7/2})\nu h_{11/2}]$ were considered, i.e. the negative parity is either due to the $p_{1/2}$ proton-hole or the $h_{11/2}$ neutron-particle. For information on the choice of single-particle energies, two-body matrix elements and effective single-particle charges and magnetic moments, see Reference [3]. Concerning the individual spin couplings of the various valence particles, these calculations resulted in fairly mixed shell model partitions, but well-defined proton and neutron seniorities, $\nu_\pi = 1$ and $\nu_\nu = 3$ or 5. At spins $I = 8^- - 12^-$, the predicted configurations are of type A $[\pi p_{1/2} \otimes \nu^5(dg)]$ and reach up to spin 12^-. On the other hand, the predicted yrast configurations in the spin range $I = 12^- - 20^-$ are of type B $[\pi g_{9/2} \otimes \nu^4(dg)\nu h_{11/2}]$, with $I = 20$ being the maximum achievable spin of this configuration. Note that two close-lying 12^- states are indeed predicted by these calculations, which is in agreement with observation. The partitions of the type-B states involve the sub-configuration $[\pi gp_{9/2} \otimes \nu h_{11/2}]$, which may be considered responsible for the shears mechanism. In Table 2, the leading type-B configurations for the yrast band are listed. It is interesting to note that at spin 16^- there occurs a change in neutron seniority from $\nu_\nu = 3$ to $\nu_\nu = 5$. This change is responsible for the small $16^- \rightarrow 15^-$ M1 strength, since none of the large components of the 16^- and 15^- states can be connected by an allowed M1 transition. On the other hand, the very similar partitions of the 16^- and 17^- states lead to a large M1-strength between them.

In conclusion, careful lifetime measurements in the negative-parity yrast band of ^{104}In revealed deviations from the general shears-band type decrease in M1 strengths for increasing spin values, which are due to a change in neutron seniority in the $d_{5/2}$, $g_{7/2}$ sub-configuration. Similar effects have recently been found in the negative-parity yrast structures of ^{101}Ag and ^{95}Ru by Galindo et al. [8,9]and in the N = 50 isotones ^{93}Tc, ^{94}Ru and ^{95}Rh by Jungclaus, Hausmann et al.[10,11].

Acknowledgments

This work has been supported by Deutsches BMBF, Bonn. The authors gratefully acknowledge the excellent measuring conditions at LNL and particularly of the GASP spectrometer.

Table 2. Main partitions of the calculated type-B wavefunctions of the states within the negative-parity dipole band in ^{104}In[3].

I^π	Partition[a]	Partition(%)	I^π	Partition[a]	Partition(%)
12^-	$\pi g\otimes\nu^3[(gd)_5h]_{21/2}$	6.8	15^-	$\pi g\otimes\nu^3[(gd)_6h]_{23/2}$	53.5
	$\pi g\otimes\nu^3[(gd)_3h]_{17/2}$	13.4		$\pi g\otimes\nu^3[(gd)_5h]_{21/2}$	9.7
	$\pi g\otimes\nu^3[(gd)_5h]_{17/2}$	8.2	16^-	$\pi g\otimes\nu^5[(gd^3)_7h]_{25/2}$	10.5
	$\pi g\otimes\nu^3[(gd)_6h]_{21/2}$	7.5		$\pi g\otimes\nu^5[(g^2d^2)_8h]_{27/2}$	18.2
13^-	$\pi g\otimes\nu^3[(gd)_5h]_{21/2}$	31.1		$\pi g\otimes\nu^5[(g^2d^2)_{10}h]_{31/2}$	12.8
	$\pi g\otimes\nu^3[(gd)_6h]_{21/2}$	11.3		$\pi g\otimes\nu^5[(gd^3)_8h]_{27/2}$	19.0
	$\pi g\otimes\nu^3[(gd)_5h]_{23/2}$	26.2	17^-	$\pi g\otimes\nu^5[(gd^3)_8h]_{27/2}$	31.1
14^-	$\pi g\otimes\nu^3[(gd)_6h]_{23/2}$	26.7		$\pi g\otimes\nu^5[(g^2d^2)_{10}h]_{31/2}$	27.8
	$\pi g\otimes\nu^3[(gd)_5h]_{21/2}$	26.2		$\pi g\otimes\nu^5[(g^2d^2)_8h]_{27/2}$	12.8

Note: a) The partition $\pi g\otimes\nu^n[(g^id^k)_{I'}h]_{I\nu}$ has a proton-hole in the $g_{9/2}$ orbit, a neutron particle in the $h_{11/2}$ orbit, i neutrons in the $d_{5/2}$ and k neutrons in the $g_{7/2}$ orbits the (i + k) neutrons being coupled to spin I'. The total angular momentum of all n neutrons is labelled I_ν. Evidently the neutron seniority is $\nu_\nu = i + k + 1$.

References

1. G. Baldsiefen et al., *Nucl. Phys.* **A574**, 521 (1994).
2. S. Frauendorf and J. Reif, *Nucl. Phys.* **A621**, 738 (1997).
3. D. Kast et al., *Eur. Phys. J.* **A3**, 115 (1998).
4. A. Johnson et al., *Nucl. Phys.* **A557**, 401c (1993); D. Sewernyak et al., *Nucl. Phys.* **A589**, 175 (1995)
5. K. P. Lieb, in *Experimental Techniques in Nuclear Physics*, D. A. Poenaru and W. Greiner, Eds. (de Gruyter, Berlin, 1997) p. 425.
6. F. Brandolini and R. V. Ribas, *Nucl. Instr. and Meth.* **A417**, 150 (1998).
7. L. C. Northcliffe and R. F. Schilling, *Nucl. Data Tabl.* **A7**, 233 (1970).
8. E. Galindo et al., *Phys. Rev.* **C64**, 034304 (2001).
9. E. Galindo et al., *Phys. Rev.* **C69**, 024304 (2004).
10. A. Jungclaus et al., *Nucl. Phys.* **A637**, 346 (1998); *Eur. Phys. J.* **A6**, 29 (1999); *Phys. Rev.* **C60**, 014309 (1999).
11. M. Hausmann et al., *Phys. Rev.* **C68**, 024309 (2003).

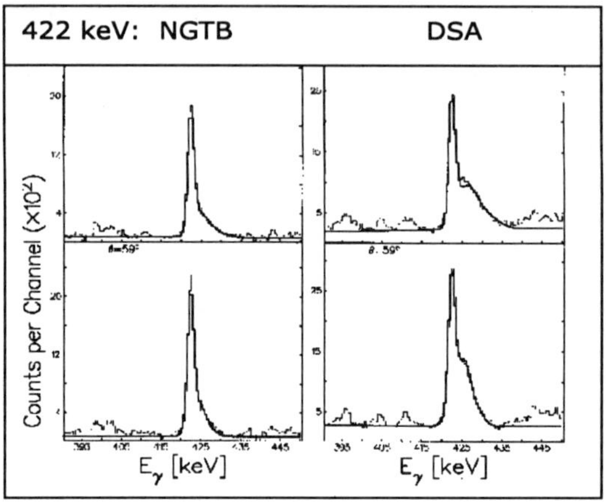

Figure 1. The 422 keV transition(NGTB, i.e. with timing condition: right) at $\theta = 35°$ (top) and 59° (bottom). The deduced lifetime is $\tau(16^-) = 0.31(2) ps$ (from NGTB).

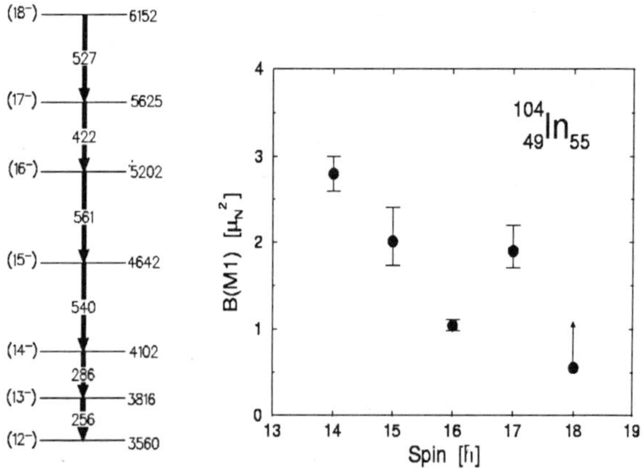

Figure 2. Negative-parity M1 cascade in ^{104}In (left) and B(M1) values (right).

KEY TOPICS IN NUCLEAR STRUCTURE
Proceedings of the 8th Int. Spring Seminar on Nuclear Physics
edited by Aldo Covello
© 2005 World Scientific Publishing Co.

STRUCTURE OF THE ^{100}SN REGION BASED ON A CORE EXCITED $E4$ ISOMER IN ^{98}CD

M. GÓRSKA, A. BLAZHEV*, H. GRAWE, J. DÖRING, C. PLETTNER
GSI, Planckstr. 1, D-64291 Darmstadt, Germany

J. NYBERG
Department of Neutron Research, Uppsala University, Uppsala, Sweden

M. PALACZ
Heavy Ion Laboratory, Warsaw University, Warsaw, Poland

E. CAURIER, D. CURIEN, O. DORVAUX, F. NOWACKI
IReS, Strasbourg, Cedex 2, France

A. GADEA, G. DE ANGELIS
INFN Laboratori Nazionali di Legnaro, Legnaro, Italy

C. FAHLANDER, D. RUDOLPH
Division of Cosmic and Subatomic Physics, Lund University, Sweden

The status of the experimental approach to ^{100}Sn is reviewed. In particular the nuclei in the closest vicinity of the doubly magic nucleus and with an isomeric high spin state are addressed. The nature of isomeric states is discussed, and the wave function analysed in terms of measured decay transition probabilities. The observation of the high spin isomer in ^{98}Cd in an experiment at EUROBALL IV yields information on single particle structure, size of the shell gap for ^{100}Sn and solves the puzzle of contradictory results of previous measurements. Experimental results are compared to the empirical and large-scale shell model calculation employing realistic interactions. The data on ^{98}Cd is discussed together with a survey of known and predicted isomeric states in this region.

*Present address: University of Sofia, Sofia, Bulgaria

1. Introduction

The region of nuclei near doubly magic ^{100}Sn has attracted recently a large experimental and theoretical interest. This is triggered by the trend of designing experiments where an efficient γ-array is combined with a highly selective method of production or separation of the nuclei of interest. This experimental development motivated an effort from the theory side employing modern shell model codes capable to deal with a large configuration space [1]. Such calculation in combination with a realistic residual interaction [2] gives a constructive input to understanding the structure of this region. ^{100}Sn as the heaviest doubly magic nucleus with $N = Z$ is a principle test ground for nuclear models to get a deeper insight into the nature of the proton-neutron interaction in nuclear medium.

The single particle structure of ^{100}Sn as extracted from the empirical shell model analysis of neighbouring nuclei [3] shows close resemblance to ^{56}Ni, one major shell lower and likewise a ls-open doubly magic core. This is corroborated by nearly the same $I^{\pi} = 2^{+}$ excitation energies in the neighbouring isotopes and isotones of ^{100}Sn and ^{56}Ni [4]. The shell gap energy is also expected to be of a similar size in both doubly magic nuclei. The correspondence of shell structure in different regions of nuclei provides a key evidence for further conclusions on the nucleon-nucleon interaction in general. As an example in Fig. 1 level schemes are shown comparing ^{54}Fe [5] and the recently measured ^{98}Cd [6,7], both two-proton hole nuclei in with respect to a doubly magic core.

The experimental approach to nuclei close to ^{100}Sn known with excited states is demonstrated by two proton-hole neighbour ^{98}Cd [6] and its neutron counterpart ^{102}Sn [8]. They are both known to have an isomeric high spin state. The observed levels below the isomeric states gave the first direct experimental information on the in-medium residual interaction in this region of nuclei. The isomeric states in this context by definition are those with lifetimes measurable by electronic means, i.e. in the ns – ms range, which allows for discrimination against prompt contamination by less exotic events. The three-particle neighbours have no isomer and hence are known only scarcely as the detection methods are less selective in such cases. They are ^{103}Sn [9] and ^{101}In [10]. From the levels observed in ^{103}Sn the more direct information on the single particle level above the ^{100}Sn core, i.e. the $g_{7/2}$-$d_{5/2}$ splitting energy was inferred.

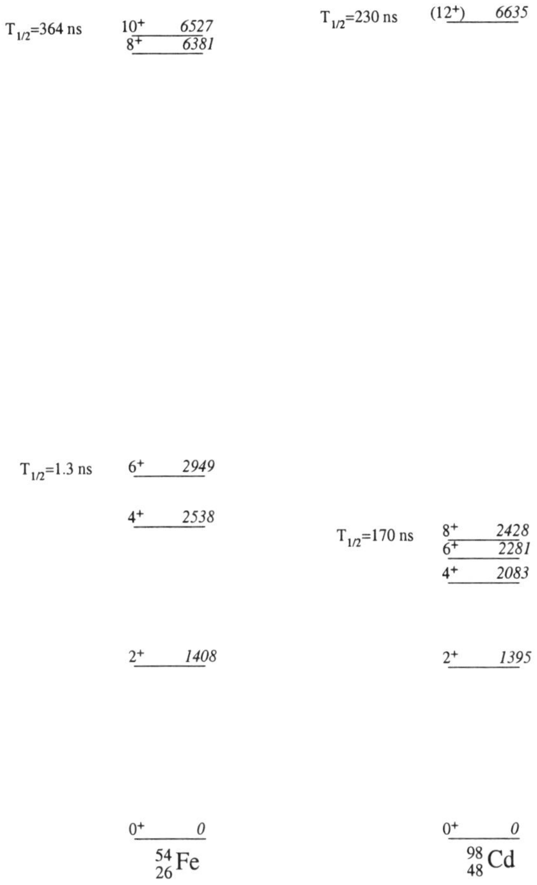

Figure 1. Partial experimental level schemes for the two-proton hole nuclei ^{54}Fe and ^{98}Cd with respect to the corresponding doubly magic nuclei.

2. Experimental Methods

Structure studies of exotic nuclei in γ-ray spectroscopy require both highly efficient and versatile γ arrays and selective ancillary devices which have a decisive role for investigation and identification of excited states in nuclei produced with the $\leq 1 \times 10^{-5}$ fraction of the total production cross section.

The most traditional method yielding highest production cross section and the highest spin population along the yrast line is the use symmetric fusion reaction in in-beam spectroscopy. The employment of large γ-ray arrays like EUROBALL [11], GAMMASPHERE [12] or GASP [13] in combination with ancillary detectors for the reaction exit channel identification as Neutron Wall [14] and Euclides charge particle ball [15] is optimal. When an isomeric state in the nucleus of interest is expected an additional catcher foil for reaction products is also placed at a certain distance behind the thin target. Additional Ge detectors are mounted around this foil, which yield only the delayed coincidence data and hence of much higher purity [6]. The prompt-delayed coincidences can be also studied as it was recently realised in a EUROBALL experiment [7] in which a core excited state in ^{98}Cd was found.

The same reactions are used in the standard β-decay experiments exploiting ISOL type facilities and the ion source chemistry [16] or resonant laser ionisation (see Ref. [17] and references therein). These are a very selective tools for the decay of high spin isomeric states, so called spin-gap isomers. They undergo β decay, or high-multipolarity γ-ray emission. The later scenario is detected by γ-γ coincidence and aniticoincidence with registered β particles [18]. This method is best applicable to the investigation of states with lifetimes $\geq 10ms$.

An alternative method for production of nuclei in this region far off stability is fragmentation. The fragments are fully identified event by event with respect to mass, charge and their trajectory. The last observable is crucial when the Doppler correction has to be done for γ rays emerging from the reaction of the fragment nuclei with a secondary target [19]. However, in the cases considered in this paper the fragment ions were implanted in the catcher and isomer deexitation γ-rays were measured [20]. It should be noted that the angular momentum gained by the fragments in these reactions is much lower on average than in the fusion reactions and not necessarily yrast states are populated [21].

The use of radioactive nuclei produced either by fragmentation [22,19]or spallation [23] for the secondary reactions like Coulomb excitation, fragmentation, transfer, knock-out etc., and further performing in-beam spectroscopy is a new field opening all over the world. New and complementary physics information is supposed to be delivered from these experiments in the future.

3. Core Excited Isomeric States and their Structure

Since the first observation of excited states in ^{98}Cd via population of an $I^{\pi} = 8^{+}$ isomer [6] in a fusion reaction , the extremely small proton polarisation charge $\delta e_{\pi} < 0.1$ e as derived, assuming a pure proton hole valence configuration, from a measured half-life, was a puzzling fact. Moreover, a follow-up experiment employing fragmentation and in-flight separation [20] yielded much shorter half-life and $\delta e_{\pi} = 0.3$ e. This value was very well reproduced by realistic shell model calculations where the two-body matrix elements were derived in a consistent way with the one-body operators describing the transition strength [24]. The right solution to this puzzle was proposed already in a previous paper by invoking the existence of a second, core excited $I^{\pi} = 12^{+}$ isomer [25]. As discussed before, the high spin isomer will be produced in fusion reaction with a higher cross section (isomeric ratio). The resulting level scheme of ^{98}Cd including the half-lives of both isomers is shown in Fig. 1. The new result on the proton polarisation charge is now consistent with Ref.[20]. The data was compared to large scale shell model (LSSM) calculations including up to 4p4h excitation of the ^{100}Sn core and using a realistic interaction which reproduced the experimental energies very well. On the other hand an empirical shell model calculation in a small configuration space were performed. The analysis of the two independent approaches provided a consistent information on the size of the ^{100}Sn shell gap [7].

In Fig. 2 a mixture of theoretical and experimental data is shown to visualize the present knowledge about isomeric states in the closest vicinity of ^{100}Sn. Until recently only the $I^{\pi} = 21/2^{+}$ spin-gap beta decaying isomer in ^{95}Pd has been observed. The information has been greatly expanded by the identification of the predicted $23/2^{+}$ and the unexpected $37/2^{+}$ isomers in ^{95}Ag[18] which becomes isomeric only by regarding excitation of the ^{100}Sn core. Using the same parameters of the LSSM calculation for the 21^{+} state in ^{94}Ag, the highest spin β decaying state ever found was very well reproduced, which could not be accounted for in the small configuration space [26]. The prediction of a further $I^{\pi} = 14^{+}$ spin trap in ^{98}Cd positioned in energy below the 12^{+} state by the LSSM calculation invokes the existence of an $E2$ isomer with spin 6^{+} in ^{100}Sn.

4. Predicted Seniority Isomers

There are several known cases of the pure configuration seniority isomers in the ^{100}Sn region as already mentioned here: ^{98}Cd, ^{102}Sn, and ^{94}Pd [27]

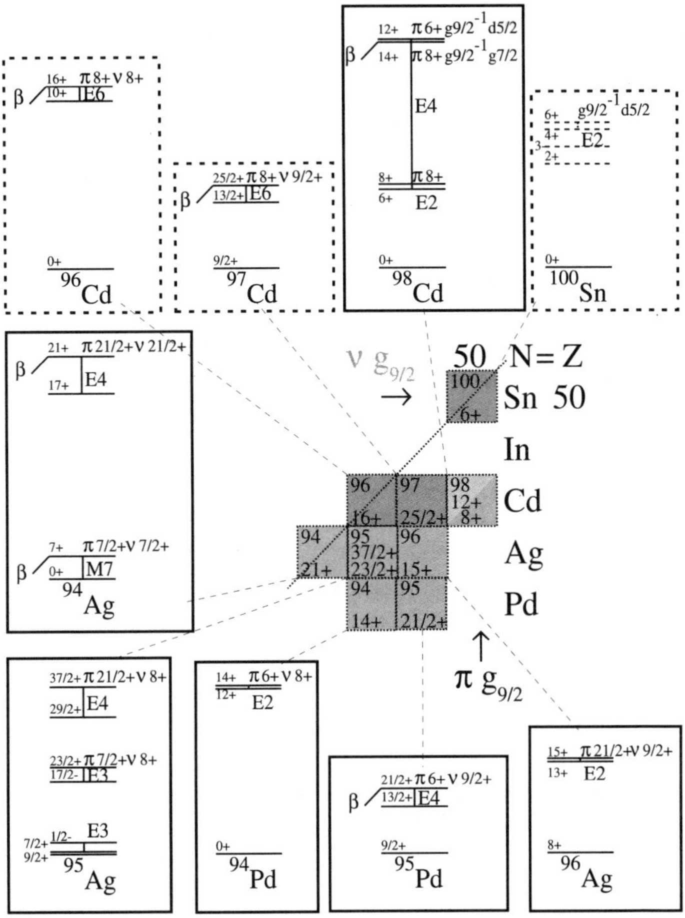

Figure 2. Simplified level schemes of nuclei near ^{100}Sn with existing (solid line rectangles) and/or predicted (dashed line rectangles) isomeric state.

which brought crucial information on the basic shell structure and residual interaction in this region. Further spin-gap isomers originating from the strong proton-neutron interaction in identical orbitals are predicted in ^{96}Ag ($E2$), ^{97}Cd ($E6$) and ^{96}Cd ($E6$). A beta-delayed proton decay observed in ^{97}Cd [28] is thought to originate from a $25/2^+$ high-spin state. Similarly, the $N = Z$ isotope ^{96}Cd is expected to reveal a highly energetically favoured beta-decaying 16^+ state. The latter nucleus was identified in projectile fragmentation reactions (see e.g., Ref.[20]), however to date no spectroscopic information is known.

5. Conclusion

Rich amount of new data available in the ^{100}Sn region shows a vivid interest in this field. The structure of observed states proved an importance of the LSSM calculation in combination with realistic residual interaction. The identification of the predicted isomeric states is a challenge for the future experiments employing most likely radioactive ion beams separated in-flight. However, the stable beam facilities will be useful to continue the research on selected non-isomeric high spin states, as e.g. magnetic rotational bands in a close vicinity of ^{100}Sn.

Acknowledgments

The authors gratefully acknowledge the comments by J. Blomqvist to the nature of core-excited isomers near ^{56}Ni and ^{100}Sn. The Swedish coauthors acknowledge the help of the Swedish Research Council and the AIM Graduate Education Programme at Uppsala University. The work was partly supported by the EU under contract HPRI-CT-1999-00078 and partly by the Polish Committee of Scientific Research (grants no. 5P03B 046 20 and 1P03B 031 26).

References

1. F. Nowacki, *Nucl. Phys.* **A704**, 223c (2002).
2. M. Hjorth-Jensen et al., *Phys. Rep.* **261**, 125 (1995).
3. H. Grawe et al., *Phys. Scr.* **T56**, 71 (1995).
4. H. Grawe et al., *Z. Phys.* **A358**, 185 (1997).
5. M.H. Rafailovich et al., *Phys. Rev.* **C27**, 602 (1983).
6. M. Górska et al., *Phys. Rev. Lett.* **79**, 2415 (1997).
7. A. Blazhev et al., *Phys. Rev.* **C69**, 064304 (2004).
8. M. Lipoglavšek et al., *Phys. Lett.* **440B**, 246 (1998).
9. C. Fahlander et al., *Phys. Rev.* **C63**, 021307(R) (2000).
10. C. Cederkäll et al., *Phys. Rev.* **C53**, 1955(R) (1996).
11. J. Simpson, *Z. Phys.* **A358**, 139 (1997).
12. I. Y. Lee, *Prog. Part. Nucl. Phys.* **38**, 65 (1997).
13. C. Rossi-Alvarez, *Nuclear Physics News* **3**, 10 (1993).
14. Ö. Skeppstedt et al., *Nucl. Instrum. Methods Phys. Res.* A**421**, 531 (1999).
15. E. Farnea et al., *Nucl. Instrum. Methods Phys. Res.* A**400**, 87 (1997).
16. R. Kirchner, *Nucl. Instrum. Methods Phys. Res.* B**70**, 186 (1992).
17. Y. Kudryavtsev et al., *Nucl. Instrum. Methods Phys. Res.* B**179**, 412 (2001).

18. J. Döring et al., *Phys. Rev.* **C 68**, 034306 (2003).
19. H.J. Wollersheim et al., *Nucl. Instrum. Methods Phys. Res.*, submitted.
20. R. Grzywacz et al., *Proc ENAM 98, AIP Conf. Proc.* **455**, 257 (1998).
21. M. de Jong et al., *Nucl. Phys* **A628**, 479 (1998).
22. T. Glasmacher, *Annu. Rev. Nucl. Part. Sci.* **48**, 1 (1998).
23. D. Habs et al., *Hyperfine Interactions* **129**, 43 (2000).
24. L. Coraggio et al., *J. Phys. G: Nucl. Part. Phys.* **26** 1697 (2000).
25. H. Grawe et al., *Nucl. Phys.* **A704**, 211c (2002).
26. C. Plettner et al., *Nucl. Phys.* **A733**, 20 (2004).
27. M. Górska et al., *Z. Phys.* **A353**, 233 (1995).
28. T. Elmroth et al., *Nucl. Phys.* **A304**, 493 (1978).

KEY TOPICS IN NUCLEAR STRUCTURE
Proceedings of the 8th Int. Spring Seminar on Nuclear Physics
edited by Aldo Covello
© 2005 World Scientific Publishing Co.

NEW YRAST STATES IN NUCLEI FROM THE ^{48}Ca REGION STUDIED WITH DEEP-INELASTIC HEAVY ION REACTIONS

R. BRODA, B. FORNAL, W. KRÓLAS, T. PAWŁAT, J. WRZESIŃSKI
Institute of Nuclear Physics PAN,
PL-31-342 Kraków, Poland
E-mail: rafal.broda@ifj.edu.pl

R.V.F. JANSSENS, M.P. CARPENTER, S.J. FREEMAN, N. HAMMOND,
T. LAURITSEN, C.J. LISTER, F. MOORE, D.SEWERYNIAK
Physics Division, Argonne National laboratory,
Argonne, Illinois 60439, USA

P.J. DALY, Z.W. GRABOWSKI
Chemistry and Physics Departments, Purdue University,
West Lafayette, Indiana 47907, USA

B.A. BROWN
National Superconducting Cyclotron Laboratory and Department of Physics and Astronomy, Michigan State University,
East Lansing, Michigan 48824, USA

M. HONMA
Center for Mathematical Sciences, University of Aizu, Tsuruga, Ikki-machi,
Aizu Wakamatsu, Fukushima 965-8580, Japan

Data from three gamma spectroscopy experiments using deep-inelastic heavy ion reactions provided new information on high-spin states in the ^{48}Ca core nucleus and in the N=30, ^{50}Ca and ^{51}Sc isotones. Shell model calculations restricted to neutron excitations only are shown to reproduce with good accuracy some of the experimental levels. It is demonstrated that proton excitations not accounted in these calculations are abundantly present in the observed yrast structures. High energy of the 4^+ state in ^{50}Ca underlines the validity of the N=32 shell closure.

1. Introduction

The nuclei located in the doubly-magic ^{48}Ca region were until recently not accessible for spectroscopic studies of yrast states and the available information was restricted to excitations populated in the inelastic scattering, light particle transfer reactions and/or beta decay.[1] The development of gamma spectroscopy techniques which exploit deep-inelastic heavy ion reactions opened new way to study higher spin excitations, including also isotopes with larger neutron excess than the ^{48}Ca closed core. Efforts to obtain experimental information on yrast states, characterized by particularly high configuration purity, is vital for the refinement of the shell model description in this region of nuclei. Theoretical approaches to describe quantitatively observed structures involve large scale shell model calculations using various sets of phenomenological interactions[2,3,4] and attempts focused mainly in other regions to use effective interactions derived from the free nucleon-nucleon (NN) potential.[5]

A fairly successful quantitative description obtained for the recently established level structures in neutron-rich nuclei with Z>20 above ^{48}Ca, specifically in the heavy Ti isotopes[6] indicates that reasonable choice of effective interactions was made for the $f_{7/2}$ protons and neutron orbitals located above the N=28 closed shell. Here e.g. the experimentally established effect of the N=32 subshell closure could be clearly accounted for by reproducing the increase of the 2^+ energy in the calculated spectrum for the ^{54}Ti isotope. In calcium isotopes the N=32 subshell closure is even more directly demonstrated by the high energy of the 2^+ state in ^{52}Ca, but generally Ca yrast excitations are dominated by the ^{48}Ca core excitations which require the knowledge of other, not so well known effective interactions. Therefore, experimental study of nuclei located in the closest neighborhood of ^{48}Ca might be crucial in providing the guidance for theoretical efforts.

Preliminary results obtained in our studies of yrast structures in the ^{48}Ca core nucleus, as well as in the closest one-particle and one-hole neighbors were already reported earlier.[7] In the present talk new results will be described concerning the high-spin excitations in the ^{48}Ca and in N=30 ^{50}Ca and ^{51}Sc isotopes.

2. Experiments and data analysis

Spectroscopic results presented below were obtained in the analysis of gamma-gamma coincidence data collected in three separate experiments performed within the recent few years. In all cases the pulsed ^{48}Ca beam was used to bombard thick targets of ^{48}Ca (backed with ^{208}Pb), ^{208}Pb and ^{238}U isotopes at energies of 210, 305 and 330 MeV correspondingly, i.e. significantly exceeding the Coulomb barrier. The first experiment was performed at the LNL Legnaro ALPI Linac accelerator for the symmetric system ^{48}Ca+^{48}Ca using the GASP multidetector array. The other two experiments, using the ^{208}Pb and ^{238}U targets, were performed at the ATLAS accelerator of the Argonne NL using the more powerful GAMMASPHERE detector array. Targets were located in the center of the detector array and multifold gamma coincidences were stored in a standard way including time parameters to separate prompt and delayed events. The final selectivity is obtained exclusively by the discrete gamma line analysis of the gamma coincidence data. Naturally the observation is restricted to gamma rays which are emitted from the recoiling products stopped in the target material, otherwise Doppler broadening prevents their detection. However, this condition is fulfilled for most of the yrast transitions as they usually involve state and/or feeding times longer than approx. 2 ps needed to stop the reaction product. In the ^{48}Ca+^{48}Ca experiment the original data were dominated by the fusion evaporation products, but spectacular enhancement of the deep-inelastic reaction products was achieved by additional gating on the low fold (1 to 7) part of the BGO multiplicity ball of the GASP array which largely rejected the high multiplicity fusion events. The data from experiments with the ^{48}Ca and ^{208}Pb targets provided clear-cut identification of the studied nuclei by the analysis of gamma cross-coincidences which correlate the presence of the corresponding partner nuclei in the reaction exit channel. In the case of the ^{238}U experiment the heavy fragment usually undergoes fission which obscures this type of analysis. It should be emphasized that the availability of the three independent sets of data helped to make many cross-checks and level schemes resulting from these complex analyses have to be considered as very solid. Moreover the production cross-section of the neutron-rich species increases with the N/Z of the target nucleus and consequently the ^{238}U target experiment provided the best data for the ^{50}Ca and ^{51}Sc isotopes discussed in Sec. IV.

3. Yrast excitations in the ^{48}Ca magic nucleus

The new data from the ^{48}Ca+^{238}U experiment revealed several high energy gamma transitions connecting earlier reported states in the ^{48}Ca isotope[7] with the sequence of lower lying states of well established spin-parity assignments. With this new input the data from previous experiments were reexamined and yielded the ^{48}Ca level scheme as shown in Fig.1. ¿From the data it could be concluded that the observed intensities of high energy transitions completely balance the feeding intensity from low energy transitions located in the upper part of the level scheme. It is obvious that the lifetimes of states which are initial for the new high energy transitions must be short and intensities reflect feeding from above, whereas any expected side feeding intensity is smeared out by the Doppler broadening and not observed. Consequently these transitions must be of an $E1$, $M1$ or $E2$ character which provides some clues for tentative spin-parity assignments of states above the 7 MeV excitation energy. In fact the suggested spin-parity assignments indicated in Fig.1 are based on such considerations, including also standard yrast population arguments. It is clear that the group of states above 7.5 MeV excitation energy represents two-particle two-hole core excitations with positive parity states arising from the two-proton or two-neutron excitations and negative parity involving combinations of proton and neutron one particle-hole excitations. The lower part of the Fig.1 level scheme, as discussed earlier[7], includes only the yrast levels populated in our experiments and does not indicate other known states e.g. those populated in the beta decay . In this part the state structure is rather transparent since for the ^{48}Ca core nucleus the proton and neutron one particle-hole excitations differ by the parity. The negative parity states arise from the proton excitation across the Z=20 gap and positive parity states are formed by the promotion of one neutron across the N=28 energy gap. Whereas the complete quadruplet of states arising from the neutron $f_{7/2}^{-1}p_{3/2}$ structure is strongly populated in the present experiments, some of the proton particle-hole excitations assuming non-yrast positions and not observed by us are known from other studies of e.g. ^{48}K beta decay[1]. Results of the shell model calculations based on effective interactions as described in Ref. 3 are shown on the right hand side of Fig.1. Calculated states, all of positive parity, correspond to neutron excitations only, since the knowledge of important proton interactions is much less advanced. For the calculated states the agreement with experiment is rather satisfactory and in the upper part of the level scheme it serves as indication of possibly predominant neutron structure of few states. As the two-proton particle-hole states of positive parity and spin up to I=8 are expected in a similar energy range, it is obvious that some of the experimental states must arise from the coupling of proton and neutron one particle- hole excitations yielding negative parity states as tentatively indicated in Fig.1.

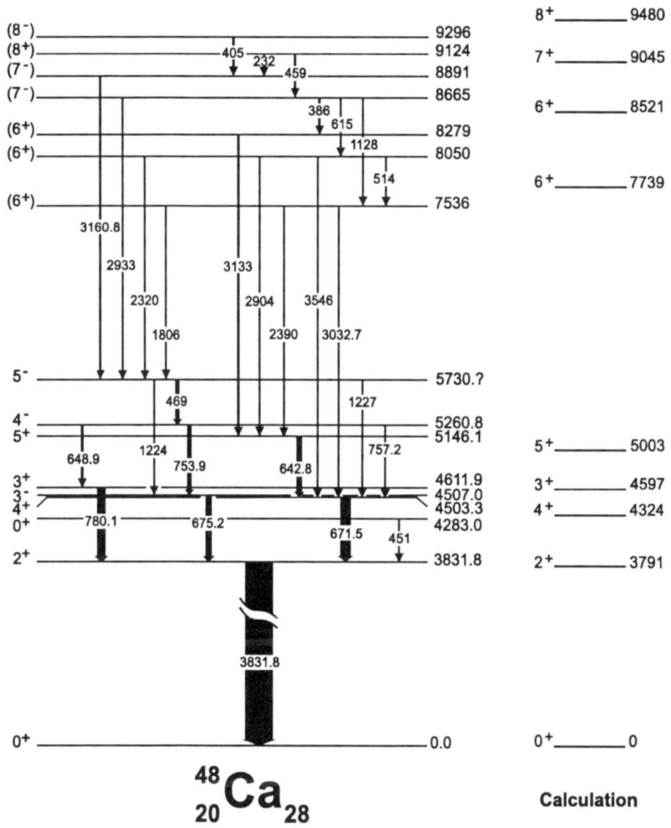

Figure 1. Yrast and near yrast levels of the ^{48}Ca doubly-magic nuleus. Shell model calculations shown to the right include neutron excitations only.

Here our reexamination of the ^{48}K (2^-) ground state beta decay revealed the existence of this type of state with $I^\pi = 3^-$ at 6.855 MeV energy. In the ^{238}U target experiment the ^{48}K isotope was produced with the satisfactory yield to enable the clean analysis of the off-beam triple gamma coincidence data. This provided new information which completes the earlier detailed study of this beta decay.[1] A new, rather strongly directly populated state identified in the ^{48}K beta decay was characterized as 3^-state. The consideration of all possible final states leads to the conclusion that apparently it must correspond to the complex proton and neutron one particle-hole excitation state populated in direct beta decay process by changing the $f_{7/2}$ neutron into $f_{7/2}$ proton.

4. The N=30 isotones – ^{50}Ca and ^{51}Sc

The production of the ^{50}Ca and ^{51}Sc isotopes could be observed in all three experiments and important initial identifications were obtained from the analysis of the ^{48}Ca and ^{208}Pb target run data. However, the much more favorable yield observed in the ^{238}U target experiment provided the highest statistics data which allowed to extract main information on yrast structures of these N=30 isotones. Before starting our experiments all available information concerning the ^{50}Ca isotope was limited to the first excited 2^+ state at 1026(1) keV and a few higher lying low-spin states with poorly defined energies, studied in the beta decay, (t,p) and $(\alpha,2p)$ reactions.[1] The first analysis of the deep-inelastic spectroscopy data revealed a strongly populated sequence of 595, 3488 and 1027 keV yrast transitions, clearly identified with the ^{50}Ca on the basis of gamma cross-coincidence analysis. The clean coincidence spectra shown in Fig.2 and obtained from two sets of data (indicated in the figure) by selecting the double 1027-3488 keV transition gates, demonstrate this identification. In the upper spectrum, besides several ^{50}Ca transitions, one observes lines corresponding to the well known gamma transitions in the ^{206}Pb and ^{205}Pb, the expected main reaction partner nuclei. In the lower spectrum a presence of weak intensity lines from the ^{236}U ground state rotational band confirms this identification and the approx. 7 times higher statistics observed for the ^{50}Ca lines shows the superior quality of the ^{238}U data in the analysis of neutron-rich species. The detailed analysis of all available data allowed to construct the ^{50}Ca level scheme as shown in Fig.3 which includes all transitions that could be safely identified with this isotope. The tentative spin-parity assignments are based on yrast population arguments, the observed gamma decay and guided by the shell model calculations made for neutron excitations only and shown to the left.

The assignment of the yrast 4^+ state at 4515 keV energy has to be considered as most likely and the calculation attributes it to the neutron particle-hole core excitation rather than involving the high energy $f_{5/2}$ orbital. Apparently the latter structure must be located at yet higher energy which reemphasizes the validity of the N=32 shell closure originating from the large energy gap between the $p_{3/2}$ and $f_{5/2}$ neutron orbitals. Originally the strongly populated yrast state at 5110 keV was anticipated to be the calculated 5^+ arising also from the neutron particle-hole excitation, however the identification of the 3^- state at 3997 keV energy and the connecting transition of 1113 keV excluded such possibility. The location of the 3997

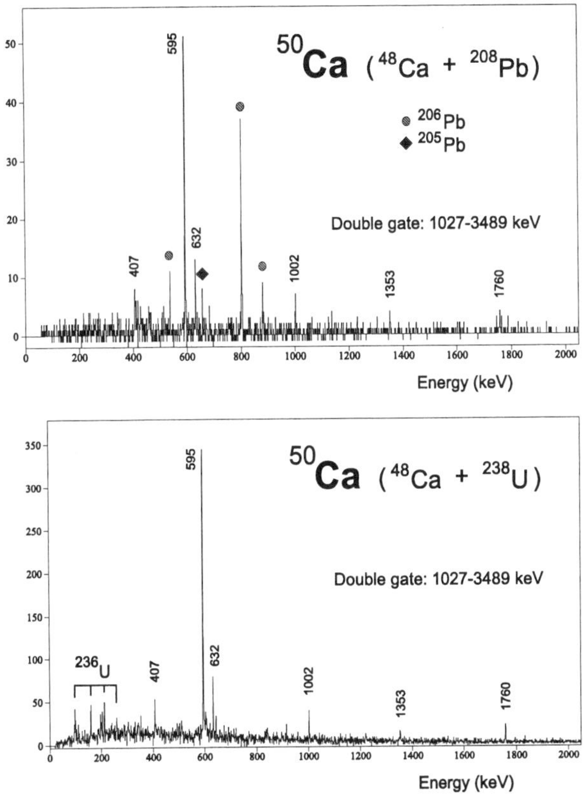

Figure 2. Gamma coincidence spectra with double gates set on 1027 - 3489 keV transitions in ^{50}Ca obtained in the ^{208}Pb (upper) and ^{238}U (lower) target experiments. Transitions with indicated energies belong to ^{50}Ca and other transitions marked arise from gamma cross-coincidences with corresponding reaction partner nuclei.

keV level assigned as 3^- was confirmed in the data of all three experiments and is consistent with the earlier identified level at 3993(15) keV and assigned as 3^- in the (t,p) reaction.[1] The most likely assignment for the 5110 keV level is therefore 5^- and the 5^+ assignment is naturally attributed to the close lying level at 5147 keV energy which fits even better the calculated 5^+ state. On the right hand side of Fig.3 the lowest ^{48}Ca core excitations are indicated to be compared with the level energies observed in ^{50}Ca. In this comparison the negative parity proton core excitations not accounted in calculations are most important. It is apparent that in the ^{50}Ca case the possible coupling with the low energy 2^+ excitation of two extra neutrons gives rise to additional negative parity states which assume near yrast po-

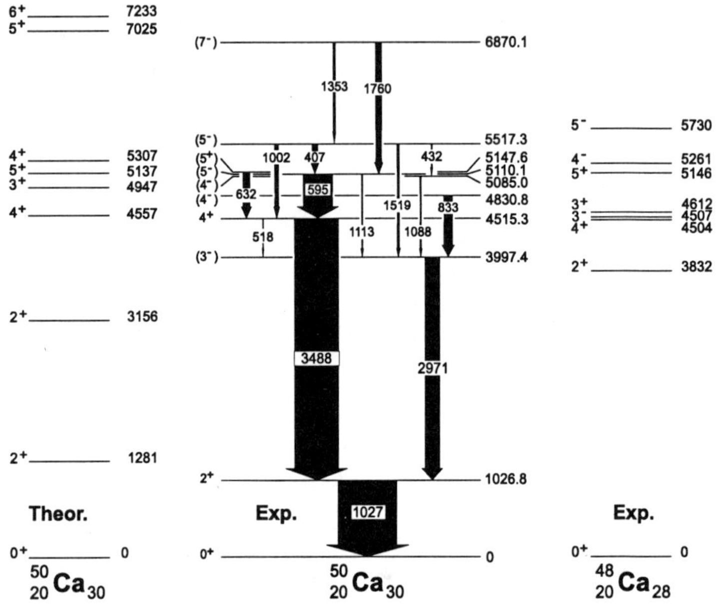

Figure 3. Yrast level scheme of the ^{50}Ca nucleus. Shell model calculated neutron excitations are shown to the left and relevant ^{48}Ca core levels are indicated to the right.

sitions. The highest energy level at 6870 keV is suggested to be the 7^- yrast state with maximum spin available for the proton particle-hole and the neutron 2^+ excitation.

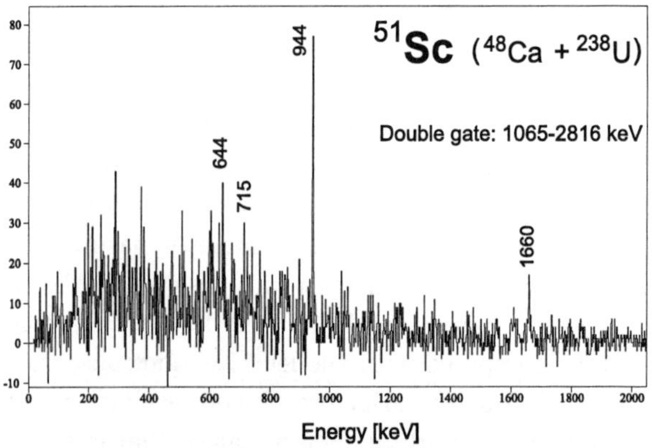

Figure 4. Gamma coincidence spectra obtained from the ^{48}Ca + ^{238}U experiment data with double gate set on the 1065 - 2816 keV yrast transitions in ^{51}Sc.

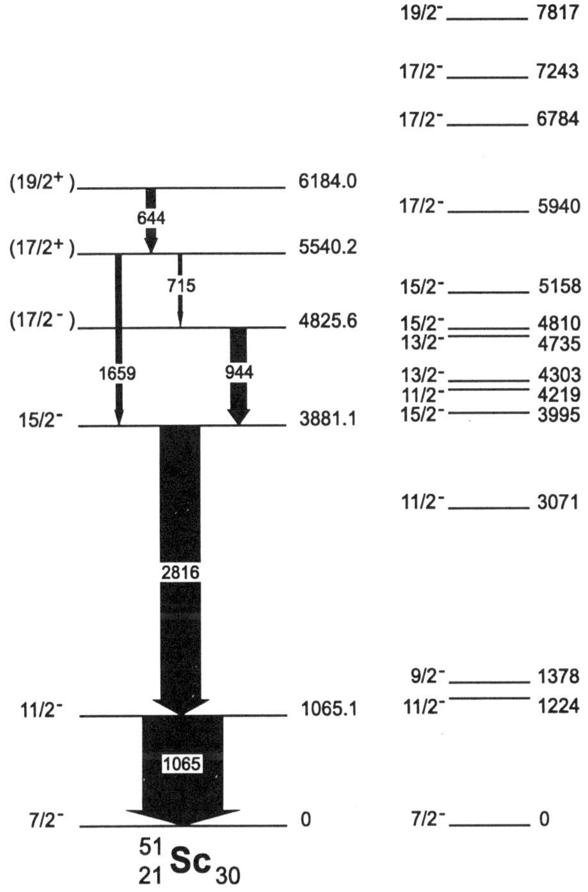

Figure 5. Level scheme of yrast excitations in the ^{51}Sc isotope. Shell model calculations shown to the right include neutron excitations only.

The odd proton ^{51}Sc N=30 isotone is even more difficult to study since in odd mass nuclei yrast excitations are less distinctly populated. From the available information the tentative location of the $11/2^-$ state at 1062(1) keV was the starting point of our analysis. The analysis involving identification procedures established the 1065 – 2816 keV coincident transitions to form the lowest yrast sequence in the ^{51}Sc. Fig.4 shows the corresponding double gate gamma coincidence spectrum displaying higher lying transitions which were located in the level scheme shown in Fig.5. The shell model calculated levels restricted to the coupling of the $f_{7/2}$ proton with neutron excitations are shown to the right of Fig.5. The $11/2^-$ and $15/2^-$

assignments for correspondingly the 1065 and 3881 keV levels are consistent with calculations and strong yrast feeding observed in experiment. Assignments for higher lying states are much less certain, however the calculated sequence does not allow to correlate the observed levels with theoretical ones and clearly some positive parity states arising from the proton core excitations must enter the consideration. The most important feature is that the observed $15/2^-$ level which corresponds to the 4^+ ^{50}Ca yrast excitation is lowered in energy by more than 600 keV and that this lowering is well accounted by interactions used in calculations.

In summary, the gamma spectroscopy experiments using deep-inelastic heavy ion reactions provided new information on high-spin states in the ^{48}Ca core nucleus and revealed yrast structures of hitherto largely unknown yrast structures in N=30 isotones.

The shell model calculations reproduce with good accuracy experimental levels for excitations involving neutron orbitals. The proton core excitations play a major role in observed yrast structures and efforts to establish appropriate effective interactions able to involve such states in calculations seem to be an urgent task for theoreticians.

Acknowledgments

This work was supported by the U.S. Department of Energy, Nuclear Physics Division, under Contracts Nos. W31-109-ENG-38 and DE-FG02-87ER40346, by National Science Foundation Grant Nos. PHY-01-01253, PHY-00-70911 and PHY-97-24299, and by Polish Scientific Committee Grant No. 2PO3B-074-18.

References

1. *Nuclear Data Sheets* **68**, 1 (1993) Updated (1995), **75**,1 (1995), **81**,183 (1997).
2. B.A.Brown et al., The comp.code OXBASH, *MSU NSCL Report* No. **524** , 1998.
3. M.Honma et al.,*Phys. Rev.* **C65**, 061301R (2002).
4. A.Poves et al., *Nucl. Phys.* **A694**, 157 (2001).
5. A.Covello et al., *Proc. of the 7th Int. Spring Seminar on Nuclear Physics*, p.139, 2002 World Scientific Publishing Co.
6. B.Fornal et al., *contribution to this conference and references therein.*
7. R.Broda et al., *Acta Phys.Pol.* **B32**, 2577 (2001) and *Proc. of the 7th Int.Spring Seminar on Nuclear Physics*, ed. by A.Covello, p.195, 2002 World Scietific Co.

KEY TOPICS IN NUCLEAR STRUCTURE
Proceedings of the 8th Int. Spring Seminar on Nuclear Physics
edited by Aldo Covello
© 2005 World Scientific Publishing Co.

YRAST STRUCTURE OF NEUTRON-RICH N=31,32 TITANIUM NUCLEI – SUBSHELL CLOSURE AT N=32

B. FORNAL, R. BRODA, W. KRÓLAS, T. PAWŁAT, J. WRZESIŃSKI

Institute of Nuclear Physics, Polish Academy of Sciences, PL-31342 Cracow, Poland

R.V.F. JANSSENS, M.P. CARPENTER, F.G. KONDEV, T. LAURITSEN, D. SEWERYNIAK, I. WIEDENHÖVER

Physics Division, Argonne National Laboratory, Argonne, Illinois 60439

M. HONMA

Center for Mathematical Sciences, University of Aizu, Tsuruga, Ikki-machi, Aizu-Wakamatsu, Fukushima 965-8580, Japan

B.A. BROWN, P.F. MANTICA

National Superconducting Cyclotron Laboratory, Michigan State University, East Lansing, Michigan 48824, USA

P.J. DALY, Z.W. GRABOWSKI

Chemistry and Physics Departments, Purdue University, West Lafayette, Indiana 47907, USA

S. LUNARDI, N. MARGINEAN, C. UR

Dipartimento di Fisica dell'Università di Padova, I-35131 Padova and, INFN, Laboratori Nazionali di Legnaro, I-35020 Legnaro, Italy

T. MIZUSAKI

Institute of Natural Sciences, Senshu University, Higashimita, Tama, Kawasaki, Kanagawa 214-8580, Japan

T. OTSUKA

Department of Physics, University of Tokyo, Hongo, Tokyo 113-0033, Japan and RIKEN, Hirosawa, Wako-shi, Saitama 351-0198, Japan

Gamma rays from neutron-rich Ti nuclei in the vicinity of $N = 32$ have been studied at Gammasphere using deep-inelastic reactions induced by a 305 MeV ^{48}Ca beam on a thick ^{208}Pb target. The yrast γ-ray cascades in ^{53}Ti were identified for the first time and the location in energy of the states with spin up to J=21/2 was determined. The yrast excitations of ^{53}Ti, together with the earlier studied yrast structure of ^{54}Ti, provided new tests of effective interactions for full pf-shell calculations. The data confirm the presence of a significant subshell gap at N=32. Comparisons between theory and experiment regarding the highest spin states located in ^{53,54}Ti suggest that energy gap at N=34 in neutron-rich nuclei is not as large as predicted by the recently proposed GXPF1 interaction.

1. Introduction

As the studies of exotic nuclei advance, unforeseen modifications to the shell structure have been observed and it is possible that traditional magic numbers may not be valid away from the valley of stability, especially in neutron-rich systems. One of the causes of the reordering of single particle orbits and, as a result, of the development of new subshell closures, could be the proton-neutron monopole interaction[1]. Examples of such structural changes, discussed recently, include the appearance of an energy gap at N=32 for neutron-rich nuclei just above ^{48}Ca. This phenomenon was first suggested by A. Huck et al.[2] who tentatively identified a candidate for the 2_1^+ state in ^{52}Ca with an excitation energy of 2563 keV that is significantly higher than the corresponding 2^+ energy in ^{50}Ca. Another evidence for a subshell closure at N=32 came from the systematic variation of the E(2_1^+) energy for the chromium (Z=24) isotopes, which was found to reach a maximum for $^{56}Cr_{32}$[3]. Little was known, however, about excited states in the Z=22 titanium isotopes close to N=32.

In a series of past experiments, it was shown that the yrast spectroscopy of hard-to-reach neutron-rich nuclei populated in heavy-ion multi-nucleon transfer reactions (at energies 15-25% above Coulomb barrier), can be studied successfully in $\gamma - \gamma$ coincidence measurements with a thick target[4,5,6]. Production of neutron-rich species in these processes is possible due to a tendency towards N/Z equilibration of the di-nuclear system formed during collision. Usually, in the studied reactions the light colliding partner had lower N/Z ratio than the heavier one. In the region of the light reaction partner, such projectile-target combinations favor production of the more neutron-rich species than the reaction partner itself. Taking into account these considerations, the system ^{48}Ca+^{208}Pb seemed to offer good prospects in reaching neutron-rich titanium isotopes.

2. Experimental procedure and results

The experiment was performed at Argonne National Laboratory using Gammasphere[7], which consisted of 101 Compton-suppressed Ge detectors. A 305 MeV ^{48}Ca beam from the ATLAS accelerator was focused on a 50 mg/cm^2 ^{208}Pb target. Gamma-ray coincidence data were collected with a trigger requiring three or more Compton-suppressed γ rays to be in prompt coincidence. Energy and timing information for all Ge detectors that fired within 800 ns of the triggering signal was stored. A total of 8.1×10^8 three-fold and higher events were recorded. The beam, coming in bursts with $\sim$0.3 ns time width, was pulsed with an $\sim$400 ns repetition time, providing clean separation of prompt and isomeric events. Conditions set on the γ-γ time parameter were used to obtain various versions of prompt and delayed $\gamma\gamma$ and $\gamma\gamma\gamma$ coincidence matrices and cubes covering γ-ray energy ranges to $\sim$4 MeV.

In γ-ray spectroscopic studies of the deep-inelastic reaction products the identification of an unknown sequence of γ rays to a specific product may be possible by using the cross-coincidences with transitions in reaction partners. In the present work, complementary products in binary reactions leading to Ti isotopes are Hg nuclei, but a given Ti product is in coincidence with several Hg partners because of neutron evaporation from the fragments after the collision. The situation is illustrated in Fig. 1a, where the spectrum arising from a sum of double gates on known transitions in ^{196}Hg is presented. This spectrum, according to expectations, displays known lines from ^{50}Ti, ^{51}Ti and ^{52}Ti, which are partners to ^{196}Hg associated with 10, 9 and 8 evaporated neutrons, respectively. Further inspection of the spectrum revealed also the presence of unknown gamma rays at energies 1002, 1237, 1495 and 1576 keV. These gamma rays were observed also in the spectra gated on transitions from the Hg isotopes with A=197-200 which indicated that they originate from the titanium products. It seemed very likely that the new transitions occur in the Ti isotopes with masses A>52. Pursuing this hypothesis we applied the identification method based on the γ-ray cross-coincidence intensities. For the ^{50,51,52}Ti reaction products, the mean mass A_{av}(Hg) of the complementary mercury fragments was determined from the Hg γ-ray intensities measured in coincidence with the γ rays of that particular Ti isotope. The same procedure of calculating the Hg mean mass was applied also to newly found 1237, 1495 and 1576 keV Ti lines. The results are illustrated in Fig. 2, where the A_{av}(Hg) as a function of A(Ti) is shown. The data points corresponding to the known gamma

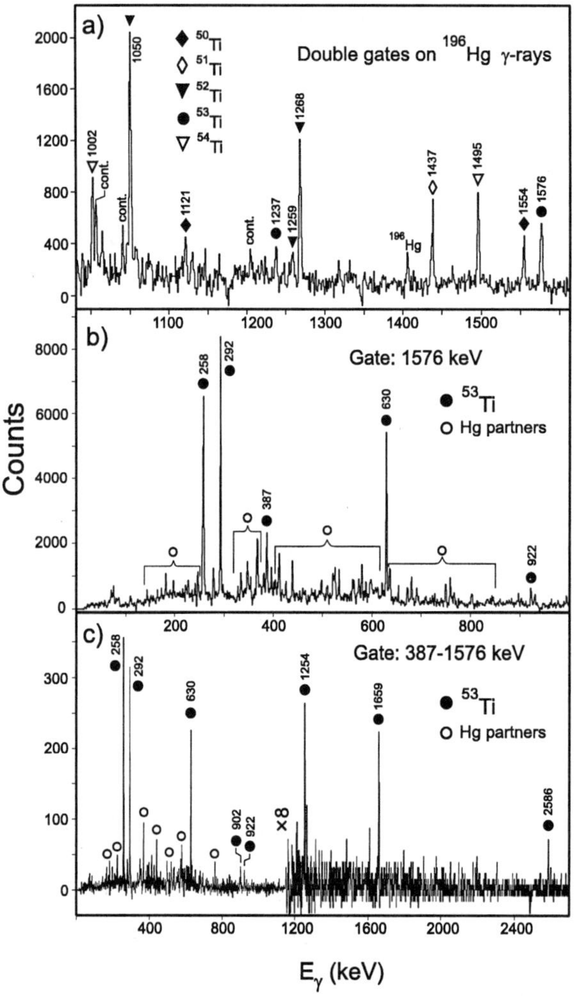

Figure 1. Representative γ-ray spectra from the Gammasphere experiment for the reaction ^{48}Ca+^{208}Pb; (a) part of a coincidence spectrum gated on pairs of yrast transitions in ^{196}Hg showing γ rays belonging to Ti partners, (b) spectrum from the $\gamma\gamma$ coincidence matrix gated on the 1576 keV line (assigned to ^{53}Ti), (c) double gate on selected ^{53}Ti two γ rays used to find higher spin transitions in the nucleus.

rays from ^{50,51,52}Ti nuclei exhibit a smooth correlation. The A_{av}(Hg) for the new transitions fit nicely into this pattern, if one assigns the 1237 and 1576 keV gamma rays to ^{53}Ti and the 1495 keV gamma ray to ^{54}Ti. The suggestion that the 1495 keV line is a transition in ^{54}Ti, was fully supported by the data from an experiment performed at Michigan State University (National Superconducting Cyclotron Laboratory), in which a beta-decay measurement of the ^{54}Sc parent produced in fragmentation of a Kr beam, was studied[8]. The MSU measurement identified the first two gamma-ray transitions in ^{54}Ti: $2^+ \rightarrow 0^+$ with energy of 1495 keV and $4^+ \rightarrow 2^+$ with energy of 1002 keV[8].

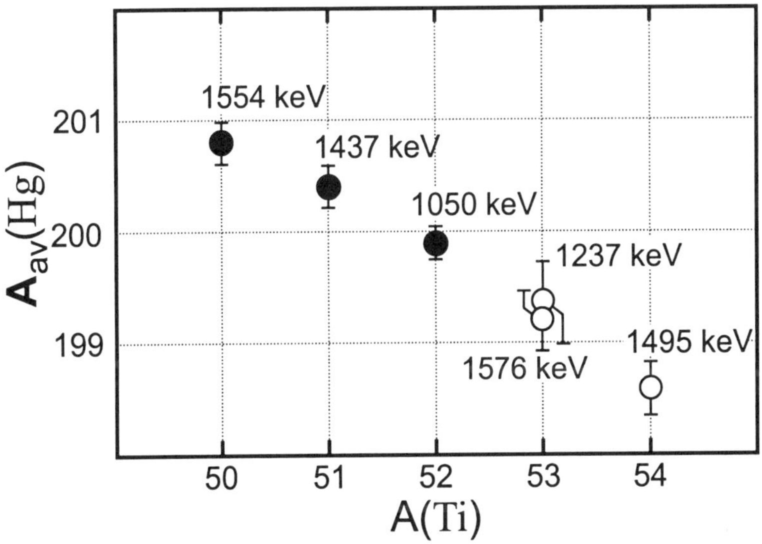

Figure 2. Plot of the average mass A_{av}(Hg) of complementary Hg fragments against the Ti product mass A(Ti) deduced from γ-ray cross-coincidence intensities. See text for further details.

Using the 1495 keV transition as a "starting point" in the analysis of $\gamma\gamma$ coincidence data from the Gammasphere experiment, we could establish the yrast structure in the ^{54}Ti nucleus up to an energy of $\sim$6.5 MeV. The details of the procedure are reported in Ref. 8 and the experimental level scheme of ^{54}Ti is shown in Fig. 3.

The correlation between A_{av}(Hg) and A(Ti) showed in Fig. 2, together with an independent identification of the ^{54}Ti 1495 keV gamma ray from the beta-decay measurement, provides persuasive support for the assignment of the 1237 and 1576 keV γ rays to the ^{53}Ti nucleus. Based on these two transitions one could start constructing the level scheme of ^{53}Ti.

A spectrum gated with the 1576 keV transition (Fig. 1b) shows four strong, mutually coincident γ rays with respective energies of 258, 292, 387 and 630 keV. The same sequence of γ rays was displayed also in a spectrum arising from the 1237 keV gate, with an additional prominent transition of 339 keV. The coincidence relationships between the newly identified gamma rays brought us to a conclusion that: a) the 339-1237 keV cascade occurs in parallel to the 1576 keV transition and deexcites a level at 1576 keV (both 1576 and 1237 keV γ rays are the ground state transitions), b) the series of transitions 630, 292, 258, 387, with the crossovers of 969 keV (339+630) and 922 keV (630+292), locate, above the 1576 keV excitation, the levels at 2206, 2498, 2756 and 3143 keV. The inspection of double gates placed on the lines identified so far (an example of which is shown in Figure 1c), revealed the presence of weaker 902, 1254, 1659 and 2586 keV γ rays belonging to ^{53}Ti. The observed coincidences and the measured intensities led to the level scheme given in Figure 4 with additional states located at 4802, 5729, 6056 and 6630 keV. A very weak 574 keV gamma ray connecting the 6630 and 6056 levels was also observed. All spin assignments are tentative - they are based on the close correspondence between established and calculated levels (see discussion below), on the fact that the reaction feeds yrast states preferentially and, on the gamma-decay pattern.

3. Discussion

The first important result from the performed measurement, already reported in Ref. 8, is an increase in the $E(2_1^+)$=1495 keV excitation energy in ^{54}Ti with respect to the $E(2_1^+)$=1050 keV in ^{52}Ti. This is consistent with the suggestion of a subshell closure at $N = 32$.

Another distinct feature of the ^{54}Ti level structure (also shown in Ref. 8) is the presence of higher-energy transitions ($E_\gamma \geq$ 2 MeV) at moderate spins. Such large jumps in yrast level energies are often signatures for excitations involving the breaking of the core, *e.g.*, requiring the promotion of nucleons across a subshell gap. In order to explore these observations further, a number of shell-model calculations was performed.

The shell-model calculations were carried out within the full pf-shell model space using the two Hamiltonians FPD6[9] and GXPF1[10]. The predicted energy levels in the ^{53,54}Ti nuclei are compared with the data in Figs. 3 and 4. The agreement with experiment is very good, especially for the GXPF1 interaction, and was used for spin-parity assignments.

The yrast states in ^{53}Ti and ^{54}Ti, calculated using the GXPF1 Hamiltonian, have a remarkably simple interpretation as in most of the cases their wavefunctions are dominated (40-70%) by a single shell-model component.

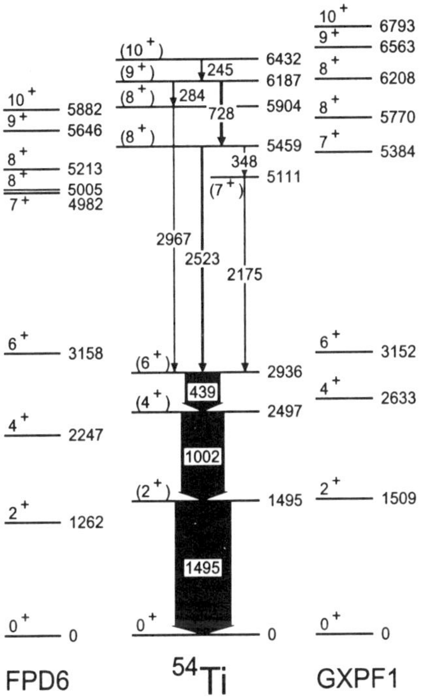

Figure 3. The level scheme of ^{54}Ti from Ref. 8 with the results of the full-pf shell model calculations using the GXPF1 and FPD6 Hamiltonians.

The ^{54}Ti level structure starts with $J^\pi = 0^+, 2^+, 4^+, 6^+$ sequence which, according to the calculations, is due to the dominance of the $\pi f_{7/2}^2$ proton configuration. Above it come the high spin 7^+, 8_1^+, 8_2^+, 9^+ and 10^+ states which arise from the coupling of two $f_{7/2}$ protons with $(\nu f_{7/2}^8 p_{3/2}^3 p_{1/2})2^+$ and $(\nu f_{7/2}^8 p_{3/2}^3 f_{5/2})4^+$ neutrons. The low lying yrast levels in ^{53}Ti, with spins up to J=15/2, are members of the $\pi f_{7/2}^2 \nu f_{7/2}^8 p_{3/2}^3$ multiplet. Next, there are the $17/2_1^-$ and $17/2_2^-$ states corresponding to a mixture of the two configurations $\pi f_{7/2}^2 \nu f_{7/2}^8 p_{3/2}^2 p_{1/2}$ and $\pi f_{7/2}^2 \nu f_{7/2}^8 p_{3/2}^2 f_{5/2}$. An excitation with the highest spin is the $21/2^-$ state dominated by the $\pi f_{7/2}^2 \nu f_{7/2}^8 p_{3/2}^2 f_{5/2}$ configuration.

The observed features of the yrast structure of ^{54}Ti such as the relatively high energy of the 2_1^+ state, the energy spacings between the $J^\pi = 0^+, 2^+, 4^+, 6^+$ excitations characteristic for a magic nucleus and, an energy gap between the yrast states involving $p_{3/2}$ neutrons and the excitations with $p_{1/2}$ and $f_{5/2}$ neutrons, fully support the existence of the N=32 subshell closure in neutron-rich species. This new subshell closure has been attributed to a decreased $\pi f_{7/2}$ - $\nu f_{5/2}$ monopole interaction as protons are removed from the $\pi f_{7/2}$ orbital[3]. The migration of the $\nu f_{5/2}$ orbital to higher energies with the removal of protons from the $\pi f_{7/2}$ orbital, in combination with the large spin-orbit splitting for $\nu p_{3/2} - \nu p_{1/2}$, gives rise to a subshell closure at N=32 for neutron-rich nuclides having Z<24. As shown in Figures 3 and 4, the full pf-shell model calculations employing the GXPF1 interactions are very successful in describing the features associated with the $\nu p_{3/2}$ sub-shell closure at N=32. On the contrary, the FPD6 predictions do not account well for these characteristics.

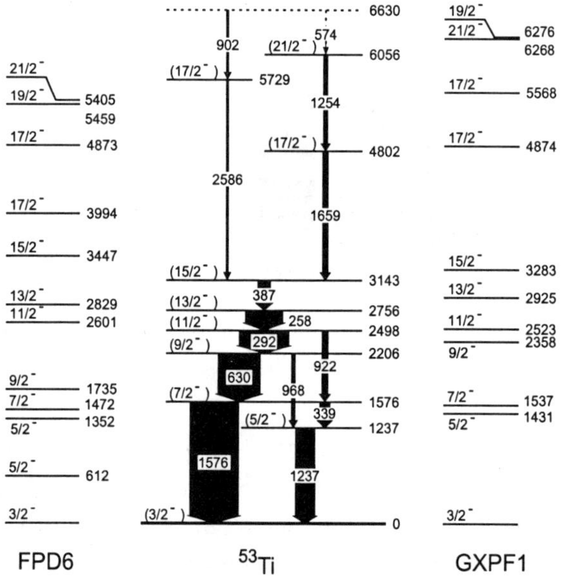

Figure 4. The proposed level scheme for ^{53}Ti. The results of the full-pf shell model calculations using the GXPF1 and FPD6 interactions are also shown for comparison.

Differences in the predictions of the two Hamiltonians regarding the low lying yrast levels in the ^{53}Ti and ^{54}Ti nuclei can be associated with different location in energy of the $\nu p_{1/2}$ and $\nu f_{5/2}$ orbitals with respect to the $\nu p_{3/2}$ state in neutron-rich nuclei for the two interactions. The FPD6 interaction is characterized by a smaller, with respect to the GXPF1 interaction, energy gap between the $\nu p_{3/2}$ state and the pair of $\nu p_{1/2}$, $\nu f_{5/2}$ levels. Comparison between the experimental data and the shell model calculations suggest that the separation between the $\nu p_{3/2}$ and the $\nu p_{1/2}$, $\nu f_{5/2}$ orbitals is well accounted for by the GXPF1 Hamiltonian, whereas this gap is too small in the FPD6 case. To explore this issue further a comparison between experimental energies of the high-spin yrast excitations in ^{53,54}Ti and the results of the shell model predictions with the two interactions has been done.

The state $21/2^-$ in ^{53}Ti and the excitations 9^+ and 10^+ in ^{54}Ti are of particular interest since their energies depend on the relative effective single-particle energies of the $\nu p_{3/2}$ and $\nu f_{5/2}$ orbitals, and the comparison with experiment provides a first crucial test of this part of the shell model Hamiltonians. The energy of the excitation involving a $f_{5/2}$ neutron in ^{53}Ti (the $21/2^-$ state) is predicted by FPD6 at an energy lower than the experimental value by ~750 keV. Also, the energies of states in ^{54}Ti involving a $f_{5/2}$ neutron (9^+ and 10^+) are underestimated by the FPD6 calculations by, on average, ~550 keV (Figure 3 and 4). The GXPF1 calculation, in turn, predicts the respective energies higher than the experimental values by ~210 keV in ^{53}Ti and by ~370 keV in ^{54}Ti. The above comparison is in line with an earlier observation that the $\nu p_{3/2}$-$\nu f_{5/2}$ energy gap is underestimated by the FPD6 interaction. On the other hand, the calculations with the GXPF1 Hamiltonian, which do very well for the low-lying yrast states, show a tendency to over predict the energies of states, for which the wave functions are dominated by the configurations with a $f_{5/2}$ neutron. It indicates that the gap between effective single-particle energies of the $\nu p_{3/2}$ and $\nu f_{5/2}$ orbitals, and, as a result, between the $\nu p_{1/2}$ and $\nu f_{5/2}$ levels, is too high in the GXPF1 Hamiltonian. This observation received a strong support from the data on ^{56}Ti, as discussed below.

A sizeable energy gap between the effective neutron single particle energies associated with the $\nu p_{1/2}$ and $\nu f_{5/2}$ orbitals in neutron-rich nuclei, predicted by the GXPF1 interaction, would lead to a sub-shell closure at $N = 34$ that should become apparent, for example, in the relatively high energy of the 2_1^+ state in ^{56}Ti. However, the energy of the ^{56}Ti 2_1^+ excitation, measured very recently in the β-decay of the ^{56}Sc parent produced in a fragmentation reaction, was reported to be only 1127 keV[11], i.e. almost 400 keV below the GXPF1 expectation. This result agrees with our observation that the $\nu f_{5/2}$ orbital is not located as high as predicted by the GXPF1 interaction, at least in Ti isotopes.

4. Summary

γ-ray coincidence measurements using Gammasphere with the ^{48}Ca + ^{208}Pb reaction at 305 MeV have provided information about the yrast excitations in neutron-rich Ti isotopes. In particular, yrast γ-ray cascades in ^{53}Ti were identified and were used to locate the states with spins up to $J^{\pi} = 21/2^{-}$. The yrast structure of ^{53}Ti, together with the earlier reported results on excitations in ^{54}Ti, was compared with the results of shell model calculations using the FPD6 and GXPF1 interactions. The experimental data are very well described by the GXPF1 calculations - this is in line with the existence of a significant energy gap between the $\nu p_{3/2}$ orbital and the pair of $\nu p_{1/2}$, $\nu f_{5/2}$ single particle states and, consequently, in line with the subshell closure at $N = 32$. On the other hand, the experimental findings suggest that the $\nu f_{5/2}$ state is not separated from the $\nu p_{3/2}$ level to the degree assumed by the GXPF1 Hamiltonian which indicates that it may be located close to the $\nu p_{1/2}$ orbital. This finding does not support the hypothesis of the appearance of another subshell closure at $N = 34$ in Ti isotopes.

Acknowledgments

This work was supported by the U.S. Department of Energy, Nuclear Physics Division, under Contracts Nos. W31-109-ENG-38 and DE-FG02-87ER40346, by National Science Foundation Grant Nos. PHY-01-01253, PHY-00-70911 and PHY-97-24299, and by Polish Scientific Committee Grant No. 2PO3B-074-18.

References

1. T. Otsuka *et al.*, *Phys. Rev. Lett.* **87**, 082502 (2001).
2. A. Huck *et al.*, *Phys. Rev.* **C31**, 2226 (1985).
3. J.I. Prisciandaro *et al.*, *Phys. Lett.* **B510**, 17 (2001).
4. R. Broda, *Eur. Phys. J.* bf A13, 1 (2002) and references therein.
5. B. Fornal *et al.*, *Acta Phys. Pol.* **B26**, 357 (1995).
6. W. Krolas *et al.*, *Nucl. Phys.* **A724**, 289 (2003).
7. I.Y. Lee, *Nucl. Phys.* **A520**, 641c (1990).
8. R.V.F. Janssens *et al.*, *Phys. Lett.* **B546**, 55 (2002).
9. W.A. Richter *et al.*, *Nucl. Phys.* **A523**, 325 (1991).
10. M. Honma *et al.*, *Phys. Rev.* **C65**, 061301 (2002).
11. S. N. Liddick *et al.*, *Phys. Rev. Lett.* **92**, 072502 (2004).

KEY TOPICS IN NUCLEAR STRUCTURE
Proceedings of the 8th Int. Spring Seminar on Nuclear Physics
edited by Aldo Covello
© 2005 World Scientific Publishing Co.

MAGNETIC MOMENT MEASUREMENTS OF NEUTRON-RICH $\nu g_{9/2}$ ISOMERIC STATES

J. M. DAUGAS, G. BÉLIER, M. GIROD, H. GOUTTE, V. MÉOT, O. ROIG
CEA/DIF/DPTA/SPN,
BP 12,
91680 Bruyères le Châtel, France

I. MATEA, G. GEORGIEV, M. LEWITOWICZ, F. DE OLIVEIRA SANTOS
GANIL,
BP 55027,
14076 Caen Cedex 5, France

M. HASS, L. T. BABY, G. GOLDRING
The Weissman Institute, Rehovot, Isarel

G. NEYENS, D. BORREMANS, P. HIMPE
IKS, KULeuven, Celestijnenlaan 200 D, 3001 Leuven, Belgium

R. ASTABATYAN, S. LUKYANOV, YU. E. PENIONZHKEVICH
FLNR-JINR, Dubna, Russia

D. L. BALABANSKI
Faculty of Physics, St. Kliment Ohridski University of Sofia, 1164 Sofia, Bulgaria

M. SAWICKA
IFD, Warsaw University, Hoża 69, 00681 Warsaw, Poland

Electromagnetic moments measurement of isomeric states produced and spin-oriented in projectile fragmentation reactions at intermediate energies have been performed using the Time Dependent Perturbed Angular Distribution (TDPAD) method. This allows the study of neutron-rich nuclei unaccessible by other kind of reaction. An important experimental achievement is presented.

1. Introduction

Magnetic moments are very sensitive probes to the detailed composition of the nuclear wave function. Measurement of the g-factor of a nuclear state provides unique information on its single particle structure and is a test ground for nuclear models. Due to its dependence of the spin and orbital angular momentum of the involved valence nucleons, magnetic moment is a rigorous probes for the spin and parity assignment of the nuclear states.

It is especially interesting to test nuclear models for nuclei far from stability in the vicinity of a spherical shell closure, where the nuclear wave functions are expected to be rather pure, and in regions where intruder states are expected, reflecting a large deformation. One current region of interest is the neutron rich species around $N = 40$ and $Z < 28$. We focus here on the role played by the $\nu g_{9/2}$ orbital in the low-energy level structure of nuclei below $N = 40$ in particular case where this orbital manifests irself as an isomeric state.

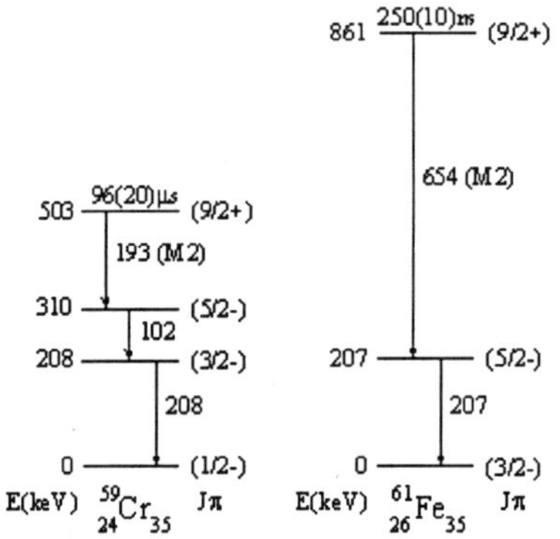

Figure 1. The level schemes for the isomeric decays of ^{59}Cr and ^{61}Fe with tentative spin and parity assignments.

The $N = 35$ istones ^{59}Cr and ^{61}Fe are the lightest nuclei exhibiting such low-lying energy isomeric states. In figure 1 we show the drafts of both level schemes. The $9/2^+$ tentatively assigned spin and parity was obtained mainly from the isomer [1] and β-decay [2] studies. Isomeric levels are at excitation energies of 861 keV and 503 keV for ^{61m}Fe and ^{59m}Cr respectively. Along the $N = 35$ isotones we clearly see the drastic lowering of the $\nu g_{9/2}$ shell in neutron-rich nuclei going from an excitation energy of 1292 keV for $Z = 28$ down to 503 keV for $Z = 24$. This reflects the possible onset of deformation induced by $\nu g_{9/2}$ intruder orbitals in this region.

We repport here an important experimental achievement in the study of magnetic moments of isomeric states produced by projectile fragmentation reactions at intermediate energy.

2. Experimental detail

2.1. *The TDPAD-method*

Magnetic moment measurements have been performed using the Time Differential Perturbed Angular Distribution (TDPAD) [3] method in combination with ion-γ correlations. The isomers are stopped in a perurbation free environment by choosing an appropriate stopper material that has cubic lattice structure. In our case we have choosen to use an annealed high-purity $500\mu m$ thick Cu foil as a stopper. Ions of Fe and Cr are expected to implant well as they have the same electronegativity and similar atomic radius. The implantation foil is placed between the poles of an electromagnet delivering an external constant magnetic field B in the vertical direction. The magnetic field causes a Larmor precession of the initially aligned isomeric nuclear spins with a frequency

$$\omega_L = -\frac{g\mu_N B}{\hbar} \tag{1}$$

where g is the nuclear gyromagnetic factor; μ_N is the nuclear magneton and $\hbar$ is Plank's constant. The monitoring of γ radiations is performed using 4 high-purity Ge detectors. For each Ge detector, energy and time signals are stored in an event by event mode. γ-ray time spectra are started by the ion implantation time, and stopped by the detection of a delayed γ-ray. To extract the precession pattern out of the individual time spectra, detectors at 90° with respect to each other are combined to generate the $R(t)$-function [3]

$$R(t) = \frac{I_1(\theta,t) - \epsilon I_2(\theta + \frac{\pi}{2},t)}{I_1(\theta,t) + \epsilon I_2(\theta + \frac{\pi}{2},t)} \sim A_2 B_2^0(t=0) cos(2(\omega_L t + \alpha - \theta)) \tag{2}$$

where I_1 and I_2 are the summed intensities of detectors placed at 180°; ϵ is the relative efficiency between the different Ge-detectors; A_2 is the radiation parameter of the γ-ray transition; B_2^0 is the second component of the orientation tensor describing the initial alignment of the isomric state; θ is the angle between the beam axis and the first Ge detector and α is the angle between the beam axis and the symmetry axis of the spin-aligned ensemble. When the produced aligned isomer is going throught the dipole of the spectrometer, the spin-orientation axis is deflected at an angle α with respect to the beam axis [4]:

$$\alpha = \theta_C(1 - \frac{gA\mu_N}{Z\hbar}) \tag{3}$$

where θ_C is the rotation angle in the spectrometer; A the mass of the implanted ion and Z its charge.

2.2. *Secondary isomeric beam production*

In the present experiment we aim to measure the g-factor of neutron-rich exotic isomers. The TDPAD is a very powerful methods which has been widely used in fusion-evaporation reactions [3]. This kind of reaction does not allow the production of the request nuclei. The neutron-rich part of the nuclides chart can be reach using projectile fragmentation reaction. Such measurement requires the combined production of spin-aligned isomeric state. The first application of the TDPAD method on isomeric states produced in high-energy projectile fragmentation has been performed on the case of ^{43m}Sc [5], where a significant amount of spin-alignment (up to 35%) has been observed. For intermediate energies calculations in the framework of a kinematical fragmentation model [6,7,8] might be used to reproduce the behavior of the angular momentum transferred during the reaction. This gives us information on the spin transferred and on the amount of spin-alignment as function of the fragment momentum distribution.

Nuclei of interest were produced by the fragmentation of a ^{64}Ni beam at an energy of 54.7 $AMeV$ impiging a 98 mg/cm^2 ^9Be target. Fragments were selected using the LISE spectrometer [10].

2.3. *Experimental set-up*

The pionnering experiment [9] in which the TDPAD technique was applied on an unknown case after projectile fragmentation was performed at GANIL facility using a ^{76}Ge primary beam at 61.4 $AMeV$. The amount of spin-alignment observed in this experiment was low, 2.6(10)% and 1.7(5)% for the ^{69m}Cu and ^{67m}Ni respectively. A significant achievement of the experimental procedure has been performed [11]. The experimental set-up is schematically presented in figure 2. The secondary beam identification and position has been performed using a removable position sensitive XY Si-detector placed in front of the vacuum chamber. The detector was removed from the beam line after the selection of the nuclei of interest. A 50 μm

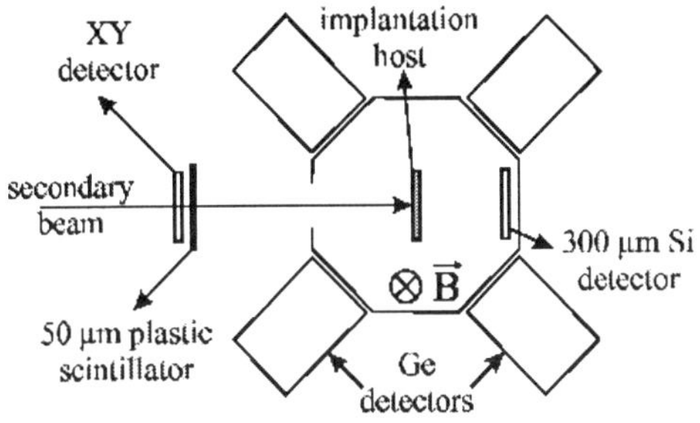

Figure 2. Schematic drawing of the experimental setup.

was used for giving the $t = 0$ signal for the time-decay curves. The use of a thin plastic instead of a 300 μm Si detector, which was used for the previous experiement, allow us to have a significant decrease of the electron pick-up of the ions passing through it. We estimated that less than 2 % of the nuclei were implanted in not fully stripped condition, instead of about 60 %. If the detected ion is not fully stripped, the interaction of the randomly oriented electron spin, coupled to the nuclear spin, with any external magnetic field can cause a significant decrease up to a complete loss of the orientation [12]. The implantation Cu foil was surrounded by 4 high purity single crystal Ge detectors positionned in the horizontal plane.

The delayed γ-rays coming from implanted isomers have been registered in an event by event mode within a time window of 3 μs. In order to reduce random coincidences in this time we have used a package suppresser (p.s.) allowing 1 out of 10 packages provided by the GANIL accelerator with a frequency of 10.5 MHz. The implantation frequency was then 950 ns instead of 95 ns without this suppression.

3. Results

3.1. *Measurement of the* ^{54m}Fe

The measurement of a known case have been first performed in order to validate and calibrate the method. We have choosen the ^{54m}Fe where the g-factor of the 10^+ isomeric state at an excitation energy of 6527 keV with a half life of $t_{1/2} = 357(4)$ ns is known to be $g = +0.7281(10)$ [13]. The knowledge of the magnetic moment allows us to extract the effective value of the applied magnetic field $B = 0.680(4)$ T. The obtained magnetic field includes all the systematical errors which are: the distribution of the magnetic field over the beam spot, the paramagnetic amplification of the

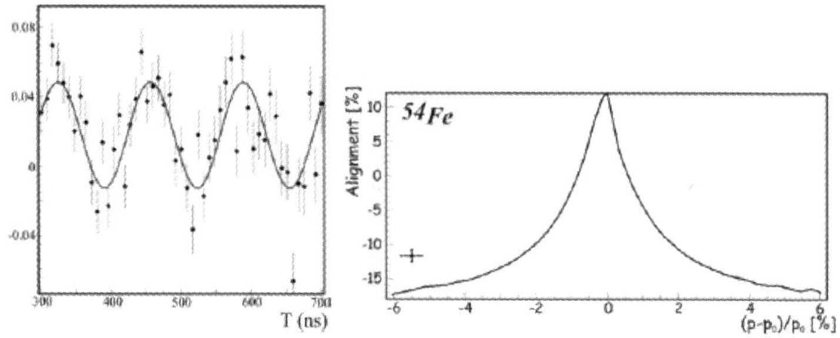

Figure 3. The left picture represents the $R(t)$ function for ^{54m}Fe and the right one the comparison between the measured and the calculated alignment, theoritical curve is scaled by a factor of 1.8.

applied field and the Knight shift. In the left part of the figure 3 is shown the $R(t)$ function corresponding to the ^{54m}Fe. We have also measured the spin alignment as a function of the momentum distribution, the comparison between the calculated and experimental alignment is presented in the right of the figure 3. The selection of the fragments in the wing of the momentum distribution $(p - p_0)/p_0 = -5.5(3)\%$ has yield to a large negative alignment of $-12.5(9)\%$. The amount of alignment is in agreement with the kinematical fragmentation model. Note that the theoretical curve have to be divided by 1.8.

3.2. *g-factor of the* ^{61m}Fe

The measurement of the ^{61m}Fe is presented in figure 4. The measured half-life of the isomeric state, $t_{1/2} = 245(5)$ ns, is in good agreement with the previous measurement [1]. A g-factor value of $-0.229(2)$ was extracted using formula 2. The error on the fitting procedure includes the errors on the effective field. The selection of the fragments have been performed in the center of the momentum distribution at $(p - p_0)/p_0 = -0.2(2)\%$ and in the outer wing at $(p - p_0)/p_0 = 2.45(70)\%$, the observed spin-alignment was $+6.2(7)\%$ and $-15.9(8)\%$ respectively. One can note that without the packages suppression the observed spin-alignment is about 3 times less. The opposite phase of the $R(t)$ function for the 2 different γ-rays indicates different multipolarities, the ratio between the amplitudes of the $R(t)$ functions is 1.43(16). Using realistic GEANT simulations [14] and assuming that the 654 keV and 207 keV transitions have pure $M2$ and $M1$ multipolarities with a level sequence of $9/2^+ \to 5/2^- \to 3/2^-$, we estimate

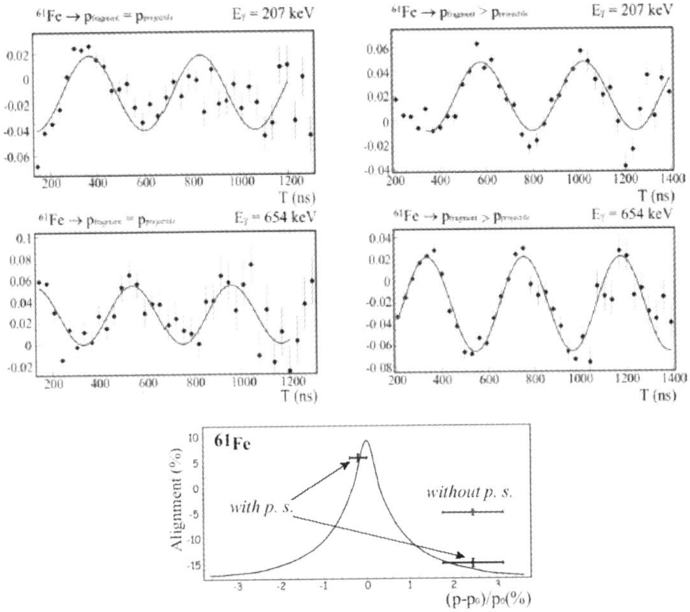

Figure 4. In the top are presented the $R(t)$ function for the 207 keV γ-ray for 2 different selections in the momentum distribution. Below are the same pictures for the 654 keV isomeric γ-ray. In the bottom is the comparison between the measured and the calculated alignment with and without packages suppression, theoritical curve is scaled by a factor of 1.8.

this ratio to be 1.30(6).

3.3. *The* ^{59m}Cr *case*

The isomeric state of the ^{59}Cr is longer than ^{54m}Fe and ^{61m}Fe. The half-life of this state has been measured more precisely and has been found to be $t_{1/2} = 86(5)$ μs. In the 3 μs time window for the acquisition of events, a much lower count rate is observed. The 2.7×10^9 fragments implanted in the crystal lead to only 5×10^4 193 keV γ-rays detected in the full time window, which represents $\approx$ 250 γ-rays par channel for a 16 ns bining. Due to this lake of statistic, the g-factor of this isomeric state has not been extracted, but we have shown that this measurement can be done.

4. Summary

Electromagnetic moments measurement of neutron-rich nuclei have been performed using projectile fragmentation reactions. The appreciably large residual alignment shows that this kind of reactions can provide a powerful tool for measuring gyromagnetic factors and quadrupole moments in neutron-rich exotic nuclei using the TDPAD method. This allows investigation of nuclear structure away from stability.

The measured g-factor for ^{61m}Fe is in very good agreement with the assigned spin and parity of the isomeric state. Large Scale Shell Model and HFB calculations have been performed and both indicates that this state is characterised by a deformed potential. Using the same method and a crystal with an electric field gradient (EFG) it will be possible to measure this deformation in order to valid theoretical calculations.

Acknowledgments

We are grateful for the technical support received from the staff of the GANIL facility. This work has been partially supported by the Access to Large Scale Facility program under the TMR program of the EU, under contract nr. HPRI-CT-1999-00019, the INTAS project nr. 00-0463 and the IUAP project P5/07 of the Belgian Science Policy Office. We are grateful to the IN2P3/EPSRC French/UK loan pool for providing the Ge detectors. The Weissman Institute group was supported by the Israel Science Foundation. Authors G.N. and D.B. acknowledge the financial support of the FWO-Vlaanderen.

References

1. R. Grzywacz *et al.*, *Phys. Rev. Lett.* **81**, 766 (1998).
2. O. Sorlin *et al.*, *Nucl. Phys.* **A660**, 3 (1999); Erratum *Nucl. Phys.* **A669**, 351 (2000).
3. G. Goldring and M. Hass, *Treaties in Heavy Ion Sciences*, D. E. Bromley ed. Plenum Press, Vol. 3, p.539.
4. G. Neyens *et al.*, *Nucl. Instr. and Meth.* **A340**, 555 (1994).
5. W. D. Schmidt-Ott *et al.*, *Z. Phys.* **A350**, 215 (1994).
6. K. Asahi *et al.*, *Phys. Rev.* **C43**, 456 (1991).
7. H. Okuno *et al.*, *Phys. Lett.* **B335**, 29 (1994).
8. J. M. Daugas *et al.*, *Phys. Rev.* **C63**, 064609 (2001).
9. G. Georgiev *et al.*, *Journ. of Physics* **G28**, 2993 (2002).
10. R. Anne *et al.*, *Nucl. Instr. and Meth.* **A257**, 215 (1987).
11. G. Georgiev *et al.*, *AIP Conf. Proc.* **701**, 169 (2004).
12. K. Vyvey *et al.*, *Phys. Rev.* **C62**, 034317 (2000).
13. M. H. Rafailovich *et al.*, *Phys. Rev.* **C27**, 602 (1983).
14. GEANT, http://wwwasdoc.web.cern.ch/wwwasdoc/

KEY TOPICS IN NUCLEAR STRUCTURE
Proceedings of the 8th Int. Spring Seminar on Nuclear Physics
edited by Aldo Covello
© 2005 World Scientific Publishing Co.

MULTINUCLEON TRANSFER REACTIONS STUDIED WITH LARGE SOLID ANGLE SPECTROMETERS

L. CORRADI

INFN - Laboratori Nazionali di Legnaro,
Via Romea 4,
35020 - Legnaro (Padova), Italy
E-mail: corradi@lnl.infn.it

A discussion is made on the status of multinucleon transfer studies performed with high resolution spectrometers and the new possible experiments to be done via γ-particle coincidences in suitable systems to elucidate the structure of interesting excited states populated via transfer reactions.

1. Introduction

Quasi-elastic and deep inelastic processes have been extensively studied in different ways during the past twenty years, both theoretically and experimentally[1,2]. However, it was not until very recently that, thanks to the development of efficient spectrometers, one could identify with high resolution multinucleon transfer channels populated in heavy-ion reactions up to the pick-up and stripping of several neutrons and protons[3,4,5,6]. Advances in theoretical calculations[7,8,9] allowed also to quantitatively study how degrees of freedom of different complexity act in the transfer process. This "transition" regime, where many nucleons are transferred and where shell effects still play a significant role in the dynamics, represents a window not well studied in its detail and where it is interesting to investigate the interplay between single nucleon and pair transfer modes and their dynamical effects as a function of energy loss, angular momentum and number of transferred nucleons. The understanding of these processes is important in view of future research to be done with radioactive beams[10,11], since one expects for instance a very different behaviour of nucleon correlation effects for neutron-rich nuclei. In this contribution examples of recent and ongoing studies performed with time-of-flight and magnetic spectrometers are discussed.

2. Inclusive measurements : main findings

From the inclusive data obtained for different heavy-ion systems[3,4,5,6] one sees a quite regular drop of the cross sections for pure neutron transfer channels indicating that neutrons seem to be transferred as independent particles. Other degrees of freedom beyond single particle transfer modes, like pair transfer, are not clearly identified in the total cross sections since several contributions are mixed up. For example it is not clear if the pair transfer itself proceeds in a cluster- or sequential-like picture. Inelastic excitations generally push the transfer strength at energies of several MeV and an interesting question is to what extent pair correlations get weaker.

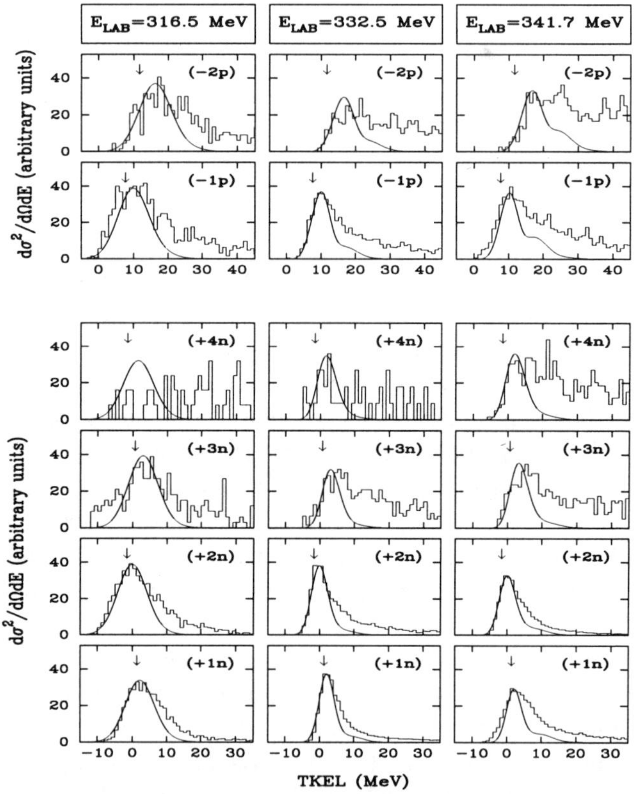

Figure 1. Experimental (histograms) and theoretical (lines) total kinetic energy loss distributions for pure neutron-pick and proton stripping channels in the reaction $^{62}Ni+^{206}Pb$ (see Ref.[5] for details).

In a recent study[5] of the system ^{62}Ni+^{206}Pb multinucleon transfer cross sections have been measured at three bombarding energies up to three nucleon pairs, focusing on pure neutron channels which all have well matched Q-values. The aim of the experiment was to observe a possible odd-even staggering in the cross sections for pure neutrons, indicating a contribution from the pair transfer mode. The measured Q-integrated cross sections drop regularly with the number of transferred neutrons, without any clear odd-even staggering, and this behaviour is almost energy independent (see Ref. [5] for details). Additional information can be gained by inspecting at the total kinetic energy loss distributions, shown in Fig. 1. One sees that the +1n and +2n channels have the main population close to the ground-ground state Q-value (indicated by the down arrows) while the more massive transfer channels display a population towards more negative Q-values, with a tail increasing with the number of transferred neutrons. The observation of these long tails in the Q-value spectra suggests that any contribution from "cold" transfers, associated with low excitation energy, is likely hidden by dominant "warm" sequential transfer processes. On the other hand a significant fraction of the flux remains close to Q=0 up to the +4n channel. This suggests that experiments with higher energy resolution able to distinguish the population to the ground state and to the higher excited states should give better signatures of the transfer contribution coming from pair modes.

A macroscopic and presently not understood problem is a different behaviour of neutrons and protons in multinucleon transfer. The comparison of cross sections measured in various systems with semiclassical Complex WKB (CWKB) calculations[7,9,12] including only single nucleon transfer modes shows that starting from the (-2p) channel, data are strongly underpredicted. This is shown in Fig. 2 for the ^{58}Ni+^{208}Pb system[6]. The discrepancies indicate that degrees of freedom beyond single-particle transfer modes have to be incorporated in the theory, or that more complex processes, i.e. deep inelastic components, play an important role. Adding into theory a new degree of freedom, namely the transfer of a pair of protons, and fixing the strength of the formfactor[13,14] to reproduce the pure -2p channel, the predictions for the other charge transfer channels are also much better indicating that the proton pair mode may be an important degree of freedom in the transfer process. Its microscopic treatment is however very difficult and to relate its strength to the pair correlations in target and in projectile (both enter in the definition of the form factor) still represents a challenge for theorists.

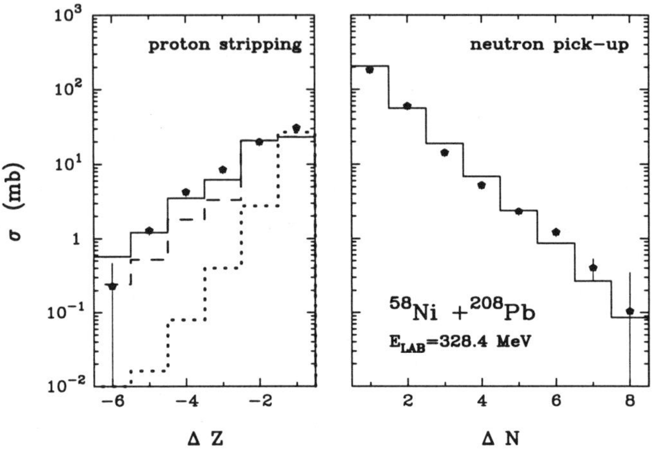

Figure 2. Total cross sections for pure proton stripping (left side) and pure neutron pick-up (right side) channels. The dotted line is the CWKB calculation including single nucleon transfer modes. The dashed line represents the result with an additional two-proton pair mode and the solid line includes also the effect of nucleon evaporation from the primary fragments. For the pure neutron transfer no pair transfer was included and the cross section represented by the dashed and the dashed-dotted lines (not shown) are very close to the full line.

3. Inclusive measurements : the ^{40}Ca+^{208}Pb system

Further interesting inputs for understanding the role of pair modes degrees of freedom comes from the study[15] of ^{40}Ca+^{208}Pb, where both projectile and target are double closed shell nuclei and therefore provide an excellent opportunity for a quantitative comparison with theoretical model. Fig.3 shows the total kinetic energy loss distributions at the three measured bombarding energies for the two neutron pick-up channel toghether with CWKB calculations. As can be appreciated, the two neutron pick-up channel displays a well defined maximum, which, within the energy resolution of the experiment, is consistent with a dominant population, not of the ground state of ^{42}Ca, but of the excitation region around 6 MeV (see below). All measured energies show the same behavior and as the beam energy decreases the distributions become narrower, the large energy loss tail tends to disappear, and the centroids slightly shift to lower energies reflecting the energy dependence of the optimum Q-value. Both features are well reproduced by the theoretical calculations indicating that the used single particle levels cover the full Q-value ranges spanned by the reaction. Systematic studies of (p,t) and (t,p) reactions performed in the past iden-

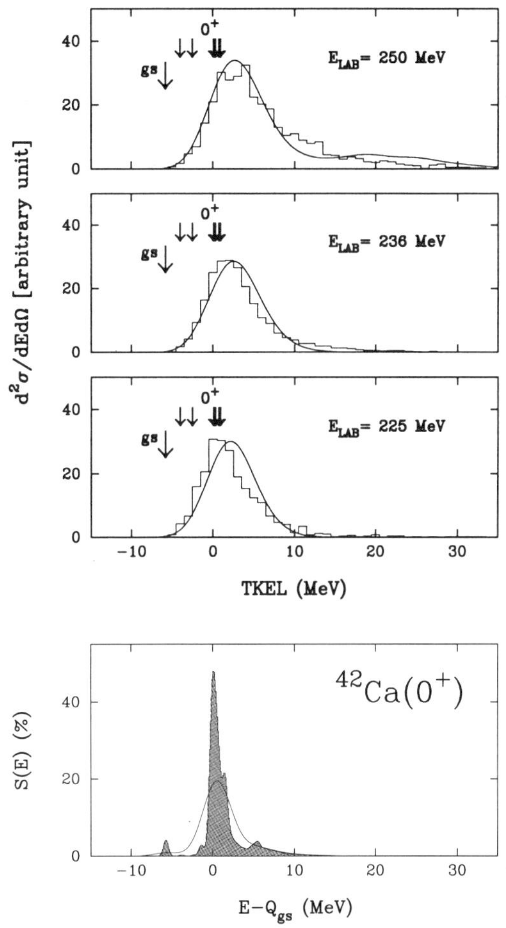

Figure 3. Experimental (histograms) and theoretical (curves) total kinetic energy loss distributions of the two neutron pick-up channels at the indicated energies. The arrows correspond to the energies of 0^+ states in ^{42}Ca with an excitation energy lower than 7 MeV. The group of states at 5.87 (collecting most of the strength), 6.02, 6.51 and 6.70 MeV were strongly populated in (t,p) reactions. Bottom panel shows the strength function $S(E)$ from SM calculations (see text) after convoluting with Gaussian of two different widths: 300 keV and 1.5 MeV (close to the experimental energy resolution - curve). The represented strength function has been obtained after 200 Lanczos iterations to allow a correct convergence of all eigenstates (see Ref.[15] for details).

tified the calcium region as the only known where the cross sections for the population of the excited 0^+ states is larger than the ground state. In most Ca isotopes the excited 0^+ strength is concentrated in one state only. Those states have been recognised as multi (additional and removal) pair-phonon states[16]. Nuclear structure and reaction dynamics studies attribute this behavior to the influence of the $p_{3/2}$ orbital that gives a much larger contribution to the two-nucleon transfer cross section than the $f_{7/2}$ orbital, which dominates the ground state wave function. Referring to Fig.3, the final population of the single particle levels used in the CWKB theory suggests that the maxima are essentially due to two neutrons in the $p_{3/2}$ orbital, i.e to the excited 0^+ states at around 6 MeV of excitation energy that were interpreted as corresponding to the pair vibrational mode[16,17]. If so, these results show that, at least in suitable cases, one can selectively populate specific Q-value ranges even in heavy-ion collisions, opening the possibility to study multi pair-phonon excitations. Since the illustrated experimental results and reaction calculations point to a selective feeding of 0^+ states dominated by a pair of neutron in the $p_{3/2}$ orbital it was computed, for ^{42}Ca, the strength distribution of these peculiar 0^+ states over all others 0^+ states. This has been performed in the framework of large scale shell model (SM) calculations by using the same model space and interaction as in a recent publication concerning various spectroscopic features of calcium isotopes[18]. To extract the channel strength function it was used the Lanczos method with a state that corresponds to the creation of two neutrons, coupled to 0, on the $|0^+\rangle$ ground state of ^{40}Ca, in turn described by a $(np - nh)$ configuration with n up to 12. The strength distribution $S(E)$ displays, clearly (see Fig.3), a strong concentration near $\sim$ 6 MeV of excitation energy, an energy very close to that of a configuration where a $p_{3/2}$ neutron pair is coupled to a closed shell of ^{40}Ca ground state ($E \sim 5.9$ MeV). This calculation demonstrates the dominant contribution of the $p_{3/2}$ orbitals and the predicted very narrow energy distribution suggests its interpretation as a pair mode.

4. Measurements with the PRISMA set-up

In order to pursue these studies in a more detailed way it would be important to distinguish the population to specific nuclear states and to determine both their strength distribution and decay pattern. This, in fact, carries information on the wavefunction components of the populated levels and therefore also on the pairing correlation. Experiments in this direction must exploit the full capability of spectrometers with solid angles much larger than the conventional ones, and with A, Z, and energy resolutions sufficient to deal also with heavier mass ions. This is possible now with the PRISMA spectrometer[19,20] recently installed at LNL and designed for the A=100-200, E = 5-10 MeV/amu heavy-ion beams of the accelerator complex of LNL. The main features of the spectrometer are its large solid angle 80 msr, wide momentum acceptance ±10%, mass resolution 1/300 via time-of-flight and energy resolution up to 1/1000. First experiments on heavy-ions grazing collisions have been already performed with different beams, with main goals to investigate the population of neutron-rich nuclei in the A=40-90 mass region by means of multinucleon transfer reactions, and to study the dynamics of such transfer processes. The spectrum in Fig.4 shows an example of the obtained A resolution in the $^{90}Zr+^{208}Pb$ reaction at E_{lab}=560 MeV and at $\theta_{lab}=54^0$. One observes events corresponding to the pick-up as well as stripping of neutrons. One of the present interests

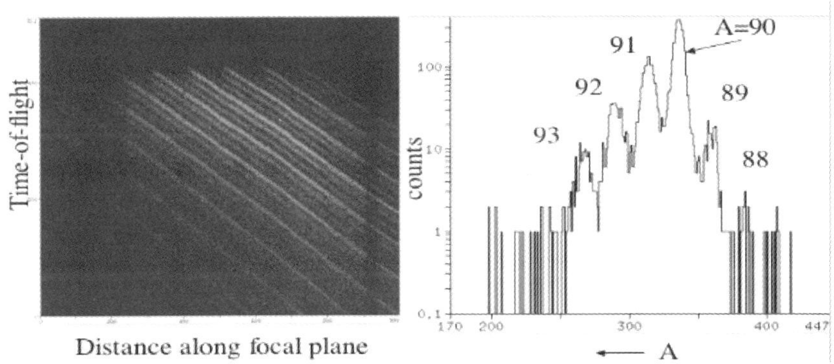

Figure 4. Left side : time-of-flight (ToF) vs. position (X) along the Prisma focal plane for Zr-like events detected in the $^{90}Zr+^{208}Pb$ reaction at E_{lab}=560 MeV and at $\theta_{lab}=54^0$. The different bands correspond to the different A/q (nuclear mass over atomic charge state) of ions produced in the reaction. A is roughly proportional to the product of ToF and X. Right side : linearized mass spectrum obtained with a gate on the most probable q and on the nuclear charge Z=40. The figures correspond to about 1% of the whole statistics.

are nuclear structure studies of moderately neutron-rich nuclei, populated at relatively high angular momentum, by means of binary reactions. Multinucleon transfer processes are in fact recognized to be a suitable tool for the production of neutron-rich nuclei, at least for certain mass regions [10,11]. These studies are being performed by combining PRISMA with the CLARA γ-array[21], recently installed close to the target point and consisting of an array of 24 Clover detectors from the Euroball collaboration.

In connection with the items previously discussed, the characteristics of the combined set-up offers new possibilities to study the population of multi-pair phonon states in heavy-ion collisions, which was difficult to do up to now. In fact, among the problems one generally faces in these experiments is that levels of interest are located at excitation energies $\geq$ 4-5 MeV and that the transfer flux is fragmented over several states. The flux fragmentation was visible in the (t,p) and (p,t) reactions performed in the A$\simeq$80-90 region[16,22]. With the use of heavy ions such an effect is likely to be even more marked, but an advantageous feature is the much higher probability to populate interesting levels at high excitation energy. Among those levels are the 0^+ states, to be investigated at first stage in nuclei near the shell closures, namely in the Ca region as well as in the Zr region. Different 0^+ levels at energies $\leq$ 2-3 MeV have main components of their wavefunctions coming from core-excited states[23], i.e they are of a multiparticle-multihole structure. Data from experiments with heavy ions specifically investigating to what extent the structure of high lying 0^+ states (i.e. at excitation energies $\geq$ 4-5 MeV) can be interpreted as pairing vibrations are extremely scarse. Recent indications in this sense come from studies of $^{90}Zr(n,n^{'})$ inelastic scattering[24], where levels at $\simeq$ 4 MeV excitation energy are suggested as having contributions from two-phonon pairing vibrations mixed with $(\pi g_{9/2})^2_{J=2^+}$ configuration coupled to $^{88}Sr_{2^+}$ core excitation.

An exploratory run with the PRISMA+CLARA set-up has been very recently done in the reaction ^{90}Zr+^{208}Pb at E_{lab}=560 MeV with the main aim of looking at the yield production of specific Q-value ranges in the Zr and Sr isotopes close to the expected region where pair vibrational modes may be excited. More in general, the Zr isotopes span a range from spherical to highly deformed shapes and it would be therefore interesting to investigate into detail the change of the population strength and decay pattern properties of specific levels populated via multinucleon transfer mechanism.

The decay properties must be studied by measuring the strength distribution of populated levels, the electromagnetic character and multipolarity in γ decay. Therefore γ-ray angular distributions and correlation of γ-rays in coincidence with PRISMA are required. A suitable candidate for next experiments is for example the ^{40}Ca+^{208}Pb system, previously discussed, where it is interesting to study the structure of the states populated in neutron transfer channels. The experimental and theoretical cross sections for the direct population of individual levels can be also compared with good precision. An example[27] of the presently achieved agreement for the +1n and -1p channels in the $^{40}Ca+^{124}Sn$ reaction is shown in Fig.5. Such comparisons are important since these channels represent the "building blocks" on which the more complex multinucleon transfer is computed.

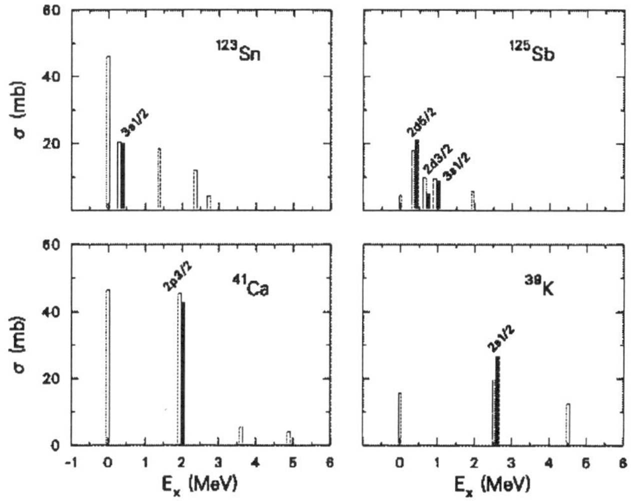

Figure 5. Experimental (full bars) and theoretical CWKB (open bars) total cross sections for the indicated channels in the reaction $^{40}Ca+^{124}Sn$ at E_{lab}=170 MeV. Measurements have been performed with the GASP γ-array in coincidence with particle detectors (see Ref.[27] for details).

A further interesting possibility, presently under discussion, is the study of $e^+ - e^-$ pair decay modes (E0) from highly excited 0^+ states populated via transfer. A two-level mixing model provides an insight into the origin of E0 transition strength in nuclei possessing shape coexistence (see Ref.[23] and references therein for a review). In general, the branching ratio between $e^+ - e^-$ and γ decays for 0^+ levels is very sensitive to the components of the wavefunction and therefore its measurement may provide cleaner signatures of pairing structures.

Acknowledgments

The material presented in this contribution is the result of a cooperative work of the following people whom I would like to thank : A.M.Stefanini, S.Szilner, G.Pollarolo, A.Gadea, F.Haas, E.Fioretto, A.Latina, S.Beghini, G.Montagnoli, F.Scarlassara, M.Trotta, A.M.Vinodkumar. I also would like to thank all the people of the CLARA collaboration who contributed to successfully perform first experiments with the new set-up.

References

1. K.E. Rehm, *Annu. Rev. Nucl. Part. Sci.* **41**, 429 (1991).
2. W.von Oertzen and A.Vitturi, *Rep. Prog. Phys.* **64**, 1247 (2001).
3. C.L.Jiang et al., *Phys. Lett.* **B337**, 59 (1994); *Phys. Rev.* **C57**, 2393 (1998).
4. L. Corradi et al., *Phys. Rev.* **C54**, 201 (1996); *Phys. Rev.* **C59**, 261 (1999).
5. L. Corradi et al., *Phys. Rev.* **C63**, 021601R (2001).
6. L.Corradi et al., *Phys. Rev.* **C66**, 024606 (2002).
7. A.Winther, *Nucl. Phys.* **A572**, 191 (1994); *Nucl. Phys.* **A594**, 203 (1995).
8. C.H.Dasso, G.Pollarolo and A.Winther, *Phys. Rev. Lett.* **73**, 1907 (1994).
9. G.Pollarolo, Proc. *XXXVII Int. Winter Meeting on Nuclear Physics*, 25-30 January 1999, Bormio (Italy), I. Iori ed., Universita' degli Studi di Milano, Vol.N.114, p.369.
10. NUPECC Report, *Nuclear Physics in Europe*, December 1997, J. Vervier et al. eds.
11. The EURISOL Report, *Key experiment task group*, J.Cornell ed., GANIL, Dec.2003; http://www.ganil.fr/eurisol.
12. E.Vigezzi and A.Winther, *Ann. of Phys.* **192**, 432 (1989).
13. C.H.Dasso and G.Pollarolo, *Phys. Lett.* **B155**, 223 (1985).
14. G.Pollarolo, R.A.Broglia and A.Winther, *Nucl. Phys.* **A406**, 369 (1983).
15. S.Szilner et al., *Eur. Phys. J.* **A**,in press;(http://arXIV.org,nuclex/0312004).
16. R.A. Broglia, O. Hansen and C. Riedel, *Advances in Nuclear Physics*, edited by M. Baranger and E. Vogt, Plenum, New York, 1973, Vol. 6, p.287.
17. M. Igarashi, K. Kubo and K. Tagi, Phys. Rep. **199**, 1 (1991).
18. E. Caurier, K. Langanke, G. Martinez-Pinedo, F. Nowacki and P. Vogel, *Phys. Lett.* **B522**, 240 (2001).
19. A.M.Stefanini et al., *Proposta di esperimento PRISMA*, LNL-INFN (Rep) - 120/97 (1997).
20. A.M.Stefanini et al., *Nucl. Phys.* **A701**, 217c (2002).
21. A.Gadea et al., *Eur. Phys. J.* **A20**, 193 (2004).
22. J.B.Ball, R.L.Auble and P.G.Roos, *Phys. Rev.* **C4**, 196 (1971).
23. J.L.Wood et al., *Phys. Rep.* **215**, 101 (1992).
24. P.E.Garrett et al., *Phys. Rev.* **C68**, 024312 (2003).
25. M.Ulrickson et al., *Phys. Rev.* **C13**, 536 (1976).
26. K.H.Souw et al., *Phys. Rev.* **C12**, 1103 (1975).
27. L.Corradi et al., *Phys. Rev.* **C61**, 024609 (2000).

KEY TOPICS IN NUCLEAR STRUCTURE
Proceedings of the 8th Int. Spring Seminar on Nuclear Physics
edited by Aldo Covello
© 2005 World Scientific Publishing Co.

STUDY OF ^{110}Sn VIA ^{112}Sn (p,t) REACTION

P. GUAZZONI, L. ZETTA

Dipartimento di Fisica dell'Università, and I.N.F.N., Via Celoria 16, I-20133 Milano, Italy
E-mail: paolo.guazzoni@mi.infn.it, luisa.zetta@mi.infn.it

A. COVELLO, A. GARGANO

Dipartimento di Scienze Fisiche, Università di Napoli Federico II, and I.N.F.N., Complesso Universitario di Monte S. Angelo, Via Cintia, I-80126 Napoli, Italy
E-mail: aldo.covello@na.infn.it, angela.gargano@na.infn.it

G. GRAW, R. HERTENBERGER, H-F. WIRTH

Sektion Physik der Universität München, D-85748, Garching, Germany
E-mail: gerhard.graw@physik.uni-muenchen.de, ralf.hertenberger@physik.uni-muenchen.de, hans-friedrich.wirth@ph.tum.de

B. F. BAYMAN

Physics and Astronomy Department, University of Minnesota, Minneapolis, U.S.A.
E-mail: bayman@umphys.spa.umn.edu

M. JASKOLA

Soltan Institute for Nuclear Studies, Hoza Street 69, Warsaw, Poland
E-mail: marian.jaskola@fuw.edu.pl

The ^{112}Sn$(p,t)^{110}$Sn reaction has been studied in a high resolution experiment at an incident energy of 26 MeV. Differential cross-sections for transitions to 27 levels of ^{110}Sn up to an excitation energy of about 4.3 MeV have been measured. Distorted wave Born approximation analysis of experimental angular distributions, assuming a dineutron cluster pickup mechanism, has been performed, allowing confirmation of previous J^{π} values as well as new assignments. Moreover, a microscopic calculation of the angular distributions of ground and first excited states of ^{110}Sn has been carried out. A shell-model study of ^{110}Sn has been also performed using a realistic effective interaction derived from the CD-Bonn nucleon-nucleon potential.

1. Introduction

The Sn isotopes have been the subject of many experimental and theoretical investigations aimed at understanding their shell model structure. The two-neutron transfer reactions as (p,t) are very sensitive to pairing correlations in the overlap between initial and final states and for this reason have been used to obtain detailed spectroscopic information on tin nuclei.

In recent years, in the framework of a systematic study of tin isotopes using (p,t) reaction in high resolution experiments, we reported the results obtained for the two nuclei ^{120}Sn and ^{114}Sn together with the predictions of realistic shell-model calculations.[1,2] In this contribution, we have extended our study to the ^{112}Sn$(p,t)^{110}$Sn reaction. The level structure of ^{110}Sn has been previously investigated by different experiments summarized in the NDS compilation,[3] where a complete list of references is reported.

Differential cross sections of 25 transitions to ^{110}Sn up to an excitation energy of 4317 keV were accurately measured, allowing to confirm or identify spin and parity of the observed levels.

In connection with the experimental work, we have performed preliminary DWBA microscopic calculations of cross section angular distributions for the ground and first 2^+ excited state of ^{110}Sn. A shell model study of ^{110}Sn has been also carried out making use of a realistic effective interaction derived from the CD-Bonn nucleon-nucleon (NN) potential.[4]

2. Experimental Procedure and Results

The ^{112}Sn$(p,t)^{110}$Sn reaction was measured in a high resolution experiment using the 26 MeV proton beam from the Munich HVEC MP Tandem accelerator. The $102\mu g/cm^2$ thick ^{112}Sn isotopic enriched (98.9%) target was evaporated on a $13\mu g/cm^2$ carbon backing. The reaction products were analyzed with the Munich Q3D spectrograph, at ten angles between 6° and 57.5^o, in two different magnetic field settings in order to reach an excitation energy of the residual nucleus of $\sim$ 4300 keV. Outgoing tritons were identified in the focal plane of the Q3D magnetic spectrograph by the 1.8 m long focal plane detector for light ions. The energy resolution was 8 keV full width at half maximum. The uncertainty in our quoted energies is estimated to be 3 keV. Absolute cross sections were estimated with an uncertainty of $\sim$ 15%.

In Table 1 all information available in the literature[3] and the results of the present transfer reaction experiment are summarized.

Table 1. Columns 1 and 2 give the adopted[3] energies, spins and parities of the ^{110}Sn levels; columns 3 and 4 the energies, spins and parities observed in the present work; column 5 gives the integrated cross sections from 5° to 60°. In column 5 absolute cross sections are reported together with the statistical errors.

Adopted		**Present Experiment**			**Adopted**		**Present Experiment**		
E_{exc} (keV)	J^π	E_{exc} (MeV)	J^π	σ_{int} (μb)	E_{exc} (keV)	J^π	E_{exc} (MeV)	J^π	σ_{int} (μb)
0.0	0^+	0.0	0^+	1309 ± 14			3.083	2^+	16 ± 1
1211.88	2^+	1.212	2^+	198 ± 6	3153	2^+			
2120.92	(2)				3182.8	$(2^+,3^+,4^+,5^+)$			
2196.92	4^+	2.197	4^+	61 ± 2			3.183	0^+	17 ± 1
2302	0^+	2.309	0^+	12 ± 1	3222.5	$(2^+,3^+,4^+,5^+)$			
2455.5	4^+				3252	4^+			
2459	(3^-)				3320	2^+			
		2.462 D	$4^+ + 3^-$	88 ± 2	3357	5^-			
2477.7	6^+	2.478	6^+	44 ± 1			3.421	2^+	6.3 ± 0.5
2545.5	2^+	2.545	2^+	36 ± 1	3446.6	$(2^+,3^+,4^+,5^+)$			
2579	0^+	2.573	0^+	7.1 ± 0.5	3540.4	(2,3,4)	3.540	4^+	3.3 ± 0.4
2694.4	4^+	2.694	4^+	11 ± 1			3.609	4^+	5.5 ± 0.5
2745	0^+	2.742	0^+	18 ± 1	3629.7	$(2^+,3^+,4^+,5^+)$			
2753.8	$(6)^+$	2.753	6^+	6.5 ± 0.5	3687.0	$(7)^-$			
2802.3							3.751	2^+	5.3 ± 0.4
2821.3	(2,3,4,5)				3765.2	$(8)^-$			
2833.4	2^+				3807				
		2.857	2^+	7.7 ± 0.5			3.812	2^+	11 ± 1
2914.7	2^+				3812.5	$(8)^+$			
2948.1	$(2^+,3^+,4^+,5^+)$						3.844	5^-	14 ± 1
2964.8		2.965	2^+	14 ± 1	3884.9	$(2^+,3^+,4^+,5^+)$	3.885	3^-	2.6 ± 0.3
2977.0					3933.1	$(9)^-$			
2983	4^+				3971				
2997	(2^+)						4.132 D	$3^- + 5^-$	5.1 ± 0.4
		3.059	4^+	36 ± 1	4158				
3060	0^+				4316.8	(10)	4.317	4^+	4.1 ± 0.4

3. DWBA Analysis

3.1. *Cluster DWBA Calculations*

Starting from a 0^+ initial state and assuming that the neutrons are transferred in a relative L=0 state with total spin S=0, only natural parity states in the final nucleus will be populated in a one-step transfer process, with a unique L-transfer. In this case the determination of the L-transfer directly gives both spin and parity of the observed level.

For the transitions populating the ^{110}Sn states, DWBA analyses have been carried out assuming a semimicroscopic dineutron cluster pickup mechanism. The calculations have been done in finite range approximation, using the computer code TWOFNR[5] and a proton-dineutron interaction potential of Gaussian form $V(r_{p2n}) = V_0 \exp -(r_{p2n}/\xi)^2$ with $\xi = 2$ fm. We have used the same set of optical model parameters employed in the ^{122}Sn(p,t)[1] and ^{116}Sn(p,t)[2] analyses. Examples of typical analyses for L=0 and L=2 transfers are reported in Fig. 1.

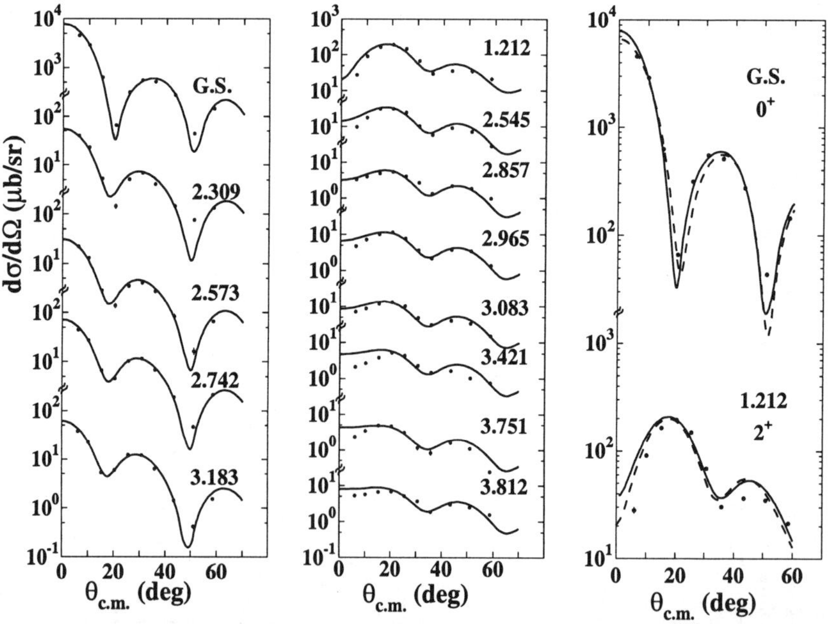

Figure 1. Experimental (*dots*) cross section angular distributions for L=0 (left) and L=2 (middle), compared with cluster DWBA calculations (*solid lines*). On the right, microscopic DWBA calculations (*dashed lines*) are compared with experimental and cluster DWBA results for 0^+ ground state and 2^+ 1.212 MeV level (see text).

Spin-parity assignments have been done for all the observed levels. In particular, 9 levels have been observed for the first time and identified in J^π, 10 levels have been confirmed with respect to the levels reported in NDS,[3] and 4 ambiguities removed. Two unresolved doublets have been observed, giving 1 confirmation, 1 removed ambiguity and 2 new assignments.

3.2. *Microscopic DWBA calculations*

The microscopic calculation of the (p,t) transfer has been also done with the reaction code TWOFNR.

The shell model description of the pair of transferred neutrons is

$$\sum_{n_1,\ell_1,j_1;n_2,\ell_2,j_2} S^J_{n_1,\ell_1,j_1;n_2,\ell_2,j_2} \left\{ \begin{array}{l} \dfrac{\left[\psi^{n_1,\ell_1,j_1}(\mathbf{r}_1,\sigma_1)\psi^{n_2,\ell_2,j_2}(\mathbf{r}_2,\sigma_2)\right]^J_M}{2(1+\delta_{n_1,n_2}\delta_{\ell_1,\ell_2}\delta_{j_1,j_2})} \\ -\dfrac{\left[\psi^{n_1,\ell_1,j_1}(\mathbf{r}_2,\sigma_2)\psi^{n_2,\ell_2,j_2}(\mathbf{r}_1,\sigma_1)\right]^J_M}{2(1+\delta_{n_1,n_2}\delta_{\ell_1,\ell_2}\delta_{j_1,j_2})} \end{array} \right\} \quad (1)$$

The spectroscopic amplitudes, $S^J_{n_1,\ell_1,j_1;n_2,\ell_2,j_2}$, are calculated from the target and residual wave-functions. To calculate the form-factor for the reaction, it is necessary to extract that part of the above two-neutron wave-function in which the neutrons have the same relative and spin wave function that they have in the outgoing triton. The (p,t) transfer calculation yields angular distributions, and relative cross section for different final states, but it is unable to predict absolute cross sections.

The ground states of ^{112}Sn and ^{110}Sn are described in term of 50 protons and 50 neutrons in filled shells, plus (N-50) active neutrons moving in the five single-particle (SP) orbits in the 50-82 shell. The SP energies are (in MeV) $\epsilon_{d_{5/2}} = 0.0$, $\epsilon_{g_{7/2}} = 0.05$, $\epsilon_{s_{1/2}} = 2.30$, $\epsilon_{d_{3/2}} = 2.45$, and $\epsilon_{h_{11/2}} = 3.25$. These active neutrons experience the shell model potential, and are assumed to interact via a pairing force. The Schroedinger equation is solved in the BCS approximation, and the spectroscopic amplitudes, calculated from the ^{112}Sn and ^{110}Sn zero quasi-particle states, are given in the following: $0g_{7/2}$ $0g_{7/2}$ 1.216; $1d_{5/2}$ $1d_{5/2}$ 1.070; $2s_{1/2}$ $2s_{1/2}$ 0.478; $1d_{3/2}$ $1d_{3/2}$ 0.642; $0h_{11/2}$ $0h_{11/2}$ -0.849.

The ^{110}Sn first excited 2^+ state is described as a mixture of the two-quasiparticle states that is favored by a quadrupole-quadrupole interaction. A random-phase-approximation calculation is performed, with the strength of the interaction taken to bring the predicted one-quadrupole-phonon state to its observed distance above the ground state (1.212 MeV).

The (p,t) spectroscopic amplitudes calculated from the ^{112}Sn zero-quasiparticle state and the one-quadrupole-phonon 2^+ state are the following: $0g_{7/2}$ $0g_{7/2}$ 0.505; $1d_{5/2}$ $0g_{7/2}$ -0.153; $1d_{5/2}$ $1d_{5/2}$ 0.365; $2s_{1/2}$ $1d_{5/2}$ 0.180; $1d_{3/2}$ $0g_{7/2}$ 0.155; $1d_{3/2}$ $1d_{5/2}$ 0.096; $1d_{3/2}$ $2s_{1/2}$ -0.196; $1d_{3/2}$ $1d_{3/2}$ -0.137; $0h_{11/2}$ $0h_{11/2}$ 0.167. These spectroscopic amplitudes are used in connection with a two-neutron-transfer code to calculate differential cross sections for the (p,t) population of the two lowest states of ^{110}Sn.

A fairly good agreement with the observed angular distributions is obtained in these very preliminary calculations, as shown in Fig. 1. However the calculated cross section for this assumed quadrupole oscillation is too small relative to the ground state cross section, by a factor 4.7. This suggest that the 2^+ level exhibits a collectivity that is not well described by the simple quadrupole oscillation model.

4. Shell Model Calculations and Results

Our shell model calculation for ^{110}Sn has been performed assuming ^{100}Sn as a closed core, with the 10 valence neutrons occupying the five SP levels $0g_{7/2}$, $1d_{5/2}$, $1d_{3/2}$, $2s_{1/2}$, and $0h_{11/2}$ of the 50-82 shell.

As mentioned in the Introduction, our two body effective interaction has been derived from the CD-Bonn NN potential. The difficulty posed by the strong short-range repulsion contained in the bare potential has been overcome by constructing a renormalized low-momentum potential, $V_{\rm low-k}$, that preserves the physics of the original NN potential up to a certain cutoff momentum Λ.[6] This is a smooth potential that can be used directly in the calculation of the shell-model effective interaction. In the present paper, we have used for Λ the value $2.1\,{\rm fm}^{-1}$ according to the criterion discussed in Ref. 6.

Once the $V_{\rm low-k}$ is obtained, the calculation of the effective interaction is carried out within the framework of the $\hat{Q}$-box plus folded diagram method, as described in Ref. 7. In our calculation of the $\hat{Q}$ box we include all diagrams up to second order and compute these diagrams by inserting intermediate states composed of particle and hole states restricted to the two major shells above and below the $N = Z = 50$ Fermi surface. A description of our derivation of the effective interaction can be found in Ref. 8. Note that the effective interaction represents the interaction between two-valence neutrons outside the doubly closed ^{100}Sn and may be not completely adequate for systems with several valence particles, as is the case of ^{110}Sn. However, we have not attempted in the present study to modify the effective interaction derived as described above. As regards the neutron SP energies, they are the same as those used in the DWBA microscopic calculations. The shell-model calculations have been performed by using the ANTOINE shell-model code.[9]

Table 2. ^{110}Sn Energy Spectrum

Positive Parity States							
NDS		Shell Model		NDS		Shell Model	
J^π	E_{exc}	J^π	E_{exc}	J^π	E_{exc}	J^π	E_{exc}
0^+	0.00	0^+	0.00	$(6)^+$	2.75	5^+	2.61
2^+	1.21	2^+	1.73		2.80	2^+	2.79
(2)	2.12	4^+	2.20	(2,3,4,5)	2.82	4^+	2.89
4^+	2.20	6^+	2.32	2^+	2.83	2^+	2.91
0^+	2.30	0^+	2.36	2^+	2.91	3^+	2.98
4^+	2.45	6^+	2.41	$(2^+,3^+,4^+,5^+)$	2.95	0^+	3.11
6^+	2.48	6^+	2.51		2.96	3^+	3.13
2^+	2.54	2^+	2.51		2.98	4^+	3.22
0^+	2.58	4^+	2.55	4^+	2.98	0^+	3.25
4^+	2.69	1^+	2.56	(2^+)	3.00	5^+	3.28
0^+	2.74	3^+	2.60			2^+	3.29
Negative Parity States							
5-	3.36	5-	3.58	8-	3.77	9-	3.77
7-	3.69	7-	3.70	9-	3.93	8-	3.78

We now compare our shell-model results with the experimental ones of Ref. 3. In Table 2 all the experimental and calculated levels of ^{110}Sn up to about 3.0 MeV are reported. We have only excluded the 3^- state, whose calculated energy is rather higher (~ 1 MeV) than the observed one (2.45 MeV). This is not surprising since configurations outside our model space are likely to be important for this state. In the higher energy region only negative-parity states are reported. As may be seen in Table 2, some ambiguities are present in the experimental spectrum. In fact, 8 out of the 20 excited states have no, or no firm spin-parity assignment.

Here, we do not try to establish a one-to-one correspondence between theory and experiment, but make only some general remarks. First, we note that the number of calculated 0^+, 2^+, 4^+, and 6^+ states compares well with the observed one, being just the same for $J^\pi = 0^+$ and 4^+ and differing by one for $J^\pi = 2^+$ and 6^+ (five and three calculated states versus the six and two observed ones). This is so with our assignment $J^\pi = 2^+$, 6^+, and 2^+ to the observed states at 2.12, 2.75 and 3.00 MeV with $J^\pi = (2)$, $(6)^+$, and

(2^+), respectively. Furthermore, from 2.6 to 3.3 MeV excitation energy we predict one 1^+ state, two 5^+ and three 3^+ states. In this energy range three states without spin and parity assignment have been observed at 2.80, 2.96, and 2.98 MeV and two states with $J^\pi = (2,3,4,5)$, and $(2^+,3^+,4^+,5^+)$ at 2.82 and 2.95 MeV, respectively.

5. Summary

In a high resolution experiment we have measured 25 transitions to levels of ^{110}Sn up to an excitation energy of 4.317 MeV via the (p,t) reaction on ^{112}Sn at an incident proton energy of 26 MeV.

A finite range DWBA analysis of the experimental angular distributions performed in the framework of a semimicroscopic dineutron cluster pickup mechanism allowed for J^π assignment to all the observed levels.

In connection with the experimental work, a very preliminary microscopic calculation of the ^{110}Sn 0^+ ground state and first 2^+ excited state angular distributions has been carried out, using spectroscopic amplitudes calculated from the target ^{112}Sn and residual ^{110}Sn wave functions in BCS approximation for 0^+ ground states, and RPA approximation for the ^{110}Sn 2^+ state.

A theoretical study of ^{110}Sn within the framework of the shell model has been performed using a two-body effective interaction derived from the CD-Bonn NN potential. We have reported here some preliminary results of this study. Our final results and a detailed comparison with the experimental data will be presented in a forthcoming publication.

References

1. P. Guazzoni, M. Jaskóla, L. Zetta, A. Covello, A. Gargano, Y. Eisermann, G. Graw, R. Hertenberger A. Metz, F. Nuoffer, and G. Staudt, *Phys. Rev.* **C60**, 054603 (1999).
2. P. Guazzoni, L. Zetta, A. Covello, A. Gargano, G. Graw, R. Hertenberger, H.-F. Wirth, and M. Jaskóla, *Phys. Rev.* **C69**, 024619 (2004).
3. D. De Frenne and E. Jacobs *Nuclear Data Sheets* **89**, 481 (2000).
4. R. Machleidt, *Phys. Rev.* **C63**, 024001 (2001).
5. M. Igarashi, computer code TWOFNR (1977) unpublished.
6. Scott Bogner, T. T. S. Kuo, L. Coraggio, A. Covello, and N. Itaco, *Phys. Rev.* **C65**, 051301(R) (2002).
7. T. T. S. Kuo, S. Y. Lee, and K. F. Ratcliff, *Nucl. Phys.* **A176**, 62 (1971).
8. M. F. Jiang, R. Machleidt, D. B. Stout, and T. T. S. Kuo, *Phys. Rev.* **C46**, 910 (1992).
9. E. Caurier, shell-model code ANTOINE, IRES, Stasbourg 1989-2002.

KEY TOPICS IN NUCLEAR STRUCTURE
Proceedings of the 8th Int. Spring Seminar on Nuclear Physics
edited by Aldo Covello
© 2005 World Scientific Publishing Co.

EXPRESSIONS FOR THE NUMBER OF PAIRS OF A GIVEN ANGULAR MOMENTUM IN THE SINGLE j SHELL MODEL: Ti ISOTOPES

L. ZAMICK[1)], A. ESCUDEROS[2)], S. J. LEE[3)], A. MEKJIAN[1)],
E. MOYA DE GUERRA[2)], A. A. RADUTA[4)] AND P. SARRIGUREN[2)]

[1)] *Department of Physics and Astronomy, Rutgers University, Piscataway, New Jersey USA 08854*

[2)] *Instituto de Estructura de la Materia, C.S.I.C., Serrano 123, E-28006 Madrid, Spain*

[3)] *Department of Physics and Institute of Natural Sciences, Kyung Hee University, Suwon, KyungGiDo, Korea*

[4)] *Department of Theoretical Physics and Mathematics, Bucharest University, P.O. Box MG11, Bucharest, Romania*

Motivated by the problem of the relative importance of $J = 1^+$ $T = 0$ pairing versus the better established $J = 0^+$ $T = 1$ pairing in nuclei, we here address the problem of the number of *np* pairs of a given angular momentum in selected nuclei, ^{44}Ti, ^{46}Ti, and ^{48}Ti. The number of pairs is obviously relevant to *np* pickup reactions such as (p,^{3}He). One can address also the *np* transfer reaction (^{3}He,p) and indeed there is a proposal by the Berkeley group to do the reaction ^{44}Ti(^{3}He,p)^{46}Sc. In this work we will consider the single j shell model for the Ti isotopes. In the near future more elaborate calculations are planned.

1. The single j model for ^{44}Ti, ^{46}Ti, and ^{48}Ti

In the single j shell model for nuclei in the $f_{7/2}$ model one can take the matrix elements from experiment. From the (j^2) configuration the states of isospin $T = 0$ have odd values of$J$ (1, 3, 5 and 7) while the states of isospin $T = 1$ have even values (0, 2, 4 and 6). We can look at the spectrum of ^{42}Sc and make the association $< (j^2)^J \, V \, (j^2)^J >= E^*(J) + constant$. For the spectra the constant does not matter. The values of $E^*(J)$ for $J = 0$ to 7 are respectively (in MeV) 0, 0.611, 1.5863, 1.4904, 2.8153, 1.5101, 3.242 and 0.6163. The constant above can be obtained as $E(^{42}\text{Sc}) + E(^{40}\text{Ca}) - E(^{41}\text{Ca}) - E(^{41}\text{Sc})$ and is equal to -3.182 MeV. It can be regarded as the pairing interaction in the ($J = 0^+ \, T = 1$) ground state. Note that the lowest levels have $J = 0^+ \, T = 1$, $J = 1^+ \, T = 0$ and $J = 7^+ \, T = 0$. The latter two are almost degenerate at an excitation energy of about 0.6 MeV.

We now consider the even-even Ti isotopes. We use the notation n for the number of $f_{7/2}$ neutrons and N for the number of valence nucleons. For Ti isotopes $N = n + 2$. In the single j shell model [1] the wave function can be written as

$$\Psi^{I\alpha} = \sum_{J_P J_N} D^{I\alpha}(J_P J_N) \left[(j^2)^{J_P} (j^n)^{J_N} \right]^I \tag{1}$$

where I is the total angular momentum, α is any additional quantum number and $D^{I\alpha}(J_P J_N)$ is the probability amplitude that the protons couple to J_P and the neutrons to J_N. For $I = 0$ states we have $J_P = J_N = J$,

$$\Psi^{I=0} = \sum_J D(JJ) \left[(j^2)^J (j^n)^J \right]^{I=0} . \tag{2}$$

The coefficients $D(JJ)$ for ^{44}Ti are shown in Table 1.

The ground states of ^{44}Ti, ^{46}Ti, and ^{48}Ti have isospins $T = 0, 1$ and 2, respectively. We define $T_{\min} = |N - Z|/2$, i.e., the minimum isospin. All the even-even ground states have $T = T_{\min}$.

In all the Ti isotopes above there is only one $I = 0$ state with isospin $T = T_{\min} + 2$, all the rest have $T = T_{\min}$. For the higher isotopic spin states the coefficients $D^{I=0}(JJ)$ are independent of what isospin conserving interaction is used. Indeed these coefficients are two particle coefficients of fractional parentage (cfp)

$$D^{I=0, T_{\min}+2}(JJ) = \left(j^n J j^2 J \,|\}\, j^{n+2} 0 \right) . \tag{3}$$

The reason for this is that the $T = T_{\min} + 2$ state is the double analog of a corresponding state in Calcium, which in this model consists of identical particles (neutrons) in the valence shell. The two particle cfp in Eq.(3) is used to separate N neutrons into n and 2 neutrons.

We have also found in the past [2] an identity to be useful, which relates a one particle cfp to a two particle cfp

$$\begin{aligned} \left(j^n J j \,|\}\, j^{n+1} j \right) &= \left(j^n J j^2 J \,|\}\, j^{n+2} 0 \right) \quad \text{i.e.,} \\ D^{T_{\min}+2}(JJ) &= \left(j^n J j \,|\}\, j^{n+1} j \right) \end{aligned} \tag{4}$$

We get two conditions from this:

a) ORTHOGONALITY of the $T_{\min} + 2$ state to all $T_{\min}$ states

$$\sum_J D^{JT_{\min}}(JJ)\,(j^n Jj\,|\}\,j^{n+1}j) = 0 \tag{5}$$

b) NORMALIZATION of the unique $T_{\min}+2$ state

$$\sum_J D^{JT_{\min}+2}(JJ)\,(j^n Jj\,|\}\,j^{n+1}j) = 1 \tag{6}$$

We define the two particle matrix element $E(J_A) =< (j^2)^{J_A} V (j^2)^{J_A} >$. The expression for the energy is

$$< \Psi H \Psi >= \sum_{J_A} N(J_A) E(J_A) \tag{7}$$

where $N(J_A)$ is the number of pairs with angular momentum J_A. Let us focus on the number of np pairs in a state of angular momentum $I = 0$. We find

$$N_{np}(J_A) = 2n \sum_{J_0} \Bigg| \sum_J D(JJ)\,(j^{n-1} J_0 j\,|\}\,j^n J) \times \sqrt{(2J+1)(2J_A+1)} \begin{Bmatrix} j & j & J \\ j & J_0 & J_A \end{Bmatrix} \Bigg|^2 \tag{8}$$

The coefficient of fractional parentage is required to separate one neutron from the others and the six-j (actually a nine-j with a zero) is required to combine the neutron with a proton in order to form a pair.

We now can obtain the number of even J_A pairs by multiplying by the factor $[1 + (-1)^{J_A}]/2$ and summing on J_A. For the odd J_A pairs we use $[1 - (-1)^{J_A}]/2$.

We find the following recursion formula in Talmi's books [3,4] useful

$$\begin{aligned} n\ & (j^{n-1}\alpha_0 J_0 j\,|\}\,j^n J)\,(j^{n-1}\alpha_1 J_1 j\,|\}\,j^n J) = \delta_{\alpha_0\alpha_1}\delta_{J_1 J_0} + (n-1) \\ & \times \sqrt{(2J_0+1)(2J_1+1)} \sum_{J_2} \begin{Bmatrix} J_2 & j & J_1 \\ J & j & J_0 \end{Bmatrix} (-1)^{J_0+J_1} \\ & \times (j^{n-2} J_2 j\,|\}\,j^{n-1}\alpha_0 J_0)\,(j^{n-2} J_2 j\,|\}\,j^{n-1}\alpha_1 J_1) \end{aligned} \tag{9}$$

We find that the total number of even/odd J_A pairs is:

$$\text{Number of } \begin{pmatrix}\text{even}\\ \text{odd}\end{pmatrix} \text{ pairs } = (n \mp 1) \pm (n+1) \left| \sum_J D(JJ)\, (j^n Jj \,|\}\, j^{n+1} j) \right|^2 \tag{10}$$

But from Eqs.(5) to (6) one sees that the second term is zero when $T = T_{\rm min}$, and it is $(n+1)$ for $T = T_{\rm min} + 2$. Hence:

$$\begin{aligned}\text{Total number of even (T = 1) } np \text{ pairs } &= (n-1) \ \text{ for } \ T = T_{\rm min}\\ &= 2n \ \text{ for } \ T = T_{\rm min} + 2 \end{aligned} \tag{11}$$

$$\begin{aligned}\text{Total number of odd (T = 0) } np \text{ pairs } &= (n+1) \ \text{ for } \ T = T_{\rm min}\\ &= 0 \ \text{ for } \ T = T_{\rm min} + 2 \end{aligned} \tag{12}$$

That we get no odd pairs for $T = T_{\rm min} + 2$ is no surprise. For a system of identical particles in the single j shell, all pairs have isospin $T = 1$ and hence have even J. Notice that the total number of even pairs is independent of the interaction, likewise odd pairs.

2. The number of $J_A = 0$ $T_A = 1$ np pairs

For a $J_A = 0$ pair Eq.(8) simplifies to

$$N_{np}(J_A = 0) = \frac{2n}{(2j+1)^2} \left| \sum_J D(JJ)\, (j^{n-1} jj \,|\}\, j^n J)\, \sqrt{(2J+1)} \right|^2 \tag{13}$$

We can develop things further by using explicit expressions for cfp's found in De Shalit and Talmi [4]:

$$\begin{aligned}
(j^{n-1} jj \,|\}\, j^n J = 0) &= 1\,,\\
(j^{n-1} jj \,|\}\, j^n Jv = 2) &= \sqrt{\frac{2(2j+1-n)}{n(2j-1)}}\,,\\
(j^n J = 0 j \,|\}\, j^{n+1} j) &= \sqrt{\frac{(2j+1-n)}{(n+1)(2j+1)}}\,,\\
(j^n Jv = 2j \,|\}\, j^{n+1} j) &= -\sqrt{\frac{2n(2J+1)}{(n+1)(2j+1)(2j-1)}}\,.
\end{aligned} \tag{14}$$

We define $M = \sum_{J\geq 2} D(JJ)\sqrt{(2J+1)}$. From Eqs.(5,6) and the above cfp's we find

$$\sqrt{\frac{2j+1-n}{(n+1)(2j+1)}}D(00) - M\sqrt{\frac{2n}{(n+1)(2j+1)(2j-1)}} = \begin{cases} 0 & (T = T_{\min}) \\ 1 & (T = T_{\min}+2) \end{cases} \tag{15}$$

More explicitly for $T = T_{\min}$ we find $D(00) = M/3$ for ^{44}Ti, $D(00) = M/\sqrt{3}$ for ^{46}Ti and $D(00) = M$ for ^{48}Ti. We also find for $T = T_{\min}$

$$D(00) = \frac{n}{2j+1}\sum_J D(JJ)\left(j^{n-1}jj\,|\}\,j^n J\right)\sqrt{(2J+1)} \tag{16}$$

Using Eq.(15) and Eq.(16) we get the main result of this section

$$N_{np}(J_A = 0,\ T = T_{\min}) = \frac{2|D(00)|^2}{n} \tag{17}$$

$$N_{np}(J_A = 0,\ T = T_{\min}+2) = 2n|D(00)|^2 = \frac{2n(2j+1-n)}{(2j+1)(n+1)} \tag{18}$$

i.e., 1, 4/5 and 3/7 for the $T = 2, 3$ and 4 states of ^{44}Ti, ^{46}Ti and ^{48}Ti, respectively.

Previously a formula for the number of np $J_A = 0$ pairs for a $J = 0$ $T = 1$ pairing interaction was obtained by Engel et al.[5]. Our results above are of course true for any isospin conserving interaction.

3. The special case of ^{44}Ti

Previous to the work discussed here we considered the number of pairs only in ^{44}Ti [6,7]. We came up with a condition on $D(JJ)$ which at first sight looks different from those of Eq.(5) and Eq.(6). The result [7] is

$$-\sum_J D(JJ)\sqrt{(2J+1)(2J_A+1)}\begin{Bmatrix} j & j & J \\ j & j & J_A \end{Bmatrix} = \begin{cases} D(J_AJ_A)/2 & (T=0) \\ -D(J_AJ_A) & (T=2) \end{cases} \tag{19}$$

Using this result we can show that for all even J_A (0, 2,4 and 6) and all $T = 0$ states:

$$\begin{aligned} &\text{Number of } nn \text{ pairs} = \text{Number of } np \text{ pairs} \\ &= \text{Number of } pp \text{ pairs} = |D(J_AJ_A)|^2 \end{aligned} \tag{20}$$

This is true for any isospin conserving interaction.

We can show however that Eq.(19) is identical to Eqs.(5) and (6) by noting that the recursion relation (9) simplifies for $n = 3$ to

$$2\sqrt{(2J+1)(2J'+1)}\begin{Bmatrix} j & j & J' \\ j & j & J \end{Bmatrix} = -\delta_{JJ'} + 3\left(j^2 J j \,|\}\, j^3 j\right)\left(j^2 J' j \,|\}\, j^3 j\right) \tag{21}$$

Amusingly Eq.(19) is an eigenvalue equation in which the operator is the unitary $6j$ symbol. ¿From the fact that the eigenvalues in Eq.(5) for $T_{\min}$ and in Eq.(6) for $T_{\min}+2$ are respectively 0 and 1, we can infer that the eigenvalues of the unitary $6j$ symbol are -1/2 and 1. The eigenvalue $-1/2$ corresponds to the $T = 0$ states of ^{44}Ti and is triply degenerate. The eigenvalue 1 corresponds to the unique $T = 2$ state and the eigenfunction $D(JJ)$ is equal to the two particle cfp $\left(j^2 J j^2 J \,|\}\, j^4 0\right)$.

It is fascinating to note that Rosensteel and Rowe [9] had already found the need to diagonalize the above unitary $6j$ symbol. They were addressing an apparently different problem, the number of seniority conserving interactions for a system of identical particles in a given j shell. We, on the other hand, are considering a system of mixed neutrons and protons. Despite these obvious differences, there might possibly be some connections that deserve further investigations. Note that in Ref.[7] and in this work we are essentially able to get the eigenvalues of the unitary $6j$ symbol without an explicit diagonalization. Rather, we show that the $6j$ operator in angular momentum space is equivalent to a simple operator in isospin space.

4. The number of J_A pairs in ^{44}Ti

In Table 1 we show the $J = 0$ wave functions of ^{44}Ti, in which the two-body matrix elements $E^J = <(j^2)^J V (j^2)^J>$ were obtained from the spectrum of ^{42}Sc. For the ground state we have: $D(00) = 0.7878$; $D(22) = 0.5616$; $D(44) = 0.2208$; $D(66) = 0.1234$.

Table 1. Excitation energies [MeV] and eigenvectors in ^{44}Ti using the *Spectrum of ^{42}Sc* as input interaction.

excitation energies		0.0000	5.4861	8.2840	8.7875
eigenvectors	D(00)	0.78776	-0.35240	-0.50000	-0.07248
	D(22)	0.56165	0.73700	0.37268	-0.04988
	D(44)	0.22082	-0.37028	0.50000	0.75109
	D(66)	0.12341	-0.44219	0.60093	-0.65432

The state at 8.284 MeV is the $T = 2$ double analog state of ^{44}Ca. The $D(JJ)$ are as mentioned before two particle cfp's. For the $T = 2$ state $D(00) = -0.5$; $D(22) = 0.37258$; $D(44) = 0.5$; $D(66) = 0.60093$.

In Table 2 we give the number of pairs of particles $(nn + np + pp)$ with angular momentum J_{12} (for $T = 1$ $J_{12} = 0, 2, 4, 6$; for $T = 0$ $J_{12} = 1, 3, 5, 7$). Let us focus on columns E and F. In F we define the no-interaction case as a basis of comparison. Here we show the average number of pairs for the three $T = 0$ states in ^{44}Ti shown in Table 1 (we take the $T = 2$ state out of the picture).

Table 2. Number of pairs for various interactions: A ($J = 0$, $T = 1$) pairing; B ($J = 1$, $T = 0$) pairing; C equal $J = 0, J = 1$ pairing; D $Q \cdot Q$ interaction; E spectrum of ^{42}Sc; F no interaction.

	A	B	C	D	E	F
$J_{12} = 0$	2.250	0.433	2.045	1.499	1.862	0.750
$J_{12} = 2$	0.139	1.420	0.492	1.413	0.946	0.861
$J_{12} = 4$	0.250	0.320	0.416	0.086	0.146	0.750
$J_{12} = 6$	0.361	0.626	0.048	0.001	0.046	0.639
$J_{12} = 1$	0.250	1.297	0.618	0.834	0.675	0.432
$J_{12} = 3$	0.583	0.388	0.165	0.156	0.271	0.902
$J_{12} = 5$	0.916	0.003	0.564	0.013	0.159	1.000
$J_{12} = 7$	1.250	1.311	1.654	1.996	1.895	0.667

Note that the number of states is 8 and the number of pairs is $n(n-1)/2 = 6$. So, on average there are 6/8=0.75 pairs per angular momentum. However even in the no-interaction case we do not always get 0.75. We do get it for a $J_{12} = 0\,T = 1$ pair but for a $J_{12} = 1\,T = 0$ pair we get only 0.432.

When the interaction is turned on we find in column E an enhancement of pairs with $J_{12} = 0, 2, 1, 7$ and a depletion of the others ($J_{12} = 3, 4, 5, 6$). Going from F to E the number of $J_{12} = 0$ pairs goes from 0.750 to 1.862, the number of $J_{12} = 1$ pairs from 0.432 to 0.675 and the number of $J_{12} = 7$ pairs from 0.667 to 1.896. In columns A, B, C and D we show results for schematic interactions. Column A is for $J = 0$ $T = 1$ pairing. Here we lower the $J = 0$ $T = 1$ state in the two particle system keeping all the others degenerate. In B we lower only $J = 1$ $T = 0$. In C we lower $J = 0\,T = 1$ and $J = 1\,T = 0$ by the same amount and in D we use a $Q \cdot Q$ interaction. We should compare these with E, the realistic case, and F, the no-interaction case.

Not surprisingly for A ($J = 0\,T = 1$ pairing) we get an enhancement of

$J = 0\ T = 1$ pairs but a depletion of $J = 1\ T = 0$ pairs. For B ($J = 1\ T = 0$ pairing) we get an enhancement of $J = 1\ T = 0$ pairs but a depletion of $J = 0\ T = 1$ pairs. What is perhaps surprising is that for both cases A and B we get an enhancement of $J = 7$ pairs even though we do not lower the $J = 7$ two particle energy.

On the other hand we have a zero sum game at hand. The total number of $T = 1$ pairs ($J = 0, 2, 4$) must equal 3, likewise the total number of $T = 0$ pairs ($J = 1, 3, 5, 7$) must also equal 3 for ^{44}Ti.

Note that in case C (equal lowering of $J = 0$ and $J = 1$ energies) we get results which are in better agreement with the realistic case E. On the other hand the number of $J = 2$ pairs is too low, but as shown in case D this can be cured by including a $Q \cdot Q$ interaction.

We feel the results in Table 2 will be a useful guide for the dynamics of two nucleon transfer reactions.

Acknowledgments

This work was supported in part by DOE grant DE-FG01-04ER04-02, by DOE grant DE-FG02-96ER-40987, by grant 2001-1-11100-005-3 from the BRP (Korea), by MCyT (Spain) under contract number BFM2002-03562, by MEC (Romania) under the grant CERES2/24 and by a NATO Linkage Grant PST 978158.

References

1. J.D. McCullen, B.F. Bayman and L. Zamick, *Phys. Rev.* **134** 515 (1964); Technical Report NYO-9891.
2. L. Zamick and Y. Durga Devi, *Phys. Rev.* **C 60** 054318 (1999).
3. A. de Shalit and I. Talmi, *Nuclear Shell Theory*, Academic Press, New York (1963).
4. I. Talmi, *Simple Models of Complex Nuclei*, Harwood Academic, Switzerland (1993).
5. J. Engel, K. Langanke and P. Vogel, *Phys. Lett.* **B 389** 211 (1996).
6. L. Zamick, A. Mekjian and S.J. Lee, LANL preprint nucl-th/0402089.
7. E. Moya de Guerra, A.A. Raduta, L. Zamick and P. Sarriguren, *Nucl. Phys.* **A727** 3 (2003).
8. L. Zamick, E. Moya de Guerra, P. Sarriguren, A. Escuderos and A.A. Raduta, LANL preprint nucl-th/0312077.
9. G. Rosensteel and D.J. Rowe, *Phys. Rev.* **C67** 014303 (2003).

KEY TOPICS IN NUCLEAR STRUCTURE
Proceedings of the 8th Int. Spring Seminar on Nuclear Physics
edited by Aldo Covello
© 2005 World Scientific Publishing Co.

A SAMPLING ALGORITHM FOR LARGE SCALE SHELL MODEL CALCULATIONS

F. ANDREOZZI, N. LO IUDICE, AND A. PORRINO

Dipartimento di Scienze Fisiche, Universitá di Napoli "Federico II"
and Istituto Nazionale di Fisica Nucleare,
Monte S Angelo, Via Cinthia I-80126 Napoli, Italy
E-mail: loiudice@na.infn.it

We show that a recently developed algorithm for generating a subset of eigenvalues and eigenvectors of large matrices may be used for an efficient definition of an importance sampling of the shell-model basis. The sampling so implemented allows for substantial reductions of configuration spaces, and, also, for extrapolations to exact eigenvalues and E2 strengths.

1. Introduction

The increasing computational power has encouraged the attemtps of facing directly the eigenvalue problem of complex quantum systems. We mention the Monte Carlo methods to compute ground states properties [1] and, more recently, to generate a truncated basis for diagonalizing the many-body Hamiltonian [2]. The first approach has to deal with the sign problem, the second with the redundancy of the basis states and the broken symmetries.

A more direct route is provided by the Lanczos [3] and Davidson [4] algorithms, which face the diagonalization of the Hamiltonian. The critical points of these methods are the amount of memory needed and the time spent in the diagonalization process.

In a recent paper [5], we have developed an iterative algorithm for generating a selected set of eigenvectors of a large matrix which is fast, *robust*, yielding always stable numerical solutions, free of *ghost* eigenvalues, and extremely simple to be implemented. Like the other methods, however, it requires the storage of at least one eigenvector, which exceeds the limits of modern computers in many complex systems. On the other hand, the algorithm can be endowed with an importance sampling which allows for a drastic reduction of the space and offers other important advantages [6], which will be discussed here.

2. The algorithm

For the sake of simplicity, we consider here a symmetric matrix $A = \{(a_{ij}) = \langle i \mid \hat{A} \mid j \rangle\}$ representing a self-adjoint operator $\hat{A}$ in an orthonormal basis $\{\mid 1\rangle, \mid 2\rangle, \ldots, \mid N\rangle\}$. The algorithm goes through several iteration loops. The first loop consists of the following steps: 1a) Diagonalize the two-dimensional matrix (a_{ij}) (i,j=1,2), 1b) select the lowest eigenvalue λ_2 and the corresponding eigenvector $\mid \phi_2\rangle = c_1^{(2)} \mid 1\rangle + c_2^{(2)} \mid 2\rangle$, 1c) for $j = 3, \ldots, N$ diagonalize the matrix

$$\begin{pmatrix} \lambda_{j-1} & b_j \\ b_j & a_{jj} \end{pmatrix} \tag{1}$$

where $b_j = \langle \phi_{j-1} \mid \hat{A} \mid j\rangle$ and select the lowest eigenvalue λ_j together with the corresponding eigenvector $\mid \phi_j\rangle$. This zero approximation loop yields the approximate eigenvalue and eigenvector

$$E^{(1)} \equiv \lambda_N, \qquad \mid \psi^{(1)}\rangle \equiv \mid \phi_N\rangle = \sum_{i=1}^{N} c_i^{(N)} \mid i\rangle. \tag{2}$$

With these new entries we start an iterative procedure which goes through $n = 2, 3, \ldots$ refinement loops, consisting of the same steps with the following modification. At each step $j = 1, 2, \ldots, N$ of the n-th loop $(n > 1)$ we have to solve an eigenvalue problem of general form, since the states $\mid \phi_{j-1}\rangle$ and $\mid j\rangle$ are no longer orthogonal. The eigenvalue $E^{(n)} \equiv \lambda_N$ and eigenvector $\mid \psi^{(n)}\rangle \equiv \mid \phi_N\rangle$ obtained after the n-th loop are proven to converge to the exact eigenvalue E and eigenvector $\mid \psi\rangle$ respectively [5].

The algorithm has been shown to be completely equivalent to the method of optimal relaxation [8] and has therefore a variational foundation. Because of its matrix formulation, however, it can be generalized with minimal changes so as to generate at once an arbitrary number n_v of eigensolutions. Indeed, we have to replace the two dimensional matrix (1) with a multidimensional one

$$\begin{pmatrix} \lambda_{j-1} & b_j \\ b_j & a_{jj} \end{pmatrix} \Longrightarrow \begin{pmatrix} \Lambda_k & B_k \\ B_k^T & A_k \end{pmatrix}, \tag{3}$$

where Λ_k is a n_v-dimensional diagonal matrix whose non-zero entries are the eigenvalues $\lambda_1^{(k-1)}, \lambda_2^{(k-1)}, \ldots, \lambda_{n_v}^{(k-1)}$, $A_k = \{a_{ij}\}$ $(i, j = (k-1)p+1, \ldots, kp)$ is a p-dimensional submatrix, B_k and its transpose are matrices composed of the matrix elements $b_{ij}^{(k)} = \langle \phi_i^{(k-1)} \mid \hat{A} \mid j\rangle$ (i = 1, $\ldots, n_v$; $j = (k-1)p + 1, \ldots, kp$). A loop procedure similar, though more general, to the one adopted in the one-dimensional case, yields a set of n_v eigenvalues $E_1, \ldots, E_{n_v}$ and corresponding eigenvectors $\psi_1, \ldots, \psi_{n_v}$.

3. Importance sampling

Since for many complex systems the dimensions of the Hamiltonian matrix become prohibitively large, one must rely on some importance sampling which allows for a truncation of the space by selecting only the basis states relevant to the exact eigensolutions. A notable example is the stochastic diagonalization method [4], which samples the basis states relevant to the ground state through a combination of plane (Jacobi) rotations and matrix inflation.

A similar, but more efficient, strategy can be implemented in the framework of our diagonalization process. Exploiting the fact that the algorithm yields quite accurate solutions already in the first approximation loop, we can devise a sampling which makes use of the first loop only. This is to be accordingly modified and through the following steps :1a) Diagonalize the v-dimensional principal submatrix $\{a_{ij}\}$ $(i, j = 1, v)$; 1b) For $j = v+1, \ldots, N$, diagonalize the $v + 1$-dimensional matrix

$$\begin{pmatrix} \Lambda_v & \vec{b}_j \\ \vec{b}_j^T & a_{jj} \end{pmatrix}, \tag{4}$$

where $\vec{b}_j = b_{1j}, b_{2j}, \cdot, \vec{b_{vj}}$; 1c) Select tha lowest v eigenvalues $\lambda'_i, (i = 1, v)$ and accept the new set only if

$$\sum_{i=1,v} \mid \lambda'_i - \lambda_i \mid > \epsilon \tag{5}$$

The outcome of this procedure is that the selected states span a $n_s(< N)$ dimensional space, so that the subsequent refinement loops iterate only on the smaller set of n_s basis vectors. The number of operations is thereby reduced by a factor N/n_s. Indeed, for large N, they are $O\left(n_s\left(N + \mu n_s\right)\right)$, to be compared with $O(\mu' N^2)$ when dealing with the full matrix, having denoted by μ and μ' the number of iterations required to reach convergence in the two cases.

4. Applications to typical nuclei

We applied the sampling algorithm to three typical nuclei, the semi-magic ^{108}Sn, the N=Z even-even ^{48}Cr and the $N > Z$ odd-even ^{133}Xe. We adopted a realistic effective interaction deduced from the Bonn-A potential [11] for

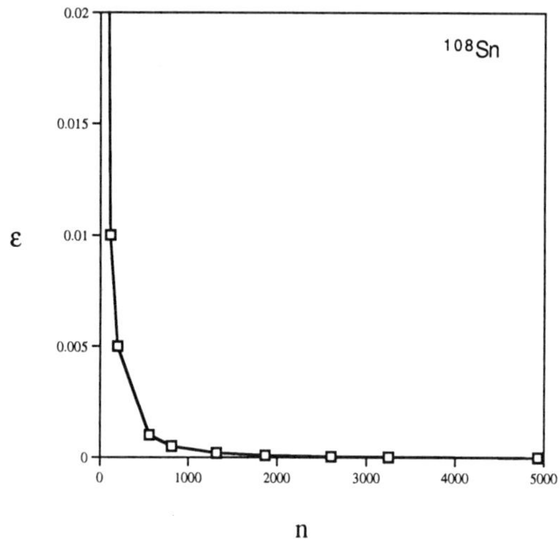

Figure 1. Importance sampling parameters versus the dimensions n of the truncated matrices resulting from the sampling.

^{108}Sn and ^{133}Xe, and the KB3 interaction [12] for ^{48}Cr. We deduced the single particle (s.p.) from a fit of properly selected experimental data.

To get a faster convergence, we generated a new correlated basis $| j\rangle$ by the multipartitioning method [7] and used this new basis to implement the importance sampling.

4.1. *Eigenvalues*

We discuss only few of the lowest states of ^{108}Sn and ^{48}Cr as illustrative examples. As the plot of Fig. 1 shows, the sampling parameter ϵ varies very smoothly with the dimensions n of the reduced matrices. In these, as in all other states considered, it scales with n according to

$$\epsilon = b\frac{N}{n^2}\exp\left[-c\frac{N}{n}\right] \tag{6}$$

As shown in Fig. 2, starting from a sufficiently small ϵ, the energies scale with ϵ according to the law

$$E = E_0 + b\frac{N}{n}\exp\left[-c\frac{N}{n}\right] \tag{7}$$

where b, c, and E_0 are constants specific of each state and the full dimension N provides the scale. This fit allows for an extrapolation to asymptotic eigenvalues which differ from the exact ones in the second or third decimal

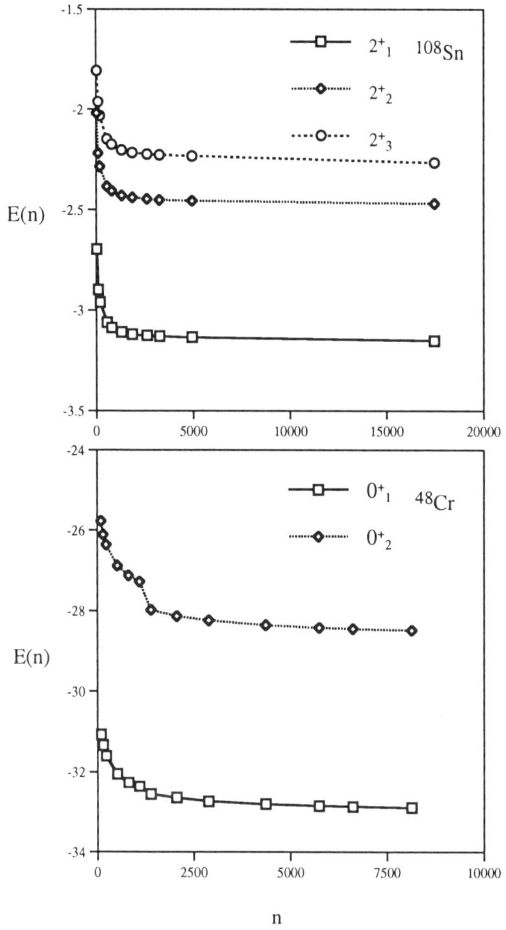

Figure 2. Eigenvalues versus the dimensions n of the truncated matrices resulting from the sampling.

digit. This law is somewhat different from the one proposed in Refs. [13]. On the other hand, it is valid for all states and nuclei examined and follows directly from the sampling [6].

4.2. *Eigenvectors and E2 transitions*

The sampling guarantees a high accuracy also for the eigenstates ψ_n As shown in Fig. 3, their overlap with the exact eigenvectors ψ for the first five $J^\pi = 2^+$ states of ^{108}Sn and $J^\pi = 0^+$ of ^{48}Cr converges rapidly to

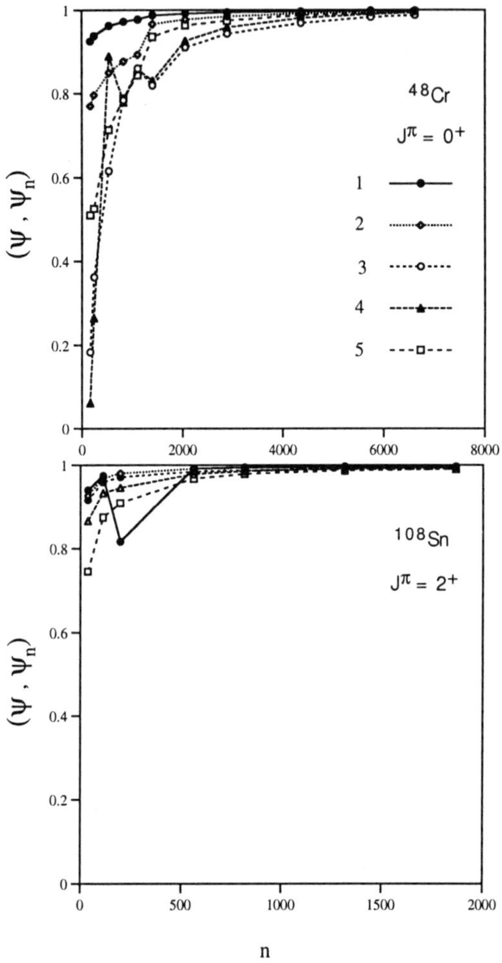

Figure 3. Overlap of sampled wave functions with the corresponding exact ones.

unity. To complete the analysis we studied the behavior versus $1/\epsilon$ of the strengths of the $E2$ transitions between some low-lying states (Fig. 4). In all cases, the strengths reach soon a plateau and, then, undergo very small variations, appreciable only on a very small scale (see inset). This fine tuning analysis shows that each strength has smooth behavior which allows for an extrapolation to asymptotic values through a formula having the same structure as the scaling law adopted for the energies (Eq. 7). In any case, the strengths computed at a relatively large ϵ differ very little from

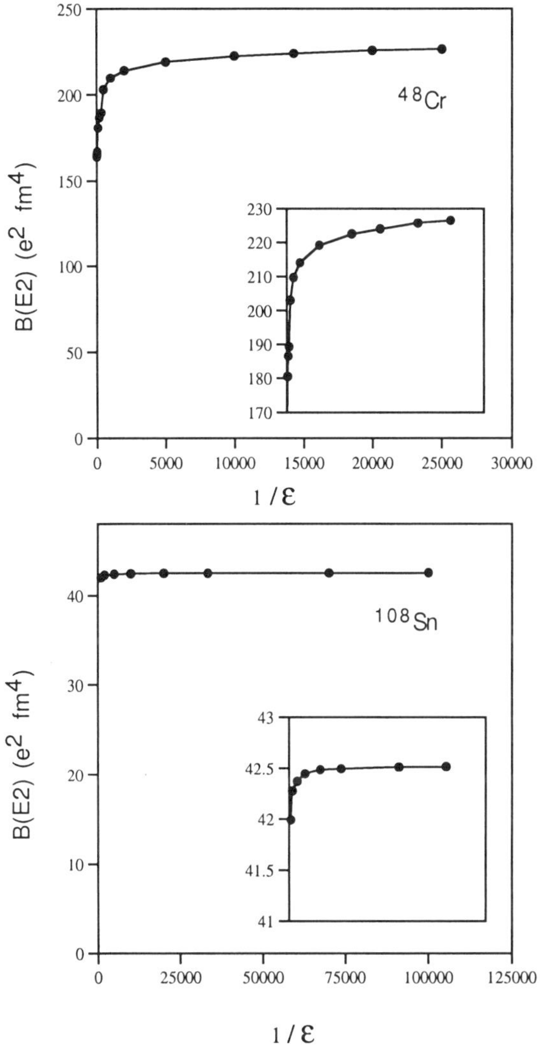

Figure 4. Strengths of some selected $E2$ transitions versus the dimensions n of the truncated matrices resulting from the sampling.

the extrapolated ones, which in turn practically coincide with the exact values. This rapid convergence is quite significant in view of the extreme sensitivity of the transition strengths to even very small components of the wave function.

5. Concluding remarks

We have shown that is easy to endow the iterative algorithm developed in Ref. [5] with an importance sampling, which allows for a drastic truncation of the matrices, while keeping it under strict control.

The important sampling has the further important virtue of providing scaling laws for energies and transition strengths which allow to extrapolate to the exact quantities.

Since the truncation is far more effective in nuclei with neutron excess [6], we feel confident that the sampling may be successfully applied to heavy nuclei.

Acknowledgements

This work is partly supported by the Ministero dell' Istruzione, Università e Ricerca (MIUR).

References

1. See for instance J.A. White, S.E. Koonin, and D.J. Dean, Phys. Rev. C **61**, 034303 (2000).
2. T. Otsuka, M. Honma, and T. Mizusaki, Phys. Rev. Lett. **81**, 1588 (1998).
3. See for instance G. H. Golub and C. F. Van Loan, *Matrix Computations*, (John Hopkins University Press, Baltimore 1996).
4. E. R. Davidson, J. Computational Phys. **17**, 87 (1975).
5. F. Andreozzi, A. Porrino, and N. Lo Iudice, J. Phys. A: Math. Gen. **35** (2002) L61.
6. F. Andreozzi, N. Lo Iudice, and A. Porrino, J. Phys. : Nucl. Part. Phys. **29** (2003) 2319.
7. F. Andreozzi and A. Porrino, J. Phys. G: Nucl. Part. Phys. **27** (2001) 845.
8. I. Shavitt, C. F. Bender, A. Pipano, and R. P. Hosteny, J. Computational Phys. **11**, 90 (1973).
9. A. Ruhe, in *Lecture Notes in Mathematics* **527**, edited by A. Dold and B. Eckmann, (Springer-Verlag, Berlin 1977) p. 130.
10. J.H. Wilkinson, The Algebraic Eigenvalue Problem, (Clarendom Press, Oxford, 1965) p. 71.
11. R. Machleidt, Adv. Nucl. Phys. 19 (1989) 189.
12. E. Caurier, A.P. Zuker, A. Poves, and G. Martinez-Pinedo, Phys. Rev. C **50** (1994) 225.
13. M. Horoi, A. Volya, and V. Zelevinsky, Phys. Rev. Lett. **82** (1999) 2064.

KEY TOPICS IN NUCLEAR STRUCTURE
Proceedings of the 8th Int. Spring Seminar on Nuclear Physics
edited by Aldo Covello
© 2005 World Scientific Publishing Co.

SHIFTED-CONTOUR MONTE CARLO METHOD FOR NUCLEAR STRUCTURE

G. STOITCHEVA AND D.J. DEAN
Physics Division, Oak Ridge National Laboratory, P.O. Box 2008, Oak Ridge, Tennessee 37831, USA

The formalism of the auxiliary-field Monte Carlo method is well suited for studying the finite-temperature properties of nuclear structure. Since this approach avoids an explicit enumeration of the many-body states, spaces larger than in the conventional methods can be reached. The nuclear implementation of the shell model Monte Carlo method is limited by the 'sign' problem associated with the Monte Carlo weight function when we use realistic effective two-body interactions. We consider a new approach for alleviating the 'sign' problem. This method is based on shifting the Monte Carlo integration to a region in the complex plane where fluctuations around the Hartree-Fock solution are minimized.

1. Introduction

The interacting nuclear shell model is a powerful approach which has been applied successfully to the microscopic description of nuclear properties. However, the application of standard diagonalization of the shell-model Hamiltonian matrix is limited by the combinatorial increase of the rank of the matrix with increasing the numbers of valence particles within a given model space. To alleviate this problem truncation techniques are used. An alternative approach requires recasting the standard diagonalization problem into a problem of multidimensional integration.

The Shell Model Monte Carlo (SMMC) method [1,2] avoids an explicit enumeration of the many body-states. In addition, it circumvents the difficulties related to the dimensions of the many-body space. Nevertheless, the power of the SMMC technique has been limited by a sign problem associated with the Monte Carlo weight function when using realistic interactions. Different attempts were pursued that alleviate the sign difficulties. One of these employed an extrapolation approach that overcomes the sign problem and was previously validated [3]. However, the approach is not exact and does rely on extrapolations.

2. Shifted-Contour Monte Carlo Method

In this work, we exploit a new method, the Shifted-Contour Shell Model Monte Carlo (SC-SMMC) method for nuclear structure. The shifted-contour method has been successfully applied for electron-structure calculations of ground and low-lying excited-states [4]. The foundation of this new method is based on shifting the Monte Carlo integration to a region in the complex plane where fluctuations around the Hartree-Fock solution are minimized.

We consider a general two-body Hamiltonian

$$\hat{H} = \sum_{\alpha\beta} \varepsilon_\alpha a_\alpha^\dagger a_\alpha + \frac{1}{2} \sum_{\alpha\beta\gamma\delta} V_{\alpha\beta\gamma\delta} a_\alpha^\dagger a_\beta^\dagger a_\delta a_\gamma, \tag{1}$$

where $a_\alpha^\dagger$ and a_α are anti-commuting fermion creation and annihilation operators associated with the single particle state α defined by the complete set of quantum numbers $(nlmjt_z)$, and ε_α and $V_{\alpha\beta\gamma\delta}$ are the single-particle energies and the two-body matrix elements of the residual interaction, respectively. The SMMC technique relies on the ability to employ the imaginary-time evolution operator, $\hat{U}$=exp$(-\beta\hat{H})$, where β is the inverse temperature, to project out the ground state. To achieve a simplification of the propagator, $\hat{U}$, the Hamiltonian in Eq. (1) can be brought into quadratic form

$$\hat{H} = \varepsilon\hat{\Theta} + \frac{1}{2}\Lambda\hat{\Theta}\hat{\Theta}, \tag{2}$$

where $\hat{\Theta}$ is a density operator, Λ is the strength of the two-body interaction, and ε is a single-particle energy. Then, by using the Hubbard-Stratonovich (HS) transformation [6]

$$e^{\frac{1}{2}\Lambda\Theta^2} = \sqrt{\frac{|\Lambda|}{2\pi}} \int d\sigma\, e^{-\frac{1}{2}\Lambda\sigma^2 + s\sigma\Lambda\Theta}, \tag{3}$$

where the phase, $s = \pm 1$ if $\Lambda \geq 0$ or $\pm i$ if $\Lambda < 0$, the two-body term of the Hamiltonian is presented as a superposition of one-body operators

$$e^{-\beta\hat{H}} = \sqrt{\frac{\beta\,|\Lambda|}{2\pi}} \int d\sigma\, e^{-\frac{1}{2}\beta|\Lambda|\sigma^2}\, e^{-\beta\hat{h}}, h = (\varepsilon + s\Lambda\sigma)\hat{\Theta}. \tag{4}$$

The integral is a weighted sum over all possible densities σ and the weight, $e^{-\frac{1}{2}\beta|\Lambda|\sigma^2}$, is Gaussian. Furthermore, fluctuating auxiliary fields which are coupled to the protons and the neutrons are generated and the exponential is split into N_t time-slices with $\beta = N_t\Delta\beta$. In these fields the fermions are considered as non-interacting particles. Thus, the interacting many-body problems is reduced by a set of one-body problems noninteracting in fluctuating fields.

The associated expectation value of a given observable is expressed by

$$\left\langle \hat{\Omega} \right\rangle_A = \frac{\int D[\sigma]G(\sigma)\Phi(\sigma)\left\langle \Omega \right\rangle_\sigma}{\int D[\sigma]G(\sigma)\Phi(\sigma)}, \tag{5}$$

where $D[\sigma] = \prod_{n,\alpha} d\sigma_\alpha(\tau_n)$ is the volume element and $G(\sigma) = \exp\left[-\frac{1}{2}\Delta\beta\sum_{\alpha,n}|V_\alpha|\,\sigma_{\alpha n}^2\right]$ is the Gaussian factor.

By using Eq. (5), in the limit of low temperature ($T \to 0$ or $\beta \to \infty$), the properties of the ground state can be extracted. Although this expression is exact, the fluctuations in $\Phi(\sigma)$, which is defined as the sign of the Monte Carlo weight function, determines how precise the Monte Carlo evaluation can be. When the average sign is smaller than its uncertainty, this leads to the sign problem: the sign of the integrand fluctuates among samples and the integral is a result of cancelations that cannot be reproduced with a finite number of samples. Some Hamiltonians are free from sign problems, e.g. pairing plus quadrupole. These Hamiltonians include correctly the dominant collective components of the effective interaction. In addition, the sign problem is related to the time-reversal properties of the one-body part of the Hamiltonian in Eq. (4) [1,7]. It has been shown that all realistic interactions suffer form the sign problem and the Monte Carlo method in this case fails unless extrapolation methods are employed.

Our goal is to modify the HS transformation so that the formation of large-amplitude oscillations in the integrand are reduced. For this purpose, let us consider the two-body Hamiltonian, Eq. (1), in which the density operator is shifted by an arbitrary density, η:

$$\hat{H} = \sum_{\alpha\beta}(\varepsilon_{\alpha\beta} + W_{\alpha\beta})a_\alpha^\dagger a_\beta + \frac{1}{2}\sum_{\alpha\beta\gamma\delta} V_{\alpha\beta\gamma\delta}(a_\alpha^\dagger a_\delta - n_{\alpha\delta})(a_\beta^\dagger a_\gamma - n_{\beta\gamma}) + const, \tag{6}$$

where $W_{\alpha\beta} = \sum_{\gamma\delta} V_{\alpha\beta\gamma\delta} n_{\alpha\beta}$ is the change in the one-body part. The Hamiltonian in Eq. (6) is equivalent to the original Hamiltonian, Eq. (1). The effect of the shift in $\hat{H}$ is equivalent to a shift of the auxiliary density σ to a line with the density η. Formally, the contour of integration passes through a stationary point. The HS transformation applied to Eq. (6) becomes:

$$e^{\frac{1}{2}\Lambda\Theta^2} \sim \int d\sigma\, e^{-\frac{1}{2}\Lambda\sigma^2 + s(\sigma - s\eta)\Lambda\Theta}. \tag{7}$$

As in case of electrons, for our calculations we also consider the arbitrary density, $\eta_{\alpha\beta}$, to be chosen as the expectation value of the Hartree-Fock density in the ground state. Due to the change in the integrand, Eq. (7), the oscillating part, $e^{s(\sigma - s\eta)\Lambda\Theta}$, is damped by η and the fluctuations are highly reduced.

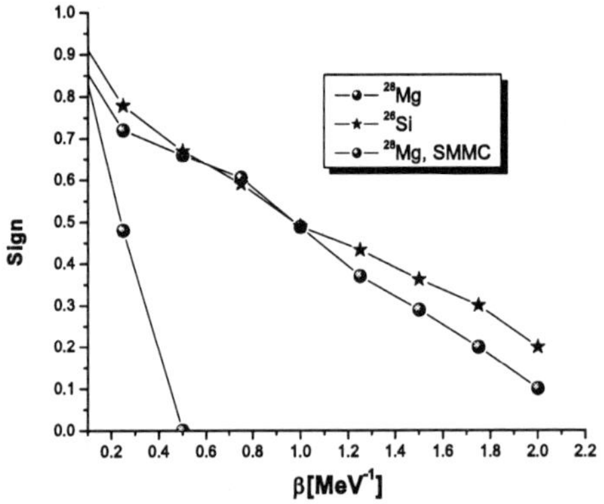

Figure 1. The sign as a function of β

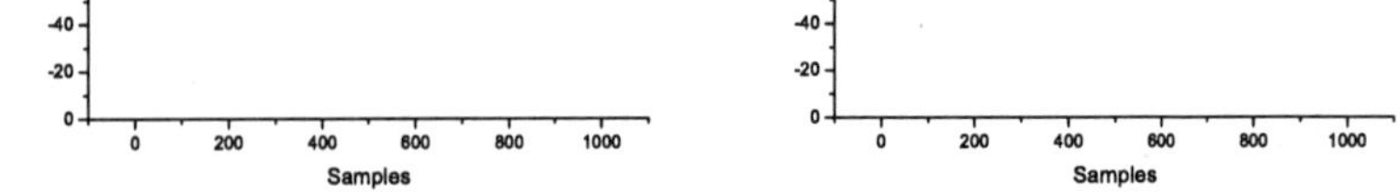

Figure 2. Energy versus samples, ^{28}Si

We applied this technique to nuclear systems in *sd*-shell and consider their properties at finite temperature (canonical ensemble). In Figure 1, we plot the behavior of the sign versus different values of β. Two examples are presented, ^{28}Mg and ^{28}Si. The change of the sign when increasing β for ^{28}Mg based on the shifted-contour method is compared with the SMMC without applying the shift. Our results show a significant delay of the sign due to the mean-field shift.

Figure 2 (left panel) shows the statistical convergence of the energy of ^{28} Si versus 1000 samples. In the case where no shifted contour method is applied, Figure 2 (right panel), the convergence degrades as β increases.

Acknowledgments

Oak Ridge National Laboratory is managed by UT-Battelle, LLC for the U.S. Department of Energy under Contract No. DE-AC05-00OR22725.

References

1. G. H. Lang, C. W. Johnson, S. E. Koonin, and W. E. Ormand, Phys. Rev. C **48**, 1518 (1993).
2. S. E. Koonin, D. J. Dean, and K. Langanke, Phys. Rep. **278**, 2 (1997).
3. Y. Alhassid, D. J. Dean, S. E. Koonin, G. Lang, and W. E. Ormand, Phys. Rev. Lett. **72**, 613 (1994).
4. N. Rom, D.M. Charutz, and D. Neuhauser, Chem. Phys. Lett. **270**, 382 (1997).
5. S. Jacobi and R. Baer, Journal of Chem. Phys. **120**, 43 (2004).
6. J. Hubbard, Phys. Rev. Lett. **3**, 77 (1959); R. L. Stratonovich, Dokl. Akad. Nauk. S.S.S.R. **115**, 1097 (1957).
7. C.W. Johnson and D.J. Dean, Phys. Rev. C **61**, 044327 (2000)

SECTION IV

COLLECTIVE ASPECTS OF NUCLEAR STRUCTURE

KEY TOPICS IN NUCLEAR STRUCTURE
Proceedings of the 8th Int. Spring Seminar on Nuclear Physics
edited by Aldo Covello
© 2005 World Scientific Publishing Co.

QUANTUM PHASE TRANSITIONS IN NUCLEI

F. IACHELLO
Center for Theoretical Physics, Sloane Physics Laboratory, Yale University, New Haven, CT 06520-8120

Shape phase transitions in nuclei are reviewed, with particular emphasis to the recent introduction of spectral signatures of critical behavior.

1. Quantum Phase Transitions

A Quantum Phase Transition (QPT) is a phase transition that occurs as a function of a coupling constant, g, (called control parameter), that appears in the quantum Hamiltonian, H, that describes the system

$$H = \alpha_1 H_1 + \alpha_2 H_2 = (\text{scale})(H_1 + gH_2). \tag{1}$$

Associated with a phase transition there is an *order parameter*, the expectation value of a suitably chosen operator that describes the state of the system $\langle P \rangle$.

Many physical systems (molecules, nuclei, atomic clusters, ...) are characterized in their equilibrium configuration by a shape. These shapes are in many case rigid. However, there are several situations in which the system is rather floppy and undergoes a phase transition between two different shapes. Shape phase transitions are examples of Quantum Phase Transitions. Challenging problems are: (i) how to study shape phase transitions and (ii) how to describe their properties in the phase transition region and in particular at the phase transition point. A method that has been found very useful to decribe shape phase transitions is that of bosonizing the shape variables and then taking the classical limit of the corresponding boson Hamiltonian [1,2,3].

2. Quantum Phase Transitions in nuclei

2.1. *Bosonization*

A bosonization of shape variables in nuclei is provided by the interacting boson model [4]. For one type of particles (one fluid system) the model is the Interacting Boson Model-1. The first step in the construction of the phase diagram is the identification of all possible dynamic symmetries of the system (phases). In IBM-1 there are three dynamical symmetries

$$\begin{aligned} &U(6) \supset U(5) \supset SO(5) \supset SO(3) \supset SO(2) \quad \text{(I)} \\ &U(6) \supset SU(3) \supset SO(3) \supset SO(2) \quad \text{(II)} \\ &U(6) \supset SO(6) \supset SO(5) \supset SO(3) \supset SO(2) \quad \text{(III)}. \end{aligned} \tag{2}$$

The next step is to write down a Hamiltonian that is a combination of Hamiltonians describing the dynamic symmetries. For the Interacting Boson Model-1 a suitable Hamiltonian is [5]

$$H = \varepsilon_0[(1-\xi)\hat{n}_d - \frac{\xi}{4N}\hat{Q}^{\chi}\cdot\hat{Q}^{\chi}] \tag{3}$$

where

$$\begin{aligned} \hat{n}_d &= (d^\dagger \cdot d) \\ \hat{Q}^{\chi} &= (d^\dagger \times s + s^\dagger \times \tilde{d})^{(2)} + \chi(d^\dagger \times \tilde{d})^{(2)}. \end{aligned} \tag{4}$$

There are here two control parameters ξ, χ. When $\xi = 0$ one has $U(5)$, when $\xi = 1, \chi = -\frac{\sqrt{7}}{2}$ one has $SU(3)$ and when $\xi = 1, \chi = 0$ one has $SO(6)$. The phase diagram is a triangle.

2.2. *Phase diagram of nuclei (one fluid)*

Shapes can be studied by introducing coherent or intrinsic states. The ground coherent state is [6]

$$\mid N;\beta,\gamma\rangle = (s^\dagger + \beta[\cos\gamma d_0^\dagger + \frac{1}{\sqrt{2}}\sin\gamma(d_{+2}^\dagger + d_{-2}^\dagger)])^N \mid 0\rangle. \tag{5}$$

The ground energy functional (potential) is

$$E(\beta,\gamma) = \frac{\langle N;\beta,\gamma \mid H \mid N;\beta,\gamma\rangle}{\langle N;\beta,\gamma \mid N;\beta,\gamma\rangle}. \tag{6}$$

By minimizing E with respect to β and γ one obtains the equilibrium configuration. For the three dynamic symmetries, one obtains: (I) $U(5)$: Spherical shape; (II) $SU(3)$: Axially deformed shape; (III) $SO(6)$: γ-unstable shape. The equilibrium values β_e and γ_e are the order parameters.

Shape phase transitions can be studied by considering Hamiltonians that are combinations of invariant operators of two chains. Minimizing the energy functional $E(N;\beta,\gamma)$ as a function of β,γ one can determine $E_{\min}$. By studying the behavior of $E_{\min}$ and its derivatives as a function of the coupling constants (control parameters, ξ,χ) one can determine the

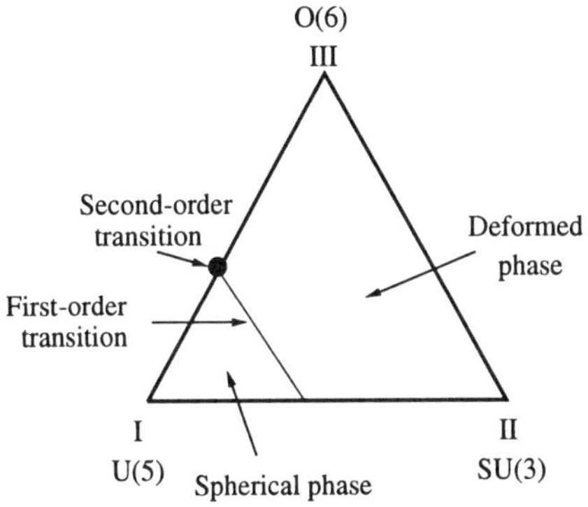

Figure 1. Phase diagram of IBM-1.

order of the phase transition (Erhenfest classification). When $N \to \infty$, discontinuities appear at the critical point. When the discontinuity is in $E_{\rm min}$, the transition is called of $0th-order$, when it is in $\frac{\partial E_{\rm min}}{\partial \xi}$ is of $1st-order$, when in $\frac{\partial^2 E_{\rm min}}{\partial \xi^2}$ of $2nd-order$, etc. For the Interacting Boson Model-1, one obtains: $U(5)-SO(6)$ (spherical to γ-unstable deformed) 2nd order; $U(5)-SU(3)$ (spherical to axially deformed) 1st order; $SO(6)-SU(3)$ (γ-unstable deformed to axially deformed) no phase transition (crossover). The phase diagram of nuclei is shown in Fig.1.

2.3. *Behavior of the order parameter around the critical point*

2.3.1. *The second order $U(5)-SO(6)$ transition*

This is characterized by the critical value ξ_c and the critical exponent α, with mean field value $\frac{1}{2}$,

$$\beta_e \propto (\xi - \xi_c)^\alpha; \qquad \alpha = \frac{1}{2}. \tag{7}$$

2.3.2. *The first order $U(5)-SU(3)$ transition*

For first order transitions, there is coexistence of two minima. There are here three values, the spinodal value, ξ^*, the critical value, ξ_c, and the antispinodal value, ξ^{**}. One can define a spinodal exponent α^* [7], with mean field value $\frac{1}{2}$

$$\beta_e \propto (\xi - \xi^*)^{\alpha^*}; \quad \alpha^* = \frac{1}{2} \tag{8}$$

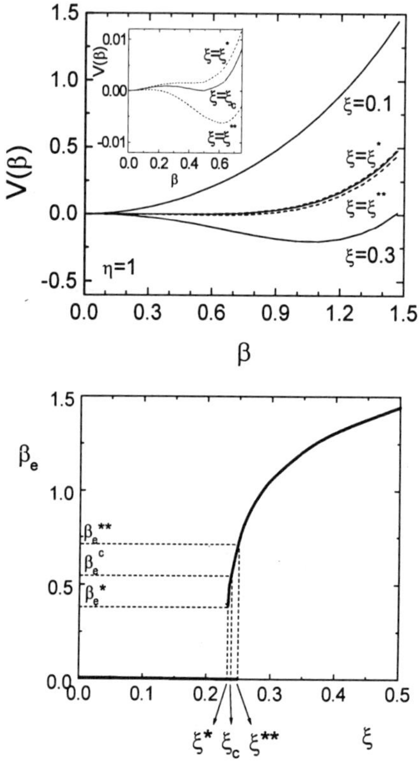

Figure 2. Behavior of the potential (top panel) and order parameter (bottom panel) around the critical point of the $U(5) - SU(3)$ transition. From [7].

The behavior of the order parameter and the potentials around the critical point of the $U(5) - SU(3)$ transition is shown in Fig.2.

2.4. *Finite N effects*

Finite N effects can be studied by a numerical diagonalization of the Hamiltonian of Eq. (3). For the $U(5) - SO(6)$ transition (second order) it is easy to define a quantum order parameter

$$\nu_1 = \langle n_d \rangle \propto \beta_e^2. \tag{9}$$

It has been found that the main features of QPT persists even for small number of particles $N \sim 10$. It has been possible to extract numerically the critical exponent α for the order parameter β_e

$$\alpha = 0.54 \pm 0.06 \tag{10}$$

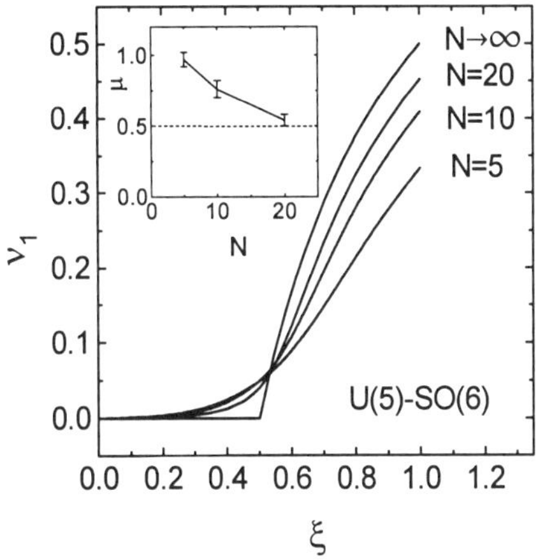

Figure 3. The order parameter ν_1 as a function of the control parameter ξ for the $U(5)-SO(6)$ transition. In the insert the critical exponent. From [7].

as shown in Fig.3. For the $U(5)-SU(3)$ transition (first order) the situation is more complex and will not be discussed here. A detailed account can be found in [7]. Finite N effects have also been studied recently by Leviatan and Ginocchio [8].

2.5. *How to detect QPT in nuclei*

In same cases it is possible to measure directly the order parameter β as a function of the control parameter ξ. In the majority of cases a direct measurement is not possible, and one must resort to an indirect measurement of a functional of the order parameter as a function of the control parameter. In nuclei, since ξ and χ are a function of proton and neutron number (or the number of bosons N), phase transitions can be studied by measuring nuclear properties as a function of N. Experimental evidence for $U(5)-SU(3)$ transition has been found in the Nd-Sm-Gd isotopes by measuring : (i) $B(E2; 2_1^+ \rightarrow 0_1^+)$ values, Fig. 4; (ii) Isomer shift $\delta\langle r^2\rangle = \langle r^2\rangle_{2_1^+} - \langle r^2\rangle_{0_1^+}$; (iii) Two-neutron separation energies $S_2(N) = E_B(N+1) - E_B(N)$, Fig.5 and (iv) Isotope shift $\Delta\langle r^2\rangle = \langle r^2\rangle^{(N+1)} - \langle r^2\rangle^{(N)}$.

3. Spectral signatures of critical points

It has been suggested recently that spectral signatures of critical points in finite quantal systems can be found by analyzing the potential at or around

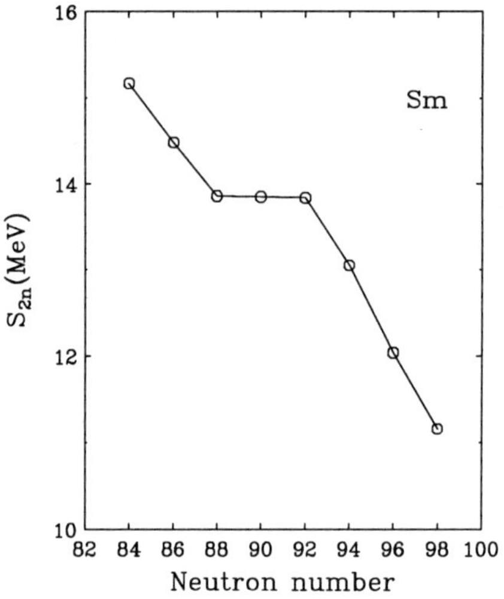

Figure 4. Two neutron separation energies in $_{62}Sm$ as a function of neutron number, showing that the $U(5) - SU(3)$ transition is first order.

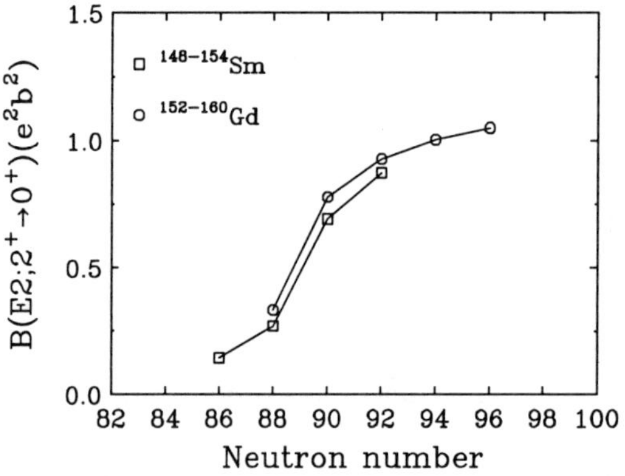

Figure 5. $B(E2; 2_1^+ \rightarrow 0_1^+)$ values in $_{62}Sm$ and $_{64}Gd$.

the critical point [9,10]. The main point is that at the critical point of a (second order) phase transition the potential is flat and can be replaced by a square well. By solving the differential equation $H\psi = E\psi$, where H is the Bohr Hamiltonian [11], one can obtain the spectrum at the critical point. Two solutions have been worked out:

(I) Critical point of the spherical-γ unstable transition ($U(5) - SO(6)$), called $E(5)$.

The potential is the 5-dimensional square well (γ-independent) potential

$$u(\beta) = 0, \quad \beta < \beta_W; \qquad u(\beta) = \infty, \quad \beta \geq \beta_W \tag{11}$$

with wave functions

$$\Psi(\beta, \gamma, \theta_i) = c_{s,\tau}\beta^{-3/2} J_{\tau+3/2}(k_{s,\tau}\beta)\Phi_{\tau,\nu_\Delta,L,M}(\gamma, \theta_i). \tag{12}$$

Here $k_{s,\tau} = \frac{x_{s,\tau}}{\beta_W}$, where $x_{s,\tau}$ is the s-th zero of $J_{\tau+3/2}(z)$. The eigenvalues are:

$$E_{s,\tau} = \frac{\hbar^2}{2B} k_{s,\tau}^2. \tag{13}$$

An experimental example of $E(5)$ symmetry has been found by Casten and Zamfir [12], in the spectrum and electromagnetic transition rates of ^{134}Ba.

(II) Critical point of spherical to axially deformed ($U(5) - SU(3)$) transition, called $X(5)$.

The situation here is much more complex. The phase transition, being of first order, has coexistence of two different shapes (spherical and axially deformed). The phase transition involves two coupled variables β, γ. The Bohr equation does not support any solution of the $SU(3)$ type. Nonetheless an approximate solution has been obtained by decoupling the two variables and replacing the potential by

$$U(\beta, \gamma) = u(\beta) + v(\gamma) \tag{14}$$

with

$$u(\beta) = 0\ , \quad \beta \leq \beta_W; \quad u(\beta) = \infty\ , \quad \beta > \beta_W\ ;$$
$$v(\gamma) = A\frac{\gamma^2}{2}. \tag{15}$$

The solutions are

$$E(s, L, n_\gamma, K, M) = E_0 + B\,(x_{s,L})^2 + An_\gamma + CK^2, \tag{16}$$

$$\Psi(\beta, \gamma, \theta_i) = c_{s,L}\beta^{-3/2} J_\nu(k_{s,L}\beta)\eta_{n_\gamma,K}(\gamma) D^L_{MK}(\theta_i), \tag{17}$$

where $x_{s,L}$ is the s-th zero of $J_\nu(z)$ with

$$\nu = \left(\frac{L(L+1)}{3} + \frac{9}{4}\right)^{1/2}, \tag{18}$$

in general, an irrational number.

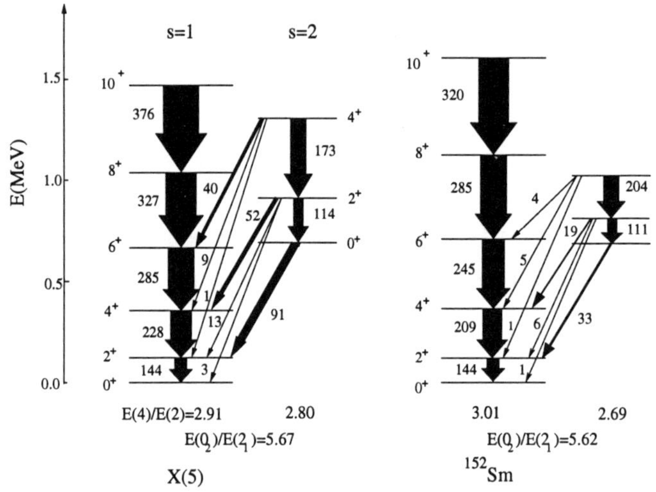

Figure 6. Comparison between the experimental spectrum of ^{152}Sm and the spectrum of $X(5)$. Adapted from [12].

After the initial discovery by Casten and Zamfir [13] in the spectrum and electromagnetic transition rates of ^{152}Sm, Fig.6, several examples have been found. A very recent example is given in [14].

In summary: (i) Approximate analytic solutions have been worked out for the critical region of the phase diagram of nuclei. (ii) These analytic solutions $E(5), X(5)$ depend only on a scale and thus provide parameter independent signatures of critical points. This is a major novel result for finite quantal systems (mesoscopic systems). The present situation is schematically shown in Fig.7. The introduction of the new solutions, called "critical point symmetries", provide a challenge for Radioactive Beam Facilities as they explore uncharted regions of the periodic table.

4. Developments in the study of QPT in nuclei

4.1. *Triaxial shapes*

These shapes cannot be obtained in IBM-1 with quadratic terms. One needs cubic or higher order terms. This is a difficult problem. The phase diagram with triaxial shapes is a tedrahedron. Triaxial QPT's are being presently investigated by Caprio [15]. Spectral signatures of the $SU(3) - SU_\gamma(3)^*$ transition (axial to triaxial transition), called $Y(5)$, have been worked out [16].

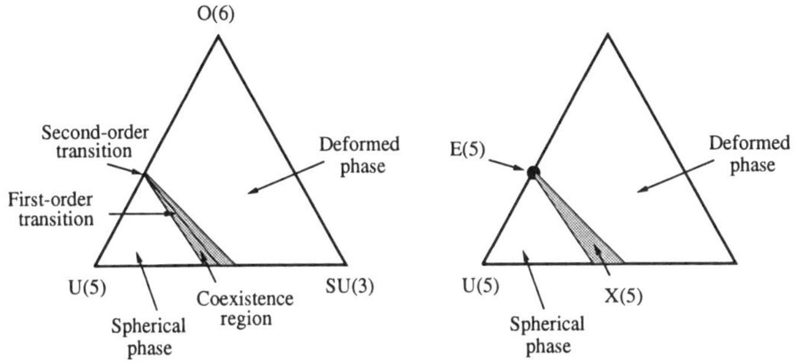

Figure 7. Phase diagram of IBM-1 showing the location of the solutions $E(5)$ and $X(5)$.

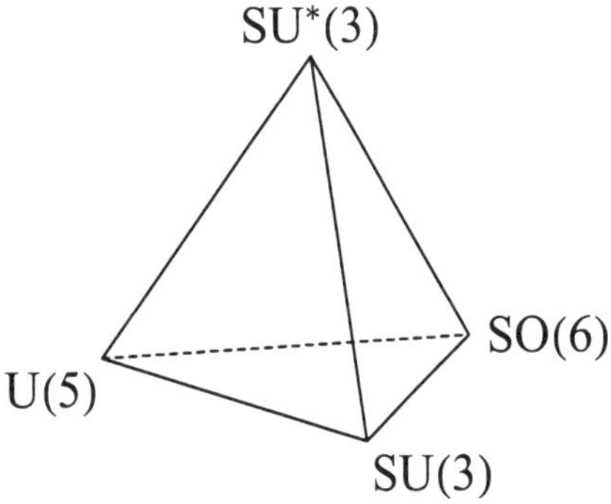

Figure 8. Phase diagram of IBM-2.

4.2. *Proton-Neutron systems (Two-fluid)*

In nuclei, one has both protons and neutrons. A much better description is in terms of proton and neutron bosons (Interacting Boson Model-2) [17]. This model has four dynamical symmetries: (I) $U(5)$ spherical shape; (II) $SU(3)$ axially deformed shape; (III) $SO(6)$ γ-unstable shape; (IV) $SU(3)^*$ triaxial shape. An Hamiltonian that contains all the features of the phase transitions is

$$H = \varepsilon_0 \left[(1-\xi)(n_{d_\pi} + n_{d_\nu}) - \frac{\xi}{N_\pi + N_\nu} (Q_\pi^{\chi_\pi} + Q_\nu^{\chi_\nu}) \cdot (Q_\pi^{\chi_\pi} + Q_\nu^{\chi_\nu}) \right]. \tag{19}$$

The phase diagram is again a tetrahedron, Fig.8, with three control parameters: ξ, $\frac{\chi_\pi+\chi_\nu}{2}$, $\frac{\chi_\pi-\chi_\nu}{2}$. This situation has been investigated by Caprio [18] and by Arias, Dukelsky and Garcia-Ramos [19].

5. Conclusions

(i) Nuclei provide one of the best grounds to study 'quantum phase transitions' in finite systems.

(ii) New insight into the structure of systems at the phase transition point can be obtained by deriving spectral signatures at the critical point of a quantum phase transition. This insight can be used to study shape transitions in a variety of quantum systems: nuclei, molecules, atomic clusters, macromolecules, finite polymers,...

(iii) Radioactive Beam Facilties are ideal to study Quantum Phase Transitions in finite systems as they will provide a much larger set of nuclear species where the general concepts of Phase Transitions and Critical Signatures can be studied.

6. Aknowledgements

This work was performed in part under DOE Contract No. DE-FG-02-91ER-40608.

I wish to thank Jose' Arias for providing me the results of his studies prior to publication, Mark Caprio for the development of the theory of phase transitions in two-fluid systems and Victor Zamfir for the study of critical exponents in the Interacting Boson Model-1.

References

1. A.E.L. Dieperink, O. Scholten, and F. Iachello, Phys. Rev. Lett. **44**, 1747 (1980).
2. D.H. Feng, R. Gilmore, and S.R. Deans, Phys. Rev. C **23**, 1254 (1981).
3. O.S. Van Roosmalen, *Algebraic Description of Nuclear and Molecular Rotation-Vibration Spectra,* Ph.D. Thesis, University of Groningen, The Netherlands (1982).
4. For a review see, F. Iachello and A. Arima, *The Interacting Boson Model,* Cambridge University Press, Cambridge, 1987.
5. D.D. Warner and R.F. Casten, Phys. Rev. C **28**, 1798 (1983).
6. J. Ginocchio and M.W. Kirson, Phys. Rev. Lett. **44**, 1744 (1980).
7. F. Iachello and N.V. Zamfir, Phys. Rev. Lett. **92**, 212501 (2004).
8. A. Leviatan and J.N. Ginocchio, Phys. Rev. Lett. **90**, 212501 (2003).
9. F. Iachello, Phys. Rev. Lett. **85**, 3580 (2000).
10. F. Iachello, Phys. Rev. Lett. **87**, 052502 (2001).
11. A. Bohr, Mat. Fys. Medd. Dan. Vid. Selsk. **26**, No. 14 (1952).
12. R.F. Casten and N.V. Zamfir, Phys. Rev. Lett. **85**, 3584 (2000).
13. R.F. Casten and N.V. Zamfir, Phys. Rev. Lett. **87**, 052503 (2001).
14. D. Tonev, A. Dewald, T. Klug, et al., Phys. Rev. C **69**, 034334 (2004).

15. M.A. Caprio, private communication (2004).
16. F. Iachello, Phys. Rev. Lett. **91**, 132502 (2003).
17. A. Arima, T. Otsuka, F. Iachello and I. Talmi, Phys. Lett. **66B**, 205 (1977).
18. M. A. Caprio and F. Iachello, submitted (2004).
19. J.M. Arias, J. Dukelsky, and J.E. Garcia-Ramos, submitted (2004).

KEY TOPICS IN NUCLEAR STRUCTURE
Proceedings of the 8th Int. Spring Seminar on Nuclear Physics
edited by Aldo Covello
© 2005 World Scientific Publishing Co.

DEVELOPMENTS OF ALGEBRAIC COLLECTIVE MODELS AND SECOND-ORDER PHASE TRANSITIONS*

D. J. ROWE
Deparment of Physics, University of Toronto,
Toronto, Ontario, M5S 1A7, Canada
E-mail: rowe@physics.utoronto.ca

This talk focuses on three topics: the development of a program to determine SO(5) spherical harmonics and SO(5) Clebsch-Gordan coefficients; efficient ways to do collective model calculations in an $SU(1,1) \times SO(5) \supset U(1) \times SO(3)$ basis; and quasi-dynamical symmetry in an IBM second-order phase transition.

1. Introduction

Attempts to understand phase transitions have profited considerably from the study of models with symmetry. Landau stated that two phases of matter with different symmetries (which cannot change continuously from one to the other) must be separated by a line of transition. Consider a system with control parameter α which is in a phase with a symmetry group G_1 when $\alpha = 0$ and in a phase with symmetry group G_2 when $\alpha = 1$. The question then is what happens when α is varied continuously from 0 to 1? It often transpires that the model exhibits a second-order phase transition from a phase characterized by one symmetry to a phase characterized by the other. However, closer examination reveals that, in the phase characterized by the G_1 symmetry, the symmetry of the system is increasingly distorted by the forces that favour the competing phase, as α is increased, until a point comes at which it can be distorted no further and a rapid change occurs to a phase dominated by the G_2 symmetry. A complementary behaviour may be observed when the critical point is approached from the other side. The distorted symmetries, called quasi-dynamical symmetries, have an elegant expression in the language of group theory and lead to new concepts in representation theory of considerable

*Work supported by the Natural Sciences and Engineering Research Council of Canada.

significance for understanding why models with symmetries are often more successful in practice than they apparently have any right to be. Such phase transitions have been examined in a variety of models.[1] A review of quasi-dynamical symmetry has been given in Ref.[2]

2. Bases for the hydrodynamic collective model

Standard basis functions are given by eigenfunctions of the harmonic vibrator Hamiltonian

$$\hat{H} = \frac{1}{2B}\nabla^2 + \frac{1}{2}B\omega^2\beta^2\,, \tag{1}$$

where $\beta^2 = q \cdot q$ is the squared length of the quadrupole tensor. This Hamiltonian is U(5), SO(5), and SO(3) invariant and an element of an $\mathrm{SU}(1,1) \times \mathrm{SO}(5)$ spectrum generating algebra. These basis functions

$$\Psi_{nv\alpha LM}(\beta,\gamma,\Omega) = R_{nv}(\beta)\mathcal{Y}_{v\alpha LM}(\gamma,\Omega)\,, \tag{2}$$

reduce the subgroup chain

$$\begin{array}{ccccccccc} \mathrm{SU}(1,1) & \times & \mathrm{SO}(5) & \supset & \mathrm{U}(1) & \times & \mathrm{SO}(3) & \supset & \mathrm{SO}(2) \\ & & v & & \alpha & v & L & & M \end{array} \tag{3}$$

The SU(1,1) $\supset$ U(1) beta wave functions are well known in terms of generalized Laguerre polynomials. A basis of SO(5) $\supset$ SO(3) $\supset$ SO(2) wave functions can be written down immediately (cf. Ref. [3]) in the form

$$\Phi_{tKLM}(\gamma,\Omega) = f_{tKL}(\gamma)\left[\mathcal{D}^L_{KM}(\Omega) + (-1)^L\mathcal{D}^L_{-K,M}(\Omega)\right], \tag{4}$$

where the functions $\{f_{tKL}\}$ are simple polynomials in $\cos\gamma$ and $\sin\gamma$ and K is an even integer. This basis is then orthonormalized sequentially to give SO(5) spherical harmonics that satisfy the familiar inner product

$$\int \mathcal{Y}^*_{v\alpha LM}(\gamma,\Omega)\mathcal{Y}_{v'\alpha'L'M'}(\gamma,\Omega)\,\sin 3\gamma\,\mathrm{d}\gamma\,\mathrm{d}\Omega = \delta_{vv'}\delta_{\alpha\alpha'}\delta_{LL'}\delta_{MM'}\,. \tag{5}$$

The integrals needed for this procedure are evaluated analytically although a computer is used to keep track of the results. The methods we use[3] make build on many of the results of Chacón *et al.*[4]

Having determined a set of SO(5) spherical harmonics, it is straightforward to compute the SO(5) CG coefficients of relevance to the collective model and the IBM1 by evaluating the integrals

$$\int \mathcal{Y}^*_{v_3\alpha_3L_3M_3}(\gamma,\Omega)\mathcal{Y}_{v_2\alpha_2L_2M_2}(\gamma,\Omega)\mathcal{Y}_{v_1\alpha_1L_1M_1}(\gamma,\Omega)\,\sin 3\gamma\,\mathrm{d}\gamma\,\mathrm{d}\Omega$$
$$\propto (v_1\alpha_1L_1M_1, v_2\alpha_2L_2M_2|v_3\alpha_3L_3M_3)\,. \tag{6}$$

CG coefficients for the couplings $v \otimes 1 \rightarrow v'$ are tabulated in Ref.[3]

3. A more efficient basis for deformed nuclei

With a basis of SO(5)-coupled wave functions, it is possible to diagonalize a general collective model Hamiltonians. However, for well-deformed nuclei, a large number of spherical vibrator basis states are needed for accurate results.

As shown by Elliott *et al.*[5], beta wave functions that are much closer to those of a deformed nucleus are given by eigenfunctions of the Hamiltonian

$$\hat{H}(\beta_0) = -\frac{1}{2B}\nabla^2 + \frac{1}{2}B\omega^2\left(\beta^2 + \frac{\beta_0^4}{\beta^2}\right), \tag{7}$$

where β_0 is a suitably chosen parameter. The potential for this Hamiltonian has a minimum value when $\beta = \beta_0$. As shown in Refs.[6,7], this Hamiltonian also defines a basis for the collective model that reduces the subgroup chain $\mathrm{SU}(1,1) \times \mathrm{SO}(5) \supset \mathrm{U}(1) \times \mathrm{SO}(3)$, where the SU(1,1) algebra is spanned by operators (with β now expressed in harmonic oscillator units)

$$\hat{X}_1 = \tfrac{1}{4}\left(-\nabla^2 - \beta^2 + \tfrac{\beta_0^4}{\beta^2}\right), \tag{8}$$

$$\hat{X}_2 = \tfrac{1}{4}\left(q \cdot \nabla + \nabla \cdot q\right), \tag{9}$$

$$\hat{X}_3 = \tfrac{1}{4}\left(-\nabla^2 + \beta^2 + \tfrac{\beta_0^4}{\beta^2}\right), \tag{10}$$

that satisfy the commutation relations

$$[\hat{X}_1, \hat{X}_2] = -i\hat{X}_3\,, \quad [\hat{X}_2, \hat{X}_3] = i\hat{X}_1\,, \quad [\hat{X}_3, \hat{X}_1] = i\hat{X}_2\,. \tag{11}$$

The energy-level spectrum for the Hamiltonian (7) is given by[5,6]

$$E_{\nu v} = (2\nu + \lambda_v)\hbar\omega, \quad \lambda_v = 1 + \sqrt{\left(v + \tfrac{3}{2}\right)^2 + \beta_0^4} \tag{12}$$

and the corresponding wave functions are again known in terms of generalized Laguerre polynomials.

As an example of the kind of calculation that can be done with the above-defined basis wave functions, Fig. 1 shows the energy-level spectrum and E2 transition rates obtained by diagonalizing the Hamiltonian

$$\hat{H}(\beta_0) = -\frac{1}{2B}\nabla^2 + \frac{1}{2}B\omega^2\left(\beta^2 + \frac{\beta_0^4}{\beta^2}\right) - \chi\cos 3\gamma \tag{13}$$

with a single beta wave function. Results obtained to a similar accuracy in a spherical vibrator basis require of the order of 100 basis wave functions. Details of the calculation are given in Ref.[7]

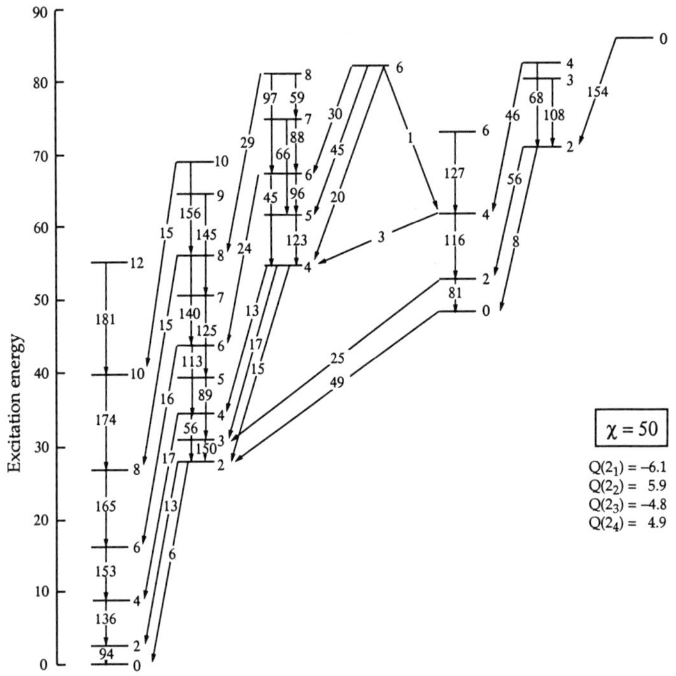

Figure 1. The low-energy spectrum and $B(\mathrm{E2})$ transition rates calculated for the Hamiltonian (7) in Ref.[7]

4. The U(5) to O(5) phase transition in the IBM

I now come to the main subject of this talk which is to review the evolution of the states of a system as it progresses from a phase with one dynamical symmetry to another with variation of a control parameter. We have studied several such systems and the results are remarkably similar.[1] Here I focus on a system of N interacting bosons having two states: a lower-energy s-boson state of angular momentum $L = 0$ and a higher-energy d-boson state of angular momentum $L = 2$. This model was developed for use in nuclear physics[8] but is of much wider interest.

Consider the Hamiltonian

$$\hat{H}(\alpha) = (1-\alpha)\hat{n} + \frac{\alpha}{N}\hat{S}_+\hat{S}_- \,, \tag{14}$$

where $\hat{n}$ is the d-boson number operator and

$$\hat{S}_+ = \tfrac{1}{2}(d^\dagger \cdot d^\dagger - s^\dagger s^\dagger), \quad \hat{S}_- = \tfrac{1}{2}(d \cdot d - ss)\,. \tag{15}$$

are the raising and lowering operators of an SU(1,1) Lie algebra. The

Hamiltonian $\hat{H}(\alpha = 0)$ has eigenstates that reduce the subgroup chain

$$U(6) \supset U(5) \supset O(5) \supset SO(3) \supset SO(2) \tag{16}$$

whereas eigenstates of $\hat{H}(\alpha = 1)$ reduce the chain

$$U(6) \supset O(6) \supset O(5) \supset SO(3) \supset SO(2)\,. \tag{17}$$

Moreover, $\hat{H}(\alpha)$ is easy to diagonalize for arbitrary values of α, because it is an element of an SU(1,1)+SU(1,1) Lie algebra. Thus, it is possible to follow the progression of its eigenstates as a function of the control parameter α.

The low energy-level spectra for $N = 20$ and $N = 40$ are shown in Fig. 2. It is seen that the system appears to hold onto its U(5) symmetry as α

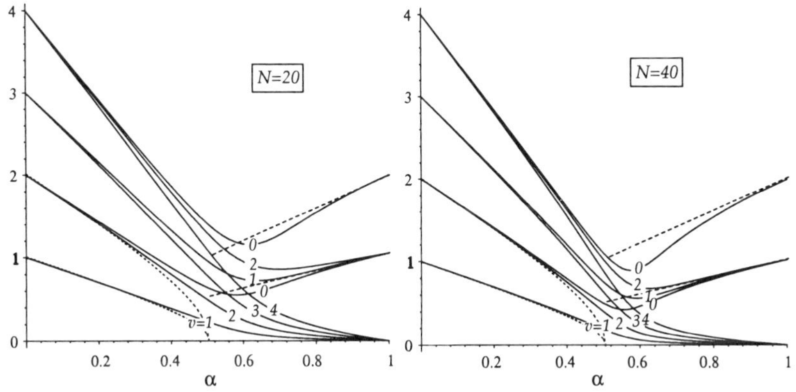

Figure 2. Spectrum of energy levels for $N = 20$ and 40 shown as a function of α for the Hamiltonian $\hat{H}(\alpha)$. Precise numerically computed energies are shown as continuous lines. The dotted lines are the results of an RPA calculation, for $\alpha < 0.5$, and the shifted harmonic approximation, for $\alpha > 0.5$.

increases until it approaches a transition region from below and similarly to holds onto its O(6) symmetry as it approaches the transition region from above. It is also seen that the transition region shrinks as N increases and, as evidenced by other calculations not shown, it approaches a singular critical point $\alpha_c = 0.5$ as $N \to \infty$.

However, a detailed inspection of the wave function shows the U(5) $\subset$ U(6) symmetry to be badly broken, well before α enters the transition region; thus, the persistent symmetries are really quasi-symmetries. For $\alpha < 0.5$, the quasi-U(5) symmetry can be understood in terms of the Random Phase Approximation. The ground state is an s-boson condensate at

$\alpha = 0$. In the RPA, the ground state becomes a quasi-s-boson condensate when $\alpha \neq 0$ in which pairs of s bosons are replaced by zero-coupled d boson pairs. The RPA predictions for excitation energies and E2 transition rates are shown in Figs. 2 and 3 which, respectively, show that the RPA

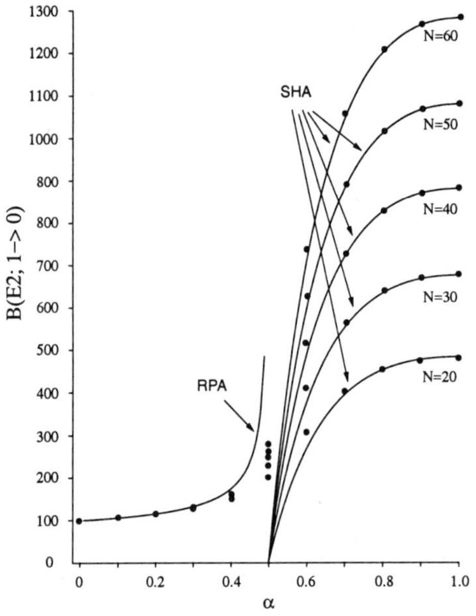

Figure 3. B(E2) transition rates for decay of the first excited $v = 1$ state to the ground state for various values of N. The continuous lines for $\alpha < 0.5$ are for the RPA and those for $\alpha > 0.5$ are for the shifted harmonic approximation (SHA).

excitation energies collapse and the E2 transition rate from the first excited state to the ground state diverges as $\alpha \to \alpha_c$. However, for the values of N shown, the RPA is a very good approximation in the region $0 \leq \alpha \lesssim 0.35$. Moreover, it becomes increasing accurate for all $\alpha < 0.5$ as N increases. The important observation for present purposes is that the RPA shows the existence of an effective Hamiltonian and effective quadrupole moment operators

$$\hat{H}_1^{\rm eff}(\alpha) = \sqrt{(1-\alpha)(1-2\alpha)}\,\hat{n}, \quad \hat{q}_m^{\rm eff} = e^{\rm eff}(\alpha)\hat{q}_m\,, \tag{18}$$

which are idential to those of the $\alpha = 0$ limit to within α-dependent normalization factors. Thus, the results of the RPA are indistinguishable from those of an effective IBM with U(5) dynamical symmetry.

Similar results hold for $\alpha > 0.5$. The coefficients in the expansion

$$\Psi_{\nu v=0} = \sum_n C_{\nu n} \varphi_{n v=0} \tag{19}$$

of the ground and first excited states of SO(5) seniority $v = 0$ are shown in the U(5) basis for $N = 60$ in Fig. 4 for two α values. The remarkable

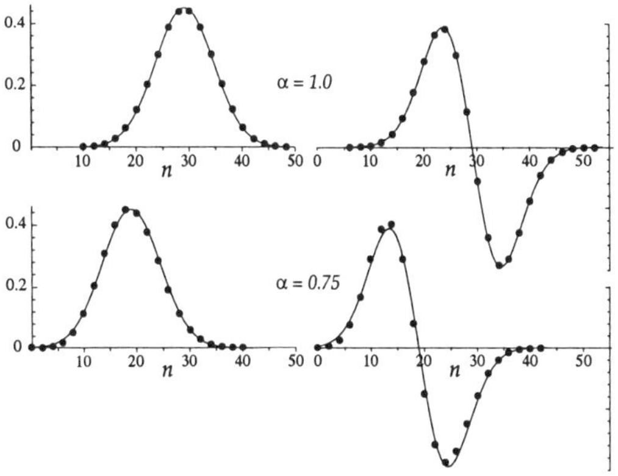

Figure 4. Coefficients of the lowest and first excited states of seniority $v = 0$ of the Hamiltonian (7) for $N = 60$ and $\alpha = 1.0$ and 0.75. It is seen that the wave functions just reach the lower $n = 0$ boundary when $\alpha = 0.75$.

fact is that the coefficients are given very precisely for large N by harmonic oscillator wave functions for $\alpha = 1.0$. Morever, the added term $(1-\alpha)\hat{n}$ in the Hamiltonian behaves as a Lagrange multiplier and simply shifts the centroid of each wave function to a smaller mean value of n but otherwise leaves it unchanged until the shifted wave function reaches the $n = 0$ boundary (n cannot take negative values). This is the point at which a shifted (coherent state) harmonic oscillator approximation starts to break down. It can be shown that the centroid of a wave functions is shifted to $n = 0$ at $\alpha = 0.5$ but because of its width, the wings of a wave function reach the $n = 0$ boundary for higher values of α for finite values of N; the width of a harmonic oscillator wave function goes to zero as $N \to \infty$. It can be shown that, when the SHA (shifted harmonic approximation) is valid, the properties of the N-boson system described by the Hamiltonian $\hat{H}(\alpha)$ are reproduced accurately by an effective Hamiltonian and effective quadrupole moment operators

$$\hat{H}_2^{\text{eff}}(\alpha) = \frac{\alpha}{N}\hat{S}_+\hat{S}_-, \quad \hat{q}_m^{\text{eff}} = \left[1 - \left(\frac{1-\alpha}{\alpha}\right)\right]^{\frac{1}{2}} \hat{q}_m \tag{20}$$

which are indistinguishable from those of an effective IBM with O(6) dynamical symmetry. However, O(6) is only a quasi-dynamical symmetry of the original N-boson model.

5. Concluding remarks

The above model analysis of a phase transition shows many properties that have been observed in several similar systems that are of wide physical significance. One is an explanation of why models with symmetry are often much more successful, than could reasonably be expected, even when there are known to be relatively strong symmetry-breaking interactions. The apparent persistence of symmetry is a wide spread phenomenon with a physically natural interpretation in terms of quasi-dynamical symmetry and the corresponding mathematical concept of embedded representations[2]. An understanding of quasi-symmetry and why and when it occurs, is of particular importance for understanding what successful models can really teach us about the systems they represent.

References

1. H. Chen, J.R. Brownstein and D.J. Rowe, Phys. Rev. **C42** 1422 (1990); H. Chen, T. Song and D.J. Rowe, Nucl. Phys. A**582** 181 (1995); D.J. Rowe, C. Bahri and W. Wijesundera, Phys. Rev. Lett. **80** (1998) 4394; C. Bahri, D.J. Rowe, and W. Wijesundera, Phys. Rev. C **58**, 1539 (1998).
2. D.J. Rowe, "Embedded representations and quasi-dynamical symmetry" in *'Computational and Group Theoretical Methods in Nuclear Physics* Proc. Symp. in Honor of Jerry P. Draayer's 60th Birthday (Eds. O. Castanos, J. Escher, J. Hirsch, S. Pittel, and G. Stoicheva. World Scientific, Singapore, 2004).
3. D.J. Rowe, P. Turner and J. Repka, *J. Math. Phys.* **45**, 2761 (2004).
4. E. Chacón, M. Moshinsky, and R.T. Sharp, J. Math. Phys. **17**, 668 (1976); E. Chacón and M. Moshinsky, J. Math. Phys. **18**, 870 (1977).
5. J.P. Elliott, J.A. Evans and P. Park, Phys. Lett. **169**B, (1986) 309; J.P. Elliott, P. Park and J.A. Evans, Phys. Lett. **171**B (1986) 145; cf. also S.G. Rohoziśki, J. Srcbrny, and K. Horbaczewska, Z. Phys. **268**, 401 (1974).
6. D.J. Rowe and C. Bahri, J. Phys. A: Math. Gen. **31**, 4947 (1998).
7. D.J. Rowe, *Nucl. Phys.* A**735**, 372 (2004).
8. A. Arima and F. Iachello, Ann. Phys. **99**, 253 (1976); **111**, 201 (1978); O. Scholten, A. Arima and F. Iachello, Ann. Phys. **115**, 325 (1978).

KEY TOPICS IN NUCLEAR STRUCTURE
Proceedings of the 8th Int. Spring Seminar on Nuclear Physics
edited by Aldo Covello
© 2005 World Scientific Publishing Co.

VARIATIONAL PROCEDURE LEADING FROM DAVIDSON POTENTIALS TO THE E(5) AND X(5) CRITICAL POINT SYMMETRIES

DENNIS BONATSOS, D. LENIS, AND D. PETRELLIS

Institute of Nuclear Physics,
National Centre for Scientific Research "Demokritos",
GR-15310 Aghia Paraskevi, Attiki, Greece
E-mail: bonat@inp.demokritos.gr

N. MINKOV, P. P. RAYCHEV, AND P. A. TERZIEV

Institute for Nuclear Research and Nuclear Energy,
Bulgarian Academy of Sciences,
72 Tzarigrad Road, BG-1784 Sofia, Bulgaria
E-mail: nminkov@inrne.bas.bg

Davidson potentials of the form $\beta^2 + \beta_0^4/\beta^2$, when used in the E(5) framework (i.e., in the original Bohr Hamiltonian after separating variables as in the E(5) model) bridge the U(5) and O(6) symmetries, while they bridge the U(5) and SU(3) symmetries when used in the X(5) framework (i.e., in the original Bohr Hamiltonian after separating variables as in the X(5) model). Using a variational procedure, we determine for each value of angular momentum L the value of the parameter β_0 at which the rate of change of the energy ratio $R_L = E(L)/E(2)$ has a maximum, the collection of the values of R_L formed in this way being a candidate for describing its behavior at the relevant critical point. This procedure leads to the E(5) ground state band in the E(5) framework and to the X(5) ground state band in the X(5) framework, thus indicating that the use of an infinite well potential in β is an optimum choice in both cases.

1. Introduction

The recently introduced E(5) [1] and X(5) [2] models are supposed to describe shape phase transitions in atomic nuclei, the former being related to the transition from U(5) (vibrational) to O(6) (γ-unstable) nuclei, and the latter corresponding to the transition from U(5) to SU(3) (rotational) nuclei. In both cases the original Bohr collective Hamiltonian [3] is used, with an infinite well potential in the collective β-variable. Separation of variables is achieved in the E(5) case by assuming that the potential is independent of the collective γ-variable, while in the X(5) case the potential is assumed to be of the form $u(\beta)+u(\gamma)$. We are going to refer to these two cases as "the E(5) framework" and "the X(5) framework" respectively. The selection of an infinite well potential in the β-variable in both cases is justified by the fact that the potential is expected to be flat around the point at which a shape phase transition occurs. Experimental evidence for the occurence of the E(5) and X(5) symmetries in some appropriate nuclei is growing ([4,5] and [6,7] respectively).

In the present work we examine if the choice of the infinite well potential is the optimum one for the description of shape phase transitions. For this purpose, we need one-parameter potentials which can span the U(5)-O(6) region in the E(5) framework, as well as the U(5)-SU(3) region in the X(5) framework. It turns out that the exactly soluble [8,9] Davidson potentials [10]

$$u(\beta) = \beta^2 + \frac{\beta_0^4}{\beta^2}, \tag{1}$$

where β_0 is the position of the minimum of the potential, do possess this property. Taking into account the fact that various physical quantities should change most rapidly at the point of the shape phase transition [11], we locate for each value of the angular momentum L the value of β_0 for which the rate of change of the ratio $R_L = E(L)/E(2)$, a widely used measure of nuclear collectivity [12], is maximized. It turns out that the collection of R_L ratios formed in this way in the case of a potential independent of the γ-variable correspond to the E(5) model, while in the case of the $u(\beta)+u(\gamma)$ potential lead to the X(5) model, thus proving, *without using any free parameter*, that the choice of the infinite well potential made in Refs. [1,2] is the optimum one. The variational procedure used here is analogous to the one used in the framework of the Variable Moment of Inertia (VMI) model [13], where the energy is minimized with respect to the (angular momentum dependent) moment of inertia for each value of the angular momentum L separately.

2. Davidson potentials in the E(5) framework

The original Bohr Hamiltonian [3] is

$$H = -\frac{\hbar^2}{2B}\left[\frac{1}{\beta^4}\frac{\partial}{\partial\beta}\beta^4\frac{\partial}{\partial\beta} + \frac{1}{\beta^2\sin 3\gamma}\frac{\partial}{\partial\gamma}\sin 3\gamma\frac{\partial}{\partial\gamma} - \frac{1}{4\beta^2}\sum_{k=1,2,3}\frac{Q_k^2}{\sin^2\left(\gamma-\frac{2}{3}\pi k\right)}\right] + V(\beta,\gamma), \tag{2}$$

where β and γ are the usual collective coordinates describing the shape of the nuclear surface, Q_k (k = 1, 2, 3) are the components of angular momentum, and B is the mass parameter.

Assuming that the potential depends only on the variable β, i.e. $V(\beta,\gamma) = U(\beta)$, one can proceed to separation of variables in the standard way [3,14], using the wave function $\Psi(\beta,\gamma,\theta_i) = f(\beta)\Phi(\gamma,\theta_i)$, where θ_i ($i = 1,2,3$) are the Euler angles describing the orientation of the deformed nucleus in space.

In the equation involving the angles, the eigenvalues of the second order Casimir operator of SO(5) occur, having the form $\Lambda = \tau(\tau+3)$, where $\tau = 0, 1, 2, \ldots$ is the quantum number characterizing the irreducible representations (irreps) of SO(5), called the "seniority" [15]. This equation has been solved by Bès [16].

The "radial" equation can be simplified by introducing [1] reduced energies $\epsilon = \frac{2B}{\hbar^2}E$ and reduced potentials $u = \frac{2B}{\hbar^2}U$, leading to

$$\left[-\frac{1}{\beta^4}\frac{\partial}{\partial\beta}\beta^4\frac{\partial}{\partial\beta}+\frac{\tau(\tau+3)}{\beta^2}+u(\beta)\right]f(\beta)=\epsilon f(\beta). \tag{3}$$

When plugging the Davidson potentials of Eq. (1) in the above equation, the β_0^4/β^2 term is combined with the $\tau(\tau+3)/\beta^2$ term appearing there and the equation is solved exactly [8,9], the eigenfunctions being Laguerre polynomials of the form

$$F_n^\tau(\beta)=\left[\frac{2n!}{\Gamma\left(n+p+\frac{5}{2}\right)}\right]^{1/2}\beta^p L_n^{p+\frac{3}{2}}(\beta^2)e^{-\beta^2/2}, \tag{4}$$

where $\Gamma(n)$ stands for the Γ-function, while p is determined by [8]

$$p(p+3)=\tau(\tau+3)+\beta_0^4, \tag{5}$$

leading to

$$p=-\frac{3}{2}+\left[\left(\tau+\frac{3}{2}\right)^2+\beta_0^4\right]^{1/2}. \tag{6}$$

The energy eigenvalues are then [8,9] (in $\hbar\omega = 1$ units)

$$E_{n,\tau}=2n+p+\frac{5}{2}=2n+1+\left[\left(\tau+\frac{3}{2}\right)^2+\beta_0^4\right]^{1/2}. \tag{7}$$

For $\beta_0 = 0$ the original solution of Bohr [3], which corresponds to a 5-dimensional (5-D) harmonic oscillator characterized by the symmetry U(5)

$\supset$ SO(5) $\supset$ SO(3) $\supset$ SO(2) [17], is obtained. The values of angular momentum L contained in each irrep of SO(5) (i.e. for each value of τ) are given by the algorithm [18] $\tau = 3\nu_\Delta + \lambda$, where $\nu_\Delta = 0, 1, \ldots$ is the missing quantum number in the reduction SO(5) $\supset$ SO(3), and $L = \lambda, \lambda+1, \ldots, 2\lambda-2, 2\lambda$ (with $2\lambda - 1$ missing).

The levels of the ground state band are characterized by $L = 2\tau$ and $n = 0$. Then the energy levels of the ground state band are given by

$$E_{0,L} = 1 + \frac{1}{2}\left[(L+3)^2 + 4\beta_0^4\right]^{1/2}. \tag{8}$$

For $u(\beta)$ being a 5-D infinite well

$$u(\beta) = \begin{cases} 0 & \text{if } \beta \leq \beta_W \\ \infty & \text{for } \beta > \beta_W \end{cases} \tag{9}$$

one obtains the E(5) model of Iachello [1] in which the eigenfunctions are Bessel functions $J_{\tau+3/2}(z)$ (with $z = \beta k$, $k = \sqrt{\epsilon}$), while the spectrum is determined by the zeros of the Bessel functions

$$E_{\xi,\tau} = \frac{\hbar^2}{2B} k_{\xi,\tau}^2, \qquad k_{\xi,\tau} = \frac{x_{\xi,\tau}}{\beta_W} \tag{10}$$

where $x_{\xi,\tau}$ is the ξ-th zero of the Bessel function $J_{\tau+3/2}(z)$. The spectra of the E(5) and Davidson cases become directly comparable by establishing the formal correspondence $n = \xi - 1$.

It is instructive to consider the ratios

$$R_L = \frac{E_{0,L} - E_{0,0}}{E_{0,2} - E_{0,0}}, \tag{11}$$

where the notation $E_{n,L}$ is used.

For $\beta_0 = 0$ it is clear that the original vibrational model of Bohr [3] (with $R_4 = 2$) is obtained, while for large β_0 the O(6) limit of the Interacting Boson Model (IBM) [18] (with $R_4 = 2.5$) is approached [8]. The latter fact can be seen in the left part of Table 1, where the R_L ratios for two different values of the parameter β_0 are shown, together with the O(6) predictions (which correspond to $E(L) = AL(L+6)$, with A constant [19]). It is clear that the O(6) limit is approached as β_0 is increased, the agreement being already quite good at $\beta_0 = 5$.

It is useful to consider the ratios R_L, defined above, as a function of β_0. As seen in Ref. [20], these ratios increase with β_0, the increase becoming very steep at some value $\beta_{0,max}$ of β_0, where the first derivative $dR_L/d\beta_0$ reaches a maximum value, while the second derivative $d^2R_L/d\beta_0^2$ vanishes. Numerical results for $\beta_{0,max}$ are shown in the right part of Table 1, together

Table 1. Left part: R_L ratios (defined in Eq. (11)) for the ground state band of the Davidson potentials in the E(5) framework (Eq. (8)) for different values of the parameter β_0, compared to the O(6) exact results. Right part: Parameter values $\beta_{0,max}$ where the first derivative of the energy ratios R_L in the E(5) framework has a maximum, while the second derivative vanishes, together with the R_L ratios obtained at these values (labeled by "var") and the corresponding ratios of the E(5) model, for several values of the angular momentum L.

L	R_L	R_L	R_L	$\beta_{0,max}$	R_L	R_L
	$\beta_0 = 5.$	$\beta_0 = 10.$	O(6)		var	E(5)
4	2.494	2.500	2.500	1.421	2.185	2.199
6	4.475	4.498	4.500	1.522	3.549	3.590
8	6.935	6.996	7.000	1.609	5.086	5.169
10	9.861	9.991	10.000	1.687	6.793	6.934
12	13.242	13.483	13.500	1.759	8.667	8.881
14	17.064	17.471	17.500	1.825	10.705	11.009
16	21.312	21.954	22.000	1.888	12.906	13.316
18	25.969	26.930	27.000	1.947	15.269	15.799
20	31.020	32.398	32.500	2.004	17.793	18.459

with the values of R_L occuring at these points, which are compared to the R_L ratios occuring in the ground state band of the E(5) model [1]. Very close agreement of the values determined by the procedure described above with the E(5) values is observed.

3. Davidson potentials in the X(5) framework

Starting again from the original Bohr Hamiltonian of Eq. (2), one seeks solutions of the relevant Schrödinger equation having the form $\Psi(\beta,\gamma,\theta_i) = \phi_K^L(\beta,\gamma)D_{M,K}^L(\theta_i)$, where θ_i ($i = 1, 2, 3$) are the Euler angles, $D(\theta_i)$ denote Wigner functions of them, L are the eigenvalues of angular momentum, while M and K are the eigenvalues of the projections of angular momentum on the laboratory-fixed z-axis and the body-fixed z'-axis respectively.

As pointed out in Ref. [2], in the case in which the potential has a minimum around $\gamma = 0$ one can write the last term of Eq. (2) in the form

$$\sum_{k=1,2,3} \frac{Q_k^2}{\sin^2\left(\gamma - \frac{2\pi}{3}k\right)} \approx \frac{4}{3}(Q_1^2 + Q_2^2 + Q_3^2) + Q_3^2\left(\frac{1}{\sin^2\gamma} - \frac{4}{3}\right). \quad (12)$$

Using this result in the Schrödinger equation corresponding to the Hamiltonian of Eq. (2), introducing reduced energies $\epsilon = 2BE/\hbar^2$ and reduced potentials $u = 2BV/\hbar^2$, and assuming that the reduced potential can be separated into two terms, one depending on β and the other depending on γ, i.e. $u(\beta,\gamma) = u(\beta) + u(\gamma)$, the Schrödinger equation can be separated into two equations [2], the "radial" one being

$$\left[-\frac{1}{\beta^4}\frac{\partial}{\partial\beta}\beta^4\frac{\partial}{\partial\beta} + \frac{1}{4\beta^2}\frac{4}{3}L(L+1) + u(\beta)\right]\xi_L(\beta) = \epsilon_\beta \xi_L(\beta). \tag{13}$$

When plugging the Davidson potentials of Eq. (1) in this equation, the β_0^4/β^2 term of the potential is combined with the $L(L+1)/3\beta^2$ term appearing there and the equation is solved exactly, the eigenfunctions being Laguerre polynomials of the form

$$F_n^L(\beta) = \left[\frac{2n!}{\Gamma\left(n+a+\frac{5}{2}\right)}\right]^{1/2} \beta^a L_n^{a+\frac{3}{2}}(\beta^2)e^{-\beta^2/2}, \tag{14}$$

where a is given by

$$a = -\frac{3}{2} + \left[\frac{1}{3}L(L+1) + \frac{9}{4} + \beta_0^4\right]^{1/2}. \tag{15}$$

The energy eigenvalues are then (in $\hbar\omega = 1$ units)

$$E_{n,L} = 2n + a + \frac{5}{2} = 2n + 1 + \left[\frac{1}{3}L(L+1) + \frac{9}{4} + \beta_0^4\right]^{1/2}, \tag{16}$$

the levels of the ground state band being characterized by $n = 0$.

It is clear that Eq. (16) leads to energies proportional to $L(L+1)$ for large values of β_0. Therefore for large values of β_0 one should expect to obtain spectra close to the rotational limit. This can be seen in the left part of Table 2, where the R_L ratios occuring for two different values of β_0 are shown, together with the predictions of the SU(3) limit of IBM, which correspond to the pure rotator with $E(L) = AL(L+1)$, where A constant [18]. The agreement to the SU(3) results is quite good already at $\beta_0 = 5$. On the other hand, the case $\beta_0 = 0$ corresponds to an exactly soluble model with $R_4 = 2.646$, which has been called the X(5)-β^2 model [21].

For $u(\beta)$ being a 5-D infinite well potential (see Eq. (9) one obtains the X(5) model of Iachello [2], in which the eigenfunctions are Bessel functions $J_\nu(k_{s,L}\beta)$ with

$$\nu = \left(\frac{L(L+1)}{3} + \frac{9}{4}\right)^{1/2}, \tag{17}$$

Table 2. Left part: R_L ratios (defined in Eq. (11)) for the ground state band of the Davidson potentials in the X(5) framework (Eq. (16)) for different values of the parameter β_0, compared to the SU(3) exact results. Right part: Parameter values $\beta_{0,max}$ where the first derivative of the energy ratios R_L in the X(5) framework has a maximum, while the second derivative vanishes, together with the R_L ratios obtained at these values (labeled by "var") and the corresponding ratios of the X(5) model, for several values of the angular momentum L.

L	R_L	R_L	R_L	$\beta_{0,max}$	R_L	R_L
	$\beta_0 = 5.$	$\beta_0 = 10.$	SU(3)		var	X(5)
4	3.327	3.333	3.333	1.334	2.901	2.904
6	6.967	6.998	7.000	1.445	5.419	5.430
8	11.897	11.993	12.000	1.543	8.454	8.483
10	18.087	18.317	18.333	1.631	11.964	12.027
12	25.503	25.968	26.000	1.711	15.926	16.041
14	34.102	34.941	35.000	1.785	20.330	20.514
16	43.839	45.233	45.333	1.855	25.170	25.437
18	54.665	56.841	57.000	1.922	30.442	30.804
20	66.530	69.760	70.000	1.985	36.146	36.611

while the spectrum is determined by the zeros of the Bessel functions, the relevant eigenvalues being

$$\epsilon_{\beta;s,L} = (k_{s,L})^2, \qquad k_{s,L} = \frac{x_{s,L}}{\beta_W}, \tag{18}$$

where $x_{s,L}$ is the s-th zero of the Bessel function $J_\nu(k_{s,L}\beta)$. The spectra of the X(5) and Davidson cases become directly comparable by establishing the formal correspondence $n = s - 1$.

It is useful to consider the ratios R_L, defined in the previous section, as a function of β_0. As seen in Ref. [20], these ratios again increase with β_0, the increase becoming very steep at some value $\beta_{0,max}$ of β_0, where the first derivative $dR_L/d\beta_0$ reaches a maximum value, while the second derivative $d^2R_L/d\beta_0^2$ vanishes. Numerical results for $\beta_{0,max}$ are shown in the right part of Table 2, together with the values of R_L occuring at these points, which are compared to the R_L ratios occuring in the ground state band of the X(5) model [2]. Very close agreement of the values determined by the procedure described above with the X(5) values is observed.

4. Discussion

It is clearly of interest to apply the variational procedure introduced here to physical quantities other than the energy ratios in the ground state band. Energy ratios involving levels of excited bands, ratios of B(E2) transition rates (both intraband and interband), and ratios of quadrupole moments are obvious choices. Work in these directions is in progress [22].

References

1. F. Iachello, *Phys. Rev. Lett.* **85**, 3580 (2000).
2. F. Iachello, *Phys. Rev. Lett.* **87**, 052502 (2001).
3. A. Bohr, *Mat. Fys. Medd. K. Dan. Vidensk. Selsk.* **26**, no. 14 (1952).
4. R. F. Casten and N. V. Zamfir, *Phys. Rev. Lett.* **85**, 3584 (2000).
5. N. V. Zamfir et al., *Phys. Rev.* **C65**, 044325 (2002).
6. R. F. Casten and N. V. Zamfir, *Phys. Rev. Lett.* **87**, 052503 (2001).
7. R. Krücken et al., *Phys. Rev. Lett.* **88**, 232501 (2002).
8. J. P. Elliott, J. A. Evans, and P. Park, *Phys. Lett.* **B169**, 309 (1986).
9. D. J. Rowe and C. Bahri, *J. Phys. A: Math. Gen.* **31**, 4947 (1998).
10. P. M. Davidson, *Proc. R. Soc. (London)* **135**, 459 (1932).
11. V. Werner, P. von Brentano, R. F. Casten, and J. Jolie, *Phys. Lett.* **B527**, 55 (2002).
12. C. A. Mallmann, *Phys. Rev. Lett.* **2**, 507 (1959).
13. M. A. J. Mariscotti, G. Scharff-Goldhaber, and B. Buck, *Phys. Rev.* **178**, 1864 (1969).
14. L. Wilets and M. Jean, *Phys. Rev.* **102**, 788 (1956).
15. G. Rakavy, *Nucl. Phys.* **4**, 289 (1957).
16. D. R. Bès, *Nucl. Phys.* **10**, 373 (1959).
17. E. Chacón and M. Moshinsky, *J. Math. Phys.* **18**, 870 (1977).
18. F. Iachello and A. Arima, The Interacting Boson Model, Cambridge University Press, Cambridge, 1987.
19. R. F. Casten, Nuclear Structure from a Simple Perspective, Oxford University Press, Oxford, 1990.
20. D. Bonatsos, D. Lenis, N. Minkov, D. Petrellis, P. P. Raychev, and P. A. Terziev, *Phys. Lett.* **B584**, 40 (2004).
21. D. Bonatsos, D. Lenis, N. Minkov, P. P. Raychev and P. A. Terziev, *Phys. Rev.* **C69**, 014302 (2004).
22. D. Bonatsos, D. Lenis, N. Minkov, D. Petrellis, P. P. Raychev, and P. A. Terziev, nucl-th/0402088.

KEY TOPICS IN NUCLEAR STRUCTURE
Proceedings of the 8th Int. Spring Seminar on Nuclear Physics
edited by Aldo Covello
© 2005 World Scientific Publishing Co.

TRANSITION PROBABILITIES: A KEY TO PROVE THE X(5) SYMMETRY

D. TONEV, G. DE ANGELIS, A. GADEA, D. R. NAPOLI, M. AXIOTIS, N. MARGINEAN AND T. MARTINEZ

INFN, Laboratori Nazionali di Legnaro,
Viale dell'Università 2, 35020 Legnaro (Padova), Italy
E-mail: mitko@lnl.infn.it

A. DEWALD, T. KLUG, J. JOLIE, A. FITZLER, O. MÖLLER, B. SAHA, P. PEJOVIC, S. HEINZE AND P. VON BRENTANO

Institut für Kernphysik der Universität zu Köln,
Zülpicherstr 77, 50937 Köln, Germany

P. PETKOV

Institute for Nuclear Research and Nuclear Energy,
Bulgarian Academy of Sciences, 1784 Sofia, Bulgaria

R. F. CASTEN

WNSL, Yale University,
New Haven, Connecticut, USA

D. BAZZACCO, E. FARNEA, S. LENZI, S. LUNARDI, R. MENEGAZZO

Dipartimento di Fisica dell' Università and INFN Sezione Padova, Padova, Italy

The lifetimes of excited states in ^{154}Gd were measured using the recoil distance Doppler-Shift (RDDS) method. The experiment was performed at Cologne FN Tandem accelerator where a beam of ^{32}S with an energy of 110 MeV was used to Coulomb excite states of ^{154}Gd. The determined transition probabilities as well as the low-spin level scheme of ^{154}Gd demonstrate a good agreement with the predictions of the critical point symmetry X(5). Comparison of specific experimental observables for the N = 90 rare earth isotones with the calculations of the X(5) model clearly show that ^{154}Gd is one of the best examples of the realization of the X(5) dynamical symmetry. In addition, the experimental data are compared to fits in the framework of the IBA and the General Collective Model (GCM).

1. Introduction

Symmetry is a universal concept in our life. There are many examples showing that symmetry plays a crucial role in the development of Physics. Some of them like rotational invariance in non-relativistic quantum mechanics and Lorentz invariance in relativistic quantum mechanics are called fundamental symmetries, since they are incredibly important.

Nuclei as dynamic systems can undergo phase transitions associated with a change of the shape of their equilibrium configuration. In the works of Iachello [1,2], it was shown that new dynamical symmetries can describe atomic nuclei at the critical points of phase transitions from spherical to deformed shapes. The new symmetries, called E(5) and X(5), are obtained within the framework of the collective model [3] under some simplifying approximations. Like the harmonic vibrator and the rigid rotor, these critical-point solutions correspond to square-well potentials in the deformation parameter β for E(5) and X(5) and make parameter-free predictions for the excitation energy ratios and decay patterns of excited states. In addition, in both cases E(5) and X(5) the β and γ degrees of freedom are considered to be decoupled.

The goal of our investigation is the nucleus ^{154}Gd, which is expected to be one of the very promising candidates for X(5) critical point symmetry. This fact is illustrated in Fig. 1, where characteristic energy ratios in some N = 90 isotones are compared to the values predicted in the framework of X(5).

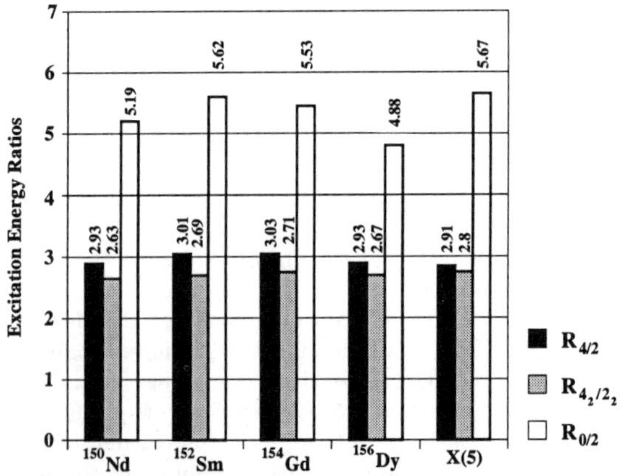

Figure 1. Comparison of the crucial experimental observables $R_{0/2}=E(0_2^+)/E(2_2^+)$, $R_{4/2}=E(4_1^+)/E(2_1^+)$, $R_{4_2/2_2}$ = $(E_{4_2^+}$ - $E_{0_2^+})$ / $(E_{2_2^+}$ - $E_{0_2^+})$ indicated on the ordinate for the N = 90 rare earth isotones ^{150}Nd, ^{152}Sm, ^{154}Gd, and ^{156}Dy with the predictions of the X(5) dynamical symmetry. Based on Ref. [4].

It turns out that a classification based only on the excitation energies and branching ratios is not sufficient for a definite assignment as was shown in the case of ^{104}Mo. P. G. Bizzeti and A. M. Bizzeti-Sona [5] demonstrated that the energy spectrum of ^{104}Mo as well as relative transition probabilities can be nicely described by the X(5) calculations. Later, however, it was established by C. Hutter et al. [6] that the absolute B(E2) values do not support an X(5)-like character of that nucleus. For this reason study of the electromagnetic properties of the first two excited bands of ^{154}Gd, based on the 0^+_1 and 0^+_2 levels, is a key to prove the realization of X(5) symmetry in ^{154}Gd.

2. Experimental details

Lifetime measurements in ^{154}Gd were performed by means of the Recoil Distance Doppler-shift method. Excited states of ^{154}Gd were populated via Coulomb excitation with a ^{32}S beam at 110 MeV delivered by the FN tandem accelerator of the University of Cologne. The target consisted of 98% enriched ^{154}Gd isotope material evaporated to a thickness of 1 mg/cm^2 onto a 2 mg/cm^2 thick Ta foil. The recoiling nuclei left the target with a mean velocity v of 2.30(3) % of the velocity of light, c, and were stopped in a 5 mg/cm^2 Nb foil mounted together with the target in the Cologne coincidence plunger apparatus [7].

The γ-rays were detected with a Euroball cluster detector placed at 0° with respect to the beam axis, and four additional high efficiency germanium detectors positioned at the backward angle of 144°. The setup is presented in details in Ref. [4].

At each distance x, the lifetimes are calculated according to the formalism of the Differential decay-curve method (DDCM) in a "singles mode" [8]:

$$\tau(x) = -\frac{R_{ij}(x) - b_{ij}\frac{W_{ij}(\theta)\epsilon_{ij}}{(1+\alpha_{ij})}\sum_h R_{hi}(x)(1+\alpha_{hi})/W_{hi}(\theta)\epsilon_{hi}}{v \cdot \frac{d}{dx}R_{ij}(x)} \tag{1}$$

Here, x is the target-to-stopper distance, $R_{ij}(x)$ is the area of the unshifted peak of the transition of interest, $t = x/v$ is the corresponding time of flight, b_{ij} is the branching ratio of the transition $i \to j$ and R_{hi} are the areas of the unshifted peaks of the direct feeding transitions. $W_{ln}(\theta)$ represents the angular distribution function for the transition $l \to n$ at the angle θ of observation with respect to the beam axis, α_{ln} is the internal conversion coefficient and ϵ_{ln} is the efficiency of the germanium detector. All details concerning the lifetime analysis are presented in Ref. [4].

The application of Eq. 1 for the lifetime determination of the 4_2^+ level is illustrated in Fig. 2 where the τ-curve, the difference in the numerator and the derivative (denominator in Eq. 1) are presented.

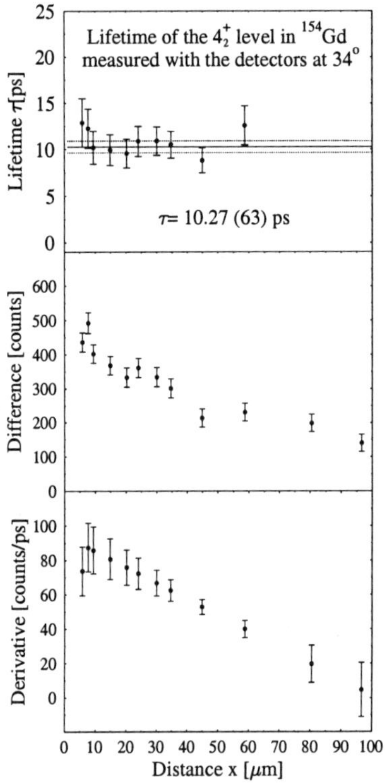

Figure 2. Lifetime determination of the 4_2^+ level of the S2 band using data collected with the cluster segments positioned at 34°. Based on Ref. [4].

This figure is also a demonstration of the quality of the data for the case of the weakest transition reported in this work. Lifetimes were derived from the data for the 4_1^+, 6_1^+, 8_1^+, 0_2^+, 2_2^+ and 4_2^+ levels in ^{154}Gd.

The quality of the experiment is demonstrated by the agreement within the error bars of the present results with literature values for levels with well-established lifetimes [9,10]. The lifetime τ= 11.00(55) ps of the 4_2^+ level is determined for the first time. This lifetime is important since it doubles the number of absolute B(E2) values now known involving the 0_2^+ band, giving several new interband transition probabilities and a second intraband B(E2) value.

In order to extract in-band transition quadrupole moments Q_t from these data, we employed the well known formula

$$B(E2, I \to I-2) = \frac{5}{16\pi} < IK20|I-2K >^2 (Q_t)^2 \tag{2}$$

in which a value of $K = 0$ was used.

3. Discussion

In Fig. 1, the crucial observables of the four N=90 isotones ^{150}Nd, ^{152}Sm, ^{154}Gd and ^{156}Dy are compared to the corresponding X(5) values. As for ^{150}Nd and ^{152}Sm, for ^{154}Gd the agreement is also remarkably good.

The absolute B(E2) values measured in this work allow now for a more stringent test of the X(5) predictions including excitation energies as well as relative and absolute transition probabilities.

Fig. 3 shows the Q_t values within the gsb of ^{154}Gd together with the theoretical values of the X(5) symmetry, the symmetric rotor and the IBA U(5) limit [11]. The four data sets are normalized to the experimental $Q_t(2_1^+ \to 0_1^+)$ value. The experimental transition quadrupole moment $Q_t(4_1^+ \to 2_1^+)$ agrees perfectly with the X(5) value. At higher spins, the experimental values lie between X(5) and the rotor values. The U(5) predictions are considerably higher for all states.

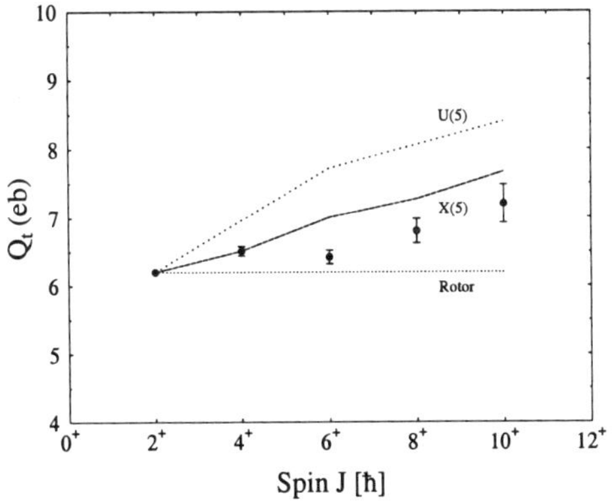

Figure 3. Comparison of the Q_t values of the gsb (band S1) in ^{154}Gd with the theoretical calculations. The lifetime of the 10^+ state necessary for derivation of the corresponding Q_t value was taken from the work of S. H. Sie et. al.[10].

Of special importance are also the transition probabilities within the first excited band (S2) and those of the inter-band transitions between the S2 and S1 bands.

In Fig.4, these quantities as well as the energy spectra of the S1 and S2 bands according to X(5) are compared to the corresponding experimental values in ^{154}Gd. Only two normalization factors, the excitation energy of the first 2^+ state and the B(E2;$2^+ \rightarrow 0^+$), are used. The overall agreement is found to be good for both energies and transition probabilities, although as generally seems to occur, the absolute X(5) energy spacings in the S2 band are larger than observed, and the intraband B(E2) values are also larger than measured.

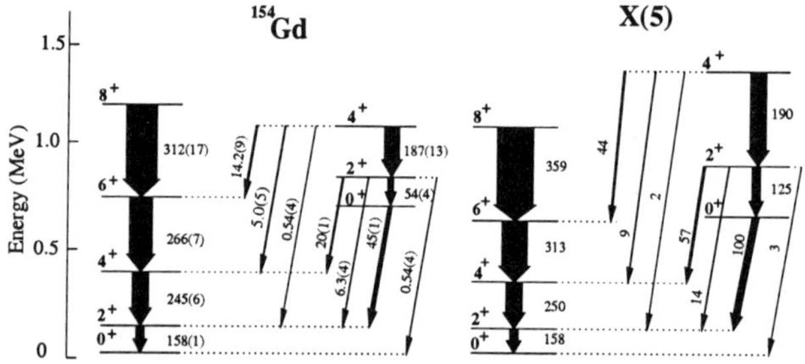

Figure 4. Comparison of the experimental level scheme of ^{154}Gd with the X(5) predictions. The B(E2) values are given next to the arrows. Based on Ref. [4].

As already mentioned when the Q_t values in Fig. 3 were discussed, the B(E2;$4_2^+ \rightarrow 2_2^+$) value is very close to the X(5) value. All other strong transitions are also well described, with the exceptions of B(E2;$2_2^+ \rightarrow 0_2^+$) and B(E2;$0_2^+ \rightarrow 2_1^+$), whose values are a factor of two smaller than the X(5) predictions. For the inter-band transitions, the measured strengths are found to be less than a factor of two smaller than the corresponding X(5) calculations which is similar to the case of ^{150}Nd [12]. A possible explanation of this fact could be found in the recent work [13]. It is shown that the evolution of the empirical correlation is well described in a parameter-free way by a new analytical solution of the Bohr-Hamiltonian's eigenvalue problem using an infinite square-well potential over a confined range of values for the parameter β. Inclusion of the second order term in the E2 operator allows for a consistent description of the $\Delta K = 0$ inter-band E2 matrix element. It is shown that the good description of the E2 rates in the ^{154}Sm results from the analytical solution of a unique Hamiltonian and a simple choice for the E2 operator.

Summarizing, the agreement between the experimental data for ^{154}Gd and the X(5) predictions is found to be good and is comparable to the quality of agreement found for ^{150}Nd and ^{152}Sm. In all cases investigated so far perfect agreement with the X(5) predictions is of course, not observed. This is not surprising since the X(5) model does not employ any fitting and the underlying nuclear potential is approximated in a rather schematic way. That is, X(5) is not a model to be fit to the data as are models such as the IBA and GCM [14] but, rather, it is an invariant (except for scale) benchmark, similar in spirit to other idealizations such as the pure rotor or harmonic vibrator.

Therefore one should expect an improved agreement between experimental and theoretical values when using other theoretical models where the calculated quantities can be adjusted by varying model parameters, such as with the IBA, GCM or the PPQ [15] models.

In the case of ^{154}Gd, we tried to reproduce the experimental observables with optimized IBA and GCM calculations. The results are shown in Fig. 5. The IBA calculation with only two scaling and two fit parameters, ξ and χ, comes very close to the experimental values including the sensitive S2 to S1 interband transition probabilities which are reproduced remarkably well. Also the GCM result can be considered to be satisfactory although 6 of the model parameters (in total 8) were employed in the calculation. The GCM Hamiltonian can be found in Ref. [14].

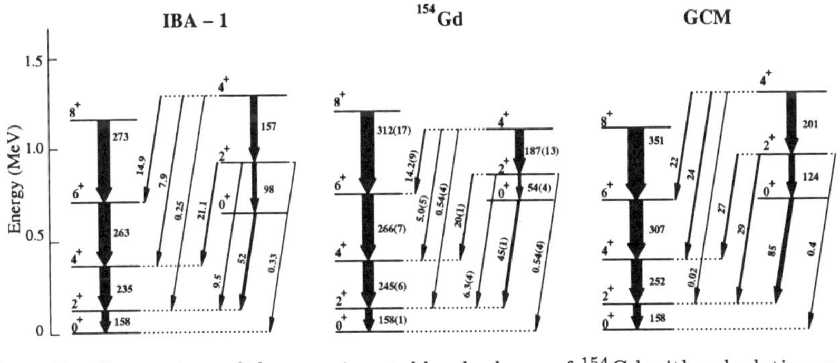

Figure 5. Comparison of the experimental level scheme of ^{154}Gd with calculations according to the GCM and IBA. The B(E2) transition strengths are given next to the arrows. Based on Ref. [4].

Finally, we want to get some idea about the perturbation of the ideal X(5) model needed to describe ^{154}Gd. It was shown [16,17] that the critical point of phase transition between SU(3) and U(5) can be associated to the parameters ξ=0.026 and χ=-$\sqrt{7/4}$ of the IBA Hamiltonian given in that reference. The actual IBA fit shown in fig. 5 was obtained with the parameter values ξ=0.028 and χ=-$\sqrt{7/4}$ which are very close to the critical point parameters. This leads to the conclusion that ^{154}Gd indeed can be located very close to the critical point of the SU(3)-U(5) phase transition for which the X(5) symmetry serves as a benchmark.

4. Conclusions

To conclude, RDDS lifetime measurements were carried out at the FN Tandem accelerator of the University of Cologne with Coulomb excitation of ^{154}Gd using a 110 MeV ^{32}S beam. From the derived lifetimes, 12 reduced E2 transition probabilities were determined whose values are in a good agreement with the predictions of the X(5) critical point symmetry as well with IBA and GCM calculations. Consistently, the different theoretical approaches show that ^{154}Gd can be associated with the critical point of a SU(3)-U(5) phase transition.

Acknowledgments

D.T. express his gratitude to Ivanka Necheva for her outstanding support. This work was funded by the BMBF under contract No. 06 OK 167. This research has been supported by a Marie Curie Fellowship of the European Community programme IHP under contract number HPMF-CT-2002-02018.

References

1. F. Iachello, *Phys. Rev. Lett.* **85**, 3580 (2000).
2. F. Iachello, *Phys. Rev. Lett.* **87**, 052502 (2001).
3. A. Bohr and B. R. Mottelson, Nuclear Structure (Benjamin, New York, 1975).
4. D. Tonev, et al., *Phys. Rev* **C69**, 034334 (2004).
5. P. G. Bizzeti and A. M. Bizzeti-Sona, *Phys. Rev.* **C 66**, 031301(R) (2002).
6. C. Hutter et al., *Phys. Rev.* **C67**, 054315 (2003).
7. A. Dewald, et al.,*Nucl. Phys.* **A545**, 822 (1992).
8. A. Dewald, S. Harisopoulos, P. von Brentano, *Z. Phys.* **A334**, 163 (1989).
9. N. Rud, et al., *Nucl. Phys.* **A191**, 545 (1972).
10. S. H. Sie, et al.,*Nucl. Phys.* **A291**, 443 (1977).
11. F. Iachello and A. Arima, The Interacting Boson Model (Cambridge University Press, Cambridge, 1987).
12. R. Krücken et al., *Phys. Rev. Lett.* **88**, 232501 (2002).
13. N. Pietralla and O. M. Gorbachenko, *Phys. Rev C* in press.
14. G. Gneuss and W. Greiner, *Nucl. Phys.* **A171**, 440 (1971).
15. K. Kumar, *Phys. Rev. Lett.* **26**, 269 (1971).
16. F. Iachello, N. V. Zamfir and R. F. Casten, *Phys. Rev. Lett.* **81**, 1191 (1998).
17. R. F. Casten, D. Kusnezov and N. V. Zamfir, *Phys. Rev. Lett.* **82**, 5000 (1999).

KEY TOPICS IN NUCLEAR STRUCTURE
Proceedings of the 8th Int. Spring Seminar on Nuclear Physics
edited by Aldo Covello
© 2005 World Scientific Publishing Co.

Z(5): CRITICAL POINT SYMMETRY FOR THE PROLATE TO OBLATE SHAPE PHASE TRANSITION

DENNIS BONATSOS, D. LENIS, AND D. PETRELLIS

Institute of Nuclear Physics,
National Centre for Scientific Research "Demokritos",
GR-15310 Aghia Paraskevi, Attiki, Greece
E-mail: bonat@inp.demokritos.gr

P. A. TERZIEV

Institute for Nuclear Research and Nuclear Energy,
Bulgarian Academy of Sciences,
72 Tzarigrad Road, BG-1784, Bulgaria
E-mail: terziev@inrne.bas.bg

A solution of the Bohr Hamiltonian is obtained by approximately separating variables for $\gamma = 30^o$. Parameter-free (up to overall scale factors) predictions for spectra and B(E2) transition rates are found to be in good agreement with experimental data for ^{194}Pt, which is supposed to be located very close to the prolate to oblate critical point, as well as for its neighbours (^{192}Pt, ^{196}Pt). Hallmarks of the model are the $R_4 = E(4)/E(2)$ ratio of 2.350, as well as the location of the γ_1- and β_1-bandheads (normalized to the 2^+ state of the ground state band) at 1.837 and 3.913 respectively, while the selection rules for B(E2) transition rates are similar to the ones of the O(6) limit of the Interacting Boson Model.

1. Introduction

Critical point symmetries in nuclear structure are recently receiving considerable attention [1,2,3], since they provide parameter-free (up to overall scale factors) predictions supported by experimental evidence [4,5,6,7]. Furthermore, it has been demonstrated [8] that experimental data in the Hf-Hg mass region indicate the presence of a prolate to oblate shape phase transition, the nucleus ^{194}Pt being the closest one to the critical point.

In the present work a new approximate solution of the Bohr Hamiltonian [9], to be called Z(5), is introduced. This solution can be used as a parameter-free (up to overall scale factors) critical point symmetry for the prolate to oblate shape phase transition, leading to parameter-free predictions which compare very well with the experimental data for ^{194}Pt. The path followed for constructing the Z(5) solution is described here:

1) Separation of variables in the Bohr equation [9] is achieved by assuming $\gamma = 30^o$. When considering the transition from $\gamma = 0^o$ (prolate) to $\gamma = 60^o$ (oblate), it is reasonable to expect that the triaxial region ($0^o < \gamma < 60^o$) will be crossed, $\gamma = 30^o$ lying in its middle. Indeed, there is experimental evidence supporting this assumption [10].

2) For $\gamma = 30^o$ the K quantum number (angular momentum projection on the body-fixed $\hat{z}'$-axis) is *not* a good quantum number any more, but α, the angular momentum projection on the body-fixed $\hat{x}'$-axis is, as found [11] in the study of the triaxial rotator [12].

3) Assuming an infinite well potential in the β-variable and a harmonic oscillator potential having a minimum at $\gamma = 30^o$ in the γ-variable, the Z(5) model is obtained.

Further comments on these choices can be found in Ref. [13].

2. The β-part of the spectrum

The original Bohr Hamiltonian [9] is

$$H = -\frac{\hbar^2}{2B}\left[\frac{1}{\beta^4}\frac{\partial}{\partial\beta}\beta^4\frac{\partial}{\partial\beta} + \frac{1}{\beta^2 \sin 3\gamma}\frac{\partial}{\partial\gamma}\sin 3\gamma\frac{\partial}{\partial\gamma}\right.$$

$$\left. -\frac{1}{4\beta^2}\sum_{k=1,2,3}\frac{Q_k^2}{\sin^2\left(\gamma - \frac{2}{3}\pi k\right)}\right] + V(\beta,\gamma), \tag{1}$$

where β and γ are the usual collective coordinates, while Q_k ($k = 1, 2, 3$) are the components of angular momentum and B is the mass parameter.

In the case in which the potential has a minimum around $\gamma = \pi/6$ one can write the last term of Eq. (1) in the form $4(Q_1^2+Q_2^2+Q_3^2)-3Q_1^2$. Using this result in the Schrödinger equation corresponding to the Hamiltonian of Eq. (1), introducing [2] reduced energies $\epsilon = 2BE/\hbar^2$ and reduced potentials $u = 2BV/\hbar^2$, and assuming [2] that the reduced potential can be separated into two terms, one depending on β and the other depending on γ, i.e. $u(\beta,\gamma) = u(\beta) + u(\gamma)$, the Schrödinger equation can be separated into two equations

$$\left[-\frac{1}{\beta^4}\frac{\partial}{\partial\beta}\beta^4\frac{\partial}{\partial\beta} + \frac{1}{4\beta^2}(4L(L+1) - 3\alpha^2) + u(\beta)\right]\xi_{L,\alpha}(\beta) = \epsilon_\beta\xi_{L,\alpha}(\beta), \tag{2}$$

$$\left[-\frac{1}{\langle\beta^2\rangle \sin 3\gamma}\frac{\partial}{\partial\gamma}\sin 3\gamma\frac{\partial}{\partial\gamma} + u(\gamma)\right]\eta(\gamma) = \epsilon_\gamma \eta(\gamma), \tag{3}$$

where L is the angular momentum quantum number, α is the projection of the angular momentum on the body-fixed $\hat{x}'$-axis (α has to be an even integer [11]), $\langle\beta^2\rangle$ is the average of β^2 over $\xi(\beta)$, and $\epsilon = \epsilon_\beta + \epsilon_\gamma$.

The total wave function should have the form

$$\Psi(\beta, \gamma, \theta_i) = \xi_{L,\alpha}(\beta)\eta(\gamma)D^L_{M,\alpha}(\theta_i), \tag{4}$$

where θ_i ($i = 1, 2, 3$) are the Euler angles, $D(\theta_i)$ denote Wigner functions of them, while M are the eigenvalues of the projection of angular momentum on the laboratory fixed $\hat{z}$-axis.

Instead of the projection α of the angular momentum on the $\hat{x}'$-axis, it is customary to introduce the wobbling quantum number [11] $n_w = L - \alpha$, which labels a series of bands with $L = n_w, n_w+2, n_w+4, \dots$ (with $n_w > 0$) next to the ground state band (with $n_w = 0$) [11].

In the case in which $u(\beta)$ is an infinite well potential

$$u(\beta) = \begin{cases} 0 & \text{if } \beta \leq \beta_W \\ \infty & \text{for } \beta > \beta_W \end{cases}, \tag{5}$$

one can use [2] the transformation $\tilde{\xi}(\beta) = \beta^{3/2}\xi(\beta)$, as well as the definitions $\epsilon_\beta = k_\beta^2$, $z = \beta k_\beta$, in order to bring Eq. (2) into the form of a Bessel equation

$$\frac{d^2\tilde{\xi}}{dz^2} + \frac{1}{z}\frac{d\tilde{\xi}}{dz} + \left[1 - \frac{\nu^2}{z^2}\right]\tilde{\xi} = 0, \tag{6}$$

with

$$\nu = \frac{\sqrt{4L(L+1) - 3\alpha^2 + 9}}{2} = \frac{\sqrt{L(L+4) + 3n_w(2L - n_w) + 9}}{2}. \tag{7}$$

Then the boundary condition $\tilde{\xi}(\beta_W) = 0$ determines the spectrum

$$\epsilon_{\beta;s,\nu} = \epsilon_{\beta;s,n_w,L} = (k_{s,\nu})^2, \qquad k_{s,\nu} = \frac{x_{s,\nu}}{\beta_W}, \tag{8}$$

and the eigenfunctions

$$\xi_{s,\nu}(\beta) = \xi_{s,n_w,L}(\beta) = \xi_{s,\alpha,L}(\beta) = c_{s,\nu}\beta^{-3/2}J_\nu(k_{s,\nu}\beta), \tag{9}$$

where $x_{s,\nu}$ is the sth zero of the Bessel function $J_\nu(z)$, while the normalization constants $c_{s,\nu}$ are determined from the normalization condition $\int_0^\infty \beta^4 \xi^2_{s,\nu}(\beta)d\beta = 1$. The notation for the roots has been kept the same as in Ref. [2], while for the energies the notation $E_{s,n_w,L}$ will be used. We shall refer to the model corresponding to this solution as Z(5) (which is not meant as a group label), in analogy to the E(5) [1], X(5) [2], and Y(5) [3] models.

3. The γ-part of the spectrum

As already mentioned, we consider a harmonic oscillator potential having a minimum at $\gamma = \pi/6$, i.e.

$$u(\gamma) = \frac{1}{2}c\left(\gamma - \frac{\pi}{6}\right)^2 = \frac{1}{2}c\tilde{\gamma}^2, \qquad \tilde{\gamma} = \gamma - \frac{\pi}{6}. \tag{10}$$

In the case of $\gamma \approx \pi/6$ Eq. (3) can be brought into the form

$$\left[-\frac{\partial^2}{\partial\tilde{\gamma}^2} + \frac{1}{2}c\langle\beta^2\rangle\tilde{\gamma}^2\right]\eta(\tilde{\gamma}) = \epsilon_{\tilde{\gamma}}\langle\beta^2\rangle\eta(\tilde{\gamma}) \tag{11}$$

which is a simple harmonic oscillator equation with energy eigenvalues

$$\epsilon_{\tilde{\gamma}} = \sqrt{\frac{2c}{\langle\beta^2\rangle}}\left(n_{\tilde{\gamma}} + \frac{1}{2}\right), \qquad n_{\tilde{\gamma}} = 0, 1, 2, \ldots \tag{12}$$

and eigenfunctions

$$\eta_{n_{\tilde{\gamma}}}(\tilde{\gamma}) = N_{n_{\tilde{\gamma}}} H_{n_{\tilde{\gamma}}}(b\tilde{\gamma}) e^{-b^2\tilde{\gamma}^2/2}, \qquad b = \left(\frac{c\langle\beta^2\rangle}{2}\right)^{1/4}, \tag{13}$$

with normalization constant

$$N_{n_{\tilde{\gamma}}} = \sqrt{\frac{b}{\sqrt{\pi}2^{n_{\tilde{\gamma}}} n_{\tilde{\gamma}}!}}. \tag{14}$$

The total energy in the case of the Z(5) model is then

$$E(s, n_w, L, n_{\tilde{\gamma}}) = E_0 + A(x_{s,\nu})^2 + Bn_{\tilde{\gamma}}. \tag{15}$$

4. B(E2) transition rates

The quadrupole operator is given by

$$T_\mu^{(E2)} = t\beta\left[D_{\mu,0}^{(2)}(\theta_i)\cos\left(\gamma - \frac{2\pi}{3}\right)\right.$$

$$\left. + \frac{1}{\sqrt{2}}(D_{\mu,2}^{(2)}(\theta_i) + D_{\mu,-2}^{(2)}(\theta_i))\sin\left(\gamma - \frac{2\pi}{3}\right)\right], \tag{16}$$

where t is a scale factor, while in the Wigner functions the quantum number α appears next to μ, and the quantity $\gamma - 2\pi/3$ in the trigonometric functions is obtained from $\gamma - 2\pi k/3$ for $k = 1$, since in the present case the projection α along the body-fixed $\hat{x}'$-axis is used. For $\gamma \simeq \pi/6$ the first term vanishes. B(E2) transition rates are calculated in the usual way [13].

The symmetrized wave function reads [11]

$$\Psi(\beta,\gamma,\theta_i) = \xi_{s,\alpha,L}(\beta)\eta_{n_{\tilde\gamma}}(\tilde\gamma)\sqrt{\frac{2L+1}{16\pi^2(1+\delta_{\alpha,0})}}(D^{(L)}_{\mu,\alpha} + (-1)^L D^{(L)}_{\mu,-\alpha}). \quad (17)$$

α has to be an even integer [11], while for $\alpha = 0$ it is clear that only even values of L are allowed, since the symmetrized wave function is vanishing otherwise.

In the calculation of the relevant matrix elements the integral over $\tilde\gamma$ leads to unity [because of the normalization of $\eta(\tilde\gamma)$], the integral over β takes the form

$$I_\beta(s_i,L_i,\alpha_i,s_f,L_f,\alpha_f) = \int \beta\xi_{s_i,\alpha_i,L_i}(\beta)\xi_{s_f,\alpha_f,L_f}(\beta)\beta^4 d\beta, \quad (18)$$

while the integral over the angles is calculated using the standard integrals involving three Wigner functions. The final result reads

$$B(E2; L_i\alpha_i \to L_f\alpha_f) = \frac{5}{16\pi}\frac{t^2}{2}\frac{I^2_\beta(s_i,L_i,\alpha_i,s_f,L_f,\alpha_f)}{(1+\delta_{\alpha_i,0})(1+\delta_{\alpha_f,0})}$$

$$\left[(L_i 2 L_f|\alpha_i 2\alpha_f) + (L_i 2 L_f|\alpha_i - 2\alpha_f) + (-1)^{L_i}(L_i 2 L_f| - \alpha_i 2\alpha_f)\right]^2. \quad (19)$$

One can easily see that the Clebsch–Gordan coefficients (CGCs) appearing in this equation impose a $\Delta\alpha = \pm 2$ selection rule. Therefore the quadrupole moments vanish, since in the relevant matrix elements of the quadrupole operator one should have $\alpha_i = \alpha_f$.

5. Numerical results

The lowest bands of the Z(5) model are given in Table 1. The notation L_{s,n_w} is used. All levels are measured from the ground state, $0_{1,0}$, and are normalized to the first excited state, $2_{1,0}$. The ground state band is characterized by $s = 1$, $n_w = 0$, while the even and the odd levels of the γ_1-band are characterized by $s = 1$, $n_w = 2$, and $s = 1$, $n_w = 1$ respectively. The β_1-band is characterized by $s = 2$, $n_w = 0$. All these bands are characterized by $n_{\tilde\gamma} = 0$, and, as seen from Eq. (15), are parameter free. The fact that the γ_1-band is characterized by $n_{\tilde\gamma} = 0$ is not surprising, since this is in general the case in the framework of the rotation-vibration model.

B(E2) transition rates, normalized to the one between the two lowest states, B(E2;$2_{1,0} \to 0_{1,0}$), are given in Table 2.

Table 1. Energy levels of the Z(5) model (with $n_{\bar{\gamma}} = 0$), measured from the $L_{s,n_w} = 0_{1,0}$ ground state and normalized to the $2_{1,0}$ lowest excited state. See Section 5 for further details.

s, n_w	1,0	1,2	2,0		1,1
L				L	
0	0.000		3.913		
2	1.000	1.837	5.697	3	2.597
4	2.350	4.420	7.962	5	4.634
6	3.984	7.063	10.567	7	6.869
8	5.877	9.864	13.469	9	9.318
10	8.019	12.852	16.646	11	11.989
12	10.403	16.043	20.088	13	14.882
14	13.024	19.443	23.788	15	18.000
16	15.878	23.056	27.740	17	21.341

Table 2. B(E2) transition rates of the Z(5) model, normalized to the transition between the two lowest states, B(E2;$2_{1,0} \to 0_{1,0}$). See Sections 4 and 5 for further details.

$L^{(i)}_{s,n_w}$	$L^{(f)}_{s,n_w}$	Z(5)	$L^{(i)}_{s,n_w}$	$L^{(f)}_{s,n_w}$	Z(5)	$L^{(i)}_{s,n_w}$	$L^{(f)}_{s,n_w}$	Z(5)
$2_{1,0}$	$0_{1,0}$	1.000						
$4_{1,0}$	$2_{1,0}$	1.590	$4_{1,2}$	$2_{1,2}$	0.736	$5_{1,1}$	$3_{1,1}$	1.235
$6_{1,0}$	$4_{1,0}$	2.203	$6_{1,2}$	$4_{1,2}$	1.031	$7_{1,1}$	$5_{1,1}$	1.851
$8_{1,0}$	$6_{1,0}$	2.635	$8_{1,2}$	$6_{1,2}$	1.590	$9_{1,1}$	$7_{1,1}$	2.308
$10_{1,0}$	$8_{1,0}$	2.967	$10_{1,2}$	$8_{1,2}$	2.035	$11_{1,1}$	$9_{1,1}$	2.665
$12_{1,0}$	$10_{1,0}$	3.234	$12_{1,2}$	$10_{1,2}$	2.394	$13_{1,1}$	$11_{1,1}$	2.952
$14_{1,0}$	$12_{1,0}$	3.455	$14_{1,2}$	$12_{1,2}$	2.690	$15_{1,1}$	$13_{1,1}$	3.190
$16_{1,0}$	$14_{1,0}$	3.642	$16_{1,2}$	$14_{1,2}$	2.938	$17_{1,1}$	$15_{1,1}$	3.392
$2_{1,2}$	$2_{1,0}$	1.620	$3_{1,1}$	$4_{1,0}$	1.243	$3_{1,1}$	$2_{1,2}$	2.171
$4_{1,2}$	$4_{1,0}$	0.348	$5_{1,1}$	$6_{1,0}$	0.972	$5_{1,1}$	$4_{1,2}$	1.313
$6_{1,2}$	$6_{1,0}$	0.198	$7_{1,1}$	$8_{1,0}$	0.808	$7_{1,1}$	$6_{1,2}$	1.260
$8_{1,2}$	$8_{1,0}$	0.129	$9_{1,1}$	$10_{1,0}$	0.696	$9_{1,1}$	$8_{1,2}$	1.164
$10_{1,2}$	$10_{1,0}$	0.092	$11_{1,1}$	$12_{1,0}$	0.614	$11_{1,1}$	$10_{1,2}$	1.069
$12_{1,2}$	$12_{1,0}$	0.069	$13_{1,1}$	$14_{1,0}$	0.551	$13_{1,1}$	$12_{1,2}$	0.984
$14_{1,2}$	$14_{1,0}$	0.054	$15_{1,1}$	$16_{1,0}$	0.507	$15_{1,1}$	$14_{1,2}$	0.910
$16_{1,2}$	$16_{1,0}$	0.043	$17_{1,1}$	$18_{1,0}$	0.459	$17_{1,1}$	$16_{1,2}$	0.846

Table 3. Comparison of the Z(5) predictions for energy levels (left part) and B(E2) transition rates (right part) to experimental data for ^{192}Pt [15], ^{194}Pt [14], and ^{196}Pt [16]. See Section 6 for further discussion.

L	Z(5)	^{192}Pt	^{194}Pt	^{196}Pt	$L^{(i)} \to L^{(f)}$	Z(5)	^{192}Pt	^{194}Pt	^{196}Pt
$4_{1,0}$	2.350	2.479	2.470	2.465	$4_{1,0} \to 2_{1,0}$	1.590	1.559	1.724	1.476
$6_{1,0}$	3.984	4.314	4.299	4.290					
$8_{1,0}$	5.877	6.377	6.392	6.333	$4_{1,2} \to 2_{1,2}$	0.736		0.446	0.715
$10_{1,0}$	8.019	8.624		8.558	$6_{1,2} \to 4_{1,2}$	1.031			1.208
$2_{1,2}$	1.837	1.935	1.894	1.936	$3_{1,1} \to 2_{1,2}$	2.171	1.786		
$4_{1,2}$	4.420	3.795	3.743	3.636					
$6_{1,2}$	7.063	5.905	5.863	5.644	$2_{1,2} \to 0_{1,0}$	0.000	0.009	0.006	0.0004
					$2_{1,2} \to 2_{1,0}$	1.620	1.909	1.805	
$3_{1,1}$	2.597	2.910	2.809	2.854	$4_{1,2} \to 2_{1,0}$	0.000		0.004	0.014
$5_{1,1}$	4.634	4.682	4.563	4.526	$4_{1,2} \to 4_{1,0}$	0.348		0.406	
$7_{1,1}$	6.869	6.677			$6_{1,2} \to 4_{1,0}$	0.000			0.012
$0_{2,0}$	3.913	3.776	3.858	3.944					

6. Comparison to experiment

Several energy levels and B(E2) transition rates predicted by the Z(5) model are compared in Table 3 to the corresponding experimental quantities of ^{194}Pt [14], which has been suggested [8] to lie very close to the prolate to oblate critical point. Its neighbours, ^{192}Pt [15] and ^{196}Pt [16], which demonstrate quite similar behaviour, are also shown. Not only the levels of the ground state band are well reproduced (below the backbending), but in addition the bandheads of the γ_1-band and the β_1-band are very well reproduced, without involving any free parameter. The staggering of the theoretical levels within the γ_1-band is quite stronger than the one seen experimentally, as it is expected [17] for models related to the triaxial rotator [11,12].

The main features of the B(E2) transition rates are also well reproduced. As far as the transitions from the γ_1-band to the ground state band are concerned, the transitions $L_{1,2} \to L_{1,0}$ are strong, while the transitions $(L+2)_{1,2} \to L_{1,0}$, which are forbidden in the Z(5) framework, are weaker by two or three orders of magnitude.

7. Discussion

In addition to the points made in the introduction, the following comments apply:

1) The β-equation [Eq. (2)] obtained after approximately separating variables in the Bohr Hamiltonian is also exactly soluble [18,19] when plugging in it the Davidson potentials [20] $u(\beta) = \beta^2 + \beta_0^4/\beta^2$, where β_0 is the minimum of the potential. In analogy to earlier work in the E(5) and X(5) frameworks [21], it is expected that $\beta_0 = 0$ should correspond to a triaxial vibrator, while $\beta_0 \to \infty$ should lead to a triaxial rotator [11].

2) Using the variational procedure developed recently in the E(5) and X(5) frameworks [21], one should be able to prove that the Z(5) model can be obtained from the Davidson potentials by maximizing the rate of change of various measures of collectivity [22] with respect to the parameter β_0, thus proving that Z(5) is also the critical point symmetry of the transition from a triaxial vibrator to a triaxial rotator.

Work in these directions is in progress.

References

1. F. Iachello, *Phys. Rev. Lett.* **85**, 3580 (2000).
2. F. Iachello, *Phys. Rev. Lett.* **87**, 052502 (2001).
3. F. Iachello, *Phys. Rev. Lett.* **91**, 132502 (2003).
4. R. F. Casten and N. V. Zamfir, *Phys. Rev. Lett.* **85**, 3584 (2000).
5. N. V. Zamfir *et al.*, *Phys. Rev.* **C65**, 044325 (2002).
6. R. F. Casten and N. V. Zamfir, *Phys. Rev. Lett.* **87**, 052503 (2001).
7. R. Krücken *et al.*, *Phys. Rev. Lett.* **88**, 232501 (2002).
8. J. Jolie and A. Linnemann, *Phys. Rev.* **C68**, 031301 (2003).
9. A. Bohr, *Mat. Fys. Medd. K. Dan. Vidensk. Selsk.* **26**, no. 14 (1952).
10. J. Gizon *et al.*, *J. Phys. G: Nucl. Phys.* **4**, L171 (1978).
11. J. Meyer-ter-Vehn, *Nucl. Phys.* **A249**, 111 (1975).
12. A. S. Davydov and G. F. Filippov, *Nucl. Phys.* **8**, 237 (1958).
13. D. Bonatsos, D. Lenis, D. Petrellis, and P. A. Terziev, *Phys. Lett.* **B588**, 172 (2004).
14. E. Browne and B. Singh, *Nucl. Data Sheets* **79**, 277 (1996).
15. C. M. Baglin, *Nucl. Data Sheets* **84**, 717 (1998).
16. C. Zhou, G. Wang, and Z. Tao, *Nucl. Data Sheets* **83**, 145 (1998).
17. N. V. Zamfir and R. F. Casten, *Phys. Lett.* **B260**, 265 (1991).
18. J. P. Elliott, J. A. Evans, and P. Park, *Phys. Lett.* **B169**, 309 (1986).
19. D. J. Rowe and C. Bahri, *J. Phys. A: Math. Gen.* **31**, 4947 (1998).
20. P. M. Davidson, *Proc. R. Soc. (London)* **135**, 459 (1932).
21. D. Bonatsos, D. Lenis, N. Minkov, D. Petrellis, P. P. Raychev, and P. A. Terziev, *Phys. Lett.* **B584**, 40 (2004).
22. V. Werner *et al.*, *Phys. Lett.* **B527**, 55 (2002).

KEY TOPICS IN NUCLEAR STRUCTURE
Proceedings of the 8th Int. Spring Seminar on Nuclear Physics
edited by Aldo Covello
© 2005 World Scientific Publishing Co.

SUPERSYMMETRY AND THE SPECTRUM OF ^{196}Au: A CASE STUDY

G. GRAW, R. HERTENBERGER, H.-F. WIRTH
Ludwig-Maximilians-Universität München, Department für Physik, Am Coulombwall 1, D-85748 Garching, Germany
E-mail: gerhard.graw@lmu.de

J. JOLIE
Institut für Kernphysik, Universität zu Köln, D-50937 Köln, Germany

J. BAREA, R. BIJKER, A. FRANK
Instituto de Ciencias Nucleares, Universidad Nacional Autonoma de Mexico, Apartado Postal 70-543, 04510 Mexico, D.F., Mexico

To provide information for comparison with predictions for a dynamical supersymmetry, the odd-odd nucleus ^{196}Au was studied via transfer reactions. With a polarized deuteron beam we measured ($\vec{d}$,t) and ($\vec{d}$,α), with unpolarized beams (p,d), (^{3}He,d), and (α,d) transfer reactions. In this way we obtain for 20 out of the 26 observed states with negative parity below 500 keV safe J^π assignments. The number of states and the safe or restricted assignments are in agreement with the predictions from the dynamical $U_\nu(6/12) \otimes U_\pi(6/4)$ supersymmetric scheme. Comparison with model predictions for spectroscopic factors will provide a further critical test to what extent this symmetry is realized in nature.

1. Introduction

The excitation energy spectra of heavy nuclei, in general, are very complex. An aim is to find "simplicity in complexity", that means to identify a few relevant degrees of freedom. Group theoretical symmetry considerations, as introduced by Arima and Iachello (1975) [1], provide a kind of order and thus a means to discuss physics. They start from the observation that valence neutrons (and protons) are especially strongly bound if they form pairs with total angular momentum $J = 0$ or $J = 2$.

1.1. *Even-even nuclei and IBA*

In their model for even-even nuclei, the Interacting Boson Approximation (IBA), these pairs, the s and d bosons, are considered as the only relevant degrees of freedom. These bosons have, because of the respective $2J+1$ magnetic substates, $\nu = 1+5 = 6$ allowed eigenstates, hence their interaction is treated within the group formalism of $U(6)$ symmetry.

The Hamiltonian, in general, is the sum of a number of terms of lower symmetry than $U(6)$. Allowed symmetries are $U(6), O(6), U(5), O(5), SU(3), O(3), O(2)$. Each term is the product of a constant giving the energy scale and a Casimir operator representing the respective group properties. Interesting are those cases where some of these terms do not appear and the remaining ones form a chain of subsequently broken symmetry (The $U(6)$ symmetry is broken in a regular way). Then we have analytical solutions both for the eigenenergies and the eigenstates, determined by a scheme of quantum numbers. The energy scale factors remain as the only values to be determined.

1.2. *Odd-even nuclei and supersymmetry*

For the description of odd-A nuclei a fermion needs to be coupled to the N boson system. Because of the interaction the fermion will create excitations of the core. This can be done within a semi-microscopical approach which relies on seniority in the nuclear shell model [3]. An alternative to this interacting boson-fermion approach is the construction of Hamiltonians exhibiting dynamical Bose-Fermi symmetries that are analytically solvable. In both approaches the boson-fermion space is spanned by the irreducible representation (irrep) $[N] \times [1]$ of $U^B(6) \otimes U^F(M)$, where M is the dimension of the single-particle space. A significant step towards unification was made in the early eighties when Iachello and coworkers embedded the Bose-Fermi symmetry into a graded Lie algebra $U(6/M)$ [4,5].

One successful case is U(6/12) in which the fermion can occupy the orbits with j = 1/2, 3/2 and 5/2. Considering those as arising from the coupling of a pseudo spin part with $s' = 1/2$ with a pseudo orbital part with $l' = 0$ and 2, the following reduction is obtained: $U^F(12) \supset U^F(6) \otimes U^F(2)$ which allows the coupling of the pseudo orbital part with the bosonic generators at the $U(6)$ level [6]. Another one is $U(6/4)$ which uses the isomorphism between the $U^F(4)$ group describing the space for a 3/2 fermion, and the bosonic $O(6)$ group [5].

1.3. *Odd-odd Nuclei and the extended supersymmetry*

A step further is the extended supersymmetry [2, 9] which deals with boson-fermion and neutron-proton degrees of freedom, allowing the description of a quartet of nuclei, using the same algebraic form of the Hamiltonian. The quartet consists of an even-even nucleus with $(N_\nu + N_\pi)$ bosons, an odd-proton and an odd-neutron nucleus with $(N_\nu + N_\pi) - 1$ bosons and an odd-odd nucleus with $(N_\nu + N_\pi) - 2$ bosons and a proton and neutron. Thus supersymmetry, if it works, relates the often very complex structure of the odd-odd nucleus to the much simpler even-even and odd-A systems.

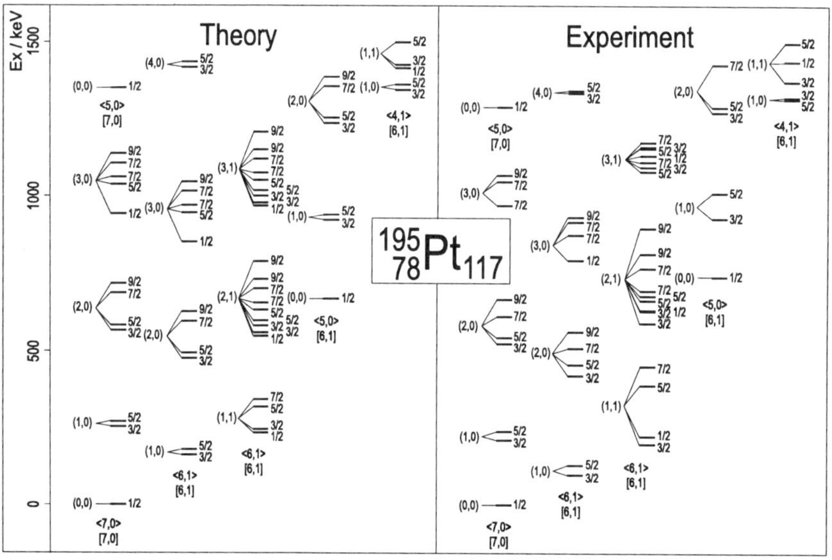

Figure 1. Supersymmetric level schemes of ^{195}Pt, as obtained by Metz et al. . The prediction of supersymmetry and respective quantum numbers (left side) is compared with experimental data (right side). For each band the SUSY quantum numbers $[N_1, N_2]$ $< \Sigma_1, \Sigma_2 >$, (τ_1, τ_2) are given.

In addition to the analytic expressions for the excitation energies the supersymmetric scheme also provides analytic results for the wave functions. These do not depend on the parameters given above and can be tested via the calculation of electromagnetic transition rates and transfer reaction amplitudes. Especially the transfer experiments shall provide a very stringent test of the existence of supersymmetry via the distribution of the transferred nucleons in the predicted wave functions.

This model, which aims to exhaust all aspects of symmetry, of course has to be considered as a minimalistic concept. A number of interactions and degrees of freedom which are expected in a fully microscopic description, are neglected. Thus the model provides a kind of benchmark one may compare with, and it is very interesting to study to what extent it is realized in nature.

With respect to the truncation of the fermion space to $j = 1/2$, $3/2$ and $5/2$ valence neutrons and $j = 3/2$ protons (or vice versa), demanded by the model, it was realized from the beginning that the ultimate candidate for the test is the low lying negative parity spectrum of the odd-odd nucleus ^{196}Au [2] as part of the quartet ^{194,195}Pt (due to the neutron $p_{1/2}$, $p_{3/2}$ or $f_{5/2}$ orbitals), and ^{195,196}Au, formed by the additional $d_{3/2}$ proton orbital. The spectrum of ^{195}Pt, compare Fig. 1, is known as the best example of the U(6/12) supersymmetry [8,11].

2. Transfer reaction studies

When we had been asked by Jan Jolie to study ^{196}Au in transfer reactions [9] no safely assigned negative-parity states in ^{196}Au except for the 2^- ground state were known. The difficulty was to relate the transitions to a level scheme for their in-beam gamma-ray and conversion electron spectroscopy experiments at the cyclotrons of the PSI and the University of Bonn and, more recently, their $\gamma\gamma$ correlation studies at the Yale accelerator [10].

Because of their excellent energy resolution (4 keV FWHM), the (p,d) transfer reactions at the Q3D magnetic spectrograph of the Munich tandem accelerator was used to provide the energy calibration of the ^{196}Au spectra, using ^{195}Pt data measured in addition to establish a correlation between measured channels and excitation energies. The achieved uncertainties of the excitation energies are less than 1 keV. These spectra establish a new and almost complete level scheme of ^{196}Au, 47 states were resolved for the first time in the energy range of 0 to 1350 keV including the resolved ground state doublet with an energy spacing of approximately 6 keV, as shown in Fig. 2. These excitation energies allowed later to set the observed γ transitions [10].

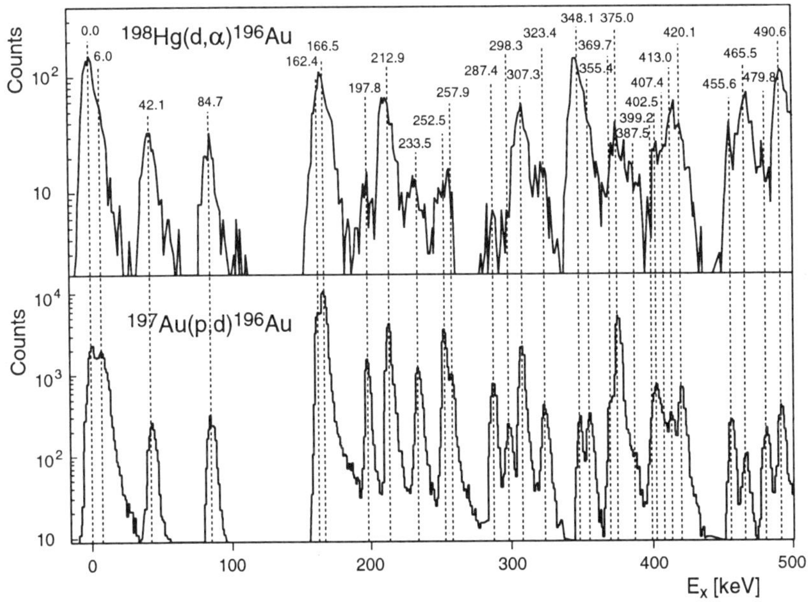

Figure 2. Part of the $^{198}Hg(\vec{d},\alpha)^{196}Au$ and the $^{197}Au(p,d)^{196}Au$ spectra at an angle of 25° measured with 18 MeV vector polarized deuterons on a 37 $\mu g/cm^2$ ^{198}Hg target and with 26 MeV protons on a 67 $\mu g/cm^2$ ^{197}Au target, respectively. Shown is the excitation energy range between 0 and 500 keV.

Below 500 keV eight new states were detected, thus we identified 26 states of negative parity in this range, in agreement with the supersymmetric model prediction, resulting from the reproduction of the $A = 194$ and 195 spectra. The next step to support the model further, is the determination of the J values. In this respect we studied three different reactions, ($\vec{d}$,t) and ($\vec{d}$,α) with polarized deuteron beam and (^{3}He,d) with unpolarized beam.

Quantum numbers and spectroscopic factors were obtained from angular distributions of single-neutron transfer in $^{197}Au(\vec{d},t)^{196}Au$, single-proton

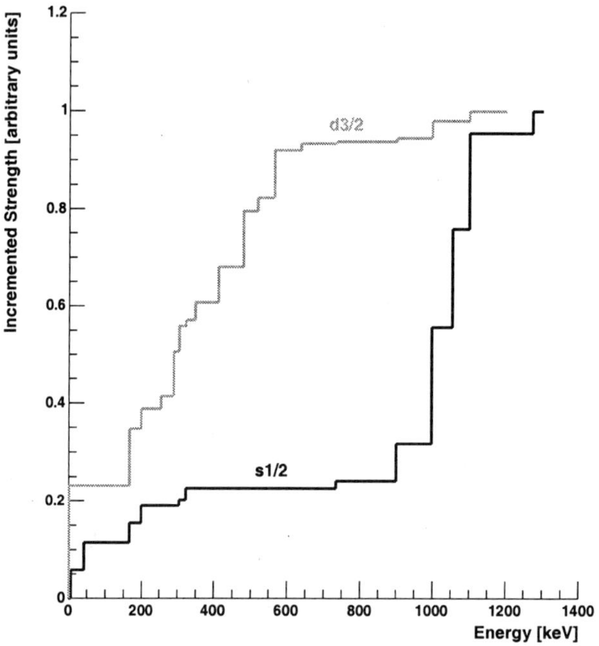

Figure 3. Incremental plot of the observed $s_{1/2}$ and $d_{3/2}$ strengths in (^{3}He,d).

transfer in ^{195}Pt(^{3}He,d)^{196}Au, and two-nucleon transfer ^{198}Hg($\vec{\rm d},\alpha$)^{196}Au. Because of the $J^\pi = 0^+$ target in the case of the ($\vec{\rm d},\alpha$) reaction identified transfer quantum numbers provide unique information about the J^π values of the excited states in ^{196}Au. This is different for the ($\vec{\rm d}$,t) and (^{3}He,d) reactions where, because of the $J^\pi = 3/2^+$ and $1/2^-$ targets, the identification of transferred lj values may provide restrictions only on the J^π values in ^{196}Au.

In (^{3}He,d) to ^{196}Au, compare Fig. 3, the $d_{3/2}$ proton strength spreads rather homogeniously over the 0 to 600 keV excitation range, in agreement with the SUSY model. Most of the $s_{1/2}$ strength is located near 1000 keV, the observed much weaker $s_{1/2}$ satellite near 200 keV results from coupling of about 25 percent of the $s_{1/2}$ strength to the $(d_{3/2}\cdot{}^{194}\mathrm{Pt}(2_1^+))_{1/2}$ state in ^{195}Au. The energy of this satellite is low also because of additional residual interactions as the quadrupole-quadrupole interaction and level repulsion.

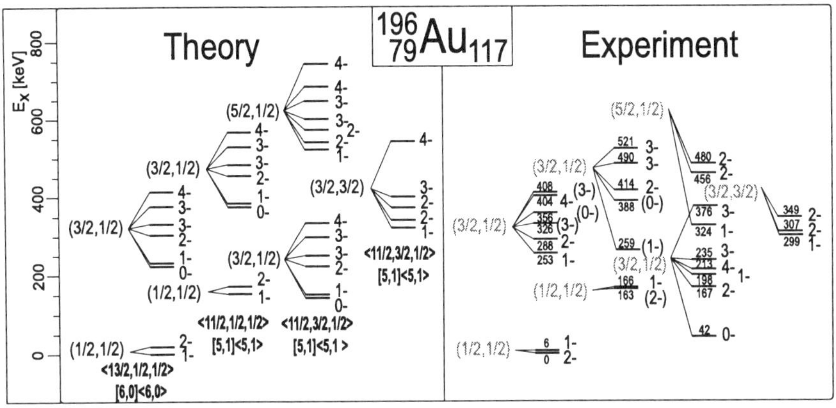

Figure 4. Level scheme of low lying negative parity states in ^{196}Au: The prediction of extended supersymmetry and respective quantum numbers (left side) is compared with experimental data (right side). Experimental J^π values are given in parentheses, if the data are consistent with this assignment but a deviating value cannot be excluded. SUSY quantum numbers $< \sigma_1, \sigma_2, \sigma_3 >$, $[N_1, N_2] < \Sigma_1, \Sigma_2 >$ are indicated below , and (τ_1, τ_2) at the left of each band. Only states with J^π lower than 5^- are shown).

Due to its nature of a comparatively weak admixture in this low energy range this coupling to the $s_{1/2}$ strength - which is outside the SUSY model - cannot cause additional "intruder" states. It will cause modifications only of the wave functions and the excitation energies. For more details, compare ref. [12].

Including also the information from γ and conversion electron spectroscopy we finally obtain for 20 out of the 26 observed states with negative parity below 500 keV safe J^π assignments. Their number and the safe or restricted assignments are in agreement with the predictions from the dynamical $U_\nu(6/12) \otimes U_\pi(6/4)$ supersymmetric scheme.

3. Spectroscopic factors and conclusions

The supersymmetric dynamical symmetry scheme for ^{196}Au derives as a prediction from a fit to the spectra of the related even-even and even-odd nuclei ^{194}Pt, ^{195}Pt and ^{195}Au. With respect to the energy range and the J^π values we observe full agreement with respect to the 20 safe and 6 tentatively assigned states. The excitation energies of the individual states

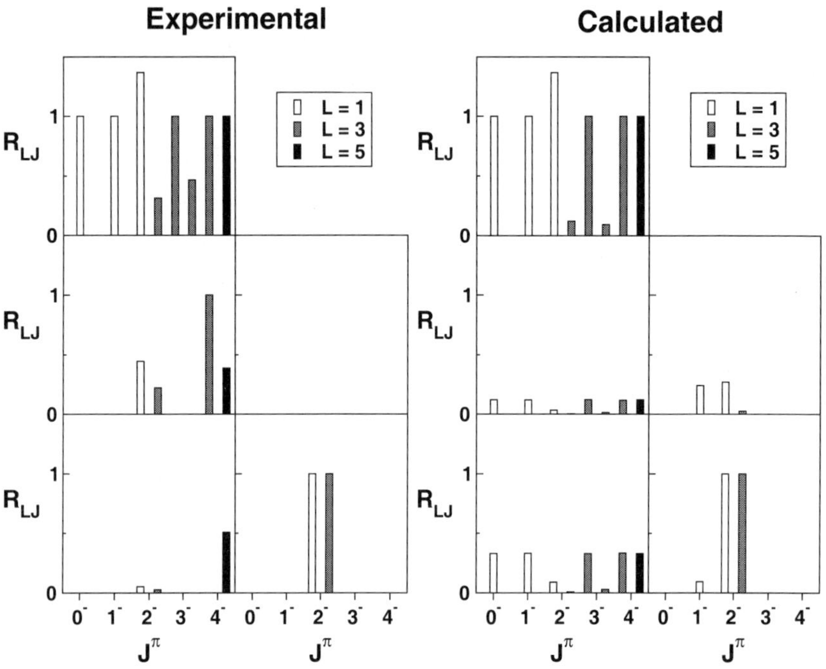

Figure 5. Comparison of experimental and calculated (d,α) spectroscopic factors for the lowest five SUSY multiplets. The normalizations had been chosen to reproduce the strongest transitions.

differ to a minor extent, for a model as schematic as SUSY this has to be expected.

A remaining test of the SUSY scheme, especially of the relation of the experimental states to SUSY bands with specific quantum numbers, is the comparison of the experimental spectroscopic factors - and also of gamma transition probabilities - with model predictions.

A first result of SUSY calculations for (d,α) spectroscopic factors by J.Barea, R.Bijker, A.Frank [13], is shown in Fig. 5. The lowest five SUSY multiplets are compared, (note, Fig. 4 shows the seven lowest SUSY multiplets) and normalize the strength of individual transfers according to the strongest transitions with respective quantum numbers. From this display it is obvious, that expected strong transitions are observed as strong ones, and expected weak transitions as weak ones, with few exceptions only.

This result is very encouraging and it will be very interesting to see respective SUSY model calculations also for the other transfer channels. This will clarify to what extent this symmetry as a scheme of order in an otherwise very complex situation is realized in nature.

On the other hand we urgently need calculations which aim to describe ^{196}Au and the neighbouring nuclei within conventionel nuclear structure models, using either geometrical collective models or the IBA for the bosonic degrees of freedom and explicit coupling of the one or two fermions. A final aim is to understand how a phenomenon as supersymmetry results as a special case of a more general description.

Acknowledgments

Our work was supported by funds of the Munich Maier Leibnitz Laboratory and the Deutsche Forschungsgemeinschaft (grant IIC4 Gr 894/2-3 and Jo391/2-1). Enlighting discussions with F. Iachello are acknowledged.

References

1. F. Iachello, A. Arima, *The Interacting Boson Model* (Cambridge University Press, Cambridge, 1987).
2. P. Van Isacker, J. Jolie, K. Heyde, A. Frank, Phys. Rev. Lett. **54**, 653 (1985).
3. F. Iachello, O. Scholten, Phys. Rev. Lett. **43**, 679 (1979).
4. F. Iachello, Phys. Rev. Lett. **44**, 772 (1980).
5. A.B. Balantekin, I. Bars, F. Iachello, Nucl. Phys. **A370**, 284 (1981).
6. P. Van Isacker, A. Frank, H.Z. Sun, Ann. of Physics **157**, 183 (1984).
7. J. Jolie, P.E. Garrett, Nucl. Phys. **A596**, 234 (1996).
8. A. Mauthofer *et al.*, Phys. Rev. **C34**, 1958 (1986).
9. A. Metz, J. Jolie, G. Graw, C. Günther, J. Gröger, R. Hertenberger, N. Warr, Y. Eisermann, Phys. Rev. Lett. **83**, 1542 (1999).
10. J. Gröger, J. Jolie, R. Krücken, C.W. Beausang, M. Caprio, R.F. Casten, J. Cederkall, J.R. Cooper, F.Corminboef, L. Gennilloud, G. Graw, C. Günther, M. de Huu, A.I. Levon, A. Metz, J.R. Novak, N. Warr, T. Wendell, Phys.Rev. **C62**, 064304 (2000).
11. A. Metz, Y. Eisermann, A. Gollwitzer, R. Hertenberger, B.D. Valnion, G. Graw, J. Jolie, Phys. Rev. **C61**, 064313 - 1 to 11 (2000). Erratum Phys.Rev. **C67**, 049901 (2003).
12. H.-F. Wirth, G. Graw, S. Christen, Y. Eisermann, A. Gollwitzer, R. Hertenberger, J. Jolie, A. Metz, O. Möller, D. Tonev, B.D. Valnion, Phys.Rev. **C70**, 014610 (2004).
13. J. Barea, R. Bijker, A. Frank, to be published (2004).

KEY TOPICS IN NUCLEAR STRUCTURE
Proceedings of the 8th Int. Spring Seminar on Nuclear Physics
edited by Aldo Covello
© 2005 World Scientific Publishing Co.

BOSONIZATION AND IBM *

FABRIZIO PALUMBO
INFN – Laboratori Nazionali di Frascati,
P. O. Box 13, I-00044 Frascati, ITALIA
E-mail: fabrizio.palumbo@lnf.infn.it

We derive a boson Hamiltonian from a Nuclear Hamiltonian whose potential is expanded in pairing multipoles and determine the fermion-boson mapping of operators. We use a new method of bosonization based on the evaluation of the partition function restricted to the bosonic composites of interest. By rewriting the partition function so obtained in functional form we get the euclidean action of the composite bosons from which we can derive the Hamiltonian. Such a procedure respects all the fermion symmetries.

1. Introduction

The IBM of Arima and Iachello [1], is most successful in describing the low energy nuclear excitations. The bosons of this model are understood as virtual pairs of nucleons, analogous to the Cooper pairs of superconductivity [2]. But no general procedure to reformulate the nuclear theory in terms of the effective bosonic degrees of freedom has been found.

The first attempt in this direction has been performed, as far as we know, by Beliaev and Zelevinsky [3]. But this work makes use of the Bogoliubov transformation which violates nucleon number conservation. Moreover the bosonization is achieved only within a perturbation scheme.

The first work which relates the IBM to a nucleon Hamiltonian is due to Otsuka, Arima and Iachello [4]. These authors got exact results for the pairing interaction in a single j-shell. Their result was somewhat generalized [5], but a full solution of the problem has not yet been achieved.

There are several recipes for bosonization [6], mostly based on the idea of mapping a fermion model space into a boson space. This requires a truncation of the nucleon space whose effect is in general not easy to control.

*This work is supported EEC under the contract HPRN-CT-2000-00131.

In order to avoid the limitations of previous works we try a different approach where we do not assume any property of the composites, other than their dominance at low energy. In particular their structure will be determined only at the end of the calculation. The problem of truncation of the nucleon space will then be traded by the problem of decoupling some bosons from the others, but in a setting where one can hopefully have a better control.

To implement Boson Dominance we perform a functional evaluation of the partition function restricted to boson composites. In this way we get the euclidean action of these composites and their coupling to external fields in closed form. All the fermion symmetries, in particular fermion number conservation, are respected. The bosonization is therefore achieved in the path integral formalism, and all physical quantities can be evaluated by standard methods. The first step, necessary also in the derivation of the Hamiltonian, is to find the minimum of the action at constant fields. Depending on the solution, one has spherical or deformed nuclei. In the latter case rotational excitations appear as Goldstone modes associated to the spontaneous breaking of rotational symmetry. The notion of spontaneous symmetry breaking survives in fact with a precise definition also in finite systems [7]. We want to emphasize that the closed form of the action opens the way to numerical simulations of fermionic systems in terms of bosonic variables, avoiding the "sign problem".

To compare with the IBM we can either write the path integral of the latter, or derive the Hamiltonian corresponding to our action. We will make here the second choice. But to derive the Hamiltonian we must perform an expansion in the inverse of the shell degeneracy.

Bosonization appears in several many-fermion systems and relativistic field theories. The effective bosons fall into two categories, depending on their fermion number. The Cooper pairs of the BCS model of superconductivity, of the IBM of Nuclear Physics, of the Hubbard model of high T_c superconductivity [8] and of color superconductivity in QCD have fermion number 2. Similar composite bosons with fermion number zero appear as phonons, spin waves and chiral mesons in QCD. They can be included in the present formalism by replacing in the composites one fermion operator by an antifermion (hole) one. Indeed the approach we are going to present can be applied, as far as we can see without any conceptual difficulty, in all the above cases, as it has been argued in a brief report of the method [9].

A different approach to bosonization which also avoids any mapping is based on the Hubbard-Stratonovich transformation. The latter renders quadratic the fermionic interaction by introducing bosonic auxiliary fields which in the end become the physical fields. The typical resulting structure is that of chiral theories [10]. In such an approach an energy scale emerges naturally, and only excitations of lower energy can be described by the auxiliary fields [7]. At present the relation with the present approach has not been fully clarified.

The paper is organized in the following way. In Sec. II we define coherent states of composites and their properties. In Sec. III we derive the path integral for the composites and find the effective bosonic action. We restrict ourselves for simplicity to a nuclear interaction given as a sum of pairing multipoles, but more general forces can easily be included and will be discussed in future works. The effective action we derive is, apart from the above limitation, general. In Sec. IV we restrict ourselves to a single j-shell with pairing multipoles and in Sec. V we determine the corresponding Hamiltonian. In Sec. VI we report our conclusions.

2. Coherent states of composites

Composites of fermion number 2 are defined in terms of the fermion creation operators $\hat{c}^\dagger$

$$\hat{b}_J^\dagger = \frac{1}{2}\hat{c}^\dagger B_J^\dagger \hat{c}^\dagger = \frac{1}{2}\sum_{m_1,m_2} \hat{c}_{m_1}^\dagger \left(B_J^\dagger\right)_{m_1,m_2} \hat{c}_{m_2}^\dagger. \tag{1}$$

In the above equation the m's represent all the fermion intrinsic quantum numbers and position coordinates and J the corresponding labels of the composites. Composites of fermion number zero can be obtained by replacing one of the fermion operators by an antifermion one. The structure matrices B_J have dimension 2Ω independent of J. Their form is determined by the fermion interaction as explained in the sequel, but we assume that they will satisfy the relations

$$\mathrm{tr}(B_J^\dagger\, B_K) = 2\,\delta_{J,K}. \tag{2}$$

We also assume them to be nonsingular. Then their dimension is twice the index of nilpotency of the composites, which is the largest integer ν such that $\left(\hat{b}_J\right)^\nu \neq 0$. It is obvious that a necessary condition for a composite to resemble a boson, is that its index of nilpotency be large. But this condition in general is not sufficient, and we must require also that

$$\det(\Omega B^\dagger B)^n \sim 1. \tag{3}$$

A convenient way to get the euclidean path integral from the trace of the transfer matrix is to use coherent states[11]. If we are interested in states with $n = \overline{n} + \nu$ bosons for an arbitrary reference number $\overline{n}$ we introduce the operator

$$\mathcal{P}_{\overline{n}} = \frac{(\Omega - \overline{n})^2}{\Omega^2} \int db^* db \langle b|b\rangle^{-1} |b\rangle\langle b| \tag{4}$$

constructed in terms of coherent states of composites

$$|b\rangle = |\exp\left(\sum_J b_J \hat{b}_J^\dagger\right)\rangle. \tag{5}$$

We would like it to be the identity in the fermion subspace of the composites. Let us see its action on composite operators. Let us first consider the case where there is only one composite with structure function satisfying the equation

$$B^\dagger B = \frac{1}{\Omega} \mathbb{1}. \tag{6}$$

In order to evaluate the matrix element $\langle b_t | b_{t-1}\rangle$ we introduce between the bra and the ket the identity in the fermion Fock space

$$\mathcal{I} = \int dc^* dc \langle c|c\rangle^{-1} |\exp(-c^* \hat{c})\rangle\langle \exp(-c\, \hat{c}^\dagger)| \tag{7}$$

where the c^*, c are Grassmann variables. We thus find

$$\langle b_t | b_{t-1}\rangle = \int dc^* dc\, E(c^*, c, b_t^*, b_{t-1}) = \left(1 + \frac{1}{\Omega} b_t^* b_{t-1}\right)^\Omega \tag{8}$$

where

$$E(c^*, c, b^*, b) = \exp\left(-c^* c + \frac{1}{2} b^*\, c\, B\, c + \frac{1}{2} b\, c^*\, B^\dagger c^*\right). \tag{9}$$

Therefore the action of $\mathcal{P}_{\overline{n}}$ on the composites

$$|\mathcal{P}_{\overline{n}} (\hat{b}^\dagger)^n\rangle = \left(1 - \frac{\nu}{\Omega - \overline{n}}\right)^{-1} \left(1 - \frac{\nu + 1}{\Omega - \overline{n}}\right)^{-1} |(\hat{b}^\dagger)^n\rangle \tag{10}$$

shows that it behaves like the identity in the neighborhood of the reference state up to an error of order $\nu/(\Omega - \overline{n})$, namely the measure $\langle b|b\rangle^{-1}$ is essentially uniform with respect to any reference state.

It is worth while noticing that in the limit of infinite Ω we recover exactly the expressions valid for elementary bosons, in particular

$$\langle b_t | b_{t-1}\rangle = \left(1 + \frac{1}{\Omega} b_t^* b_{t-1}\right)^\Omega \to \exp(b_t^* b_{t-1}), \;\; \Omega \to \infty. \tag{11}$$

In the general case of many composites the above equations become

$$\langle b_t | b_{t-1} \rangle = [\det (\mathbb{1} + \beta_t^* \beta_{t-1})]^{\frac{1}{2}}, \tag{12}$$

where $\beta_t^* = \sum_J (b_J)_t^* B_J$. Then using the condition 3 we find again that $\mathcal{P}$ approximates the identity with an error of order $1/\Omega$

$$\mathcal{P}|(\hat{b}_{I_0}^\dagger)^{n_0} ... \hat{b}_{I_i}^\dagger)^{n_i}\rangle = |\left((\hat{b}_{I_0}^\dagger)^{n_0} ... \hat{b}_{I_i}^\dagger)^{n_i} + O(1/\Omega)\right)\rangle. \tag{13}$$

Identifying the operator $\mathcal{P}$ with the identity in the subspace of the composites is the only approximation we will make in the derivation of the effective boson action.

3. Composites path integral

Now we are equipped to realize our program. The first step is the evaluation of the partition function Z_c restricted to fermionic composites. To this end we divide the inverse temperature in N_0 intervals of spacing τ

$$\tau = \frac{1}{N_0 T} \tag{14}$$

and write

$$Z_c = \mathrm{tr}\,(\mathcal{P} \exp(-H\tau))^{N_0}. \tag{15}$$

We will restrict ourselves to a Hamiltonian with interactions which can be written as a sum of pairing multipoles

$$\hat{H} = \hat{c}^\dagger h_0\, \hat{c} - \sum_K g_K \frac{1}{2}\, \hat{c}^\dagger F_K^\dagger \hat{c}^\dagger\, \frac{1}{2}\, \hat{c}\, F_K\, \hat{c}. \tag{16}$$

The single particle term includes the single particle energy with matrix e, any single particle interaction with external fields described by the matrix $\mathcal{M}$ and the chemical potential μ

$$h_0 = e + \mathcal{M} - \mu. \tag{17}$$

Therefore we will be able to solve the problem of fermion-boson mapping by determining the interaction of the composite bosons with external fields. We assume for the potential form factors the normalization

$$\mathrm{tr}(F_K^\dagger F_K) = 2\,\Omega. \tag{18}$$

For the following manipulations we need the Hamiltonian in antinormal form

$$\hat{H} = H_0 - \hat{c}\, h^T \hat{c}^\dagger - \sum_K g_K \frac{1}{2}\, \hat{c} F_K \hat{c}\, \frac{1}{2}\, \hat{c}^\dagger F_K^\dagger \hat{c}^\dagger \tag{19}$$

where the upper script T means "transposed" and

$$h = h_0 - \sum_K g_K F_K^\dagger F_K, \quad H_0 = \frac{1}{2}\mathrm{tr}(h + h_0). \tag{20}$$

Now we must evaluate the matrix element $\langle b_t | \exp(-\tau \hat{H}) | b_{t-1} \rangle$. To this end we expand to first order in τ (which does not give any error in the final $\tau \to 0$ limit) and insert the operator $\mathcal{P}$ between annihilation and creation operators

$$\begin{aligned} \langle b_t | \exp(-\tau \hat{H}) | b_{t-1} \rangle &= \exp(-H_0 \tau) \langle b_1 | \mathcal{P} - \hat{c}\, h^T \tau\, \mathcal{P} \hat{c}^\dagger \\ &\times \sum_k g_k \tau \frac{1}{2} \hat{c} F_K \hat{c} \mathcal{P} \frac{1}{2} \hat{c}^\dagger F_K^\dagger \hat{c}^\dagger | b \rangle. \end{aligned} \tag{21}$$

Using the identity in the fermion Fock space we find

$$\begin{aligned} \langle b_t | \exp(-\tau \hat{H}) | b_{t-1} \rangle^{-1} &= \int dc^* dc\, E(c^*, c, b_t^*, b_{t-1}) \\ &\times \exp(-H_0 \tau - c^* h\, \tau c) \exp\left(\sum_K g_K \tau \frac{1}{2} c\, F_K\, c \frac{1}{2} c^* F_K^\dagger c^* \right) \end{aligned} \tag{22}$$

where the function $E(c^*, c, b^*, b)$ is defined in (9). By means of the Hubbard-Stratonovich transformation we can make the exponents quadratic in the Grassmann variables and evaluate the Berezin integral

$$\begin{aligned} \langle b_t | \exp(-\tau \hat{H}) | b_{t-1} \rangle &= \det R \, \exp(-H_0 \tau) \int \prod_K da_K^* da_K \\ &\times \exp(-a^* \cdot a) \exp\left\{ \frac{1}{2} \mathrm{tr} \ln \left[\mathbb{1} + \left(\beta_t^* + \sum_{K_1} \sqrt{g_{K_1} \tau}\, a_{K_1}^* F_{K_1} \right) \right.\right. \\ &\left.\left. \times R^{-1} \left(\beta_{t-1} + \sum_{K_2} \sqrt{g_{K_2} \tau}\, a_{K_2} (F_{K_2})^\dagger \right) (R^T)^{-1} \right] \right\}, \end{aligned} \tag{23}$$

where $R = \mathbb{1} + h\,\tau$. Setting $\Gamma_t = \left(\mathbb{1} + \beta_t^* \beta_{t-1}\right)^{-1}$ and performing the integral over the auxiliary fields $a_K{}^*, a_K$ we get the final expression of the euclidean action

$$\begin{aligned} S(b^*, b) &= \tau \sum_t \left\{ H_0 - \mathrm{tr}\, h + \frac{1}{2\tau} \mathrm{tr}[\ln(\mathbb{1} + \beta_t^* \beta_t) + \ln \Gamma_t] \right. \\ &- \frac{1}{4} \sum_K g_K \left[\mathrm{tr}(\Gamma_t \beta_t^* F_K^\dagger)\, \mathrm{tr}(\Gamma_t F_K \beta_{t-1}) + 2\, \mathrm{tr}\left(\Gamma_t F_K^\dagger F_K \right) \right. \\ &\left.\left. - \mathrm{tr}[\Gamma_t \beta_t^* F_K^\dagger, \Gamma_t F_K \beta_{t-1}]_+ \right] + \frac{1}{2} \mathrm{tr}\left[\Gamma_t \beta_t^* \left(\beta_{t-1}\, h^T + h\, \beta_{t-1} \right) \right] \right\} \end{aligned} \tag{24}$$

where $[..,..]_+$ is an anticommutator. This action differs from that of elementary bosons because

i) the time derivative terms are non canonical. Indeed expanding the logarithms we get

$$\mathrm{tr}\,[\ln(\mathbb{1} + \beta_t^* \beta_t) + \ln \Gamma_t\,] = \frac{1}{2}\left\{\beta_t^* \nabla_t \beta_t - \frac{1}{2}[(\beta_t^*)^2 \nabla_t (\beta_t^*)^2] + ...\right\}, \quad (25)$$

where $\nabla_t f = \frac{1}{\tau}(f_{t+1} - f_t)$. The first term is the canonical one, while the others contain the derivative of powers of the boson variables. The canonical form of the first term is due to the normalization of Eq.(2) of the structure functions, otherwise β_t and β_{t-1} would not have the same coefficient. Note the difference of the noncanonical terms with respect to the chiral expansions, where there are powers of derivatives, rather than derivatives of powers.

ii) the coupling of the chemical potential (which appears in h) is also noncanonical. Indeed expanding Γ_t we get $\mu\,\mathrm{tr}\,(\beta_t^* \beta_{t-1} - \beta_t^* \beta_t \beta_t^* \beta_{t-1} + ...)$, and only the first term is canonical

iii) the function Γ becomes singular when the number of bosons is of order Ω, as it will become clear in the sequel. This reflects the Pauli principle.

We remind the reader that the only approximation done concerns the operator $\mathcal{P}$. Therefore these are to be regarded as true features of compositeness.

The bosonization of the system we considered has thus been accomplished. In particular the fermionic interactions with external fields can be expressed in terms of the bosonic terms which involve the matrix $\mathcal{M}$ (appearing in h). The dynamical problem of the interacting (composite) bosons can be solved within the path integral formalism. This includes the new interesting possibility of a numerical simulation of the partition function which could now be performed with bosonic variables avoiding the sign problem.

Part of the dynamical problem is the determination of the structure matrices B_J. This can be done by expressing the energies in terms of them and applying a variational principle which gives rise to an eigenvalue equation.

4. The action in a single j-shell

In this paper we restrict ourselves to a system of nucleons of in a single j-shell. Then we identify the quantum number K with the boson angular momentum, $K = (I_K, M_K)$, so that the form factors of the potential are proportional to Clebsh-Gordan coefficients

$$(F_{IM})_{m_1,m_2} = \sqrt{2\Omega}\, C^{IM}_{jm_1 jm_2}, \quad \Omega = j + \frac{1}{2}. \tag{26}$$

In such a case the structure matrices are completely determined by the angular momentum of the composites and the normalization conditions (2) $B_J = \Omega^{-\frac{1}{2}} F_J$. The points i) and ii) following Eq. (24) are the only difficulties in the derivation of the Hamiltonian which could be otherwise read from the action. We can overcome them by performing an expansion in inverse powers of Ω. We will retain only the first order corrections, which are of order Ω^0, with the exception of the coupling with external fields where they are of order Ω^{-1}. In this approximation the first difficulty is overcome because noncanonical time derivatives are of order $1/\Omega$ and the second one because the only noncanonical coupling of the chemical potential of order Ω^0 comes from the only term of the chemical potential of order Ω, which can be shown to be $\mu \sim -\frac{1}{2}\, g_0\, \Omega$, independent of the number of bosons.

The resulting action is

$$S(b^*, b) = \sum_t \left\{ \sum_{K_1 K_2} b^*_{K_1} \left[(\nabla_t - 2\mu) + \omega \right]_{K_1 K_2} b_{K_2} + \sum_{I_1 I_2 I_3 I_4} \sum_{IM} W^I_{I_1 I_2 I_3 I_4} \left(b^*_{I_1}\, b^*_{I_2} \right)_{IM} \left(b_{I_3}\, b_{I_4} \right)_{IM} \right\} \tag{27}$$

where all the b^*'s and all the b's are at time $t, t-1$ respectively, and

$$\begin{aligned} \omega_{K_1 K_2} &= \frac{1}{\Omega} \mathrm{tr} \left(F_{K_1} F^\dagger_{K_2}\, e \right) - g_{I_1} \Omega\, \delta_{K_1 K_2} \\ W^I_{I_1 I_2 I_3 I_4} &= \left(-2g_0 + \sum_{i=1}^{4} g_{I_i} \right) \Pi^4_{i=1} [(2I_i + 1)]^{1/2} \Omega \begin{Bmatrix} j & j & I_1 \\ j & j & I_2 \\ I_3 & I_4 & I \end{Bmatrix} \\ (b_{I_3}\, b_{I_4})_{IM} &= \sum_{M_3, M_4} C^{I,M}_{I_3, M_3, I_4, M_4}\, b_{I_3 M_3} b_{I_4 M_4}. \end{aligned} \tag{28}$$

Notice the factor 2 in front of the chemical potential due to the fact that the composites have fermion number 2.

5. The Hamiltonian

The Hamiltonian is obtained[11] by omitting the time derivative and chemical potential terms, and replacing the variables b^*, b by corresponding creation-annihilation operators $\hat{a}^\dagger, \hat{a}$, satisfying canonical commutation relations

$$\hat{H} = \sum_{I_1 M_1 I_2 M_2} \omega_{I_1 M_1 I_2 M_2} \hat{a}^\dagger_{I_1 M_1} \hat{a}_{I_2 M_2} + \sum_{I_1 I_2 I_3 I_4} \sum_{IM} \left\{ W^I_{I_1 I_2 I_3 I_4} \left(\hat{a}^\dagger_{I_1} \hat{a}^\dagger_{I_2} \right)_{IM} (\hat{a}_{I_3} \hat{a}_{I_4})_{IM} \right\}. \tag{29}$$

It is easy to check that, due to the symmetries of the 9j symbols, it is hermitian.

From the interaction with external fields we get the fermion-boson mapping of other operators

$$\hat{c}^\dagger \mathcal{M} \hat{c} \rightarrow \sum_{I_1 M_1 I_2 M_2} \frac{2}{\Omega} \mathrm{tr} \left(F_{I_1 M_1} \mathcal{M} F^\dagger_{I_2 M_2} \right) \hat{a}^\dagger_{I_1 M_1} \hat{a}_{I_2 M_2} + \sum_{\mathrm{all}\, I, M} \left(\frac{2}{\Omega} \right)^2 \mathrm{tr} \left(F_{I_1 M_1} \mathcal{M} F^\dagger_{I_4 M_4} F_{I_2 M_2} F^\dagger_{I_3 M_3} \right) \times \hat{a}^\dagger_{I_1 M_1} \hat{a}^\dagger_{I_2 M_2} \hat{a}_{I_3 M_3} \hat{a}_{I_4 M_4}. \tag{30}$$

We remind the reader that the above Hamiltonian has been derived under the condition $n << \Omega$ in a single subshell. Therefore if we further assume $e_{m_1 m_2} = \bar{e}\, \delta_{m_1 m_2}$, the single boson energy matrix is diagonal $\omega_{I_1 I_2} = (2\,\bar{e} - g_{I_1} \Omega) \delta_{I_1 I_2}$. But the bosonic interactions couple all the bosons with angular momenta for which the $9j$ symbols do not vanish, even if the corresponding potentials do vanish.

6. Summary

We have developed a general approach to the problem of bosonization where we introduce fermionic composites without any preliminary mapping of the fermion model space into a bosonic one. Restricting the trace in the partition function of the system to the composites we get the euclidean action of the effective bosons in closed form. The only approximation made concerns the identity operator in the space of the composites.

It is perhaps worth while to spend a few words about the nature of this approximation. Indeed it might appear that two are the approximations involved. The first one is the restriction of the partition function to composites. This is the fundamental physical assumption of Boson Dominance. Then we replace the identity in the composite subspace by the operator $\mathcal{P}$, which seems a further approximation. But $\mathcal{P}$ differs appreciably from the identity only for states with many bosons, states which cannot resemble elementary bosons because of the Pauli principle. We therefore deem that the two approximations are essentially one and the same.

The nuclear dynamics can be studied by the methods of path integrals, including numerical simulations which now are not affected by the sign problem.

To derive the Hamiltonian of the IBM we must make recourse to an expansion in the inverse of the index of nilpotency of the composites. In the present work we restricted ourselves to a nucleon model space of a single j-shell and to a number of bosons much smaller than the index of nilpotency. Both limitations can easily be removed. Concerning the second one, some care must be exercized to respect particle number conservation, as done for instance in [7]. The first one requires a parametrization of the structure functions according to

$$(B_{J,M})_{m_1,m_2} = \sum_{j_1 j_2} p_{J j_1 j_2} C^{JM}_{j_1 m_1 j_2 m_2}. \tag{31}$$

Now the energies of the bosons are functions of the parameters p. A variational principle applied to these energies generates an equation for these parameters. The solution to this equation can in general be found only numerically, but the Hamiltonian and the other operators retain their analytic expressions.

References

1. F.Iachello and A.Arima, *The Interacting Boson Model*, Cambridge University Press, Cambridge, 1987.
2. Y. Nambu and M. Mukherjee, *Phys. Lett.* **B209** 1 (1988). ; M. Mukherjee and Y. Nambu, *Ann. Phys.* **191** 143 (1991).
3. S.T.Beliaev and V.G.Zelevinsky, *Nucl.Phys.* **39** 582 (1962).
4. T.Otsuka, A.Arima and F.Iachello, *Nucl.Phys.* **A309** 1 (1978).
5. T.Otsuka, A.Arima, *Phys. Lett.* **77B** 1 (1978).
6. A.Klein and E.R.Marshalek, *Rev. Mod. Phys.* **63** 375 (1991).
7. M.B.Barbaro, A.Molinari, F.Palumbo and M.R.Quaglia, *nucl-th/0304028.*
8. M.Cini and G.Stefanucci, *cond-mat/0204311 v1.*
9. F.Palumbo, *nucl-th/0405045* .
10. V.A.Miransky, *Dynamical symmetry breaking in quantum field theories*, World Scientific, 1993.
11. J. W. Negele and H. Orland, *Quantum Many-Particle Systems*, Addison-Wesley Publishing Company, 1988.

KEY TOPICS IN NUCLEAR STRUCTURE
Proceedings of the 8th Int. Spring Seminar on Nuclear Physics
edited by Aldo Covello
© 2005 World Scientific Publishing Co.

STUDY OF β DECAY IN THE As-Ge ISOTOPES IN THE INTERACTING BOSON-FERMION MODEL

N. YOSHIDA

Faculty of Informatics, Kansai University, Takatsuki 569-1095, Japan
E-mail: yoshida@res.kutc.kansai-u.ac.jp

L. ZUFFI

Dipartimento di Fisica dell'Università di Milano and
Istituto Nazionale di Fisica Nucleare, Sezione de Milano,
Via Celoria 16, Milano 20133, Italy
E-mail: zuffi@mi.infn.it

S. BRANT

Department of Physics, Faculty of Science, University of Zagreb,
10000 Zagreb, Croatia
E-mail: brant@sirius.phy.hr

The β decay from As to Ge isotopes is studied together with the energy levels and the electromagnetic properties in the interacting boson-fermion model (IBFM). The effects of the higher-order terms in the particle transfer operators that form the β-decay operators will be discussed.

1. Introduction

The interacting boson-fermion model has been successful in describing the structures of the nuclei with collective motion.[1] The model has been applied to the analyses of various nuclear properties like energy levels, electromagnetic properties, transfer reactions, as well as β decays.

The study of β decay is important in the view points of (i) testing the nuclear models by analyzing the existing data, because the β-decay rates are very sensitive to the wave functions, and (ii) providing reliable information for astronomical research. The application of the IBFM to β decay was initiated by Dellagiacoma *et al.* for $52 \leq Z \leq 58$.[2] and has been extended to various regions.[3] In fact we have already published the results in the Rh-Pd region[4] and the Xe-Cs region.[5] We noticed that once the wave functions are determined in IBFM calculations of the energy levels, the β-decay ratios are obtained in a parameter free calculation.

In the present report, we will show the preliminary results of the study of the β decay from the odd-mass As to the Ge isotopes, in addition to the energy levels and the electromagnetic properties of the negative-parity states in the As and the Ge isotopes with $A = 69$, 71 and 73. This is the first analysis of β-decay in the Ge-As region in the IBFM. In addition to the calculations with the Fermi and the Gamow-Teller operators conventionally taken, we have tried also the β-decay operators with the additional terms proposed by Barea *et al.*[6] which take account of the fermion states with the seniority v higher than 2.

2. Calculations and results

2.1. *Energy levels*

The isotopes: ^{69,71,73}As and ^{69,71}Ge are described with the even-even core of ^{68,70,72}Ge and an odd nucleon while ^{73}Ge is described with the core of ^{74}Ge with a hole neutron. The hamiltonian is written as

$$H = H^{\mathrm{B}} + H^{\mathrm{F}} + V^{\mathrm{BF}} \tag{1}$$

where H^{B} is the hamiltonian of the proton-neutron interacting boson model (IBM2).[7] The interaction parameters are taken from Ref. 8. The hamiltonian of the odd fermion consists of the energies of the single-particle states:

$$H^{\mathrm{F}} = \sum_i \epsilon_i n_i. \tag{2}$$

The BCS calculation is performed in the orbitals $f_{7/2}$, $p_{3/2}$, $f_{5/2}$, $p_{1/2}$, $g_{9/2}$ and $d_{5/2}$ with the condition on the gap: $\Delta = 12/\sqrt{A}$. The single-particle energies similar to those in Ref. 9 are taken, and then adjusted for better description of the levels. The quasi-particle energies ϵ_j of the negative-parity orbitals are used in Eq. (2). The interaction V^{BF} between the bosons and the odd particle consists of the quadrupole, the monopole and the exchange interactions:

$$\begin{aligned} V^{\mathrm{BF}} = & \sum_{i,j} \Gamma_{ij} \left([a_i^\dagger \tilde{a}_j]^{(2)} \cdot Q_{\rho'}^{\mathrm{B}} \right) + \sum_{i,j} \Gamma'_{ij} \left([a_i^\dagger \tilde{a}_j]^{(2)} \cdot Q_{\rho}^{\mathrm{B}} \right) \\ & + A \sum n_i n_{d_{\rho'}} \\ & + \sum_{i,j} \Lambda_{ki}^{j} \Big\{ : \left[[d_\rho^\dagger \tilde{a}_j]^{(k)} a_i^\dagger s_\rho \right]^{(2)} : \cdot \left[s_{\rho'}^\dagger \tilde{d}_{\rho'} \right]^{(2)} \\ & \qquad + \text{Hermitian conjugate} \Big\} \end{aligned} \tag{3}$$

where we include the quadrupole interaction between like particles too. The modified annihilation operator $\tilde{d}$ is defined by $\tilde{d}_m = (-1)^m d_{-m}$. The symbols ρ and ρ' denote π (ν) and ν (π) if the odd fermion is a proton (a neutron). The creation operator of the odd particle is written as $a^{\dagger}_{jm}$, while the modified annihilation operator is defined as $\tilde{a}_{jm} = (-1)^{j-m} a_{j-m}$. The orbital dependence of the quadrupole and the exchange interactions are:[1]

$$\Gamma_{i,j} = (u_i u_j - v_i v_j)\, Q_{i,j}\, \Gamma \tag{4}$$

$$\Gamma'_{i,j} = (u_i u_j - v_i v_j)\, Q_{i,j}\, \Gamma' \tag{5}$$

$$\Lambda^{j}_{k,i} = -\beta_{k,i}\beta_{j,k} \left(\frac{10}{N_\rho (2j_k + 1)} \right)^{1/2} \Lambda \tag{6}$$

where $\beta_{i,j} = (u_i v_j + v_i u_j) Q_{i,j}$ and $Q_{i,j} =< j_i || Y^{(2)} || j_j >$. The parameters Γ, Γ', A and Λ are varied so as to produce good agreement with the experimental energy levels.

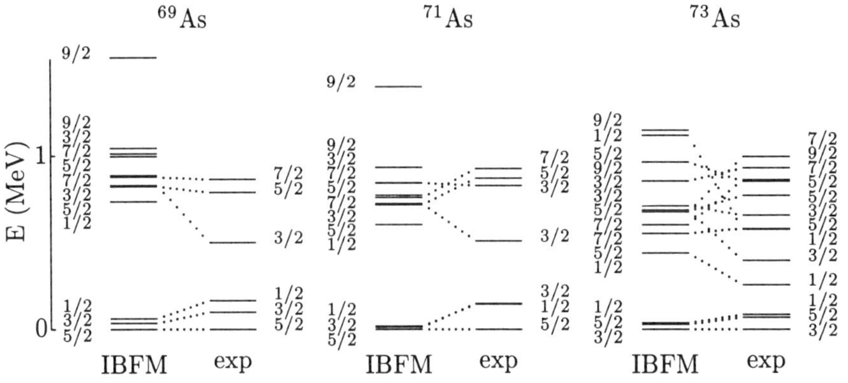

Figure 1. The Energy levels the negative-parity states in the As isotopes.

The calculated energy levels are shown in Figs. 1 and 2. The experimental data are taken from Refs. 10, 11, 12. Reasonable agreement is obtained.

2.2. *Electromagnetic properties*

The electric quadrupole transition operator is

$$T^{(\mathrm{E2})} = e^{\mathrm{B}}\, Q^{\mathrm{B}}_{\pi} + e^{\mathrm{B}}\, Q^{\mathrm{B}}_{\nu} + \sum_{i,j} e'_{i,j}\, [a^{\dagger}_i \tilde{a}_j]^{(2)} \tag{7}$$

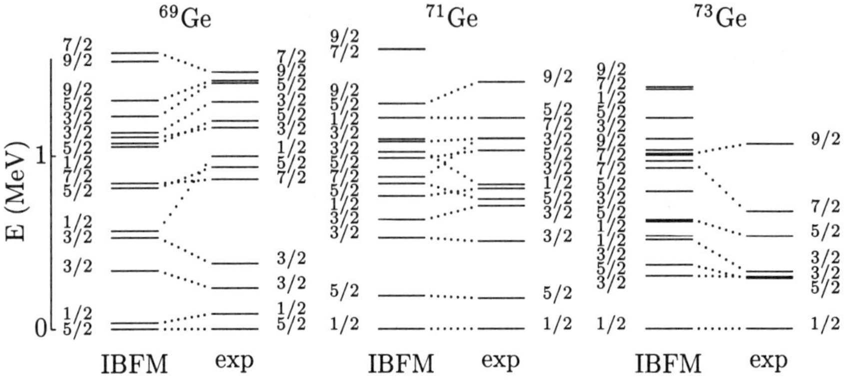

Figure 2. The Energy levels of the negative-parity states in the Ge isotopes.

where $e'_{i,j} = -e^{\rm F}_{\rho}(u_i u_j - v_i v_j) < i||r^2 Y^{(2)}||j > /\sqrt{5}$. For the magnetic dipole transitions we take,

$$T^{(\rm M1)} = \sqrt{\frac{3}{4\pi}} \left(g^{\rm B}_{\pi} L^{\rm B}_{\pi} + g^{\rm B}_{\nu} L^{\rm B}_{\nu} + \sum_{i,j} e^{(1)}_{i,j} [a^{\dagger}_i \tilde{a}_j]^{(1)} \right) \quad (8)$$

where $e^{(1)}_{i,j} = -(u_i u_j + v_i v_j) < i||g_l \mathbf{l} + g_s \mathbf{s}||j > /\sqrt{3}$. We use the boson effective charge fixed to $e^{\rm B} = 0.0631$ eb, which is determined so that the observed $B(\rm E2; 0^+_1 \to 2^+_1)$ in ^{70}Ge is reproduced. For the odd proton in As $e^{\rm F}_{\pi} = 1.5$ e, while for the odd neutron in Ge $e^{\rm F}_{\nu} = 0.5$ e. For the magnetic dipole operator, the boson g-factors for all the isotopes are: $g^{\rm B}_{\nu} = 0$, $g^{\rm B}_{\pi} = 1\ \mu_{\rm N}$. For the odd particle (proton or neutron), the spin part of the spin g-factor is reduced by the factor of 0.7.

The calculated electric quadrupole moments and the magnetic dipole moments of the lowest $5/2^-$ states are compared with the experimental values in Fig. 3. The agreement between the experiment and the calculation is generally good when the experimental values are present. For the quadrupole moment of $5/2^-_1$ in ^{73}As, only the magnitude is known experimentally. If we assume negative sign, as shown in the figure, then the calculated value agrees very well with the experiment. For ^{71}As, the calculated quadrupole moment is much larger in magnitude than the observed moment. Apart from this case, the calculated electric and magnetic moments of the ground states are consistent with the existing experimental data. We have also calculated the B(E2) values, the B(M1) values, and the branching ratios, and have obtained reasonable agreement with the experimental data. More details will be published elsewhere.

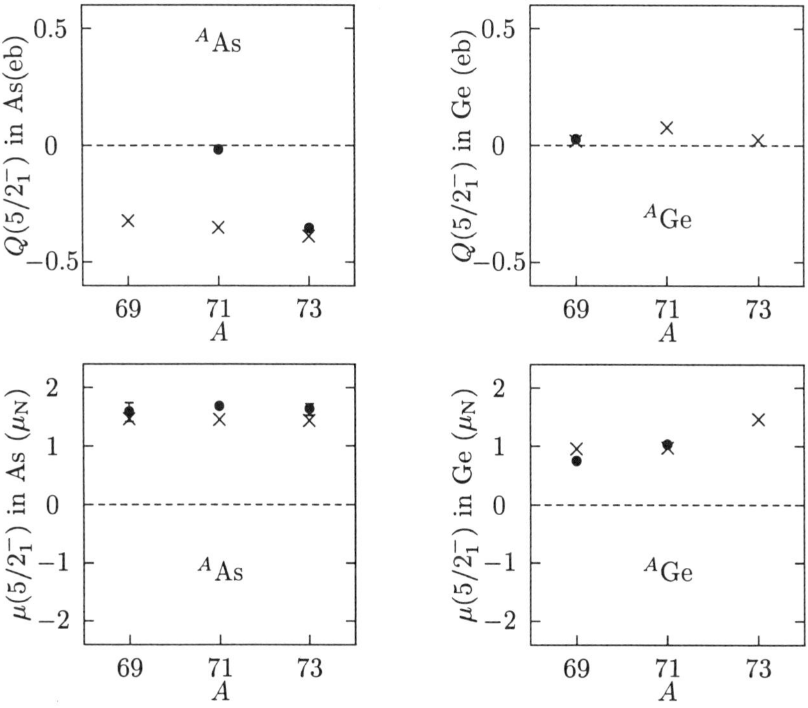

Figure 3. The calculated electric quadrupole moments and the magnetic dipole moments are compared with the experimental data. The symbol • shows the experimental values while × shows the results of the calculations.

2.3. *β-decay*

The β-decay operators, i.e., the Fermi and the Gamow-Teller, are written as

$$O^{\mathrm{F}} = \sum_{j} -\sqrt{2j+1}\left[P_{\nu}^{\dagger(j)}\tilde{P}_{\pi}^{(j)}\right]^{(0)}, \tag{9}$$

$$O^{\mathrm{GT}} = \sum_{j'j} \eta_{j'j}\left[P_{\nu}^{\dagger(j')}\tilde{P}_{\pi}^{(j)}\right]^{(1)} \tag{10}$$

where $\eta_{j'j} = - < j'||\sigma||j > \sqrt{3}$. The IBFM images of the particle transfer operators are

$$P_{\nu}^{\dagger(j)} = A_{\nu}^{\dagger(j)}, \qquad \tilde{P}_{\pi}^{(j)} = \tilde{A}_{\pi}^{(j)}, \tag{11}$$

where

$$A_{m}^{\dagger(j)} = \frac{u_j}{K'_j} a_{jm}^{\dagger} - \sum_{j'} \frac{v_j \beta_{j'j}}{K'_j K} \left(\frac{10}{N(2j+1)} \right)^{1/2} s^{\dagger} [\tilde{d} a_{j'}^{\dagger}]_{m}^{(j)}$$
$$+ \frac{1}{K'_j} \sum_{j'J} \phi_{jj'}^{J} [[a_{j'}^{\dagger} d^{\dagger}]^{(J)} \tilde{d}]_{m}^{(j)}. \tag{12}$$

The last summation, i.e., the d-boson number conserving terms with $\phi_{jj'}^{J}$ were introduced by Barea *et al.*,[6] where the coefficients $\phi_{jj'}^{J}$ are determined from the matrix elements between the states with seniority $v = 2$ and the states with $v = 3$. The normalization factors K and K'_j are determined by

$$K = \left(\sum_{j'j} \beta_{j'j}^{2} \right)^{1/2} \tag{13}$$

and

$$\sum_{\alpha J} \left\langle \mathrm{odd}; \alpha J || A^{\dagger(j)} || \mathrm{even}; 0_1^{+} \right\rangle^{2} = (2j+1) u_j^{2}. \tag{14}$$

Although the normalization factor K'_j was not adopted in Ref. 6, we use this factor in order to be consistent with the other works in β-decay. The transition between ^{73}As and ^{73}Ge requires another treatment because the creation of a neutron involves the annihilation of a neutron boson. For this case, we need to use $\tilde{B}_{\nu}^{(j)} = \theta_j a_j^{\dagger} s + \ldots$ for the neutron instead of $A_{\nu}^{\dagger(j)}$. This cases was not included in the present report.

Figure 4 shows both the results without the d-boson number conserving terms ($\phi_{jj'}^{J} = 0$) and the results with the d-boson number conserving terms ($\phi_{jj'}^{J} \neq 0$). The additional terms show some improvement in the agreement with the experimental data. But, the effect is small.

3. Discussion

In our previous works in Rh-Pd[4] and Cs-Xe[5], we obtained log-ft values comparable to the experimental ones without any normalization factor. In the present work, however, the calculated log-ft values are smaller than the experimental values in most cases. Namely, the actual β-decay rates are

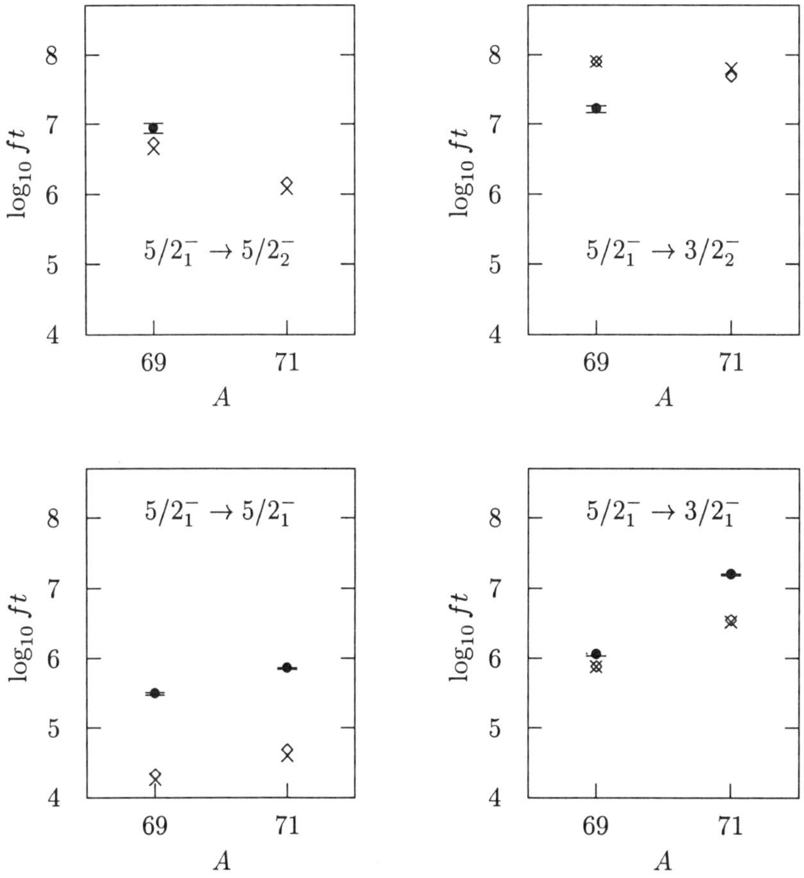

Figure 4. The $\log_{10} ft$ values of the β-decay from As to Ge isotopes. The symbol • shows the experimental values with their errors, while the symbol × shows the results of calculation the conventional operators ($\phi^J_{jj'} = 0$). The symbol ⋄ shows the results of calculation with the additional d-boson number conserving terms ($\phi^J_{jj'} \neq 0$).

appreciably lower than the calculated ones. This may be attributed to some mixture of the components of the wave functions that are not considered of the calculation. Actually, we did not include the intruder 0^+ states that are observed in the even-even core.

The additional d-boson number conserving terms produce only a little improvement. The overall normalization factor K'_j may be reducing the effects of these terms. In fact, this factor was introduced in phenomenological application in order to compensate for the effect of truncation the higher-order terms.[13]

4. Conclusions

We have analyzed the energy levels, the electromagnetic properties and β-decay in ^{69,71,73}As and ^{69,71,73}Ge. The energy levels and the electromagnetic moments reasonably agree with the experimental data.

For the β-decay, some quenching factor will be needed to explain the magnitudes of the experimental decay rates. The influence of the intruder states will be one origin of the quenching and will be a subject of a further study. It will be interesting to study in what situations the higher-order terms in the particle transfer operator play an important role. The extension to other regions, and the consistent description with even-even and odd-odd nuclei will be also a future problem.

Acknowledgment

This work was partly supported by Gakujutsu-Frontier project, Kansai University.

References

1. F. Iachello and P. Van Isacker, *The Interacting Boson-Fermion Model*, (Cambridge University Press, Cambridge, 1991).
2. F. Dellagiacoma, Ph. D. thesis, Yale University, 1988; F. Dellagiacoma and F. Iachello, *Phys. Lett.* B **218**, 299 (1989).
3. G. Maino and L. Zuffi, in *Proceedings of the 5th International Spring Seminar on Nuclear Physics, Ravello, 1995*, edited by A. Covello (World Scientific, Singapore, 1996), p. 611.
4. N. Yoshida, L. Zuffi, and S. Brant, *Phys. Rev.* **C66**, 014306 (2002).
5. L. Zuffi, S. Brant, and N. Yoshida, *Phys. Rev.* **C68**, 034308 (2003).
6. J. Barea, C. E. Alonso and J. M. Arias, *Phys. Rev.* **C65**, 034328 (2002).
7. F. Iachello and A. Arima, *The Interacting Boson Model*, (Cambridge University Press, Cambridge, 1987).
8. N. Yoshida and A. Arima, *Phys. Lett.* **B165**, 231 (1985)
9. K. Langanke, D. J. Dean and W. Nazarewicz, *Nucl. Phys.* **A728**, 109 (2003).
10. M. R. Bhat and J. K. Tuli, *Nucl. Data Sheets* **90**, 269 (2000).
11. M. R. Bhat, *Nucl. Data Sheets* **68**, 579 (1993).
12. Balraj Singh, *Nucl. Data Sheets* **101**, 193 (2004).
13. O. Scholten, *Prog. Part. Nucl. Phys.* **14**, 189 (1985).

KEY TOPICS IN NUCLEAR STRUCTURE
Proceedings of the 8th Int. Spring Seminar on Nuclear Physics
edited by Aldo Covello
© 2005 World Scientific Publishing Co.

ENERGY DISTRIBUTION OF COLLECTIVE STATES WITHIN THE FRAMEWORK OF SYMPLECTIC SYMMETRIES

A. I. GEORGIEVA*, V. P. GARISTOV, H. GANEV
Institute of Nuclear Research and Nuclear Energy,
Bulgarian Academy of Sciences, Sofia 1504, Bulgaria

AND J. P. DRAAYER
Department of Physics and Astronomy,
Louisiana State University, Baton Rouge, Louisiana, 70803-4001 USA

In a new application of the algebraic Interacting Vector Boson Model (IVBM), we exploit the reduction of its $Sp(12, R)$ dynamical symmetry group to $Sp(4, R) \otimes SO(3)$, which defines basis states with fixed values of the angular momentum L. The energy distribution of collective positive parity states with $L = 0, 2, 4, 6...$ states is studied, using the correspondence of the new reduction chain to the rotational limit of the model. Results for low-lying spectra of rare-earth nuclei show that the energies of the states with fixed L lie on second order curves with respect to the number of collective phonons n or vector bosons $N = 4n$ from which they are built. The analysis of this behavior leads to insight regarding the common nature of collective states, tracking vibrational as well as rotational features.

1. Introduction

Contemporary experimental techniques have advanced to the place where information on large sequences of collective states with $L = 0, 2, 4, 6...$ in a given nucleus is now common place, particularly in the rare-earth region.[1] Ideally, a successful theoretical model should be able to classify, explain and correctly describe all such data. Algebraic approaches attempt to relate the structure of the collective states to basis states of a dynamical symmetry group and its subgroup chains. In this regard, the symplectic model provides a natural and very general framework for investigating the nature of collective excitations in many-body systems.[2] The symplectic model incor-

*Acknowledges the support from the organizers to present this work at the conference porates "elementary" excitations into the structure of the collective states. The phenomenological Interacting Vector Boson Model (IVBM) has been shown to yield a rather accurate description of the low-lying spectra of even-even, well-deformed nuclei.[3] The spectrum generating algebra of the symplectic model is that of the $Sp(12, R)$ group, which has a very rich subgroup structure.[3,4]

Our aim is to study the behavior of the energies of sequences of collective states with fixed angular momentum L in the spectra of a given even-even nucleus. The energy distribution of these states with respect to the number of "elementary excitations" (phonons) that build them is best reproduced and interpreted within the framework of a algebraic approach that involves symplectic symmerties. In the present work we introduce the reduction of $sp(12, R)$ to the noncompact direct product $sp(4, R) \otimes so(3)$, which isolates sets of states with given L. This permits an investigation into the behavior of low-lying collective states with the same angular momentum L with respect to the number of excitations N that build these states.

2. Group theoretical background for the IVBM

We consider $Sp(12, R)$ – the group of linear canonical transformation in a 12-dimensional phase space – to be the dynamical symmetry group of the model.[3,5] Its algebra is realized in terms of creation (annihilation) operators $u_m^+(\alpha)(u_m(\alpha) = (u_m^+(\alpha))^\dagger)$, of *two* types of bosons with "pseudospin" projection $\alpha = p = 1/2$ (proton) and $\alpha = n = -1/2$ (neutron) in a 3-dimensional oscillator potential with $m = 0, \pm 1$. The bilinear products of the creation and annihilation operators of the two vector bosons generate the noncompact symplectic group $Sp(12, R)$:[3]

$$F_M^L(\alpha, \beta) =_{k,m} C_{1k1m}^{LM} u_k^+(\alpha) u_m^+(\beta),$$

$$G_M^L(\alpha, \beta) =_{k,m} C_{1k1m}^{LM} u_k(\alpha) u_m(\beta), \tag{1}$$

$$A_M^L(\alpha, \beta) =_{k,m} C_{1k1m}^{LM} u_k^+(\alpha) u_m(\beta),$$

where C_{1k1m}^{LM} are the usual Clebsch-Gordon coefficients and $L = 0, 1, 2$ with $M = -L, -L+1, ...L$ define the transformation properties of these operators under rotations.

The operators $A_M^L(\alpha, \beta)$ from (1) generate the algebra of the maximal compact subgroup $U(6)$ of $Sp(12, R)$. The rotational limit of the IVBM is further defined by the chain

$$\begin{array}{ccccccccc} U(6) & \supset & SU(3) & \times & U(2) & \supset & SO(3) & \times & U(1) \\ [N] & & (\lambda,\mu) & & (N,T) & K & L & & T_0 \end{array} \quad (2)$$

where the labels below the subgroups are the quantum numbers corresponding to their irreducible representations (IR).[6] In this limit, the Hamiltonian is expressed in terms of the first and second order invariant operators of different subgroups in (2). The *complete* spectrum of the system is calculated through the diagonalization of the Hamiltonian in the subspaces of *all* the unitary irreducible representations (UIR) of $U(6)$, belonging to a given UIR of $Sp(12,R)$.[4] Since the reduction from $U(6)$ to $SO(3)$ is carried out by the mutually complementary groups $SU(3)$ and $U(2)$,[7] their Casimir operators as well as their quantum numbers are related to one another: $T = \frac{\lambda}{2}$ and $N = 2\mu + \lambda$.[8,9] Making use of the latter, the model Hamiltonian is

$$H = aN + bN^2 + \alpha_3 T^2 + \beta_3 L^2 + \alpha_1 T_0^2 \quad (3)$$

which is obviously diagonal in the basis

$$|[N]_6;(\lambda,\mu);K,L,M;T_0\rangle \equiv|\,(N,T);K,L,M;T_0\rangle. \quad (4)$$

2.1. *Reduction through the noncompact Sp(4,R)*

We now introduce another possible reduction of $sp(12,R)$ algebra – through its noncompact subalgebra $sp(4,R)$:[5]

$$sp(12,R) \supset sp(4,R) \otimes so(3). \quad (5)$$

The generators of the $sp(4,R)$ – $F_0^0(\alpha,\beta), G_0^0(\alpha,\beta)$ and $A_0^0(\alpha,\beta)$ – are obtained from the vector addition to $L=0$ (scalar products) of the different pairs of vector bosons $u_m^+(\alpha), (u_m(\alpha))$ $m = 0, \pm 1$ representing the $sp(12,R)$ generators (1). Hence by construction, all these operators are scalars with respect to 3-dimensional rotations and commute with the components of the angular momentum $L_M = -\sqrt{2}\sum_\alpha A_M^1(\alpha,\alpha)$ that generate the $so(3)$ algebra. In this way, the direct product of the two algebras (5) is realized. Hence the quantum number L of the angular momentum algebra $so(3)$ can be used to characterize the representations of $sp(4,R)$.

The maximal compact subalgebra of $sp(4,R)$ is $u(2)$, generated by the Weyl generators $A^0(p,n) = \sqrt{\frac{2}{3}}T_+$, $A^0(n,p) = -\sqrt{\frac{2}{3}}T_-$, and $A^0(p,p) = \sqrt{\frac{1}{3}}N_+$, $A^0(n,n) = \sqrt{\frac{1}{3}}N_-$. $N_+(N_-)$ that count the number of particles of each kind. Here we use the equivalent set of infinitesimal operators with the Cartan operators $N = N_+ + N_-$ and $T_0 = \frac{1}{2}(N_+ - N_-)$. The operator N is the first-order invariant of $u(2) \supset su_T(2) \otimes u_N(1)$ and reduces the infinite dimensional spaces of the boson representations of $sp(4,R)$ to an infinite sum of finite representations of $u(2)$. The operators $T_0, T_\pm$ close to form the pseudospin algebra $su(2)$. The standard labelling of the $su_T(2)$ basis states is by means of the eigenvalues of the second order Casimir operator of $SU(2)$: $T(T+1) = \frac{N}{2}(\frac{N}{2}+1)$. From the last equation it follows that $T = \frac{N}{2}, \frac{N}{2}-1, ..., 1$ or 0, for each fixed N in the reduction $sp(4,R) \supset u(2)$. The other label of the $su_T(2)$ basis states is provided by the eigenvalues of the operator $T_0 = -T, -T+1, \ldots, T-1, T$. The following correspondence exists between chains 2 and 5 of the subalgebras of $sp(12,R)$:[10]

$$\begin{array}{ccccc} sp(12,R) & \supset & sp(4,R) & \otimes & so(3) \\ \cup & & \cup & & \cap \\ u(6) & \supset & u(2) & \otimes & su(3). \end{array} \tag{6}$$

This correspondence (6) is a result of the equivalent pseudospin $u(2) \supset sp(4,R)$ algebra in both chains, which is complementary to $su(3) \supset u(6)$.

A basis for the $sp(4,R)$ representations in the even H_+ (N-even) spaces of $sp(12,R)$ is generated by a consecutive application of the different degrees of the operators $F^0(\alpha,\beta); \alpha,\beta = \pm 1/2$ to the lowest weight state (lws) with angular momentum L that labels the $Sp(4,R)$ irrep under consideration.[10] Each starting $u(2)$ configuration is characterized by a totally symmetric representation $[L]_2$ formed by $N_{min} = L$ vector bosons. The $r/2$ degree of $F^0(\alpha,\beta)$ is obtained by the decomposition of all the even numbers $r = 0,2,4,6,...$, into a direct sum of $u(2)$ irreps $[r_1, r_2]$, where r_1, r_2 are both even and $r = r_1 + r_2$: $[r]_2 = \bigoplus_{i=0}^{\langle r/4 \rangle} [r-2i, i]$ where $\langle \frac{r}{4} \rangle$ denotes the integer part of the ratio. The action of these operators is given by the inner products $[k]_2 = [L]_2 \otimes [r]_2$, which correspond to $N = N_{min} + r$, $r = 0,2,4,6,...$ and $T = \frac{k}{2} = \frac{N}{2}, \frac{N}{2} - 1, ..., 0$. We illustrate this technique for the cases $L = 0$ and $L = 2$ in **Tables 1** and **2**. The columns are defined by the pseudospin quantum number $T = k/2$ and the rows by the eigenvalues of $N = k_{\max} = L + r$ for L even and $N = k_{\max} + 2 = 2L + r$ for $r = 0,2,4,6,....$ The symbol ρ is a multiplicity label that counts how many times the respective irrep $[k]_2$ appears for the specified value of N.

Table 1. $L = 0$

...	$T=4$	$T=3$	$T=2$	$T=1$	$T=0$	T/N
					$[0]_2(0,0)$	$N=0$
				$[2]_2(2,0)$		$N=2$
			$[4]_2(4,0)$		$[0]_2(0,2)$	$N=4$
		$[6]_2(6,0)$		$[2]_2(2,2)$		$N=6$
	$[8]_2(8,0)$		$[4]_2(4,2)$		$[0]_2(0,4)$	$N=8$
	...	...	...	...	...	...

Table 2. $L = 2$

...	$T=4$	$T=3$	$T=2$	$T=1$	$T=0$	T/N
				$[2]_2(2,0)$		$N=2$
			$[4]_2(4,0)$	$[2]_2(2,1)$	$[0]_2(0,2)$	$N=4$
		$[6]_2(6,0)$	$[4]_2(4,1)$	$2\times[2]_2(2,2)$	$[0]_2(0,3)$	$N=6$
	$[8](8,0)$	$[6]_2(6,1)$	$2\times[4]_2(4,2)$	$2\times[2]_2(2,3)$	$[0]_2(0,4)$	$N=8$
...	...	...	....	...	...	...

3. Distribution of low-lying collective states

Because of the correspodence (6) and the relation between the $SU(3)$ and $SU(2)$ second order Casimir operators, we use the same Hamiltonian (3) as in previous work.[4] Even more, as established above, the bases in both cases are equivalent and as a result the Hamiltonian (3) eigenvalues for the states with a fixed L are the energies

$$E((N,T);KLM;T_0) = aN + bN^2 + \alpha_3 T(T+1) + \alpha_1 T_0^2 + \beta_3 L(L+1). \quad (7)$$

In (7) the dependence of the energies of the collective states on the number of phonons (vector bosons) N is parabolic. All the rest of the quantum numbers definning the states T, T_0, and L are expressed in terms of N by means of the reduction procedure described above. This result confirms the empirical investigation of the states with fixed angular momentum.[11] Their energies are well described by the simple phenomenological formula $E_L(n) = An - Bn^2$, where $A > 0$ and $B > 0$ are fitting parameters and n is an integer number corresponding to each one of the states with given L. We have estabished a relation $N = 4n$ between the quantum number N and number of ideal bosons n.[12] In the present application of the IVBM, the parity of the states is defined as $\pi = (-1)^T$, so the corresponding basis states have a fixed T value in the $sp(4,R)$ representations labelled by

L.[4] We start with an evaluation of the inertia parameter β_3 in front of the term $L(L+1)$ from fitting the energies of the ground band states (gbs) with $J^\pi = 0_1^+, 2_1^+, 4_1^+, 6_1^+, ...$ to their experimental values in each nuclei. Further, the values of N_{L_i} corresponding to the experimentally observed $E_{L_i}^{\exp}$ and the values of the parameters in (7) are evaluated in a multi-step χ-square fitting procedure. The set of N_{L_i} with minimal value of χ^2 determines the distribution of the L_i^+ states energies (the parameters of the Hamiltonian) with respect to the number of bosons N_{L_i} that build the states. We fix the model parameters with respect to the three sets of states 0^+, 2^+ and 4^+, as there is usually enough of them to get good statistics in the fit and they are predominantly band heads (all the 0^+ and some of the 2^+ and 4^+). Sets of states with other values of $L = 1, 3, 6$ or with negative parity (T-odd) can be included in the consideration only by determining in a convenient way the values of T, T_0 and finding the sequences of N corresponding to the observed experimental energies. As all the parameters of Hamiltonian are evaluated from the distribution of 0^+ (a and b for $T = T_0 = 0$), 2^+ (α_3 for $T > 0, T_0 = 0$) and 4^+ (α_1 for $T > 0, T_0 > 0$) states, for any additional sets of states, we introduce a free additive constant c_L to the eigenvalues $E((N,T); KLM; T_0)$ which is evaluated in the same way as the other parameters of (7).

3.1. *Analysis of the results*

Results for an application of the theory to the collective spectra of 4 even-even rare earth nuclei are shown in **Figures 1-2**. The theoretical distribution of the energies with respect to the N_{L_i} values and the experimental numbers can be clearly seen. Values for $N_{\min}$, the T, T_0 for states with given L as well as the values for the Hamiltonian parameters β_3, a, b, α_3, α_1 and c_L that were obtained with their respective χ^2 are presented in **Table 3.** The s in the first column gives the number of the experimentally observed states with that L value.

The examples choosen are nuclei for which there is experimental data on energies for more than 5 states with angular momenta $L = 0, 2, 4$ in the low-lying spectra. Two of these nuclei have typical vibrational spectra[13] – Nd^{144} and Sm^{148} – and the rest – Gd^{154}, Hf^{178} – have typical rotational character. This is confirmed by the values obtained for the inertia parameter β_3 that is given in **Table 3**. It is well known that the main distinction of these two types of spectra is the position of the first excited 2_1^+ state of the gsb, which for vibrational nuclei is around $1 MeV$ but for the well-deformed nuclei it lies an order of magnitude or so lower, typically around $0.07MeV$.

Table 3. Parameters of the theory

Nucleus	s	L	$N_{\min}$	T	T_0	χ^2	parameters
Sm^{148}	5	0	0	0	0	0.0005	$a = 0.03096$ $b = -0.00010$
	7	2	8	4	0	0.0002	$\alpha_3 = -0.00187$
	10	4	16	8	8	0.0003	$\alpha_1 = -0.00285$
	3	6	20	10	10	0.0023	$\beta_3 = 0.03929$
Nd^{144}	7	0	0	0	0	0.0001	$a = 0.02389$ $b = -0.00003$
	11	2	12	6	0	0.0008	$\alpha_3 = 0.00309$
	7	4	20	10	10	0.0004	$\alpha_1 = -0.00450$
							$\beta_3 = 0.04074$
Hf^{178}	7	0	0	0	0	0.0010	$a = 0.02666$ $b = -0.00007$
	9	2	8	2	0	0.0008	$\alpha_3 = 0.02677$
	9	4	12	2	2	0.0016	$\alpha_1 = 0.05274$
	7	6	16	2	2	0.0008	$\beta_3 = 0.01482$
Gd^{154}	11	0	0	0	0	0.0023	$a = 0.05218$ $b = -0.00019$
	17	2	4	2	0	0.0482	$\alpha_3 = 0.0400$
	8	4	4	2	2	0.0027	$\alpha_1 = 0.09574$
	6	6	8	4	2	0.0023	$\beta_3 = 0.01634$
	3	3	6	2	0	0.0033	$c_3 = 0.05$
	2	5	10	2	0	0.00008	$c_5 = 0.09$

For the nuclei with vibrational spectra, Nd^{144} and Sm^{148}, we applied the procedure described above with values of T-even differing quite a lot ($\Delta T = 4$) for the sets with different L. This corresponds to rather large changes in the values of the initial $N_{min} = 2T$. Most of the states with fixed L are placed on the left-hand-side of symmetric parabolas so the values of N_{L_i} increase with an increase of the energy of these states.

With the procedure outlined above, the ordering of states into different bands is easy to recognize. The gsb is formed from the lowest states with $L = 0^+, 2^+, 4^+$ which are almost equally spaced the case of vibrational

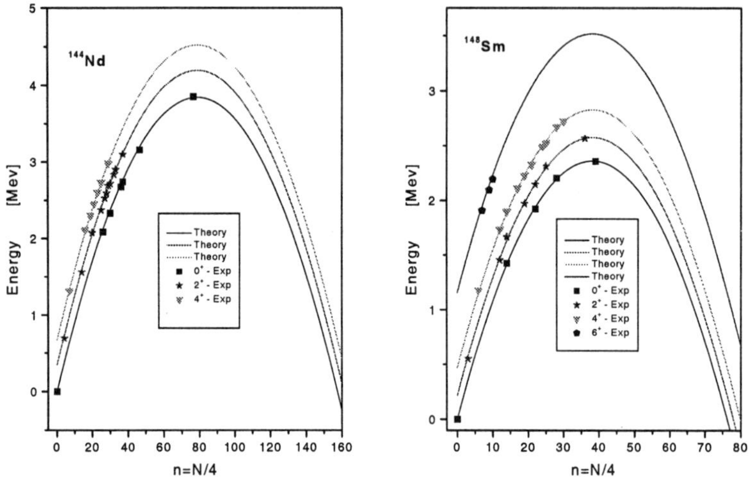

Figure 1. (color online) Comparison of theoretical and experimental energy distributions of states with $J^{\pi} = 0^{+}, 2^{+}, 4^{+}$ in ^{144}Nd (left) and $J^{\pi} = 0^{+}, 2^{+}, 4^{+}, 6^{+}$ in ^{148}Sm (right)

nuclei with very close or even equal values of n for the states in the other excited bands. The almost degenerate $0^{+}, 2^{+}, 4^{+}$ triplet of states, which is characteristic of harmonic quadrupole vibrations, can be observed in the theoretical energy curves and are characterized by almost equal differences between their corresponding values of N.

The rotational character of Gd^{154} and Hf^{178} requires very small differences in the values of N and T for the $0_1^{+}, 2_1^{+}, 4_1^{+}$ states of the gsb.[14] In order to avoid a degenaracy of the energies with respect to N_{L_i} for rotational spectra, we use the symmetric feature of the second order curves. This corresponds to the second solution, namely $N''_{0_1} = -\frac{b}{a}$, for the equation (7) for the ground state, with the maximum N_{0_1} associated with the ground state. This can be used as a restriction on the values of N_{L_i}.[15] The states with a given L in the rotational spectra are placed on the right-hand-side of the theoretical curves. On a parabola that is specified by a fixed L, the number of bosons that build the states decreases with an increase in their energes. Hence, if the number of quanta that build a collective state is taken as a measure of collectivity, the states from a rotational spectra are more collective than the vibrational ones, which is to be expected.

One can also see the structure of collective bands that are formed by

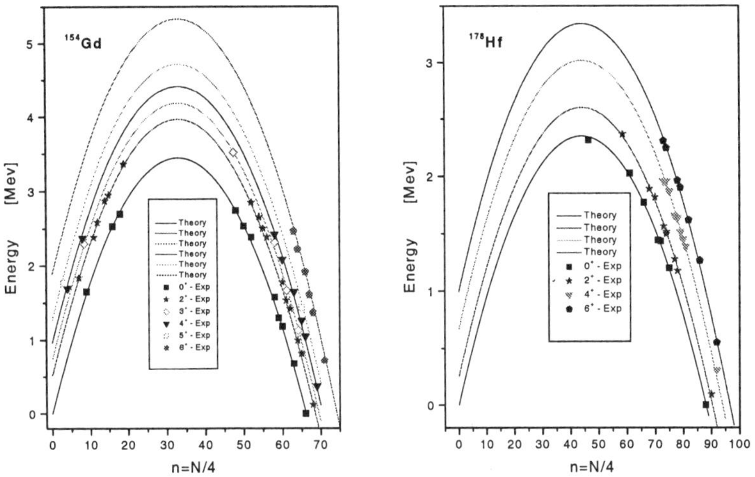

Figure 2. (color online) The same as on Figure 1 for states with $J^{\pi} = 0^{+}, 2^{+}, 3^{+}, 4^{+}, 5^{+}, 6^{+}$ in ^{154}Gd (left) and $J^{\pi} = 0^{+}, 2^{+}, 4^{+}, 6^{+}$ in ^{178}Hf (right)

sets of states from different curves. The best examples are for gsb and the first excited β- and γ-bands of deformed nuclei (see **Figure 2**). The spectrum of Gd^{154}, in addition to the $L = 0^{+}, 2^{+}, 4^{+}$, include states with $L = 3^{+}, 5^{+}, 6^{+}$. This example illustrates the $K^{\pi} = 2^{+}$ and 4^{+} bands and shows that our method works for these sets of collective states as well. The respective values of c_L are given in **Table 3**. In this case the second 0_2^{+} band is below the γ-band which is in contrast with the Hf^{178} spectra.[14]

4. Conclusion

In this paper we introduce a theoretical framework that serves to underpin the empirical observation that collective states can be accurately placed on two parametric second order curves with respect to a variable n that counts the number of collective phonons (bosons) that build the states.

The examples introduced confirm that the theory is both reliable and appropriate for all know types of collective states observed in atomic nuclei. In particular, the theory can be used to accurately describe the main types of collective vibrational and rotational spectra. These two are clearly distinguished using the symmetry property of the second order curves. The vibrational nuclei are placed on the left-hand-side of the parabolas with clearly distinguished values of T with N that increases with increasing energies. For rotational nuclei the situation is different, which confirms the traditional treatment of vibrational states as few phonon states and rotational states as having higher degrees of collectivity. The band structure and energy degeneracies in both cases are clearly observed.

The results introduced here to illustrate the theory demonstrate that the IVBM can be used to reproduce reliably empirical observations of the energy distribution of collective states. Such a demonstration can be provided for any collective model that includes one- and two-body interactions in the Hamiltonian. The main feature that leads to our parameterization is the symplectic dynamical symmetry of the IVBM. This allows for a change in the number of "phonons" that are required to build the states. This investigation provides also for insight in the structure of collective states, revealing the similar origin of vibrational and rotational spectra, but at the same time yielding information about unique features that distinguish the two cases.

References

1. Mitsuo Sacai, Atomic Data and Nuclear Data Tables, 31, 399 (1984); Level Retrieval Parameters http://iaeand.iaea.or.at/nudat/levform.html
2. D. J. Rowe, Rep. Prog. Phys., **48**, 1419 (1985).
3. A. Georgieva, P. Raychev and R. Roussev, J. Phys. G: Nucl. Phys., **8**, 1377 (1982).
4. H. Ganev, V. Garistov and A. Georgieva, Phys. Rev. **C 69**, 014305 (2004).
5. M. Moshinsky and C. Quesne, J. Math. Phys., **12**, 1772 (1971).
6. A. Georgieva, P. Raychev and R. Roussev, J. Phys. G: Nucl. Phys., **9**, 521 (1983).
7. C. Quesne, J. Phys., **A18**, 2675 (1985).
8. A.Georgieva, M.Ivanov, P. Raychev and R. Roussev, Int. J. Theor. Phys., **25**, 1181 (1986).
9. M. Moshinsky, Rev. Mod. Phys., **34**, 813 (1962).
10. V. V. Vanagas, *Algebraic foundations of microscopic nuclear theory,* Nauka, Moscow, (in russian) (1988).
11. V. Garistov, *Rearrangement of the Experimental Data of Low Lying Collective Excited States,* Proceedings of the XXII International Workshop on Nuclear Theory, ed. V. Nikolaev, Heron Press Science Series, Sofia, 305 (2003).
12. V. Garistov, On Description of the Yrast Lines in IBM-1, Proceedings of the XXI International Workshop on Nuclear Theory, Rila Mountains, ed. V. Nikolaev, Heron Press Science Series, Sofia (2002).
13. R. K. Sheline, Rev. Mod. Phys., **32**, 1 (1960).
14. A. Aprahamian, Phys. Rev. **C 65**, 031301(R) (2002).
15. J. P. Draayer and G. Rosensteel, Phys. Lett., **124B**, 281 (1983); G. Rosensteel and J. P. Draayer, Nucl. Phys., **A436**, 445 (1985).

KEY TOPICS IN NUCLEAR STRUCTURE
Proceedings of the 8th Int. Spring Seminar on Nuclear Physics
edited by Aldo Covello
© 2005 World Scientific Publishing Co.

RECENT RESULTS FROM SPECTROSCOPIC STUDIES OF EXOTIC HEAVY NUCLEI AT JYFL*

R. JULIN†

Department of Physics, University of Jyväskylä (JYFL)
PB 35 (YFL), FIN-40014 University of Jyväskylä, Finland
E-mail: rauno.julin@phys.jyu.fi

Recoil-Decay-Tagging (RDT) experiments for studies of shape coexistence in neutron deficient nuclei near Z = 82 have been continued at JYFL by employing the JUROGAM gamma-ray detector array and the GREAT spectrometer at the RITU gas-filled separator. A new non-yrast band has been observed in ^{186}Pb and tentatively associated with oblate shape. New experiments for ^{254}No were also carried out with the same set-up revealing feeding via highly coverted M1 transitions.

1. Instrumentation

The RITU separator at JYFL combined with detector systems for prompt radiation at its target area and for delayed radiation at its focal plane, forms one of the most efficient facilities for spectroscopy of exotic heavy nuclei [1]. By employing this system and the RDT technique it has been possible to observe in-beam γ-rays and conversion electrons from fusion-evaporation residues down to a level of 100 nb in production cross-section.

The gas-filled recoil separator RITU (Recoil Ion Transport Unit) was designed to separate residues of fusion-evaporation reactions from beam particles and other reaction products, especially fission [2]. Recently a new focal plane spectrometer GREAT funded by the UK institutes has been constructed for RITU [3]. In the GREAT spectrometer the fusion evaporation residues and their particle decay (so far α-decay) are detected by a double-sided silicon strip detector (DSSD). The strip pitch of the DSSD is 1mm in both directions resulting in a total of 4800 pixels and enabling

*This work is supported by the EU 5th framework IHP – Access to Research Infrastructure (HPRI-CT-1999-00044) and IHP – RTD (HPRI-CT-1999-50017) programmes and the Academy of Finland under the Finnish Centre of Excellence Programme 2000-2005.
†On behalf of the JUROGAM collaboration.

an identification of residues on the basis of their decay properties. Consequently, prompt γ-rays emitted by the fusion residues detected at the target area can be identified with high sensitivity (RDT technique). A transmission multiwire proportional counter (MWPC) in front of the DSSD is used for furher cleaning of the recoil- and α-particle spectra of the DSSD. Behind the DSSD a planar double-sided germanium strip detector is used for detection of delayed γ-rays from isomers and decay products.

Since April 2003 the JUROGAM array has been used to detect prompt γ rays at the target area. It consists of 43 Eurogam Phase 1 type of Compton suppressed detectors and has a photo-peak efficiency of 4% for 1.3 MeV γ rays.

The SACRED spectrometer was used to obtain in-beam electron spectra from heavy nuclei in RDT measurements at RITU [4]. In SACRED, electrons emitted from the target into backward angles are guided by the solenoid field and distributed over a Si detector (diameter 2 cm) which is divided into 25 independent pixels enabling to detect $e^- - e^-$ coincidences from cascades of converted transitions.

As a part of the GREAT project a new type of data acquisition system, known as Total Data Read out (TDR) [5], has been developed. It operates without any hardware trigger, and is designed to minimise dead time in the acquisition process. All detector electronic channels run independently and are associated in software, the data words all being time-stamped from a global 100 MHz clock.

In 2003 an RDT campaign of 13 experiments was carried out with the upgraded RITU + GREAT + JUROGAM system. Two of the main physics cases to be studied were the shape coexistence in the proton-drip line nuclei near Z = 82 and structure of very heavy nuclei near Z = 102. Examples of these topics are presented and discussed in the present contribution.

2. Probing the three shapes of ^{186}Pb

In our earlier in-beam spectroscopic RDT studies we have been able to extend information about yrast states in even-mass Pb isotopes down to ^{182}Pb. The ground state of light even-mass $^{182-188}$Pb nuclei is still spherical but the yrast line above the ground state is formed by a collective band, usually assumed to be based on 4p - 4h (or multiproton-multihole) intruder exitations and a prolate minimum [1].

In α-decay studies the first two lowest excited states in ^{186}Pb were observed to be 0^+ states [6]. On the basis of α-decay hindrance factors the

532 keV state was associated with an oblate minimum (2p-2h) and the 650 keV state with a prolate minimum (4p-4h). Consequently, with the spherical ground state, the three lowest 0^+ states represent triple shape coexistence in ^{186}Pb, a phenomenon which has been one of the highlights of the last decade in nuclear structure physics.

Only in ^{186}Pb the bandhead of the prolate intruder band can be associated with a 0^+ state (the 650 keV state). The oblate 0^+ state has been found to intrude down in energy with decreasing neutron number from ^{202}Pb to ^{186}Pb but only in ^{188}Pb a clear band structure has been associated with this state [7].

It is important to confirm the triple shape coexistence in ^{186}Pb by identifying members of the oblate intruder band. Observation of this band may also shed light on mixing of shapes and evolution of shape at higher spin. Yrast bands associated with an oblate minimum of proton 4p-2h intruder structures have been observed in ^{192}Po, ^{194}Po and ^{196}Rn nuclei [1]. Obviously due to smaller deformation, moments of inertia extracted from these bands are smaller than those for the prolate bands in even-mass $^{182-188}$Pb nuclei. Consequently, in spite of the lower energy of the oblate 0^+ state in ^{186}Pb the oblate band is expected to lie well above the yrast line and is therefore weakly populated in available fusion evaporation reactions. Moreover, $\gamma\gamma$ coincidence information is needed for the identification of the non-yrast states. Finally, decay tagging needed to identify ^{186}Pb γ rays has so far been difficult due to the relatively long half-life of 4.8 s of the ^{186}Pb α decay.

The new JUROGAM+RITU+GREAT+TDR system enabled us to carry out a successful RDT γ-ray experiment for ^{186}Pb at JYFL. We used the ^{106}Pd(^{83}Kr,3n)^{186}Pb reaction and recorded a total of $\sim 10^6$ α particles representing a cross-section of 185 μb for this reaction. A singles spectrum of γ- rays tagged with ^{186}Pb α decays is shown in Fig. 1a. This spectrum is dominated by the transitions of the yrast prolate band and the $2_1^+ \rightarrow 0_1^+$ transition.

Fig. 1b shows a sample RDT γ- ray spectrum obtained by gating on non-yrast transitions. This spectrum together with other coincidence information reveals a level scheme of Fig. 2 with the yrast band extended up to 16^+ and a new non-yrast band. The tentative spin and parity asignments are based on intensity, braching ratio and angular distribution information.

It is intriguing to question whether the observed non-yrast band in ^{186}Pb can be associated with the oblate minimum. Unfortunately, the low-energy

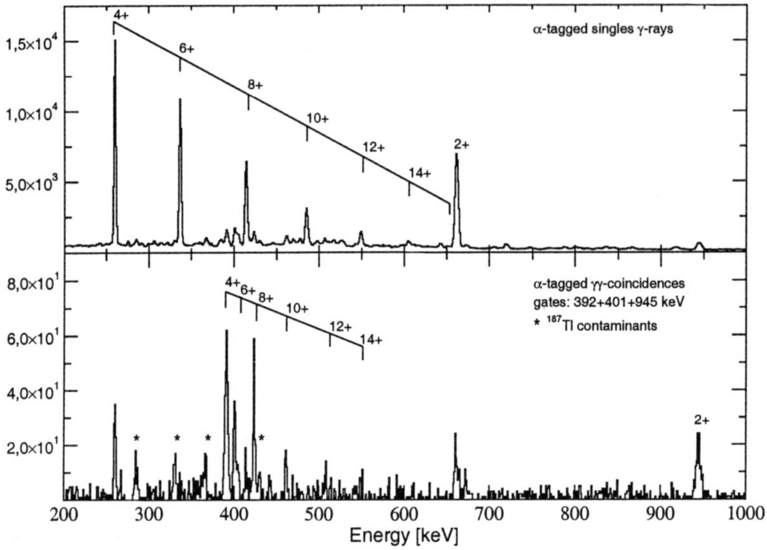

Figure 1. a) Singles γ-ray energy spectrum gated with fusion evaporation residues and tagged with ^{186}Pb α decays. b) Recoil-gated, α-tagged γ-γ coincidence spectrum with a sum gate on the three lowest non-yrast transition.

transitions to the 2_1^+ , 0_2^+ and 0_3^+ states are too weak to be observed and therefore cannot be utilised in the discussion of bandheads and mixing of shapes. Kinematic moments of inertia $J^{(1)}$ extracted from the experimental γ-ray energy and spin information for both of the observed bands in ^{186}Pb are plotted in Fig. 3 together with those for the prolate yrast bands in ^{184,186,188}Pb and the oblate yrast bands in ^{192,194}Po.

Fig. 3 also shows the $J^{(1)}$ values for a positive-parity, even-spin non-yrast band in ^{188}Pb [7]. Up to the 8^+ state this band seems to follow the pattern of the oblate bands in ^{192,194}Po and therefore could be associated with the oblate minimum. However, above the 8^+ state the band upbends with $J^{(1)}$ approaching the values of the prolate bands. The energy of the 2_2^+ state in ^{186}Pb is close to that in ^{188}Pb and therefore the band on top of it could be build on similar structures. This band reveals a strong upbend already at spin 6 and extends to $J^{(1)}$ values above those for the prolate bands. It is difficult to associate the upbends of the non-yrast bands in ^{186}Pb and ^{188}Pb with alignments of valence nucleon as they appear at very low spin. However, they could be due to a change towards more deformed structures. Indeed, more deformed oblate and prolate shapes are predicted to appear in light Pb isotopes [8,9,10]. As the oberved nonyrast states favour

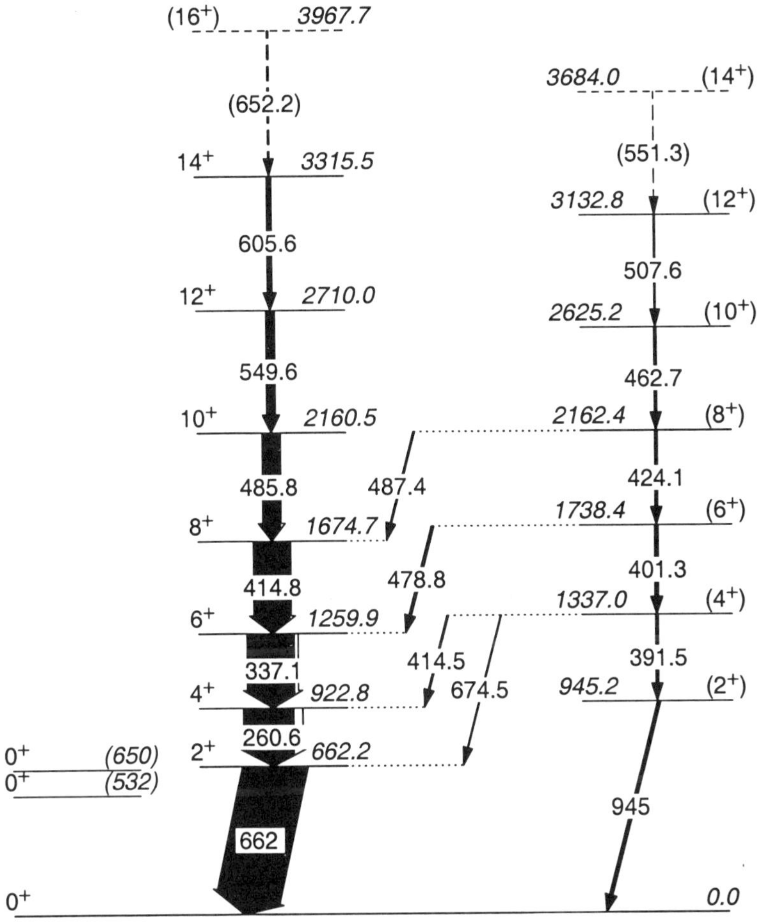

Figure 2. Level scheme of ^{186}Pb deduced from the present data (the two 0^+ states on the left side are taken from ref. [6]).

de-excitations towards the assumed oblate states it is plausible to associate the upbends in ^{186}Pb and ^{188}Pb to be due to a shape change towards more deformed oblate shapes.

3. Spectroscopic studies of very heavy elements

Heavy nuclei with $Z > 100$ can be stable against fission only due to shell effects. The shell correction energy should be large enough for creating an island of spherical superheavy elements around $Z = 114$ and $N = 184$.

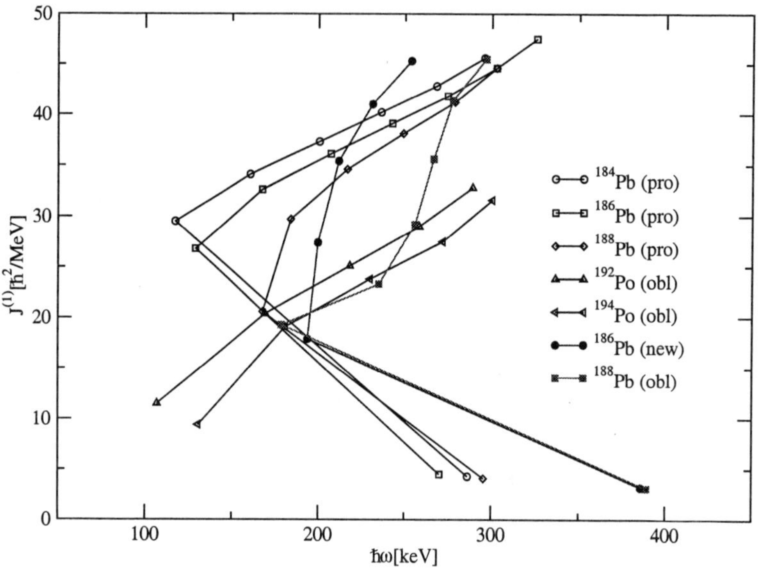

Figure 3. Kinematic moment of inertia as a function of rotational frequency for nuclei in the vicinity of ^{186}Pb.

Stability of known α-decaying nuclei with $Z > 100$ is supposed to be originated from shell effects in a deformed nucleus. It is important to verify experimentally the predicted deformation and moment of inertia as well as alignments in these nuclei. Moreover, information about single-particle orbitals near the Fermi surface of these deformed trans-fermium nuclei can be used in estimates of single particle energies in spherical super-deformed nuclei.

Most of the scarce experimental information on the structure of heaviest nuclei has been obtained in α-decay studies [11]. Important information from single-particle properties has been obtained by using transfer reactions [12]. Excited states up to $I^{\pi} = 8^{+}$ in ^{256}Fm have been seen in the β decay of the isomeric 8^{+} state of ^{256}Es [13]. Coulomb excitation has been used to populate excited states of ^{248}Cm up to a 5.1 MeV 30^{+} state [13].

The small production cross-sections make any kind of detailed spectroscopic studies of heavy elements extremely difficult. Exceptionally high cross-sections of 300 - 2000 nb in cold fusion-evaporation reactions are obtained for the production of transfermium nuclei when using the doubly magic ^{48}Ca projectile and target nuclei near ^{208}Pb. This fact has enabled us to probe structures of these nuclei in in-beam γ-ray and electron spectroscopic experiments when employing the RDT method.

In our earlier experiments with the JUROSPHERE and SACRED spectrometers we used the ^{48}Ca beam on ^{208}Pb, ^{206}Pb and ^{204}Hg targets to study ^{254}No [14], ^{252}No [15] and ^{250}Fm, respectively. The observation of discrete γ-ray lines from a rotational cascade of transitions up to $I = 20$ in ^{254}No, ^{252}No and ^{250}Fm reveals that these trans-fermium nuclei are deformed and can in rotation compete against fission up to at least that spin. The kinematic moment of inertia values for these nuclei derived from the observed transition energies are about half of the rigid rotor value and are slightly increasing with spin (Fig.4), obviously due to gradual alignment of quasi-particles. For ^{252}No the extracted values increase more rapidly at high spin indicating a more dramatic alignment of quasi-particles. The kinematic moment of inertia values for ^{250}Fm are almost identical to the ^{254}No ones at low spin but then follow the alignment pattern of ^{252}No at higher spin.

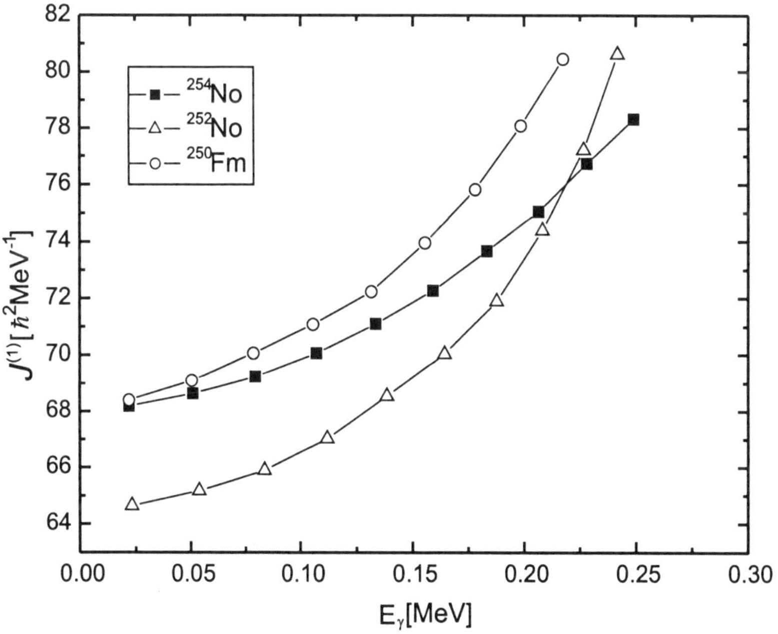

Figure 4. Kinematic moments of inertia for ^{250}Fm, ^{252}No and ^{254}No extracted from the measured γ-ray energies.

It is possible to extract the ground state deformation parameter β_2 from the extrapolated energy of the 2_1^+ state using global systematics [16,17]. The value we derived for ^{254}No is $\beta_2 \approx 0.27$, which is in good agreement with the values calculated using the macroscopic-microscopic method [18,19]. A β_2 value similar to ^{254}No is obtained for ^{250}Fm. The extracted value of $\beta_2 = 0.26$ for ^{252}No indicated that ^{252}No is less deformed than ^{254}No and ^{250}Fm.

The ^{254}No experiment was recently repeated by employing the 43 JUROGAM array in connection with the upgraded RITU separator. A preliminary recoil gated spectrum is shown in Fig. 5. The yrast line is seen up to $I = 20$ and there are peaks in the spectrum representing depopulation of sidebands in ^{254}No. The energy spacing between the lines in the 800 - 1000 keV region is the energy spacing between the 2^+ and 4^+ states of the ground state band. Therefore, the corresponding transitions must de-excite a strongly fed non-yrast state of spin 3 or 4. Obviously feeding of such a state takes place via a fast highly concerted M1 cascade not visible in the gamma-ray spectrum.

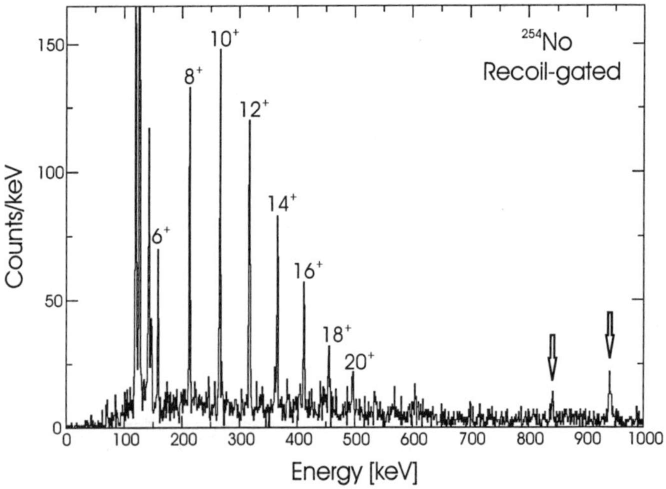

Figure 5. A preliminary recoil gated γ-ray spectrum for ^{254}No detected by the JUROGAM array at RITU.

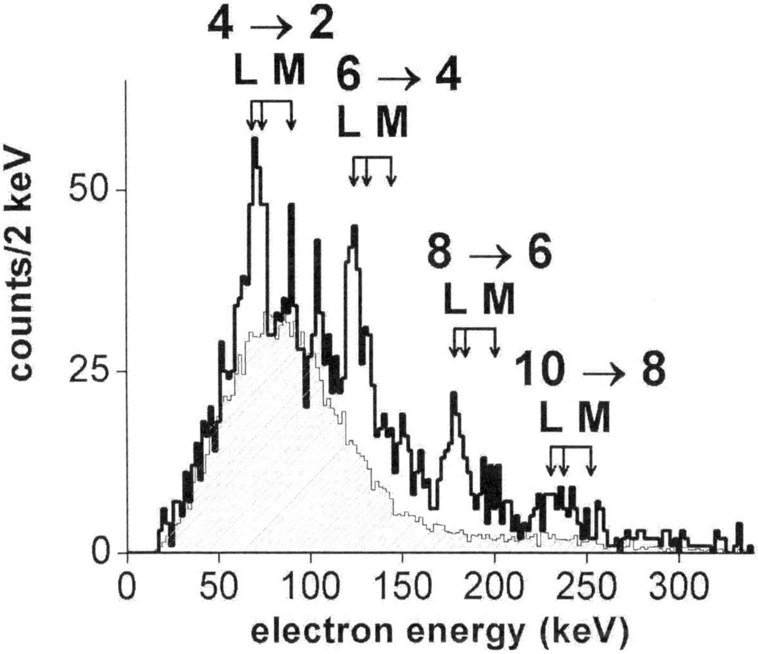

Figure 6. A conversion-electron spectrum tagged by ^{254}No recoils. The hashed area shows a simulated spectrum of electrons from M1 transitions of high K bands in ^{254}No [20].

More information about highly converted M1 cascades was obtained in an experiment, where the SACRED conversion-electron spectrometer was used to measure prompt conversion electrons from the ^{208}Pb(^{48}Ca,2n)^{254}No reactions. In a resulting recoil gated spectrum shown in Fig. 6, electron peaks originating from transitions between the low-spin yrast states in ^{254}No are seen. In a careful analysis of the prompt recoil-gated electron-electron coincidence spectra it was found out that the broad distribution under these electron peaks is not due to random events but consists of high-multiplicity events, obviously originating from cascades of highly converted M1 transitions within rotational bands built on high-K states in ^{254}No [20].

References

1. R. Julin *et al.*, *J. Phys. G: Nucl. Part. Phys.* **27**, R109 (2001).
2. M. Leino *et al.*, *Nucl. Instr. Meth.* **B99**, 653 (1995).
3. R. D. Page *et al.*, *Nucl. Instr. Meth.* **B 204**, 634 (2003).
4. P. A. Butler *et al.*,*Nucl. Instr. Meth.* **A 381**, 433 (1996).

5. I. H. Lazarus *et al.*, *IEEE Trans. on Nucl. Sci.* **48**, N:o 3,(2001)
6. A. N. Andreyev *et al.*, *Nature* **405**, 430 (2000).
7. G. D. Dracoulis *et al.*, *Phys. Rev.* **C69**, 054318 (2004).
8. W. Nazarewicz, *Phys. Lett.* **B305**, 195 (1993).
9. M. Bender *et al.*, *Phys. Rev.* **C69**, 064303 (2004).
10. J. L. Egido *et al.*, *Phys. Rev. Lett.* **93**, 082502 (2004).
11. S. Hofmann, *Rep. Prog. Phys.* **61**, 639 (1998).
12. I. Ahmad *et al.*, *Phys. Rev. Lett.* **39**, 12 (1977).
13. M. R. Schmorak, *Nucl. Data Sheets* **57**, 515 (1989).
14. M. Leino *et al.*, *Eur. Phys. J.* **A6**, 63 (1999).
15. R.-D. Herzberg *et al.*, *Phys. Rev.* **C65**, 014303 (2001).
16. L. Grodzins, *Phys. Lett.* **2** 88 (1962).
17. S. Raman, *At. Data Nucl. Data Tables* **42** 1 (1989).
18. Z. Patyk *et al.*, *Nucl. Phys.* **A533** 132 (1991).
19. S. Cwiok *et al.*, *Nucl. Phys.* **A573** 356 (1994).
20. P. A. Butler *et al.*, *Phys. Rev. Lett.* **89** 202501 (2002).

KEY TOPICS IN NUCLEAR STRUCTURE
Proceedings of the 8th Int. Spring Seminar on Nuclear Physics
edited by Aldo Covello
© 2005 World Scientific Publishing Co.

DIPOLE STRENGTH DISTRIBUTIONS IN ^{124,126,128,130,132,134,136}Xe: A SYSTEMATIC STUDY IN THE MASS REGION OF A NUCLEAR SHAPE TRANSITION

U. KNEISSL

Institut für Strahlenphysik, Universität Stuttgart
Allmandring 3, D-70569 Stuttgart, Germany
E-mail: KNEISSL@ifs.physik.uni-stuttgart.de

Systematic nuclear resonance fluorescence experiments (NRF) on all 7 stable even-even Xe isotopes have been performed at the bremsstrahlung facility of the 4.3 MV Stuttgart Dynamitron accelerator. For the first time thin-walled, high-pressure gas targets (about 70 bar) were used in NRF experiments. Precise excitation energies, transition strengths, spins, and decay branching ratios were obtained for numerous states, most of them unknown so far. The systematics of the observed $E1$ two-phonon excitations $(2^+ \otimes 3^-)$ and $M1$ excitations to 1^+ mixed symmetry states are discussed with respect to the new critical point symmetry E(5).

1. Motivation

Systematic investigations of nuclei within isotopic chains undergoing a shape or phase transition are of particular current interest in nuclear structure physics. In the framework of algebraic models the dynamical symmetry limits U(5), SU(3), and O(6) correspond to spherical, axially deformed, and γ-soft nuclear shapes. Recently Iachello [1] proposed two new so-called critical point symmetries E(5) and X(5) which apply for nuclei at the critical points of phase transitions from spherical vibrators to deformed γ-soft nuclei and to deformed rotors, respectively. Whereas the X(5) first order phase/shape transition is clearly seen in the sudden onset of the deformation splitting of the higher-lying $E1$ Giant Dipole Resonance (GDR) in the chains of Sm and Nd isotopes [2] and is indicated in the abrupt concentration of the $M1$ *Scissors Mode* strength in ^{150}Nd [3], there is, at least up to now, no evidence in dipole strength distributions for the predicted second order E(5) phase transition. Therefore, the aim of the present present work was to study the influence of the E(5) shape or phase transition on low-lying $E1$ or $M1$ strength distributions.

In Fig.1 the ratios of the excitation energies of the first 4^+ and 2^+ states in even-even nuclei near the N=82 shell closure are plotted together with the values expected for U(5), E(5), O(6), and SU(3) nuclei. The values for the Xe isotopes are shown by full circles. Obviously, the Xe isotopic chain with 7 stable even-even isotopes crosses the U(5), E(5) values and reaches the O(6) limit. Therefore, this chain provides an unique case to investigate systematically the changes of spectroscopic observables expected for shape or phase transitions near the critical points.

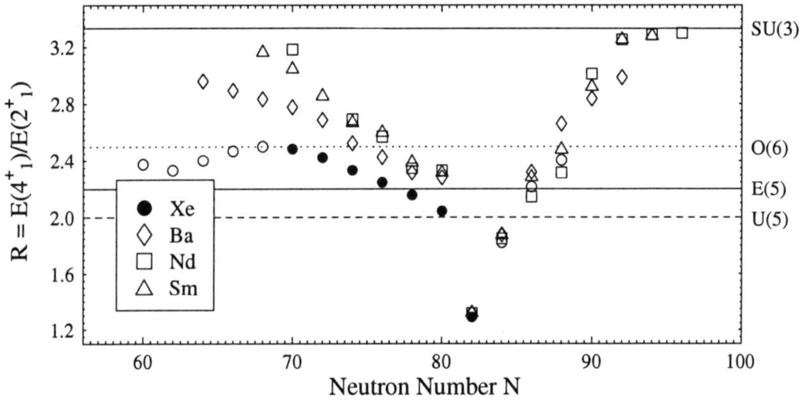

Figure 1. Ratios of the excitation energies of the first 4^+ and 2^+ states in even-even nuclei near the N=82 shell closure together with the values expected for U(5), E(5), O(6), and SU(3) nuclei.

2. Experiments

Photon scattering, nuclear resonance fluorescence (NRF), represents the most sensitive tool to investigate low-lying dipole excitations (see Ref.[3]). The present NRF studies[4] have been performed at the well-established bremsstrahlung facility of the 4.3 MV Stuttgart Dynamitron accelerator described in more detail in Ref.[3]. For the first time thin-walled, high-pressure gas targets (about 70 bar) [5], developed at the Forschungszentrum Karlsruhe, were used in the present NRF experiments. The total masses of the available, highly enriched target material were about 0.3-0.7 gram. The scattered photons are detected by three carefully shielded Ge(HP)-γ-spectrometers, with efficiencies ϵ of 100% (relative to a 3"×3" NaI/Tl detector) in each case, placed at scattering angles of 90°, 127°, and 150°

with respect to the incident beam. The detector at 127° was surrounded additionally by a BGO anti-Compton shield.

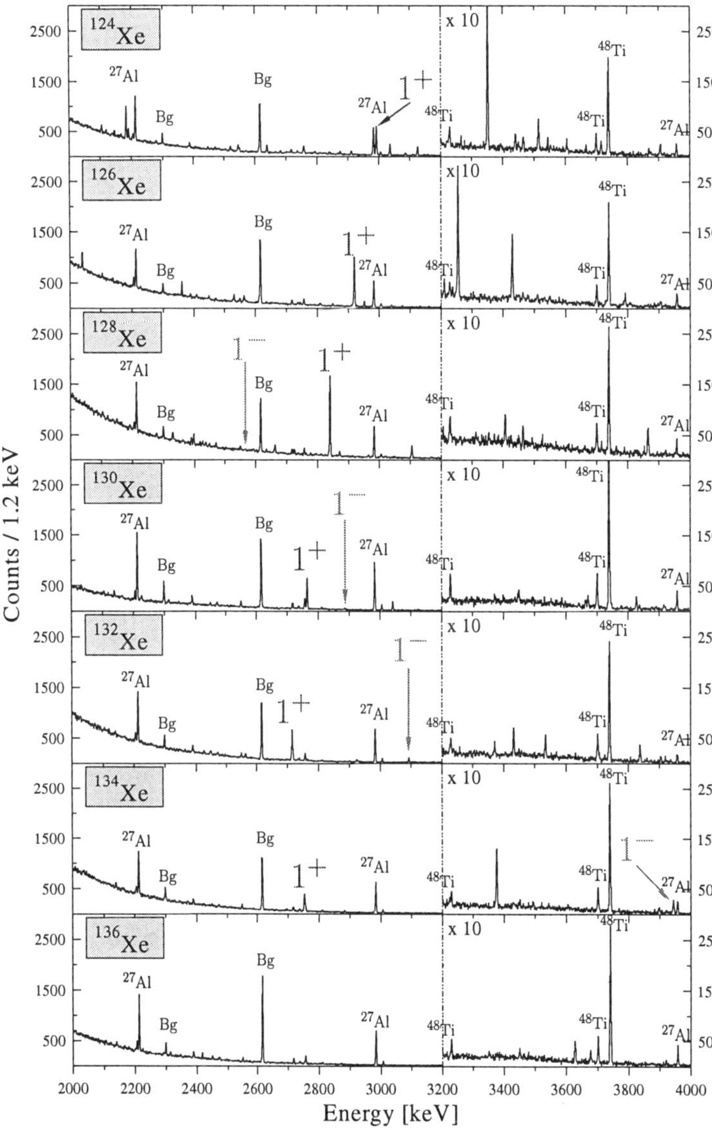

Figure 2. Spectra of photons scattered off ^{124,126,128,130,132,134,136}Xe in the energy range from 2 to 4 MeV. Above 3.2 MeV the scale is stretched by a factor of 10. For details see text.

3. Results

Precise excitation energies, transition strengths, spins, and decay branching ratios were obtained for numerous states, most of them unknown so far. Unfortunately, no parity assignments via Compton-polarimetry[6] were possible in reasonable measuring times due to the low target quantities of less than 1 gram.

Fig.2 gives on overview on the observed very clean spectra of scattered photons. Marked peaks correspond to background lines (Bg), transitions in the photon flux monitor ^{27}Al, and to excitations in ^{48}Ti (container material), which are well known from previous NRF studies [7]. The spectra are dominated by strong excitations around 2.7 – 3.0 MeV marked by (1^+) which are ascribed to $M1$ excitations to 1^+ mixed symmetry states. Candidates for $E1$ excitations to the spin 1^- member of the $(2^+ \otimes 3^-)$ two-phonon quintuplet are labeled by (1^-).

4. Discussion

In heavy nuclei two general low-lying dipole excitation modes are well established, $E1$ excitations to the 1^- member of the two-phonon quintuplet of the type $2^+ \otimes 3^-$ in nuclei near closed shells (see compilation in Ref.[8]) and $M1$ excitations to 1^+ mixed symmetry states in transitional nuclei [9] which in deformed nuclei correspond to the well known *Scissors Mode* excitations [10,11]. The aim of the present present work was to study the influence of a shape or phase transition on low-lying dipole strength distributions. The following discussion is restricted to these fundamental dipole modes and their dependence on a possible phase transition.

Fig. 3 shows the obtained dipole strengths distributions. Since no parity determination could be performed, the reduced ground-state transition widths $\Gamma_0^{red}=\Gamma_0/E_\gamma^3$ are plotted as a function of the excitation energy, which are proportional to the reduced excitation probabilities $B(E1)\uparrow$ and $B(M1)\uparrow$, respectively. For even-even nuclei a value of $\Gamma_0^{red}=1$ meV/MeV^3 corresponds to reduced excitation probabilities of $B(E1)\uparrow= 2.866\cdot 10^{-3}\mathrm{e}^2\,\mathrm{fm}^2$ or $B(M1)\uparrow=0.259\,\mu_N^2$, respectively. The strong excitations between 2.7 and 3.0 MeV marked by full circles are ascribed to $M1$ excitations to 1^+ mixed-symmetry states (see subsec. 4.2). Excitations in ^{134,132,130,128}Xe labeled by open rhombs are attributed to the $E1$ two-phonon excitations (see subsec. 4.1). In the O(6) candidates ^{124,126}Xe rather strong low-lying dipole excitations emerge between 2.0-2.5 MeV which may be due to the low-energy octupole strengths expected for these isotopes [12].

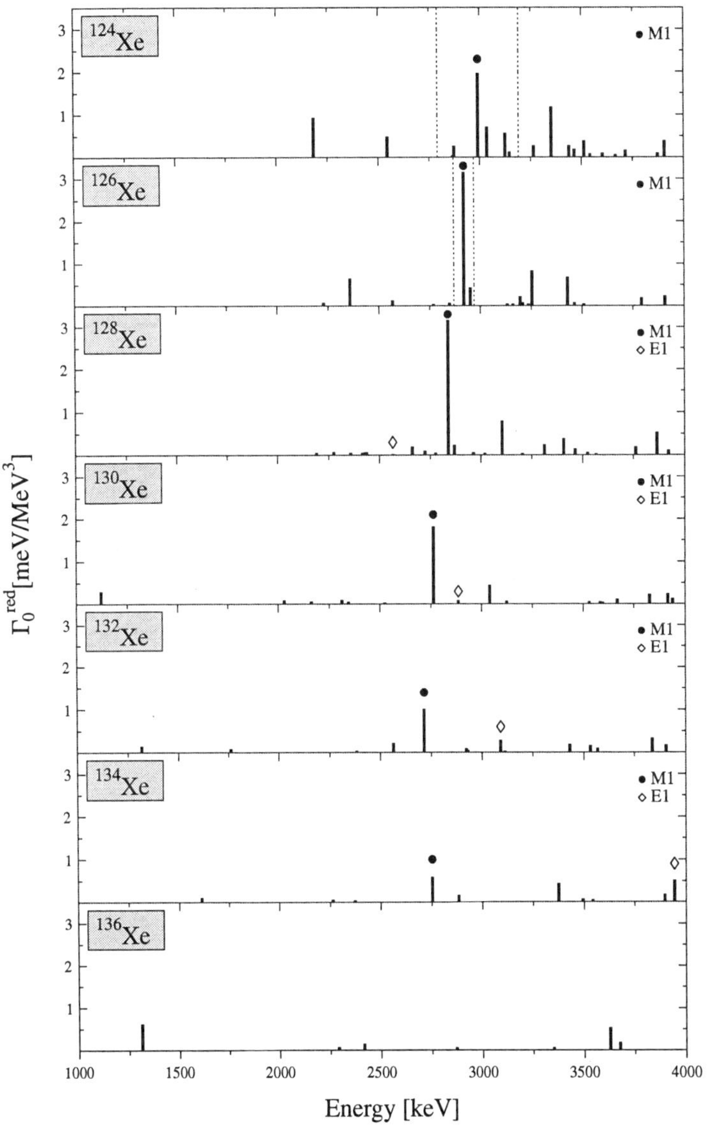

Figure 3. Dipole strength distributions observed for the even-even Xe isotopes. Plotted are the reduced ground-state transition widths Γ_0^{red} which are proportional to the reduced excitation probabilities $B(E1)\uparrow$ and $B(M1)\uparrow$, respectively. For details see text.

4.1. *E1 two-phonon excitations*

In Fig.4 the data for the ascribed two-phonon excitations are summarized. Their excitation energies, shown in the upper part by full circles, lie nearly

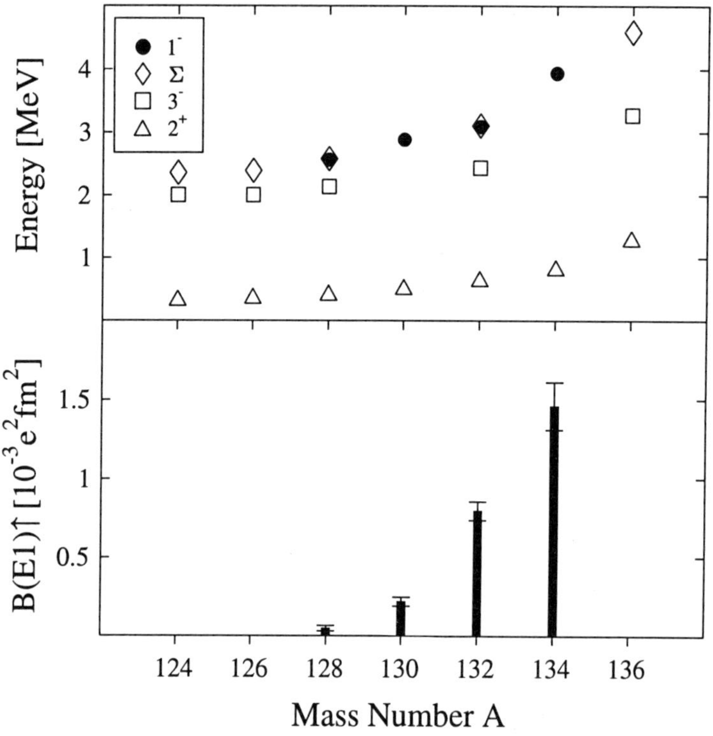

Figure 4. $E1$ two-phonon excitations in the even-even Xe isotopes. Upper part: Energies of the 2_1^+ (open triangles), 3_1^- (open squares) one-phonon excitations and of the observed 1^- two-phonon excitations (full circles) in ^{128,130,132,134}Xe compared to the expected sum energies $\Sigma = E_{2+} + E_{3-}$ (open rhombs). Lower part: Experimental $B(E1) \uparrow$ values for the two-phonon excitations (full bars).

exactly at the sum energy $\Sigma = E_{2+} + E_{3-}$ of the corresponding one-phonon excitations. This documents a rather harmonic coupling. In the lower part of the figure the $B(E1) \uparrow$ values are depicted. A steep decrease of the strengths is observed when moving away from the closed shell N=82. The same behavior was seen in recent Stuttgart NRF experiments on the Ba-isotopes [13]. The two-phonon excitations were observed only in ^{128,130,132,134}Xe. For the magic isotope ^{136}Xe (N=82) the excitation is expected at about 4.6 MeV an energy which is not accessible at the Stuttgart facility; in the lighter isotopes the expected strengths are too low to be detected (see Fig.4).

The reduced transition probabilities $B(E1, 1 \to 0)$ observed for the Xe isotopes (shown by full bars) fit nicely into the systematics o the two-phonon excitation in nuclei near the N=82 as can be seen in Fig.5. Here the $B(E1, 1 \rightarrow 0)$ values are plotted as a function of the neutron and and proton numbers, respectively. (For even-even nuclei holds: $B(E1, 1 \rightarrow 0) = 1/3 \cdot B(E1, 0 \rightarrow 1)$). Obviously, the strengths are maximal for magic nuclei and drop steeply when moving away from the shell closure marked by the bold face line at N=82. This behavior was observed at all shell closures and can be explained by the dipole core polarization effect (see Ref.[8]).

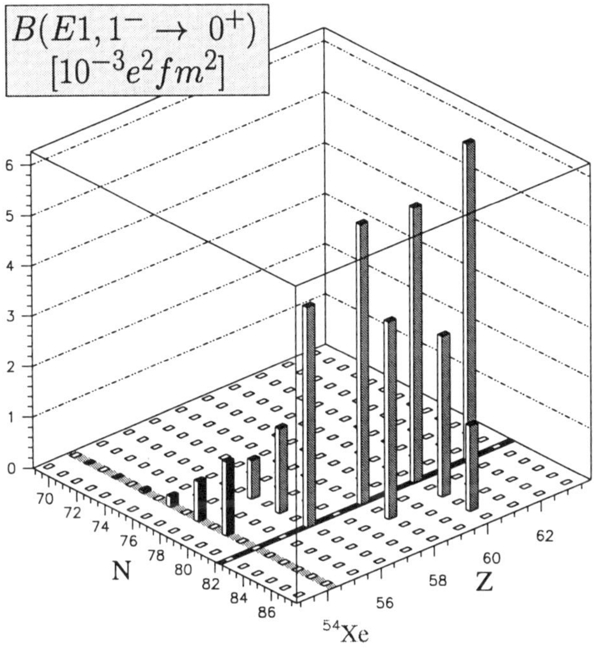

Figure 5. Strength systematics of $E1$ two-phonon excitations in the even-even nuclei near the N=82 shell closure. Plotted are the reduced transition probabilities $B(E1, 1 \rightarrow 0)$ as a function of the neutron and proton numbers N and Z, respectively. For details see text.

4.2. *M1 excitations to 1⁺ mixed-symmetry states*

Fig.6 shows the excitation energies (upper part) and $B(M1) \uparrow$ values (lower part) of the strong excitations ascribed to $M1$ excitations to 1^+ mixed symmetry states. The energies vary only smoothly between between 2.7 and 3.0 MeV. The total strengths increase when moving from the closed shell nucleus ^{136}Xe to the O(6) candidates ^{126,124}Xe were the full strength was observed as predicted within the O(6) limit. These expectation values are shown in Fig.6 as a dashed line.

Even if no direct parity determinations were possible, this interpretation is on rather safe grounds. First of all, the excitation energies and strengths are as expected from the systematics[14,15]. Furthermore, these states show strong decay branchings to the <u>second</u> $2+_2$ state, a characteristic for the 1^+ mixed symmetry states[9]. Last not least, the observed strengths scale linearly with β_2^2, the square of the deformation parameter, which is proportional to the $B(E2)$ value. This so-called β_2^2- or δ^2-law (when using an alternative definition of a deformation parameter) was first experimentally observed in the Sm[16,17] and Nd[18,19] isotopic chains as a basic property of the *Scissors Mode.* This dependence can be explained in several theoretical approaches, see, e.g., Ref.[20]. In Fig.7 the $B(M1)\uparrow$ values for the seven even-even Xe isotopes are plotted against the corresponding β_2^2 data taken from the compilation of Raman et al.[21]. For the lighter isotopes ^{126,124}Xe, which approach the O(6) limit, the $M1$ strengths seems to be somewhat fragmented, as expected. Therefore, the strengths were summed up in narrow energy intervalls as indicated by dashed lines in Fig.3.

4.3. *Hints for the E(5) Phase Transition ?*

The first order X(5) phase transition was observed experimentally in the N=90 isotones[22,23,24]. The sudden jump of increased $B(E2)$ values causes, as already discussed, the deformation splitting of the higher-lying electric GDR[2]. Due to the proportionality of $B(E2)$ and $B(M1)$ values[15] this also influences the $M1$ *Scissors Mode* strength distributions. On the other hand, the E(5) phase transition is expected to be of second order. The $B(E2)$ values in the Xe and Ba isotopes vary only smoothly and increase nearly linearly with decreasing neutron numbers (see, e.g., Ref.[13]). Therefore, it is not astonishing that no drastic changes were observed in the present dipole strength distributions in the Xe isotopes. Nevertheless, regarding the strengths of the $M1$ excitations to the mixed symmetry states (Fig.6), a change in the slope can be stated near A=128-130. These findings may be interpreted as a hint for a possible influence of the E(5) phase transition, since also the E_4/E_2 ratios for ^{128,130}Xe lie near to the E(5) expectation value of 2.2 (see Fig.1). However, it should be emphasized that for a clear manifestation of the E(5) phase transition, more spectroscopic information on the complete low-lying level scheme including transition probabilities and the various branching ratios is needed, as it is available for the best E(5) candidate up to now, the neighboring isotope ^{134}Ba [25].

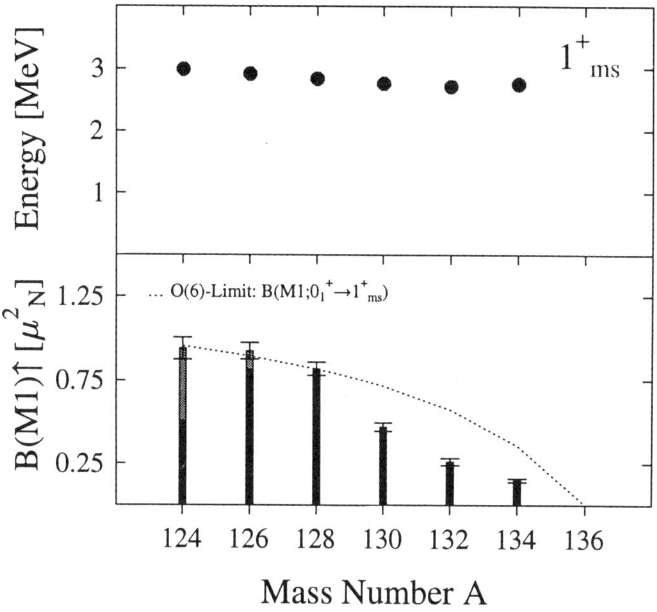

Figure 6. Experimental results for $M1$ excitations to 1^+ mixed-symmetry states in the even-even Xe isotopes. Upper part: Excitation energies. Lower part: Total $B(M1)\uparrow$ values. The dashed line corresponds to the O(6) expectation values.

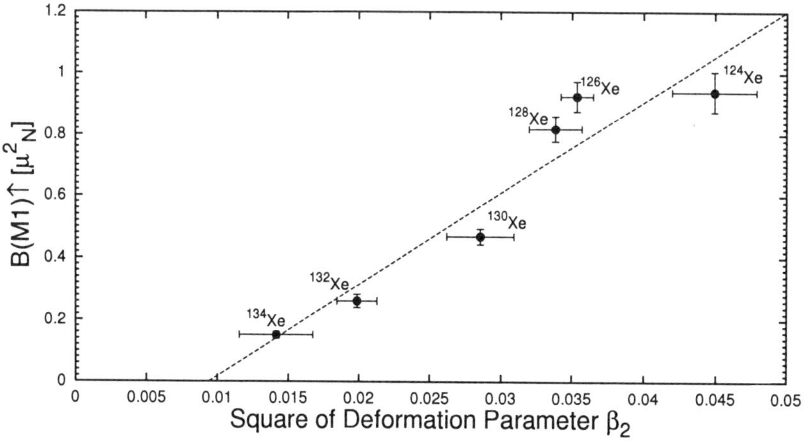

Figure 7. Observed total B(M1)↑ values as a function of the square of the deformation parameter β_2.

Acknowledgments

The author thanks P. von Brentano, C. Fransen, G. Friessner, H. von Garrel, N. Hollmann, J. Jolie, F. Käppeler, L. Käubler, C. Kohstall, L. Kostov, A.Linnemann, D. Mücher, N. Pietralla, H.H. Pitz, G. Rupp, G. Rusev, R. Schwengner, M.Scheck, F. Stedile, S. Walter, V. Werner, and K.Wisshak, all the members of the STUTTGART-KARLSRUHE-KÖLN-ROSSENDORF collaboration, who participated in the NRF experiments on the Xe isotopes, for the pleasant and fruitful collaboration. Special thanks are due to H. von Garrel, who analyzed the data within his PhD thesis. Stimulating discussions and suggestions by F. Iachello and N.V. Zamfir are gratefully acknowledged. Thanks are due to the Deutsche Forschungsgemeinschaft (DFG) for the longstanding financial support. Last not least, the author thanks Aldo Covello and his organizing team for the kind hospitality and warm atmosphere at the Seminar in Paestum.

References

1. F. Iachello, *Phys. Rev. Lett.* **85**, 3580 (2000) and ibid **87**, 052502 (2001).
2. P. Carlos *et al*, *Nucl. Phys.* **A1 172**, 437 (1971) and ibid **A 225**, 171 (1974).
3. U. Kneissl, H. H. Pitz, and A. Zilges, *Prog. Part. Nucl. Phys.* **37**, 349 (1996).
4. H. von Garrel, Doctoral Thesis, Stuttgart 2004, in preparation.
5. R. Reifrath *et al*, *Phys. Rev.* **C 66** , 064603 (2002).
6. B. Schlitt *et al*, *Nucl. Instr. a. Meth. in Phys. Res.* **A 337**, 416 (1994).
7. A. Degener *et al*, *Nucl. Phys.* **A 513**, 29 (1990).
8. W. Andrejtscheff *et al*, *Phys. Lett.* **B 506**, 239, (2001).
9. N. Pietralla *et al*, *Phys. Rev. Lett.* **83**, 1303 (1999).
10. N. Lo Iudice and F. Palumbo *Phys. Rev. Lett.* **41**, 1532 (1978).
11. D. Bohle *et al*, *Phys. Lett.* **137B**, 27 (1984).
12. N. V. Zamfir *et al*, *Phys. Rev.* **C 55**, R1007 (1997).
13. M. Scheck *et al*, *Phys. Rev.* **C**, to be published (2004).
14. N. Pietralla *et al*, *Phys. Rev.* **C 58**, 184 (1998).
15. N. Pietralla *et al*, *Phys. Rev.* **C 52**, R2317 (1995).
16. W. Ziegler *et al*, *Phys. Rev. Lett.* **65**, 2515 (1990).
17. W. Ziegler *et al*, *Nucl. Phys.* **A564**, 366 (1993).
18. J. Margraf *et al*, *Phys. Rev.* **C 47**, 1474 (1993).
19. T. Eckert *et al*, *Phys. Rev.* **C 56**, 1256 (1997) and ibid **C 57**, 1007 (1998).
20. N. Lo Iudice *et al*, *Phys. Lett.* **B 304**, 193 (1993).
21. S. Raman *et al*, *At.Data and Nucl.Data Tables* **78**, 1 (2001).
22. R. F. Casten and N. V. Zamfir, Phys. Rev. Lett. **87**, 052503 (2001).
23. R. Krücken *et al*, *Phys. Rev. Lett.* **88**, 232501 (2002).
24. D. Tonev *et al*, *Phys. Rev.* **C 69**, 034334 (2004).
25. R. F. Casten and N. V. Zamfir, *Phys. Rev. Lett.* **85**, 3584 (2000).

KEY TOPICS IN NUCLEAR STRUCTURE
Proceedings of the 8th Int. Spring Seminar on Nuclear Physics
edited by Aldo Covello
© 2005 World Scientific Publishing Co.

ROLE OF THERMAL PAIRING IN REDUCING THE GIANT DIPOLE RESONANCE WIDTH AT LOW TEMPERATURE

N. DINH DANG

RI-beam factory project office, RIKEN
2-1 Hirosawa, Wako city, 351-0198 Saitama, Japan
E-mail: dang@postman.riken.go.jp

A. ARIMA

House of Councilors, Nagatacho 2-1-1, Chyodaku, 100-8962 Tokyo, Japan
E-mail: akito_arima@sangiin.go.jp

The neutron thermal pairing gap determined from the modified BCS theory is included in the calculation of the width of the giant dipole resonance (GDR) at finite temperature T in ^{120}Sn within the Phonon Damping Model. The results obtained show that thermal pairing causes a smaller GDR width at $T \leq 2$ MeV as compared to the one obtained neglecting pairing. This effect improves significantly the agreement between theory and experiment including the most recent data point at $T = 1$ MeV.

Intensive experimental studies of highly-excited nuclei during the last two decades have produced many data on the evolution of the giant dipole resonance (GDR) as a function of temperature T and spin. The data show that the GDR width increases sharply with increasing T from $T \geq 1$ MeV up to $\simeq 3$ MeV. At higher T a width saturation has been reported (See [1] for the most recent review). The increase of the GDR width with T is described within the thermal shape fluctuation model (TSFM) [2,3] and the phonon damping model (PDM) [4] as shown in Fig. 1. The TSFM incorporates the thermal fluctuation of nuclear shapes, which accounts for the width increase. The PDM considers the coupling of the GDR to *pp* and *hh* configurations at $T \neq 0$ as the mechanism of the width increase and saturation. In general, pairing was neglected in the calculations for hot GDR as it was believed that the gap vanishes at $T = T_c < 1$ MeV according to the temperature BCS theory. However, it has been shown in [5] that thermal fluctuations smear out the superfuid-normal phase transition in finite systems so that the pairing gap survives up to $T \gg 1$ MeV. This has been confirmed microscopically in the recent modified Hartree-Fock-Bogoliubov (MHFB) theory at finite T [6], whose limit is the modified BCS (MBCS) theory [7,8]. Other approaches such as the static-path approximation [9], shell model calculations [10], as well as the exact solution of the pairing problem [11] also show that pairing correlations do not abruptly disappear at $T \neq 0$. It was suggested for the first time in [12] that the decrease of the pairing gap with increasing T, which is also caused by *pp* and *hh* configurations at low T, may slow down the increase of the GDR width. By including a simplified T-dependent pairing gap in the CASCADE calculations using the PDM strength functions, Ref. [13] has improved the agreement between the calculated GDR shapes and experimental ones. Very recently the γ decays

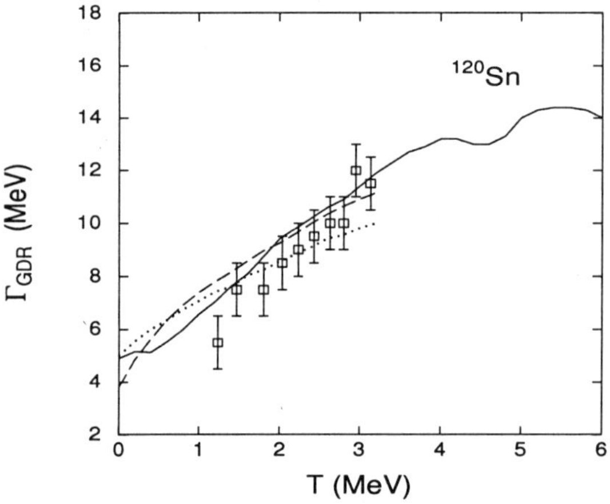

Figure 1. GDR width Γ described within the PDM (solid line) and TSFM (dotted [2] and dashed [3] lines) as a function of T for ^{120}Sn. Pairing is not included.

were measured in coincidence with ^{17}O particles scattered inelastically from ^{120}Sn. A GDR width of around 4 MeV has been extracted at $T = 1$ MeV [14], which is smaller than the value of ~ 4.9 MeV for the GDR width at $T = 0$. This result and the existing systematic for the GDR width in ^{120}Sn up to $T \simeq 2$ MeV are significantly lower than the prediction by the TSFM. Based on this, Ref. [14] concluded that the narrow width observed in ^{120}Sn at low T is not understood. This talk will show that thermal pairing causes the narrow GDR width in ^{120}Sn at low T.

For this purpose we include the thermal pairing gap obtained from the MBCS theory [6,7,8] in the PDM [4], and carry out the calculations for the

GDR width in ^{120}Sn at $T \leq 5$ MeV. The GDR width is presented as the sum of quantal (Γ_Q) and thermal (Γ_T) widths as [4,15]

$$\Gamma = \Gamma_Q + \Gamma_T \ , \tag{1a}$$

$$\Gamma_Q = 2\pi F_1^2 \sum_{ph} [u_{ph}^{(+)}]^2 (1 - n_p - n_h) \delta(E_{\rm GDR} - E_p - E_h) \ , \tag{1b}$$

$$\Gamma_T = 2\pi F_2^2 \sum_{ss'} [v_{ss'}^{(-)}]^2 (n_{s'} - n_s) \delta(E_{\rm GDR} - E_s + E_{s'}) \ , \tag{1c}$$

where $(ss') = (pp')$ and (hh') with p and h denoting the orbital angular momenta j_p and j_h for particles and holes, respectively. The quantal and thermal widths come from the couplings of quasiparticle pairs $[\alpha_p^\dagger \otimes \alpha_h^\dagger]_{LM}$ and $[\alpha_s^\dagger \otimes \tilde{\alpha}_{s'}]_{LM}$ to the GDR, respectively. At zero pairing they correspond to the couplings of ph $[a_p^\dagger \otimes \tilde{a}_h]_{LM}$ and pp (hh) $[a_s^\dagger \otimes \tilde{a}_{s'}]_{LM}$ pairs to the GDR, respectively (The tilde ~ denotes the time-reversal operation). The quasiparticle energies $E_j = [(\epsilon_j - \bar{\mu})^2 + \bar{\Delta}^2]^{1/2}$ are found from the MBCS equations (39) and (40) of [8], which determine the modified thermal gap $\bar{\Delta}$ and chemical potential $\bar{\mu}$ from the single particle energies ϵ_j and particle number N. From them one defines the Bogoliubov coefficients u_j, v_j, and combinations $u_{ph}^{(+)} = u_p v_h + v_p u_h$, and $v_{ss'}^{(-)} = u_s u_{s'} - v_s v_{s'}$. The quasiparticle occupation number n_j is given by the Fermi-Dirac distrbution as $n_j = [\exp(E_j/T) + 1]^{-1}$. The GDR energy $E_{\rm GDR}$ is found as the solution of the equation $\omega - \omega_q - P(\omega) = 0$, where ω_q is the unperturbed phonon energy, and $P(\omega)$ is the polarization operator:

$$P(\omega) = F_1^2 \sum_{ph} \frac{[u_{ph}^{(+)}]^2 (1 - n_p - n_h)(E_p + E_h)}{\omega^2 - (E_p + E_h)^2}$$

$$-F_2^2 \sum_{ss'} \frac{[v_{ss'}^{(-)}]^2 (n_s - n_{s'})(E_s - E_{s'})}{\omega^2 - (E_s - E_{s'})^2} \ . \tag{2}$$

The PDM has three T-independent parameters. The parameters ω_q and F_1 are chosen to reproduce the quantal width Γ_Q and GDR energy at $T = 0$, while F_2 is chosen such as the GDR energy does not change appreciably with T. Their values for ^{120}Sn are given in [4] for the zero-pairing case. At $\bar{\Delta} = 0$, one has $u_p = 1$, $v_p = 0$, $u_h = 0$, $v_h = 1$ so that $[u_{ph}^{(+)}]^2 = 1$, $[v_{ss'}]^2 = 1$. As for the single-particle occupation number $f_j = u_j^2 n_j + v_j^2(1 - n_j)$, one obtains $f_h = 1 - n_h$ and $f_p = n_p$ at zero pairing. The PDM equations for $\bar{\Delta} = 0$ [4] are then easily recovered from Eqs. (1) – (2).

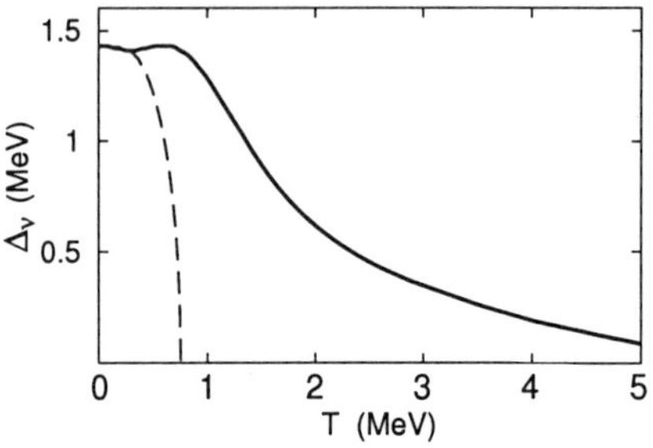

Figure 2. Neutron pairing gap as a function of T. Solid and dashed lines show the MBCS gap $\bar{\Delta}$ and BCS gap, respectively [7,8].

Shown in Fig. 2 is the temperature dependence of the neutron pairing gap $\bar{\Delta}_\nu$ for ^{120}Sn, which is obtained from the MBCS equations [6,7,8] using the single-particle energies determined within the Woods-Saxon potential at $T = 0$. They span a space from $\sim$ -40 MeV up to $\sim$ 17 MeV including 7 major shells and $1j_{15/2}$, $1i_{11/2}$, and $1k_{17/2}$ levels. The pairing parameter G_ν is chosen to be equal to 0.13 MeV, which yields $\bar{\Delta}(T=0) \equiv \bar{\Delta}(0) \simeq 1.4$ MeV. In difference with the BCS gap, which collapses at $T_c \simeq 0.79$ MeV, the gap $\bar{\Delta}_\nu$ does not vanish, but decreases monotonously with increasing T at $T \geq 1$ MeV resulting in a long tail up to $T \simeq 5$ MeV due to thermal fluctuation of quasiparticle number in the MBCS equations [6,7,8].

The single-particle occupation number f_j is plotted as a function of single-particle energy ϵ_j for the neutron levels around the chemical potential in Fig. 3. It is seen that, in general, the pairing effect always goes counter the temperature effect on f_j, causing a steeper dependence of f_j on ϵ_j. Decreasing with increasing T, this difference becomes small at $T \geq 3$ MeV. Since a smoother f_j enhances the pp and hh transitions leading to the thermal width $\Gamma_{\rm T}$ [4], pairing should reduce the GDR width, and the effect is expected to be stronger at a lower T, provided the GDR energy $E_{\rm GDR}$ is the same. A deviation from this general rule is seen at very low $T \simeq 0.1$ MeV, where the temperature effect is still so weak that f_j obtained at $\bar{\Delta} \neq 0$ (solid line) is smoother than that obtained at zero pairing (dotted line).

The GDR width Γ was calculated from Eq. (1) for ^{120}Sn using the same set of PDM parameters ω_q, F_1, and F_2, which have been chosen for the zero-pairing case [4] (set A). The result is shown as the thin solid line in Fig. 4. The oscillation of GDR energy $E_{\rm GDR}$ as T varies occurs within the range of $\sim \pm 1.5$ MeV, which is wider compared with that obtained neglecting pairing. As expected from the discussion above, the GDR energy

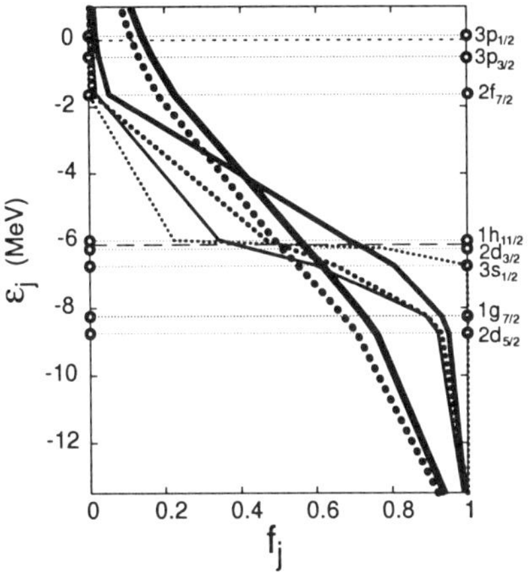

Figure 3. Single-particle occupation number f_j as a function of ϵ_j for the neutron levels around the chemical potential at T = 0.1, 1, and 3 MeV. Results obtained including and without pairing are shown by solid and dotted lines, respectively (A thicker line corresponds to a higher T). The lines are drawn just to guide the eyes. The horizontal dashed line at ~ -6 MeV shows the chemical potential at $\bar{\Delta} = 0$ and $T = 0$.

$E_{\rm GDR}$ at $T = 0.1$ MeV drops to 14 MeV, i.e. by 1.4 MeV lower than the GDR energy measured on the ground state. The GDR width increases to 5.3 MeV compared to 4.9 MeV on the ground state. At $0 \leq T \leq 0.5$ MeV, the above-mentioned competition between the decreasing quantal and increasing thermal widths makes the total width decrease first to reach a minimum of 3.4 MeV at $T \simeq 0.2$ MeV then increase again with T. At $T \geq T_{\rm c}$ the width only increases with T. At $1 \leq T \leq 3$ MeV the GDR width obtained including pairing is smaller than the one obtained neglecting pairing (dashed line), but this difference decreases with increasing T so that at $T > 3$ MeV, when the gap $\bar{\Delta}$ becomes small, both values nearly coincide. This improves significantly the agreement with the experimental systematic at $1 \leq T \leq 2.5$ MeV. In order to have the same value of 4.9 MeV for the GDR width at $T \simeq 0$, we also carried out the calculation using slightly readjusted values $F_1' = 0.96F_1$ and $F_2' = 1.03F_2$ while keeping the same ω_q (set B). The result obtained is shown in the same Fig. 4 as the thick solid line. The GDR energy $E_{\rm GDR}$ moves up to 16.6 MeV at $T = 0.1$ MeV in agreement with the value of 16.5 $\pm$ 0.7 MeV extracted in [14]. The width at $T = 1$ MeV also becomes slightly smaller, which agrees quite well with the latest experimental point [14]. At $T > 1$ MeV the results obtained using two parameter sets, A and B, are nearly the same. The effect of

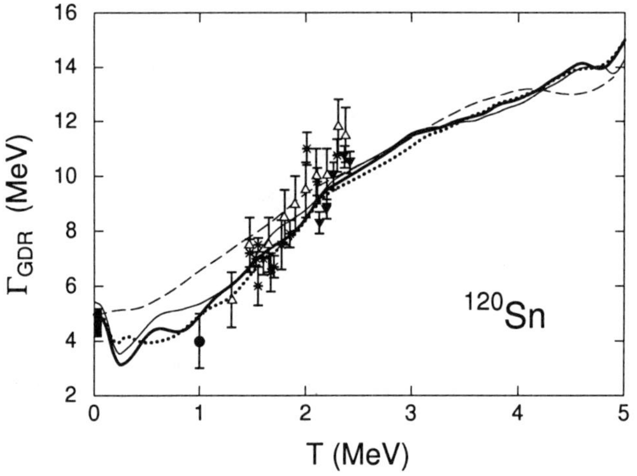

Figure 4. GDR width Γ as a function of T for ^{120}Sn. The dashed line shows the PDM result obtained neglecting pairing (the solid line in Fig. 1). The thin and thick solid lines are the PDM results including the gap $\bar{\Delta}$, which are obtained using the parameter sets A and B, respectively. The dotted line is the PDM result including the renormalized gap $\widetilde{\Delta}$ (See text). The solid circle is the low-T data point from Ref. [14]. Crosses and open triangles are from Fig. 4 of Ref. [3]. Solid upside-down triangles are data from Ref. [16]. The rectangle at $T = 0$ is GDR widths on the ground state for tin isotopes with masses $A = 116 - 120$ from Ref. [17].

quantal fluctuations ΔN^2 of particle number within the BCS theory at $T = 0$, however, is neglected in these results. To be precise, this effect should be included using the particle-number projection method at finite T. The latter is so computationally intensive that the calculations were carried out so far only within schematic models (See, e.g. [18]), or one major shell for nuclei with A $\leq$ 60 as in the Shell Model Monte Carlo method [19]. For the limited purpose of the present study, assuming that $\Delta N^2 \gg 1$, we applied the approximated projection at $T = 0$ proposed in [20], which leads to the renormalization of the gap as $\widetilde{\Delta}(T) = [1 + 1/\Delta N^2]\bar{\Delta}$ with $\Delta N^2 = \bar{\Delta}(0)^2 \sum_j (j+1/2)/[(\epsilon_j - \bar{\mu})^2 + \bar{\Delta}(0)^2]$ [21]. This yields $\widetilde{\Delta}(T=0) \simeq$ 1.5 MeV ($\Delta N \simeq \pm$ 4). The PDM result obtained using the gap $\widetilde{\Delta}$ and the parameter set B is shown in the same Fig. 4 as the dotted line. The GDR width becomes 5 MeV with $E_{\rm GDR} = 15.3$ MeV at $T = 0$ in good agreement with the GDR parameters extracted on the ground state. It is seen that the fluctuation of the width at $T \leq 0.5$ MeV is largely suppressed by using this renormalized gap $\widetilde{\Delta}$.

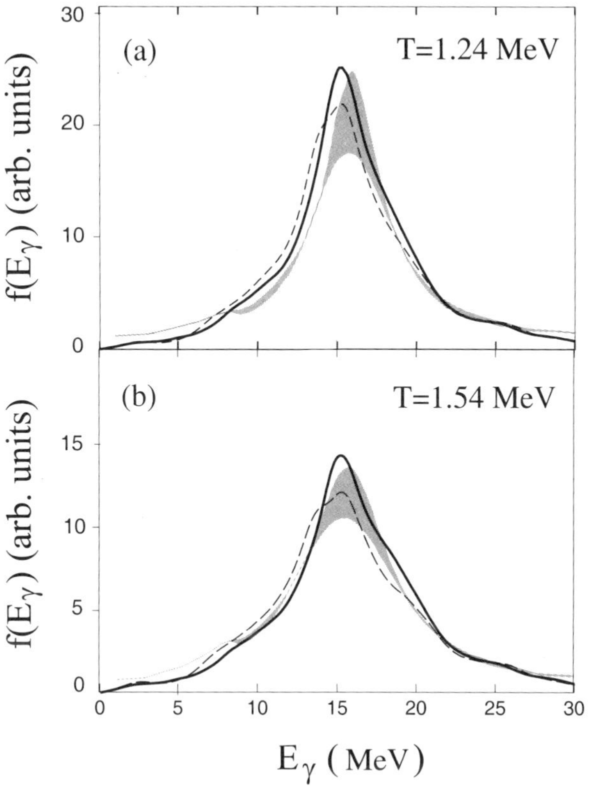

Figure 5. Experimental (shaded areas) and theoretical divided spectra obtained within PDM without pairing (dashed lines) and including the gap $\bar{\Delta}$ (thick solid lines) as for the thick solid line in Fig. 4.

Shown in Fig. 5 are GDR cross-sections obtained within PDM for ^{120}Sn using Eq. (1) of Ref. [13]. The experimental cross-section are taken from Fig. 2 of Ref. [13]. They have been generated by CASCADE at excitation energy $E^* = 30$ and 50 MeV, which correspond to $T_{\rm max} = 1.24$, and 1.54 MeV, respectively. The theoretical cross-sections have been obtained using the PDM strength function $S_{\rm GDR}(E_\gamma)$ [4] at $T = T_{\rm max}$. This is the low temperatures region, at which discrepancies are most pronounced between theory and experiment. (A divided spectrum free from detector response at $T = 1$ MeV is not available in Ref. [14]). From this figure it is seen that thermal pairing clearly offers a better fit to the experimental line shape of the GDR at low temperature.

In conclusion, we have included the pairing gap, determined within the MBCS theory [7,8], in the PDM to calculate the GDR width in ^{120}Sn at $T \leq 5$ MeV. In difference with the gap given within the conventional BCS theory, which collapses at $T_c \simeq 0.79$ MeV, the MBCS gap never vanishes, but monotonously decreases with increasing T. The results obtained show that thermal pairing indeed plays an important role in lowering the width at $T \leq 2$ MeV as compared to the value obtained without pairing. This improves significantly the overal agreement between theory and experiment. Hence, the small GDR width at $T = 1$ MeV extracted in the latest experiment [14] is explained as a manisfestation of thermal pairing at low temperature.

References

1. M.N. Harakeh and A. van der Woude, Giant resonances - Fundamental high-frequency modes of nuclear excitation (Clarendon Press, Oxford, 2001).
2. W.E. Ormand, P.F. Bortignon, and R.A. Broglia, Phys. Rev. Lett. **77**, 607 (1966); W.E. Ormand *et al.*, Nucl. Phys. A **614**, 217 (1997).
3. D. Kusnezov, Y. Alhassid, and K.A. Snover, Phys. Rev. Lett. **81**, 542 (1998) and references therein.
4. N.D. Dang and A. Arima, Phys. Rev. Lett. **80**, 4145 (1998); Nucl. Phys. A **636**, 427 (1998); N. Dinh Dang, K. Tanabe, and A. Arima, Nucl. Phys. A **645**, 536 (1998).
5. L.G. Moretto, Phys. Lett. B **40**, 1 (1972).
6. N. Dinh Dang and A. Arima, Phys. Rev. C **86**, 014318 (2003).
7. N. Dinh Dang and V. Zelevinsky, Phys. Rev. C **64**, 064319 (2001).
8. N. Dinh Dang and A. Arima, Phys. Rev. C **67**, 014304 (2003).
9. N.D. Dang, P. Ring, and R. Rossignoli, Phys. Rev. C **47**, 606 (1993).
10. V. Zelevinsky, B.A. Brown, N. Frazier, and M. Horoi, Phys. Rep. **276**, 85 (1996).
11. A. Volya, B.A. Brown, and V. Zelevinsky, Phys. Lett. B **509**, 37 (2001).
12. N. Dinh Dang, K. Tanabe, and A. Arima, Nucl. Phys. A **675**, 531 (2000).
13. N. Dinh Dang, K. Eisenman, J. Seitz, and M. Thoennessen, Phys. Rev. C **61**, 027302 (2000).
14. P. Heckman *et al.*, Phys. Lett. B **555**, 43 (2003).
15. N.D. Dang, V.K. Au, T. Suzuki, and A. Arima, Phys. Rev. C **63**, 044302 (2001).
16. M.P. Kelly *et al.*, Phys. Rev. Lett. **82**, 3404 (1999).
17. B.L. Berman, At. Data Nucl. Data Tables **15**, 319 (1975).
18. R. Rossignoli, N. Canosa, and P. Ring, Phys. Rev. Lett. **80**, 1853 (1998).
19. Y. Alhassid, S. Liu, and H. Nakada, Phys. Rev. Lett. **83**, 4265 (1999).
20. I.N. Mikhailov, Sov. Phys. JETP **18**, 761 (1964).
21. N.D. Dang, Z. Phys. A **335**, 253 (1990).

KEY TOPICS IN NUCLEAR STRUCTURE
Proceedings of the 8th Int. Spring Seminar on Nuclear Physics
edited by Aldo Covello
© 2005 World Scientific Publishing Co.

COLLECTIVITY IN LIGHT NUCLEI AND THE GDR

A. MAJ, J. STYCZEŃ, M. KMIECIK, P. BEDNARCZYK*, M. BREKIESZ, J. GRĘBOSZ, M. LACH, W. MĘCZYŃSKI, M. ZIĘBLIŃSKI AND K. ZUBER
The Niewodniczański Institute of Nuclear Physics, PAN
ul. Radzikowskiego 152, 31-342 Kraków, Poland
E-mail: Adam.Maj@ifj.edu.pl

A. BRACCO, F. CAMERA, G. BENZONI, S. LEONI, B. MILLION AND O. WIELAND
Dipartimento di Fisica, Universitá di Milano and INFN Sez. Milano
Via Celoria 16, 20133 Milano, Italy

The results are presented from the experiments using the EUROBALL and RFD/HECTOR arrays, concerning various aspects of collectivity in light nuclei. A superdeformed band in ^{42}Ca was found. A comparison of the GDR line shape data with the predictions of the thermal shape fluctuation model, based on the most recent rotating liquid drop LSD calculations, shows evidence for a Jacobi shape transition in hot, rapidly rotating ^{46}Ti and strong Coriolis effects in the GDR strength function. The preferential feeding of the SD band in ^{42}Ca by the GDR low energy component was observed

1. Introduction

The excited states in nuclei are traditionally interpreted either as the collective excitations (e.g. in models based on rotating liquid drop), or as many single-particle excitations (e.g. in shell models), or as the coupling of single-particle excitations to the collective modes. Moreover, collective oscillations as the Giant Dipole Resonance (GDR) were found in hot nuclei and satisfactory interpreted within the statistical decay models and shape fluctuation models.

The light nuclei, with mass ranging from A=30 to A=70, open new horizons for studying the excited states in nuclei. On one hand, the number of nucleons is here large enough to align and to form high spins. Because

*Present address: GSI Darmstadt, Germany

of, however, the relatively light mass and consequently small value of the moment of inertia, the angular velocities associated with high spins are extremely large. One would, therefore, expect to observe much easier effects which are related to the rapid rotation, e.g. change in the deformation, Coriolis effects, etc. On the other hand, the number of nucleons is small enough so that the band termination is easily reached, and the related single-particle effects are expected to prevail at highest spins. Also the shape fluctuations with such low number of nucleons are predicted to be sizeable. All this means that in light nuclei one should not expect to have clearly separated collective and single particle excitation modes, but rather a strong mixture of those two extreme approaches

2. Low-T regime: superdeformation in light nuclei

A textbook manifestation of the collective rotation in nuclei is the observation of the superdeformed (SD) bands. Such bands are characterized by a long cascade of γ- transitions, following the pattern of a rotor with constant moment of inertia, and with a value reflecting the 2:1 axis ratio of the nucleus. In light nuclei, the superdeformation is additionally associated with the extremely rapid rotation. Rotational frequency deduced for the SD bands known in this region can reach even 2 MeV. A typical example of the SD band in such nuclei is the one observed[1] in ^{40}Ca. The kinematical moment of inertia of that SD band (shown in Fig. 1 with open circles) is constant as a function of rotational frequency, which proofs a very stable superdeformed configuration. Similar SD bands were observed in several other nuclei in this mass region. Some of them, however, display less constant behavior in function of the rotational frequency. For example[2] in ^{61}Cu, the SD band shows gradual decrease of the moment of inertia as a function of rotational frequency (Fig. 1, full diamonds). This may indicate an important contribution of non collective degrees of freedom in building states with high angular momentum when the band termination[3] is approached.

The main experimental difficulty in studying the high spin phenomena in the light mass region is an excessive Doppler broadening of lines in in-beam γ-spectra. This is due to a high recoil velocity of residues produced in fusion-evaporation reactions and, of course, due to high energy of γ-rays expected to occur in light nuclei. This constrain can be minimized by making use of the Recoil Filter Detector (RFD)[4], when coupled to the Ge-array. The RFD is a system of 18 heavy ion detectors distributed around the beam axis placed downstream. It measures a time of flight of incoming residual nuclei produced in the reaction with respect to a beam pulse, as well as their flight direction. In this way the RFD enables a complete determination of the velocity vector of every recoiling nucleus, thus the event by event Doppler correction can be performed.

To study the high spin states in ^{42}Ca nucleus in the experiment at VIVITRON accelerator of the IReS Laboratory of Strasbourg (France), the RFD was coupled to the EUROBALL. The resulting complex level scheme[5] was dominated by the single-particle excitation and by rotational bands of normal deformation. Nevertheless it was possible, due to the improved resolution, to single out a band showing the superdeformed character. The extracted moment of inertia of this band is plotted with full squares also in Fig. 1. It shows somewhat irregular behavior at low rotational frequencies, but at higher the behavior becomes smoother and the value of the relative kinematical moment of inertia is the same as for ^{40}Ca and ^{61}Cu. This band and other discrete transitions in this nucleus were used to analyze the properties of the GDR.

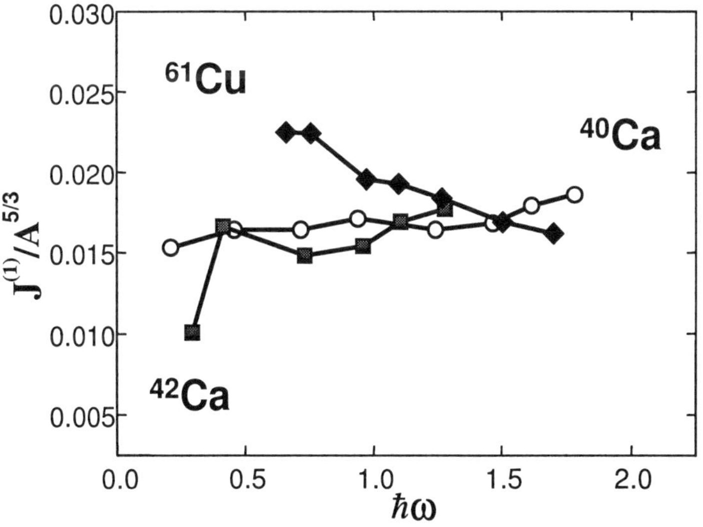

Figure 1. Relative kinematical moments of inertia for ^{40}Ca (open circles), ^{61}Cu (full diamonds) and ^{42}Ca (full squares) as a function of rotational frequency.

3. High-T regime: Jacobi shape transitions and Coriolis effects in the GDR in light nuclei

The Jacobi shape transition, an abrupt change of nuclear shape from an oblate ellipsoid non-collectively rotating around its symmetry axis to an elongated prolate or triaxial shape, rotating collectively around the shortest axis has been predicted to appear in many nuclei at angular momenta close to the fission limit. In particular, recently developed LSD (Lublin-Strasbourg Drop) model[6,7] has been used to calculate the Jacobi transition mechanism in ^{46}Ti nucleus. The results of these calculation[8] have shown, that the equilibrium shape of the nucleus (in this liquid drop approximation) is spherical at I=0 and nearly spherical for I<$10\hbar$, becomes oblate at larger spin value and the size of the oblate deformation becomes larger when the spin increases. At around I=$28\hbar$ the Jacobi shape transition sets in: the nucleus becomes unstable towards triaxial and then (for I>$34\hbar$) it follows prolate shape configurations, with rapidly increasing size of the deformation up to the fission limit (around I=$40\hbar$).

Following these prediction, an another experiment at the VIVITRON accelerator was performed, using the EUROBALL phase IV Ge-array coupled to the HECTOR array[9]. In this experiment the ^{46}Ti compound nucleus was populated in the ^{18}O+^{28}Si reaction at E_B=105 MeV (for details see Ref. 10).

The GDR spectrum, gated on known, well resolved low energy γ-ray transitions of ^{42}Ca and on high angular momentum region of the decaying compound nucleus, is shown in Fig. 2 (left panel) together with the best fit Monte Carlo Cascade calculations. The quality of the fit can be judged more clearly by inspecting the right panel of Fig. 2, where the GDR line shape, i.e. the extracted absorption cross-section using the method described in e.g. Ref. 11, is shown. One feature of the obtained GDR line shape is a broad high-energy component centered at around 25 MeV. Much more pronounced, however, is a narrow low-energy component at 10.5 MeV. This fit shows also that the average GDR line shape has to be approximated with at least 3 components.

In order to interpret this GDR line shape, we use the same approach as has been adopted in many studies concerning the GDR in hot and rotating nuclei (see e.g. Ref. 9), namely the thermal shape fluctuation model (see Ref. 12 and references therein) which assumes that the average GDR line shape is the weighted sum of individual (i.e. at given deformation) GDR line shapes. The weighting factors (the probability of finding the

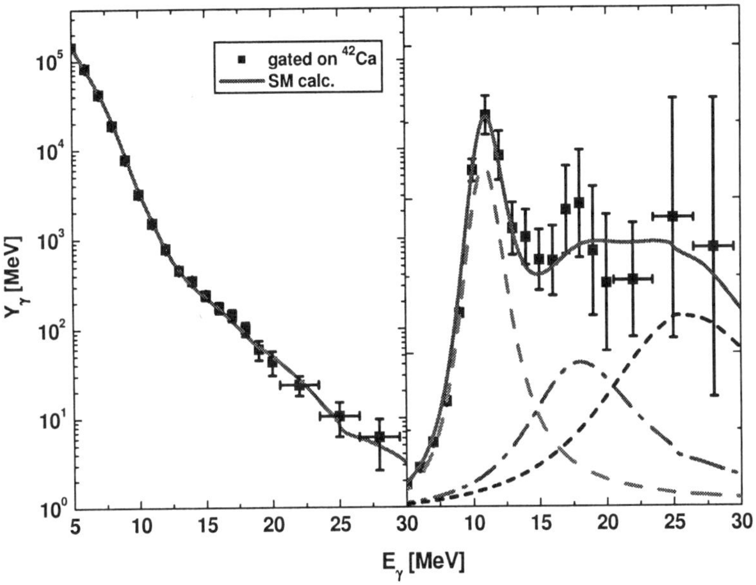

Figure 2. Left panel: The high-energy γ-ray spectrum gated by the ^{42}Ca transitions and by high fold region, in comparison with the best fitting statistical model calculations (full drawn line) assuming a 3-Lorentzian GDR line shape. Right panel: The deduced experimental GDR strength function (full drawn line) together with best fitting 3-Lorentzian function and its individual components.

nucleus at a given deformation value) are calculated by using the macroscopic deformation-dependent LSD energies. In the calculations, we include the possibility of Coriolis splitting of the GDR strength function for given spin and deformation value using the rotating harmonic oscillator model[13]. Thus, for each deformation point the GDR line shape consists in general of 5-Lorentzian parametrization. The results for spin region I=28-34 are presented together with the experimental GDR strength function in Fig. 3a.

A noteworthy good agreement between the theoretical predictions and the present experimental results can be observed. For comparison, the calculated averaged GDR line shape for I=24, i.e. in the oblate regime, is shown with dashed line in the same figure. This agreement might be an evidence of observation of the predicted Jacobi shape transition. Additional signature for the Jacobi transition may come from the presence of two other broad components in the strength function at higher energies both in the experiment and calculations. One component, at around 17 MeV, is clearly seen in the present data, and the other which is very broad (20-30 MeV) can also be identified

To demonstrate the importance of the Coriolis effect, Fig. 3b shows the same calculations of the average GDR line shape when neglecting the Coriolis splitting. As can be seen, in this case the low-energy component has higher energy than the experimental one, and also the entire GDR line shape does not reproduce the experimental data. The predictions for the oblate regime (I=24) are very weakly sensitive to the Coriolis effect.

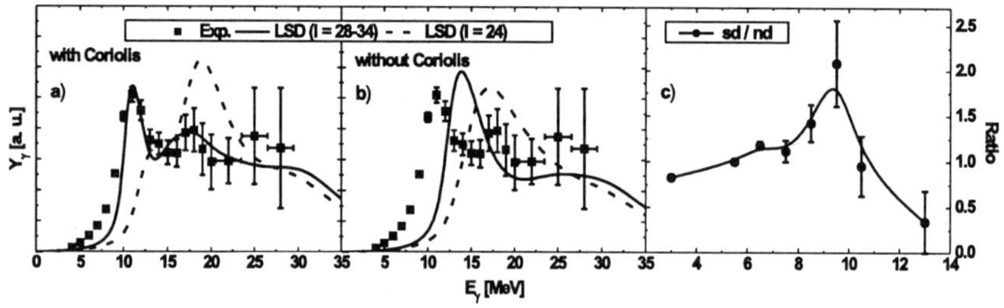

Figure 3. a) The full drawn line shows the theoretical prediction for the spin region 28-34 of the GDR line shape in ^{46}Ti obtained from the thermal shape fluctuation model (including the Coriolis splitting of the GDR components) based on free energies from the LSD model calculations. The dashed line shows similar prediction for I=24. The filled squares are the experimental GDR strength function; b) The same, as in a), but in the calculations the Coriolis splitting was not taken into account; c) The ratio of the γ-ray intensity in the superdeformed band to the intensity in the normal deformed band in ^{42}Ca, as a function of the associated high energy γ-rays from the GDR decay of ^{46}Ti.

4. Link between high-T and low-T regimes: GDR feeding of the SD band in ^{42}Ca

To see how the different regions of high-energy γ-rays feed the discrete lines in ^{42}Ca residual nucleus, the gates (1 MeV wide) were set on the GDR spectrum and with such a condition the discrete line intensities were analyzed. The ratio of the intensity within the SD band in ^{42}Ca (the one shown with full squares in Fig. 1) to the intensity of transition between states with normal deformation, is plotted in Fig. 3c. The ratio was normalized arbitrarily to 1 at 6 MeV. As can be seen, in the region 8-10 MeV the ratio is larger by a factor of 2 as compared to the low energy (statistical) region.. Considering that the gates were set on the raw spectrum (left panel of Fig. 2), not corrected for the detector's response function, this 8-10 MeV bump in the ratio corresponds to the 10.5 MeV low energy component of the GDR strength function shown in Fig. 3a.

This might indicate that the low energy component of the GDR in the compound nucleus ^{46}Ti feeds preferentially the SD band in the ^{42}Ca evaporation residue. One can also speculate that the highly deformed shapes created by the Jacobi shape transition at rapidly spinning *hot* light nuclei persist the evaporation process and this results in population the superdefomed bands in *cold* high-spin residua.

One should note that similar preferential feeding of the SD-bands by the low energy component of the GDR has been observed[14] in the case of ^{143}Eu.

5. Summary

A collective rotation forming the superdeformed band has been observed in ^{42}Ca, with a moment of inertia similar to the other measured in this mass region. High-energy γ-ray spectrum from the hot ^{46}Ti compound nucleus measured in coincidence with discrete transition in the ^{42}Ca residues shows highly fragmented GDR strength function with a broad 15-25 MeV structure and a narrow low energy 10.5 MeV component. This can be interpreted as the result of Jacobi shape transition and strong Coriolis effects. In addition the low energy GDR component seems to feed preferentially the superdeformed band in ^{42}Ca. This suggests that the very deformed shapes after the Jacobi shape transition in hot compound nucleus persist during the evaporation process. Thus the Jacobi shape transition in the compound nucleus might constitute kind of a gateway to very elongated, rapidly rotating cold nuclear shapes.

Acknowledgments

We appreciate very much the help of B. Herskind from NBI Copenhagen; E. Farnea, G. de Angelis and D. Napoli from LNL Legnaro; S. Brambilla, M. Pignanelli and N. Blasi from Milano; M. Kicińska-Habior from Warsaw; J. Nyberg from Uppsala; C.M. Petrache from Camerino; D. Curien, J. Dudek and N. Dubray from Strasbourg; and K. Pomorski from Lublin. A financial support from the Polish State Committee for Scientific Research (KBN Grant No. 2 P03B 118 22), the European Commission contract EUROVIV and the Italian INFN is acknowledged.

References

1. E. Ideguchi, D.G. Sarantites, W. Reviol, A.V. Afanasjev, M. Devlin, C. Baktash, R.V.F. Janssens, D. Rudolph, A. Axelsson, M.P. Carpenter, A. Galindo-Uribarri, D.R. LaFosse, T. Lauritsen, F. Lerma, C.J. Lister, P. Reiter, D. Seweryniak, M. Weiszflog, and J.N. Wilson, *Phys. Rev. Lett.* **87**, 222501 (2001).
2. J. Dobaczewski, J.P.Vivien, K. Zuber, P.Bednarczyk, T.Byrski, D.Curien, G. de Angelis, O. Dorvaux, G. Duchene, E. Farnea, A. Gadea, B. Gall, J. Grębosz, R. Isocrate, A. Maj, W. Męczyński, J.C. Merdinger, A. Prevost, N. Redon, J. Robin, O. Stezowski, J. Styczeń, M. Ziębliński, *AIP Conf. Proc.* **701**, 273 (2004).
3. A.V. Afanasjev, I. Ragnarsson and P. Ring, *Phys.Rev.* **C59**, 3166 (1999).
4. P. Bednarczyk, W. Męczyński, J. Styczeń, J. Grębosz, M. Lach, A. Maj, M. Ziębliński, N. Kintz, J.C. Merdinger, N. Schulz, J.P. Vivien, A. Bracco, J.L. Pedroza, M.B. Smith, K.M. Spohr, *Acta Phys. Pol.* **B32**, 747 (2000).
5. M. Lach, J. Styczeń, W. Męczyński, P. Bednarczyk, A. Bracco, J. Grębosz, A. Maj, J.C. Merdinger, N. Schulz, M.B. Smith, K.M. Spohr, J.P. Vivien, and M. Ziębliński, *Eur Phys J.* **A12**, 381 (2001).
6. K. Pomorski and J. Dudek, *Phys. Rev.* **C67**, 044316 (2003).
7. J. Dudek, K. Pomorski, N. Schunck and N. Dubray, *Eur. Phys. J* **A20**, 15 (2004).
8. A. Maj, M. Kmiecik, M. Brekiesz, J. Grębosz, W. Męczyński, J. Styczeń, M. Ziębliński, K. Zuber, A. Bracco, F. Camera, G. Benzoni, B. Million, N. Blasi, S. Brambilla, S. Leoni, M. Pignanelli, O. Wieland, B. Herskind, P. Bednarczyk, D. Curien, J.P. Vivien, E. Farnea, G. De Angelis, D.R. Napoli, J. Nyberg, M. Kicińska-Habior, C.M. Petrache, J. Dudek, and K. Pomorski, *Eur Phys J.* **A20**, 165 (2004).
9. A. Maj, J.J. Gaardhøje, A. Atac, S. Mitarai, J. Nyberg, A. Virtanen, A. Bracco, F. Camera, B. Million and M. Pignanelli, *Nucl. Phys.* **A571**, 185 (1994).
10. A. Maj, M. Kmiecik, A. Bracco, F. Camera, P. Bednarczyk, B. Herskind, S. Brambilla, G. Benzoni, M. Brekiesz, D. Curien, G. De Angelis, E. Farnea, J. Grębosz, M. KicińskaHabior, S. Leoni, W. Męczyński, B. Million, D.R. Napoli, J. Nyberg, C.M. Petrache, J. Styczeń, O. Wieland, M. Ziębliński, K. Zuber, N. Dubray, J. Dudek and K. Pomorski, *Nucl. Phys.* **A731**, 319 (2004).
11. M. Kicińska-Habior, K.A. Snover, J.A. Behr, C.A. Gossett, Y. Alhassid and N. Whelan, *Phys. Lett.* **B308**, 225 (1993).
12. P.F. Bortignon, A. Bracco and R.A. Broglia, *Giant Resonances: Nuclear Structure at Finite Temperature, Gordon Breach, New York*, 1998.
13. K. Neergård, *Phys. Lett.* **B110**, 7 (1982).
14. G. Benzoni, A. Bracco, F. Camera, S. Leoni, B. Million, A. Maj, A. Algora, A. Axelsson, M. Bergstrom, N. Blasi, M. Castoldi, S. Frattini, A. Gadea, B. Herskind, M. Kmiecik, G. Lo Bianco, J. Nyberg, M. Pignanelli, J. Styczeń, O. Wieland, M. Ziębliński, A.Zucchiatti, *Phys. Lett.* **540B**, 199 (2002).

KEY TOPICS IN NUCLEAR STRUCTURE
Proceedings of the 8th Int. Spring Seminar on Nuclear Physics
edited by Aldo Covello
© 2005 World Scientific Publishing Co.

SOFT DIPOLE EXCITATIONS NEAR THRESHOLD *

MOSHE GAI

Laboratory for Nuclear Science at Avery Point, University of Connecticut,
1084 Shennecossett Rd., Groton, CT 06340-6097, USA
moshe.gai@uconn.edu - http://www.phys.uconn.edu

A dipole degree of freedom, a most natural concept in molecular physics, is also essential for describing certain states near threshold in light and heavy nuclei. Much like for the Ar-benzene molecule, it is shown that molecular configurations are important near threshold as exhibited by states with a large halo and strong electric dipole transitions. After its first observation in ^{18}O it was also shown to be relevant for the structure of heavy nuclei (e.g. ^{218}Ra). The Molecular Sum Rule derived by Alhassid, Gai and Bertsch (AGB) is shown to be a very useful model independent tool for examining such dipole molecular structure near threshold. Accordingly, the dipole strength observed in the halo nuclei such as ^{6}He, ^{11}Li, ^{11}Be,^{17}O, as well as the N=82 isotones is concentrated around threshold and it exhausts a large fraction (close to 100%) of the AGB sum rule, but a small fraction (a few percent) of the TRK sum rule. This is suggested as an evidence for a new soft dipole Vibron like oscillations in nuclei.

1. Molecular Dipole Excitation

A molecular degree of freedom is characterized by excitations that involves the relative motion of two tightly bound constituents and not the excitation of the objects themselve. Hence it is associated with a polarization vector known as the separation vector. Such a vector can be classicaly described in a geometrical model in three dimensions or by using the corresponding group U(4) [1] and the very succesful Vibron model of molecular Physics [2]. This model has two symmetry limits that correspond to the geometrical description of Rigid Molecules, the O(4) limit, or Soft Molecules, the U(3) limit.

A most comprehensive discussion of such molecular structure and the Vibron model can be found in Iachello-Levine book [2] on "Algebraic Thoery of Molecules". In Figure 1 taken from that book we show the characteristic

*work supported by usdoe grant no. de-fg02-94er40870.

dimensions of the Ar-benzen molecule. The argon atom is losely bound to the (tightly bound) benzen molecule by a van der Waals polarization and thus this molecular state lies close to the dissociation limit. We note that the relative dimension and indeed the very polarization phenomena are reminscent of a halo structure where the argon atom creates a "halo" around the benzen molecule.

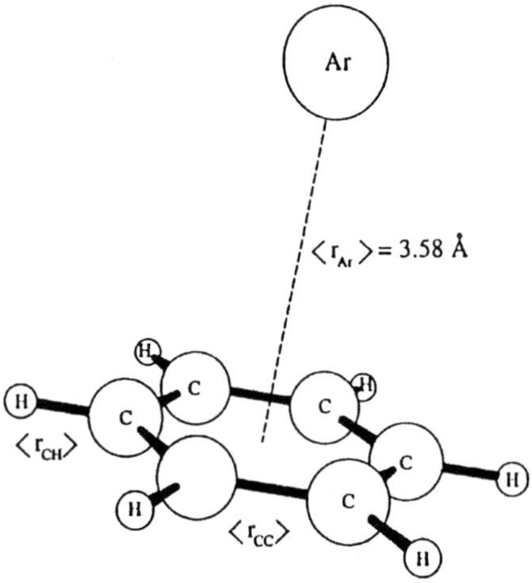

Figure 1: Characteristic dimensions of the Ar-benzen molecule, adopted from Iachello and Levine [2].

2. The AGB Molecular Sum Rule

The polarization phenomena associated with a molecular state implies that it should be associated with dipole excitations of the separation vector. In this case expectation values of the dipole operator do not vanish as the center of mass and center of charge of the polarized molecular state do not coincide [3,4]. Hence molecular states give rise to low lying dipole excitations. While the high lying Giant Dipole Resonace (GDR) is associated with a Goldhaber-Teller [5] excitation of the entire neutron distribution against the proton distribution, a molecular excitation involves a smaller fraction of the nucleus at the surface and is expected to occur at lower excitation than the GDR; i.e. a soft dipole mode [6,7].

The GDR exhausts the Thomas-Reiche-Kuhn (TRK) [8] Energy Weighted Dipole Sum Rule as applied to nuclei:

$$S_1(E1; A) = \Sigma_i \; B(E1 : 0^+ \;\rightarrow\; 1_i^-) \times E^*(1_i^-)$$

$$= \frac{9}{4\pi} \frac{NZ}{A} \times \frac{e^2\hbar^2}{2m} \qquad \text{(equ. 1)}$$

And for a molecular state Alhassid, Gai and Bertsch [9] derived sum rules by subtracting the individual sum rules of the contituents from the total sum rule:

$$S_1(E1; A_1 + A_2) = S_1(A) - S_1(A_1) - S_1(A_2)$$
$$= \frac{9}{4\pi} \frac{(Z_1 A_2 - Z_2 A_1)^2}{A A_1 A_2} \times \frac{e^2\hbar^2}{2m} \qquad \text{(equ. 2)}$$

$$S_1(E1; \alpha + A_2) = \frac{9}{4\pi} \frac{(N-Z)^2}{A(A-4)} \times \frac{e^2\hbar^2}{2m} \qquad \text{(equ. 3)}$$
$$S_1(E1; n + A_2) = \frac{9}{4\pi} \frac{Z^2}{A(A-1)} \times \frac{e^2\hbar^2}{2m} \qquad \text{(equ. 4)}$$
$$S_1(E1; 2n + A_2) = \frac{9}{4\pi} \frac{2Z^2}{A(A-2)} \times \frac{e^2\hbar^2}{2m} \qquad \text{(equ. 5)}$$

The ratio of the TRK/AGB sum rules is given by:

$$\text{TRK/AGB} = NZA_1A_2/(Z_1A_2 - Z_2A_1)^2 \qquad \text{(equ. 6)}$$
$$(\alpha) \quad = (N-Z)^2/NZ(A-4)$$
$$(1n) \quad = N(A-1)/Z$$
$$(2n) \quad = N(A-2)/2Z$$

The Molecular Sum Rule, equ (2), was shown to be useful in elucidating molecular (cluster) states in ^{18}O where the measured B(E1)'s and B(E2)'s exhaust 13% and 23%, respectively, of the Molecular Sum Rule [10]. Similarily, these molecular states in ^{18}O have alpha widths that exhaust 20% of the Wigner sum rule. The branching ratios for electromagnetic decays in ^{18}O were also shown to be consistent with predictions of the Vibron model in the U(3) limit [11]. Indeed the manifestation of a molecular structure in ^{18}O has altered our undertsanding of the coexistence of degrees of freedoms in ^{18}O [12]. Similar observations were also made in the heavy nucleus ^{218}Ra [13]. However we emphasize that the prevailing use of the term Cluster Sum Rule when discussing the AGB Sum Rule is inappropriate. As we demonstrate below even one neutron can lead to a nuclear molecular structure, much in the same way that the hydrogen atom plays a major role in the structure of molecules.

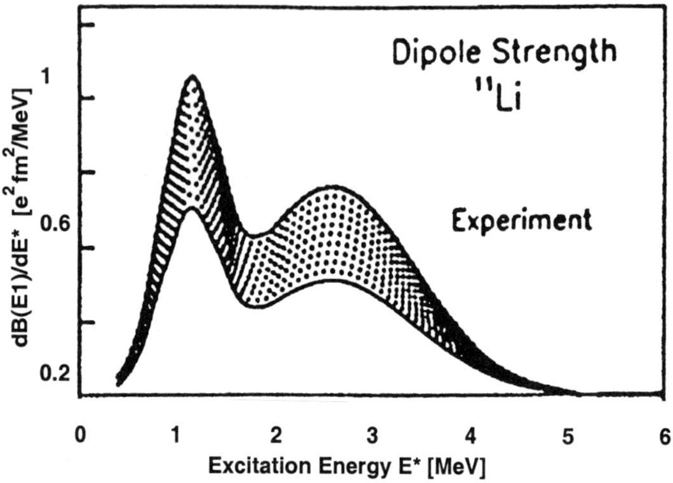

Figure 2: Dipole strength measured in ${}^{11}Li$ [14].

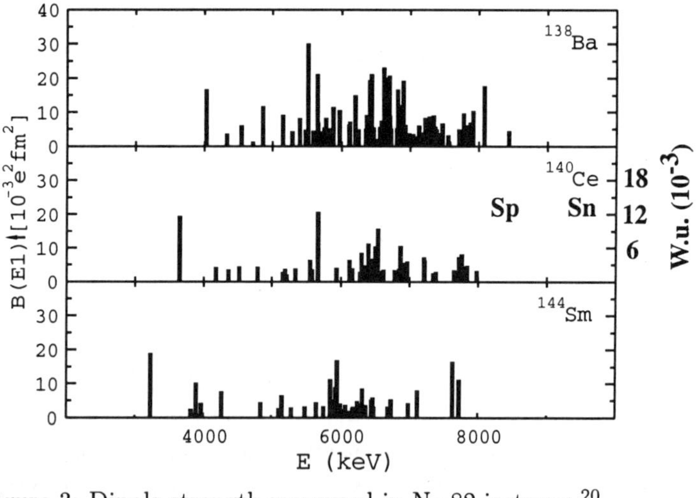

Figure 3: Dipole strength measured in N=82 isotones [20].

The dipole strength at approximately 1.2 MeV in ${}^{11}Li$ [14], shown in Figure 2, exhausts approximately 20% of the Molecular Sum Rule, and the total strength integrated up to 5 MeV exhausts approximately 100% of the AGB sum rule [15,16], but it only exhausts approximately 8% of the TRK sum rule, see Table 1. We emphasize that the experimental efficiency at for example 6 MeV is very large (30%), but no strength is found at higher energies beyond 100% of the Molecular Sum Rule. These two facts strongly suggest the existence of a low lying soft dipole mode in ${}^{11}Li$. In fact the data shown in Figure 1 suggest that two soft dipole states are observed near threshold, as would be expected for a linear triatomic molecule such as the CH_2 molecule. We also find similar low lying dipole strength in ^{11}Be [17], oxygen isotopes [18] and ^{6}He [19], that are also known to to exhibit a halo structure. Even the N=82 isotones that are not considered to exhibit exotic structure show a diople strength near threshol as shown in Figure 3 [20]. These results are summarized in Table 1.

Nucleus	$< E^* >$	TRK	TRK/AGB	AGB
^{11}Li [14,15]	1.2 MeV	$8.0 \pm 2.0\%$	(2n) 12	$96 \pm 24\%$
^{11}Be [17]	1.0 MeV	5.0%	(1n) 18	90%
^{17}O [18]	< 15 MeV	4%	(1n) 18	72%
^{138}Ba [20]	6.5 MeV	$0.78 \pm 0.15\%$	(1n) 200	$156 \pm 30\%$

3. Conclusions

In conclusions we demonstrate that molecular configurations play a major role in the structure of light and heavy nuclei. Unlike the Giant Dipole Resonance that involves oscillation of the entire neutron-proton distributions, these Vibron states involve only oscillations of the surface of the nucleus, and hence they lie at lower energies than the GDR. Similarly, while the GDR exhausts the TRK sum rule, the Vibron states exhausts the ABG Molecular Sum Rule.

References

1. F. Iachello, and A.D. Jackson; Phys. Lett. **108B**(1982)151.
2. F. Iachello and R.D. Levine, Algebraic Theory of Molecules; Oxford University Press, 1995.
3. L.A. Radicati; Phys. Rev. **87**(1952)521.
4. M. Gell-Mann and V.L. Telegdi; Phys. Rev. **91**(1953)169.
5. M. Goldhaber and E. Teller; Phys. Rev. **74**(1948)1046.
6. K. Ikeda Nucl. Phys. **A538**(1992)355c.
7. P.G. Hansen; Nucl. Phys. **A588**(1995)1c.
   P.G. Hansen and A.S. Jensen; Annu. Rev. Nucl. Part. Sci. **45**(1995)591.
8. W. Kuhn; Zeit. f. Phys. **33**(1925)408. F. Reiche, W. Thomas; Zeit. f. Phys. **34**(1925)510.
9. Y. Alhassid, M. Gai, and G.F. Bertsch ; Phys. Rev. Lett. **49**(1982)1482.

10. M. Gai, M. Ruscev, A.C. Hayes, J.F. Ennis, R. Keddy, E.C. Schloemer, S.M. Sterbenz and D.A. Bromley; Phys. Rev. Lett. **50**(1983)239.
11. M. Gai *et al.*; Phys. Rev. **C43**(1991)2127.
12. M. Gai *et al.*; Phys. Rev. Lett. **62**(1989)874.
13. M. Gai *et al.*; Phys. Rev. Lett. **51**(1983)646.
14. M. Zinser *et al.*; Nucl. Phys. **A619**(1997)151.
15. G.F. Bertsch and J. Foxwell; Phys. Rev. **C41**(1990)1300.
16. M. Gai; Rev. Mex. Fis. Supp. **45**(1999)106.
17. T. Nakamura *et al.*; Phys. Lett. **B331**(1994)296. N. Gan *et al.* http://www.phy.ornl.gov/progress/ribphys/reaction/rib023.pdf.
18. T. Aumann *et al.*; Nucl. Phys. **A649**(1999)297c. A. Leistenscheneider *et al.*; Acta. Phys. Pol. **B32**(2001)1095.
19. S. Nakayama *et al.*; Phys. Rev. Lett. **85**(2000)262.
20. A. Zilges *et al.* Phys. Lett. **B542**(2002)43.

KEY TOPICS IN NUCLEAR STRUCTURE
Proceedings of the 8th Int. Spring Seminar on Nuclear Physics
edited by Aldo Covello
© 2005 World Scientific Publishing Co.

UNIFIED SEMICLASSICAL APPROACH TO ISOSCALAR COLLECTIVE MODES IN HEAVY NUCLEI

V. I. ABROSIMOV

Institute for Nuclear Research, 03028 Kiev, Ukraine
E-mail: abrosim@kinr.kiev.ua

A. DELLAFIORE AND F. MATERA

Istituto Nazionale di Fisica Nucleare and Dipartimento di Fisica, Università di Firenze, via Sansone 1, 50019 Sesto F.no (Firenze), Italy
E-mail:della@fi.infn.it; matera@fi.infn.it

A semiclassical model based on the solution of the Vlasov equation for finite systems with a sharp moving surface has been used to study the isoscalar quadrupole and octupole collective modes in heavy spherical nuclei. Within this model, a unified description of both low-energy surface modes and higher-energy giant resonances has been achieved by introducing a coupling between surface vibrations and the motion of single nucleons. Analytical expressions for the collective response functions of different multipolarity can be derived by using a separable approximation for the residual interaction between nucleons. The response functions obtained in this way give a good qualitative description of the quadrupole and octupole response in heavy nuclei. Although shell effects are not explicitly included in the theory, our semiclassical response functions are very similar to the quantum ones. This happens because of the well known close relation between classical trajectories and shell structure. The role played by particular nucleon trajectories and their connection with various features of the nuclear response is displayed most clearly in the present approach, we discuss in some detail the damping of low-energy octupole vibrations and give an explicit expression showing that only nucleons moving on triangular orbits can contribute to this damping.

1. Introduction

It is well known that the isoscalar quadrupole and octupole response of nuclei displays both low- and high-energy collective modes [1]. Also known is that semiclassical models have difficulties in describing both these systematic features of the isoscalar response, in particular, models based on fluid dynamics, see e.g. [2], can explain the giant resonances, but fail to describe the low-energy collective modes. On the other hand it is known from quantum studies that the coupling between the motion of individual nucleons and surface vibrations plays an essential role in low-energy nuclear collective modes, see e.g. [3,4,5]. Semiclassical models of the fluid-dynamical type do not contain explicitly the single-particle degrees of freedom, so they can not describe the coupling between individual nucleons and surface motion.

In the present contribution we review a study the isoscalar collective modes in nuclei [6,7] made by using a semiclassical approach that includes the single-particle degrees of freedom explicitly and thus allows for an account the coupling between individual nucleons and surface motion. Our model is based on the linearized Vlasov kinetic equation for finite systems with moving surface [8,9]. The coupling between the motion of individual nucleons and the surface vibrations is obtained by treating the nuclear surface as a collective dynamical variable, like in the liquid drop model. Here we concentrate our attention on the isoscalar quadrupole and octupole collective modes in heavy spherical nuclei, an application of the same model to the compression dipole modes has been discussed in a previous meeting of this series [10].

2. Reminder of formalism

This Section recalls briefly the formalism of References [8,9] which is at the basis of the present approach.

The fluctuations of the phase-space density induced by a weak external force can be described by the linearized Vlasov equation, which is usually a differential equation in seven variables. For spherical systems this equation can be reduced to a system of two (coupled) differential equations in the radial coordinate alone [8]. This is achieved by means of a change of variables and a partial-wave expansion:

$$\delta f(\mathbf{r},\mathbf{p},\omega) = \sum_{LMN} [\delta f^{L+}_{MN}(\epsilon,\lambda,r,\omega) + \delta f^{L-}_{MN}(\epsilon,\lambda,r,\omega)] \times \left(\mathcal{D}^L_{MN}(\alpha,\beta,\gamma)\right)^* Y_{LN}(\frac{\pi}{2}\frac{\pi}{2}). \quad (1)$$

The functions $\delta f^{L\pm}_{MN}(\epsilon,\lambda,r,\omega)$ are partial-wave components of the (Fourier transformed in time) density fluctuations for particles with energy ϵ, magnitude of angular momentum λ and radial position r, the $\pm$ sign distinguishes between particles having positive or negative components of the radial momentum p_r.

The other terms in the expansion are Wigner matrices and spherical harmonics.

In order to solve the one-dimensional linearized Vlasov equation for the $\delta f^{L\pm}_{MN}$ functions we must specify the boundary conditions satisfied by these functions. Different boundary conditions allow us to study different physical properties of the system, so the fixed-surface boundary conditions employed in [8] were adequate to study giant resonances, but different (moving-surface) boundary conditions [9] must be introduced in order to study surface modes. We assume that the external force can also induce oscillations of the system surface according to the usual liquid-drop model expression

$$R(\vartheta,\varphi,t) = R + \sum_{LM} \delta R_{LM}(t) Y_{LM}(\vartheta,\varphi) \tag{2}$$

and the boundary condition satisfied by the functions $\delta f^{L\pm}_{MN}$ at the nuclear surface is taken as

$$\delta f^{L+}_{MN}(R) - \delta f^{L-}_{MN}(R) = 2F'(\epsilon) i\omega p_r \delta R_{LM}(\omega)\,. \tag{3}$$

This equation has been derived with the assumption that the equilibrium phase-space density is a function $F(\epsilon)$ of the particle energy alone, $F'(\epsilon)$ is its derivative. The boundary condition (3) corresponds to a mirror reflection of particles in the reference frame of the moving nuclear surface, it provides a coupling between the motion of nucleons and the surface vibrations. A self-consistency condition involving the nuclear surface tension is then used to determine the time (or frequency) dependence of the additional collective variables $\delta R_{LM}(\mathrm{t})$ [9].

Now, assuming a simplified residual interaction of separable form,

$$v(r_1, r_2) = \kappa_L r_1^L r_2^L\,, \tag{4}$$

the moving-surface isoscalar collective response function of a spherical nucleus, described as a system of A interacting nucleons contained in a cavity of equilibrium radius $R = 1.2A^{\frac{1}{3}}$ fm, is given by

$$\tilde{\mathcal{R}}_L(s) = \mathcal{R}_L(s) + \mathcal{S}_L(s)\,. \tag{5}$$

Instead of the frequency ω, as independent variable we have used the more convenient dimensionless quantity $s = \omega/(v_F/R)$ (v_F is the Fermi velocity). The response function $\mathcal{R}_L(s)$, given by

$$\mathcal{R}_L(s) = \frac{\mathcal{R}^0_L(s)}{1 - \kappa_L \mathcal{R}^0_L(s)}\,, \tag{6}$$

describes the collective response in the fixed-surface limit. The response function $\mathcal{R}^0_L(s)$ is analogous to the quantum single-particle response function and it is given explicitly by [6,7],

$$\mathcal{R}_L^0(s) = \frac{9A}{8\pi}\frac{1}{\epsilon_F}\sum_{n=-\infty}^{+\infty}\sum_{N=-L}^{N=L}(C_{LN})^2\int_0^1 dx x^2\; s_{nN}(x)\frac{(Q_{nN}^{(L)}(x))^2}{s+i\varepsilon - s_{nN}(x)}, \tag{7}$$

where ϵ_F is the Fermi energy and the quantity ε is a vanishingly small parameter that determines the integration path at poles.

The functions $s_{nN}(x)$ are defined as

$$s_{nN}(x) = \frac{n\pi + N\arcsin(x)}{x}. \tag{8}$$

The variable x is related to the classical nucleon angular momentum λ. The quantities C_{LN} in Eq. (7) are classical limits of the Clebsh-Gordan coefficients coming from the angular integration. In principle the integer N takes values between $-L$ and L, however only the coefficients C_{LN} where N has the same parity as L are nonvanishing. The coefficients $Q_{nN}^{(L)}(x)$ appearing in the numerator of Eq. (7) have been defined in Ref. [8], they are essentially the classical limit of the radial matrix elements of the multipole operator r^L and can be evaluted analytically for $L = 2, 3$.

The function $\mathcal{S}_L(s)$ in Eq. (5) gives the moving-surface contribution to the response. With the simple interaction (4) this function can be evaluated explicitly as [6,7]

$$\mathcal{S}_L(s) = -\frac{R^6}{1-\kappa_L\mathcal{R}_L^0(s)}\frac{[\chi_L^0(s) + \kappa_L\varrho_0 R^L\mathcal{R}_L^0(s)]^2}{[C_L - \chi_L(s)][1-\kappa_L\mathcal{R}_L^0(s)] + \kappa_L R^6[\chi_L^0(s)+\varrho_0 R^L]^2}, \tag{9}$$

with $C_L = \sigma R^2(L-1)(L+2) + (C_L)_{coul}$, $\sigma \approx 1\mathrm{MeV\,fm}^{-2}$ is the surface tension parameter obtained from the mass formula, $(C_L)_{coul}$ gives the Coulomb contribution to the restoring force and $\varrho_0 = A/\frac{4\pi}{3}R^3$ is the equilibrium density. The functions $\chi_L^0(s)$ and $\chi_L(s)$ are given by [11]

$$\chi_L^0(s) = \frac{9A}{4\pi}\frac{1}{R^3}\sum_{n=-\infty}^{+\infty}\sum_{N=-L}^{N=L}(C_{LN})^2\int_0^1 dx x^2\; s_{nN}(x)\frac{(-)^n Q_{nN}^{(L)}(x)}{s+i\varepsilon - s_{nN}(x)}, \tag{10}$$

and

$$\chi_L(s) = -\frac{9A}{2\pi}\epsilon_F\,(s+i\varepsilon)\sum_{nN}(C_{LN})^2\int_0^1 dx x^2\;\frac{1}{s+i\varepsilon - s_{nN}(x)}, \tag{11}$$

their structure is similar to that of the zero-order propagator (7).

We refer to the papers [6,7] for further details on the formalism and discuss here only the main points.

Equation (9) is the main result in the present context. Together with Eqs.(5) and (6), this equation gives a unified expression of the isoscalar response function, including both the low- and high-energy collective excitations. By comparing the fixed- and moving-surface response functions, we can appreciate the effects due to the coupling between the motion of individual nucleons and the surface vibrations.

3. Fixed- vs. moving-surface strength distributions

The strength function $S_L(E)$ associated with the response function (5) is defined as $(E = \hbar\omega)$

$$S_L(E) = -\frac{1}{\pi}\mathrm{Im}\,\tilde{\mathcal{R}}_L(E)\,. \tag{12}$$

We discuss here the isoscalar quadrupole and octupole strength distributions. The strength κ_L of the residual interaction (4) can be estimated in a self-consistent way, giving ([12], p. 557),

$$\kappa_{BM} = -\frac{4\pi}{3}\frac{m\omega_0^2}{AR^{2L-2}}, \tag{13}$$

with the parameter ω_0 given by $\omega_0 \approx 41A^{-\frac{1}{3}}\mathrm{MeV}$. Since this estimate is based on a harmonic oscillator mean field and we are assuming a square-well potential instead, we expect some differences. Hence we determine the parameter κ_L phenomenologically, by requiring that the peak of the high-energy resonance agrees with the experimental value of the giant multipole resonance energy. This requirement implies $\kappa_L \approx 2\kappa_{BM}$ for $L = 2, 3$.

In Fig.1 we display the quadrupole strength function (L=2 in Eq. (12)) obtained for $A = 208$ nucleons by using different approximations. The dotted curve is obtained from the zero-order response function (7), it is similar to the quantum response evaluated in the Hartree-Fock approximation. The dashed curve is obtained from the collective fixed-surface response function (6). Comparison with the dotted curve clearly shows the effects of collectivity. The collective fixed-surface response has one giant quadrupole peak. Our result for this peak is very similar to that of the recent random-pase approximation (RPA) calculations of [13] (cf. Fig.5 of [13]). However, contrary to the RPA calculations, there is no signal of a low-energy peak in the fixed-surface response function. The solid curve instead shows the moving-surface response given by Eqs. (5) and (9). Now a broad bump appears in the low-energy part of the response and a narrower peak is still present at the giant resonance energy. Of course the details of the low-energy excitations are determined by quantum effects, nonetheless the present semiclassical approach does reproduce the average behaviour of this systematic feature of the quadrupole response.

We finally notice that the width of the giant quadrupole resonance is underestimated by our approach, this is a well known limit of all mean-field calculations that include only Landau damping. A more realistic estimate of the giant-resonance width would require including a collision term into our kinetic equation.

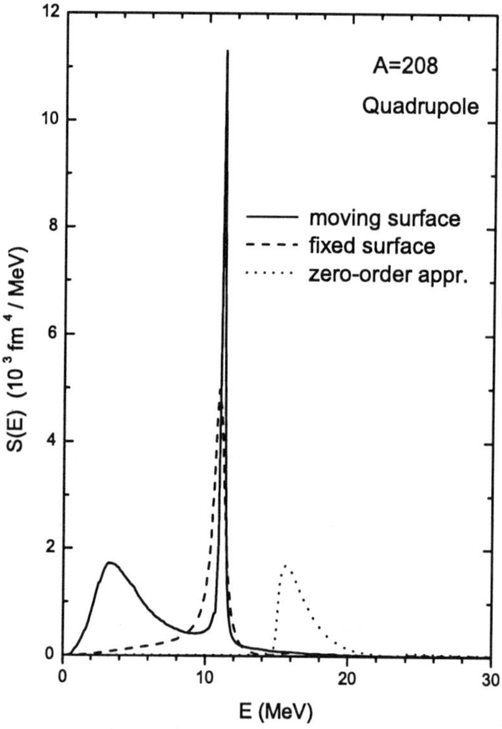

Figure 1. Quadrupole strength function for a hypothetical nucleus of $A = 208$ nucleons. The dotted curve shows the zero-order aproximation, the dashed curve instead shows the collective response evaluated in the fixed-surface approximation. The full curve gives the moving-surface response.

In Fig.2 we show the octupole strength function ($L = 3$ in Eq. (12)).

The zero-order octupole strength function (dotted curve) is concentrated in two regions around 8 and 24 MeV. In this respect our semiclassical response is strikingly similar to the quantum response, which is concentrated in the $1\hbar\omega$ and $3\hbar\omega$ regions. This concentration of strength is quite remarkable because our equilibrium phase-space density, which is taken to be of the Thomas-Fermi type, does not include any shell effect, however we still obtain a strength distribution that is very similar to the one usually interpreted in terms of transitions between different shells.

We can clearly see that the collective fixed-surface response given by Eq. (6) (dashed curve) has two sharp peaks around 20 Mev and 6-7 Mev. The experimentally observed [1] concentration of isoscalar octupole strength in the two regions usually denoted by HEOR (high energy octupole resonance) and LEOR (low energy octupole resonance) is qualitatively reproduced, however the considerable strength experimentally observed at lower energy (low-lying collective states) is absent from the fixed-surface response function. The most relevant change induced by the moving surface (solid curve in Fig.2) is the large double hump appearing at low energy. This feature is in qualitative agreement both with experiment [1] and with the result of RPA-type calculations (see e.g. [14]). We interpret this low-energy double hump as a superposition of surface vibration and LEOR.

The moving-surface octupole response of Fig. 2 displays also a novel resonance-like structure between the LEOR and the HEOR (at about 13 MeV for a system of $A = 208$ nucleons).

We also find [15] that the parameters $\delta R_{3M}(t)$, describing the octupole surface vibrations in Eq. (2), approximately satisfy an equation of motion of the damped oscillator kind:

$$D_3 \delta \ddot{R}_{3M}(t) + \gamma_3 \delta \dot{R}_{3M}(t) + C_3 \delta R_{3M}(t) = 0\,. \tag{14}$$

The friction coefficient γ_L can be evaluated analyticallty in the low-frequency limit, giving (for a generic L)

$$\gamma_L = \gamma_{wf}\, 2\frac{(4\pi)^2}{2L+1} \sum_{N=1}^{L} \frac{1}{N} \mid Y_{LN}(\frac{\pi}{2},\frac{\pi}{2}) \mid^2 \sum_{n=1}^{+\infty} \cos\alpha_{nN} \sin^3\alpha_{nN} \Theta(\frac{\pi}{2} - \alpha_{nN})\,, \tag{15}$$

with $\gamma_{wf} = \frac{3}{4}\varrho_0 p_F R^4$ and $\alpha_{nN} = \frac{n}{N}\pi$. The angles α_{nN} are related to the nucleon trajectories. In the octupole case the coefficient $\gamma_{L=3}$ gets a contribution only from the term with $n = 1$ and $N = 3$, thus we see that only nucleons moving along closed triangular trajectories can contribute to the damping of surface octupole vibrations.

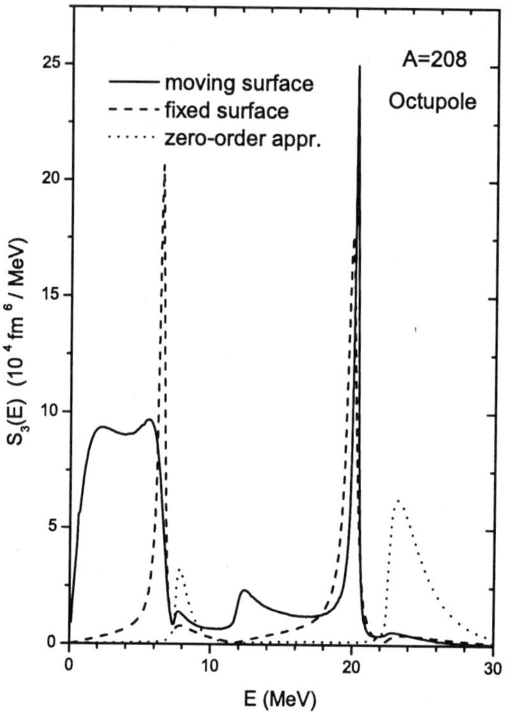

Figure 2. The same as in Fig.1 for the octupole strength function.

4. Conclusions

A unified description of the low- and high-energy isoscalar collective quadrupole and octupole response has been achieved by using appropriate boundary conditions for the fluctuations of the phase-space density described by the linearized Vlasov equation. The response functions obtained in this way give a good qualitative description of all the main features of the isoscalar response in heavy nuclei, i. e. low-lying quadrupole and octupole collective modes, plus quadrupole and octupole giant resonances. In our model the low-energy modes are surface oscillations and the coupling between single-particle motion and surface vibrations is described by simple analytical expressions.

References

1. A. van der Woude, *Progr. Part. Nucl. Phys.* **18**, 217 (1987).
2. G. Holzwarth and G. Eckart, *Nucl. Phys.* **A325**, 1 (1979).
3. P. F. Bortignon and R. A. Broglia, *Nucl. Phys.* **A371**, 405 (1981).
4. G. F. Bertsch and R. A. Broglia, *Oscillations in finite quantum systems, ch.6* (Cambridge University Press, Cambridge, UK, 1994).
5. D. Lacroix, S. Ayik and Ph. Chomaz, *Phys. Rev.* **C63**, 064305 (2001).
6. V. I. Abrosimov, A. Dellafiore and F. Matera, *Nucl. Phys.* **A717**, 44 (2003).
7. V. I. Abrosimov, O. I. Davidovskaya, A. Dellafiore and F. Matera, *Nucl. Phys.* **A727**, 220 (2003).
8. D. M. Brink, A. Dellafiore and M. Di Toro, *Nucl.Phys.* **A456**, 205 (1986).
9. V. Abrosimov, M. Di Toro and V. Strutinsky, *Nucl.Phys.* **A562**, 41 (1993).
10. V. I. Abrosimov, A. Dellafiore and F. Matera, in *Proc. of the 7th Intern. Spring Seminar on Nucl. Physics*, edited by A.Covello (World Scientific, Singapore, 2002), p.481.
11. V. I. Abrosimov, A. Dellafiore and F. Matera, *Nucl. Phys.* **A697**, 748 (2002).
12. A. Bohr and B. M. Mottelson, *Nuclear Structure, Vol. 2* (W.A. Benjamin, Inc.: Reading, Massachussets, 1975).
13. I. Hamamoto, H. Sagawa and X. Z. Zhang, *Nucl. Phys.* **A648**, 203 (1999).
14. K. F. Liu, H. Luo, Z. Ma, Q. Shen and S. A. Moszkowski, *Nucl. Phys.* **A534**, 1 (1991).
15. V. Abrosimov, A. Dellafiore and F. Matera, *Nucl. Phys.* **A653**, 115 (1999).

KEY TOPICS IN NUCLEAR STRUCTURE
Proceedings of the 8th Int. Spring Seminar on Nuclear Physics
edited by Aldo Covello
© 2005 World Scientific Publishing Co.

MICROSCOPIC DESCRIPTION OF MULTIPLE GIANT RESONANCES IN HEAVY ION COLLISIONS*

E. G. LANZA
I.N.F.N. and Dipartimento di Fisica e Astronomia dell'Universitá, Catania, Italy
and
Departamento de FAMN, Universidad de Sevilla, Spain
E-mail: Edoardo.Lanza@ct.infn.it

The excitation of a multiple giant resonance is studied within the framework of a semiclassical model. We make use of an extended RPA to treat anharmonicities and non liner terms in the external field. Successful applications to double giant resonances to both relativistic and lower incident energies are reviewed. We have extended this method to the calculation of the triple giant resonance. We show that large amplitude motion induce a strong coupling to giant monopole and quadrupole vibrations.

1. Introduction

Giant resonances (GR) are well described within the Random Phase Approximation (RPA) where the fermionic Hamiltonian can be mapped into a sum of independent harmonic oscillators corresponding to each mode. Then the RPA states are pure one-phonon or multiphonon states and their energy is the sum of the energies of the single phonons. Experiments have clearly shown the existence of double GR. For a review on this subject see ref.[1]. In relativistic Coulomb excitations some anomalies in the inelastic cross section have shown up: the experimental cross section is larger than the calculated one by a factor ranging from 1.3 to 2. To overcome this discrepancy we have introduced anharmonic terms in the internal Hamiltonian and non-linear terms in the external field. In this contribution, after a brief review of the model employed and of the results obtained for the double GR, we will describe the new calculations for the triple GR.

*This work is the result of a long standing collaboration with M.V. Andrés (Sevilla), F. Catara (Catania), Ph. Chomaz (GANIL), M. Fallot (Nantes), J.A. Scarpaci (Orsay) and C. Volpe (Orsay).

2. Beyond the standard model

Great successes have been achieved in the study of excitation processes of heavy nuclei at relatively low energy by using semiclassical methods techniques. These methods are based on the assumption that nuclei move on classical trajectories, while the internal degrees of freedom are treated quantum mechanically. These assumptions usually are well justified for grazing collisions(see ref.[2]). This methods have also been applied to relativistic Coulomb excitations[3]. The standard approach to study the double excitation of giant resonances implies the use of an internal Hamiltonian which is of harmonic type and an external excitation field which is linear in the phonon creation operators. The analysis of inelastic cross section for the excitation of double Giant Dipole Resonance (DGDR) in a relativistic Coulomb experiment has shown some differences between the theoretical standard approach and the experimental results. These differences can go from a 30 % up to a factor 2, depending on the excited nucleus (see last paper of LAND Collaboration[4]).

In order to overcome this discrepancy we have included corrections to the harmonic approximations, like anharmonicities in the internal Hamiltonian and non-linearities in the external field. In the microscopic theory of RPA the Hamiltonian can be written as

$$H_{RPA} = \sum_{\nu} E_\nu Q^\dagger_\nu Q_\nu \tag{1}$$

where the phonon creation operator

$$Q^\dagger_\nu = \sum_{p,h} (X^\nu_{ph} B^\dagger_{ph} - Y^\nu_{ph} B_{ph})$$

is defined in terms of the bosonic operators B that are the lowest order terms of the bosonic expansion of the fermionic operators [5]

$$a^\dagger_p a_h \to B^\dagger_{ph} + (1-\sqrt{2}) \sum_{p'h'} B^\dagger_{p'h'} B^\dagger_{p'h} B_{ph'} + \; ... \tag{2}$$

Here, the index p (h) labels the particle (hole) states with respect to the Hartree-Fock ground state. The other terms after the first one correct for the Pauli principle. In the space spanned by one- and two-phonon states the bosonic Hamiltonian is

$$\begin{aligned} H = \sum_{\nu} E_\nu Q^\dagger_\nu Q_\nu + [\sum_{\nu_1\nu_2\nu_3} V^{21}_{\nu_1\nu_2\nu_3} Q^\dagger_{\nu_1} Q^\dagger_{\nu_2} Q_{\nu_3} \\ + \sum_{\nu_1\nu_2\nu_3\nu_4} V^{22}_{\nu_1\nu_2,\nu_3\nu_4} Q^\dagger_{\nu_1} Q^\dagger_{\nu_2} Q_{\nu_3} Q_{\nu_4}] + h.c. \end{aligned} \tag{3}$$

where V^{21} (V^{22}) are the matrix elements connecting one- with two-phonon states (two- with two-phonon states). The eigenstates of the Hamiltonian are mixed states of one- and two-phonon states and their corresponding eigenvalues are not harmonic. Both novel aspects may increase the excitation probability for the DGDR.

In the semiclassical models of heavy ion reactions the excitation of one of the partners of the collision is due to the mean field of the other. In standard models the excitation operator is of one-body type and it is assumed to be linear in the phonon operator Q because only terms of ph type are taken into account. Applying the same boson expansion described above and taking also the contribution pp and hh terms, we obtain a non-linear excitation field

$$W = W^{00} + \sum_{\nu} W^{10}_{\nu} Q^{\dagger}_{\nu} + h.c. + \sum_{\nu\nu'} W^{11}_{\nu\nu'} Q^{\dagger}_{\nu} Q_{\nu'} + \sum_{\nu\nu'} W^{20}_{\nu\nu'} Q^{\dagger}_{\nu} Q^{\dagger}_{\nu'} + h.c. \quad (4)$$

The first term in eq. (4) represents the interaction of the two colliding nuclei in their ground state. The W^{10} part connects states differing by one phonon, the W^{11} term couples excited states with the same number of phonons, while W^{20} allows transitions from the ground state to two-phonon states. These new routes of excitation may increase again the excitation probability of the DGDR.

We write the Schrödinger equation as a set of linear differential coupled equations for the time dependent amplitude probabilities for each eigenstate $|\Phi_{\alpha}>$ of the Hamiltonian (3). The cross section is then calculated by integrating the excitation probability for each $|\Phi_{\alpha}>$ over the whole impact parameters range. For more details see ref.[6].

Calculations within this framework have been done both for a schematic model[7], where we confirm successfully our implementations to the standard model, as well as for a more realistic microscopic model[6]. In the latter case we run a self-consistent HF+RPA code with a SGII interaction, we include all one-phonon states with angular momentum less or equal to 3 and with an EWSR more than 5%. Then we construct all possible two-phonon states out of them and in the space of one- and two-phonon states we diagonalize the Hamiltonian (3). For the case of relativistic Coulomb excitation for the reaction ^{208}Pb + ^{208}Pb at 641 MeV/A, making use of the result of ref.[3], we found a value for the inelastic cross section for the DGDR excitation very close to the experimental values[6]. We want to stress the fact that the increase in the cross section with respect to the standard approaches is due to the fact that we consider the excitation of several states of different multipoles lying in the energy region of the DGDR. The population of these states is strongly suppressed by selection rules when the anharmonicities and non-linearities are not taken into account.

Application of this model to the reaction $^{40}Ca + ^{40}Ca$ at 50 MeV/A has also been done. For this reaction the existence of double Giant Quadrupole Resonance (DGQR) has been established since many years by the N. Frascaria group in Orsay[8]. In this case the nuclear contribution is important and the calculation has been carried out in the same fashion as in the previous one except for the determination of the form factors where a double folding procedure has been employed[9]. The calculations, also in this case, show a satisfactory agreement with the experimental results and strengthen once more the importance of the anharmonic and non linear terms[10].

These results have motivated the extension of the microscopic calculation to the study of the excitation of the three phonon states also because experiments have already been done in both ranges of energies discussed before[11,12]. Of course a simple extension of the just presented model to the case of the triple giant resonance will be impossible for numerical problems. We need then some approximation which make feasible the calculation without loosing the main physical properties. A help on this side comes from an analysis of few years ago where we studied an extended Lipkin-Meshow-Glick (LMG) model[13]. In that paper, the original LMG model has been extended in order to include terms that play the same role than the anharmonic terms of our microscopic Hamiltonian (3). The Hamiltonian of such extended LMG model is still exactly solvable. We then apply boson expansion methods including terms up to the forth order, obtaining in this way a quartic anharmonic bosonic Hamiltonian which corresponds to the one used in our microscopic model. We can apply then all the approximations done in the realistic calculation and comparing the results with the exact ones. The relevant results can be summarized as follow: The main approximation done in the microscopic calculation are well justified. Furthermore, the quartic Hamiltonian diagonalized in an enlarged space including up to three-phonon states produces results which are very close to the exact ones[13]. In the next section we will show that following this approach, we diagonalize a microscopic quartic Hamiltonian in the space of one-, two- and three-phonon states. The results so obtained show that a correct description of the states whose main component is a two-phonon configuration requires the inclusion of one- and three-phonon ones. We will see also the importance of the role played by the breathing mode in nuclear anharmonicity. More details and deeper discussions about these calculations can be found in ref.[14].

3. Three-phonon states: Calculations and results

The calculations were done following the model described above: We make use of a mapping of the fermion particle-hole operators into boson operators (eq. (2)). Then we construct a boson image of the Hamiltonian, truncated at the fourth order and we express it in terms of the collective operators Q. We use then the same one-phonon basis as in previous microscopic calculations[6,9,10,14], we construct all two- and three-phonon configuration out of them without energy cut-off, with both natural and unnatural parity. Then the Hamiltonian is diagonalized in the space spanned by such states. The eigenstates are mixing states whose components are of one-, two- and three-phonon kind

$$|\Phi_\alpha>=\sum_{\nu_1} c^\alpha_{\nu_1}|\nu_1> + \sum_{\nu_1\nu_2} d^\alpha_{\nu_1\nu_2}|\nu_1\nu_2> + \sum_{\nu_1\nu_2\nu_3} e^\alpha_{\nu_1\nu_2\nu_3}|\nu_1\nu_2\nu_3> . \quad (5)$$

Calculations have been done for the two nuclei ^{40}Ca and ^{208}Pb. The main result is that the spectrum of the two-phonon states is strongly modified by their coupling to the three-phonon ones. Indeed, the diagonalization in the three-phonon space produces very large shifts in the energies, in almost all the cases more than one MeV (for ^{40}Ca) and always downward. This can be understood in second order perturbation which gives a good estimate in most cases. In this case the correction to the energy is

$$\Delta E_i =< \varphi_i|V|\varphi_i> + \sum_{j\neq i} \frac{|<\varphi_j|V|\varphi_i>|^2}{E^0_i - E^0_j} \quad (6)$$

where $|\varphi_i>$ is the considered unperturbed state, $|\varphi_j>$ all the other states and E^0 the corresponding unperturbed energies. The diagonal contribution (first order) is small in most cases. If φ_i is a two-phonon state the contribution from the three-phonon states is negative since most of them lye above the two-phonon ones.

Moreover, whenever a GMR is added on top of any state, the corresponding matrix elements coupling states with one- and two-phonon states are large, of the order of 1 to 2 MeV in ^{40}Ca. This strong coupling of all collective vibrations with the breathing mode comes from the fact that in a small nucleus such as ^{40}Ca any large amplitude motion affects the central density. Therefore, surface modes cannot be decoupled from a density variation in the whole volume as clearly seen in recent TDHF simulations[15]. Equivalent role is played by the GQR although with a slighter effect.

Similar results are obtained for ^{208}Pb where the role played by the GMR and GQR are inverted. In this case the anharmonicities are of the order of hundreds of KeV. In large nuclei the surface vibration may occur without changing the volume.

Concluding about the energy of the two-phonon states, one can see that the inclusion of the three phonon configurations induces an anharmonicity of more than 1 MeV in ^{40}Ca but only of a few hundred keV in ^{208}Pb. This is related to the fact that collectivity is more pronounced in ^{208}Pb than in ^{40}Ca. Because of the location at high energy of the three phonon states, the observed shift is systematically downward. It is important to stress that the considered residual interaction only couples states with a number of phonon varying at maximum by one unit. Therefore, the energy variation of the two-phonon spectrum induced by inclusion of four and more phonon states would be small since it corresponds to a third order perturbation involving two large energy differences in the denominator.

4. Double giant resonances cross section

A question raises spontaneously: How can the new findings affect the double giant resonances cross section? In order to answer this question one should perform a complete calculation including the three-phonon states. This is under study and it will require a strong numerical effort. At the moment

Table 1. Relativistic Coulomb excitation for the system ^{208}Pb + ^{208}Pb at 641 MeV/A. Summed cross sections (in mb) in the DGDR region ($E \geq 22$ MeV)

Phonons	har. & lin.	anh. & non-lin.	anh. & non-lin. (E_{3pho})
L=0	42.66	43.46	45.49
L=1	13.93	27.00	26.23
L=2	207.30	211.33	218.01
L=3	63.26	77.53	76.30
L=4	6.53	7.90	7.88
L=5	0.38	0.50	0.50
L=6	0.02	0.03	0.01
total	334.08	367.75	374.42

we have done a very simple calculation which may shed some light on the effect of the stronger anharmonicities reported here. We have repeated the calculation for the relativistic Coulomb excitation for the system ^{208}Pb + ^{208}Pb at 641 MeV/A where we have simply substitute the old energies (that is the ones obtained by diagonalizing the Hamiltonian (3) in the space of one- and two-phonon states) with the new ones (that is the ones obtained in the basis including up to three-phonon states). The mixing coefficients have been taken equal to the old ones[6]. In table 1 we show the summed inelastic cross section for the excitation of the phonons whose angular momentum is reported in the first column. In the fourth column is presented the summed cross section for the anharmonic and non-linear case with the energies obtained including the three-phonon states. In the other columns we show the old calculation[6], namely the standard and the anharmonic and non-linear cases. The results reported in table 1 show an enhancement with respect to the old calculations which is close to the 2%.

Table 2. Coulomb plus nuclear excitation for ^{40}Ca + ^{40}Ca at 50 MeV/A. The cross sections (in mb) are summed over the energy region 24 MeV $\leq E \leq$ 40 MeV. The values in parentheses correspond to the double ISGQR states.

Phonons	har. & lin.	anh. & non-lin.	anh. & non-lin. (E_{3pho})
L=0	0.18 (0.15)	0.25 (0.21)	0.33 (0.24)
L=1	0.08	0.10	0.13
L=2	0.89 (0.33)	1.82 (0.53)	2.01 (0.57)
L=3	2.58	3.24	3.31
L=4	0.98 (0.90)	1.82 (1.73)	1.85 (1.74)
L=5	0.25	0.45	0.45
L=6	0.16	0.20	0.25
total	5.12 (1.38)	7.88 (2.47)	8.33 (2.55)

Similar calculation has been performed for the system ^{40}Ca + ^{40}Ca at 50 MeV/A. As in the previous case we have changed only the excitation energies leaving unchanged the old mixing coefficients. The results are shown in table 2 together with the old calculations. As one can see the major increase is obtained when we consider the old case: the introduction of the only the new energies does not change too much the cross sections. Of course this is only part of the story: a complete calculation should be done by introducing the states of three-phonon components and the correct mixing coefficients. Only in this case we will have a complete and satisfactory answer to the question whether the inclusion of the three-phonon states should affect and in what measure the double giant resonances cross sections. Calculations in this direction are in progress.

5. Summary

We have reviewed calculations for the inelastic excitation of a double giant resonance in heavy ions collisions within a semiclassical model, performed by using an extended RPA which includes anharmonicity. Applications to various energy regimes have been shown to describe successfully the existing experimental data. We have extended this method to the calculation of the triple giant resonance. We have shown that the spectrum of the two-phonon states are strongly modified by their coupling to the three-phonon ones. Furthermore, large amplitude motion induce a strong coupling to giant monopole and quadrupole vibrations. Calculation of the inelastic cross sections which include these findings are in progress.

Acknowledgments

I would like to thank the *Secreteria de Estado de Educación y Universidades* of the Spanish *Ministerio de Educación, Cultura y Deporte*, which under the contract SAB2002-0070 has allowed my sabbatical leave from Catania.

References

1. M.N. Harakeh and A. van der Woude *Giant Resonances*, Clarendon Press, Oxford, 2001.
2. R. A. Broglia and A. Winther, Heavy ion reactions, Addison-Wesley, 1991.
3. A. Winther and K. Alder, *Nucl. Phys.* **A319**, 518 (1979).
4. K. Boretzky et al. (LAND Collaboration), *Phys. Rev.* **C68**, 024317 (2003).
5. M. Hage-Hassan and M. Lambert, *Nucl. Phys.* **A188**, 545 (1972).
6. E. G. Lanza, M. V. Andrés, F. Catara, Ph. Chomaz and C. Volpe, *Nucl. Phys.* **A 613**, 445 (1997); *Nucl. Phys.* **A 654**, 792c (1999).
7. C. Volpe, F. Catara, Ph. Chomaz, M.V. Andrés and E.G. Lanza, *Nucl. Phys.* **A 589**, 521 (1995); *Nucl. Phys.* **A 599**, 347c (1996).
8. J. A. Scarpaci et al., *Phys. Rev.* **C56**, 3187 (1997); J. A. Scarpaci et al., *Phys. Rev. Lett.* **71**, 3766 (1993).
9. E. G. Lanza, M. V. Andrés, F. Catara, Ph. Chomaz and C. Volpe, *Nucl. Phys.* **A636**, 452 (1998).
10. M.V. Andrés, F. Catara, E.G. Lanza, Ph. Chomaz, M. Fallot and J. A. Scarpaci, *Phys. Rev.* **C65**, 014608 (2001).
11. J. A. Scarpaci, *Nucl. Phys.* **A731**, 175 (2004).
12. S. Ilievsky et al. (LAND Collaboration), *Phys. Rev. Lett.* **92**, 112502 (2004).
13. C. Volpe, Ph. Chomaz, M.V. Andrés, F. Catara, and E.G. Lanza, *Nucl. Phys.* **A 647**, 246 (1999).
14. M. Fallot, Ph. Chomaz, M.V. Andrés, F. Catara, E.G. Lanza and J. A. Scarpaci, *Nucl. Phys.* **AA 729**, 699 (2003).
15. Ph. Chomaz and C. Simenel, *Nucl. Phys.* **A731**, 188 (2004).

KEY TOPICS IN NUCLEAR STRUCTURE
Proceedings of the 8th Int. Spring Seminar on Nuclear Physics
edited by Aldo Covello
© 2005 World Scientific Publishing Co.

SHAPE EVOLUTION AND TRIAXIALITY IN NEUTRON RICH Y, Nb, Tc, Rh AND Ag

Y.X. LUO[1,2,3], J.O. RASMUSSEN[3], J.H. HAMILTON[1], A.V. RAMAYYA[1], J.K. HWANG[1], S.J. ZHU[1,4], P.M. GORE[1], E.F. JONES[1], S.C. WU[3,5], J. GILAT[3], I.Y. LEE[3], P. FALLON[3], T.N. GINTER[3,6], G. TER-AKOPIAN[7], A.V. DANIEL[7], M.A. STOYER[8], R.DONANGELO[9], AND A. GELBERG[10]

[1]*Physics Department, Vanderbilt University, Nashville, TN 37235 USA*
[2]*Institute of Modern Physics, CAS, Lanzhou 730000, China*
[3]*Lawrence Berkeley National Laboratory, Berkeley, CA 94720 USA*
[4]*Physics Department, Tsinghua University, Beijing 100084, China*
[5]*Department of Physics, National Tsinghua University, Hsinchu, Taiwan*
[6]*National Superconducting Cyclotron Laboratory, Michigan State University, East Lansing, MI 48824 USA*
[7]*Flerov Laboratory for Nuclear Reactions, JINR, Dubna, Russia*
[8]*Lawrence Livermore National Laboratory, Livermore, CA 94550 USA*
[9]*Universidade Federal do Rio de Janeiro, CP 68528, RG Brazil*
[10]*Inst. für Kernphysik, Universitöt zu Köln, 50937 Köln, Germany*

New level schemes to high spin are extracted for ^{99,101}Y, ^{101,105}Nb, ^{105,107,109}Tc, ^{111,113}Rh and ^{115,117}Ag from prompt $\gamma - \gamma - \gamma$ coincidences studies of ^{252}Cf with Gammasphere. The $\pi 5/2^+[422]$ bands in ^{99,101}Y and ^{101,105}Nb and $7/2^+[413]$ bands in ^{105,107,109}Tc, ^{111,113}Rh and ^{115,117}Ag exhibit a smooth evolution from strong prolate deformation with very little signature splittings in ^{99,101}Y to triaxial deformation with large and near maximum splitting in the Tc, Rb and Ag nuclei. These splitting data along with the presence of an excited $11/2^+$ bands with almost no decay to the $7/2^+$ levels in the Tc and Rh nuclei indicate the increasing importance of rigid triaxial shapes in these nuclei. The Nb isotopes having intermediate splitting between Y and Tc, Rh isotopes may suggest that they are transitional nuclei with regard to triaxial deformation. Triaxial-rotor-plus-particle model calculations yielded best fits to the energies, signature splitting and transition probabilities for $\epsilon(\approx \beta_2)\approx 0.32$ and γ = -22.5^o for ^{107}Tc and β_2 = 0.28 and near maximum triaxiality, γ = -28°, for ^{111,113}Rh. A K=1/2 band based on the $1/2^+[431]$ intruder orbital with large deformation is seen in ^{105,107}Tc and ^{111,113}Rh with anomalous spaces where the 1/2, 5/2, ... are above the 3/2, 7/2, ... levels. Band crossings are seen in ^{105,107,109}Tc and ^{111,113}Rh with spin alignments that indicate the alignment of a $h_{11/2}$ neutron pair.

1. Introduction

Studies of shape transitions and shape coexistence in neutron rich nuclei with A $\approx$ 100 have long been of major interest [1]. The sudden onset of superdeformed ground states and its rapid decrease with Z, identical bands and shape coexistence are known in the even-even Sr(Z=38), Zr(Z=40) and Mo(Z=42) nuclei [1,2] with triaxial shapes reported in Mo and Ru nuclei [3,4]. The sudden onset of super-deformation in the ground state of N=60,62 Sr nuclei as well as in ^{74}Kr and ^{76}Sr was explained in terms of the reinforcement of the proton and neutron shape driving forces when the protons and neutrons have shell gaps at the same large deformation [5,6]. Such gaps at the same deformation occur when N=Z=38 and Z=38, N=60,62 ($\beta_2\approx$0.4) as well as other cases where N and Z shell gaps can reinforce each other to drive nuclei to super-deformation (see Ref. 7).

Spectroscopic information of the odd-Z neutron-rich neighbors in this region can provide significant insight for developing a better understanding of shape transitions and the importance of triaxial shapes in this region. Until recently, little was known about the level structures of the odd-Z neutron-rich nuclei above the first few levels. We have investigated the level structures of ^{99,101}Y, ^{101,105}Nb [8], ^{105,107,109}Tc [8,9], ^{111,113}Rh [10], and ^{115,117}Ag [11] populated in the spontaneous fission (SF) of ^{252}Cf. Experimental details are found elsewhere [10]. Clear evidence for the role of triaxial shapes is found including the smooth increase in level splittings in the ground bands and an excited 11/2$^+$ band with strongly hindered decay to the 7/2$^+$ level of the ground band. Rigid triaxial rotor plus one particle calculations reproduce the energies, signature splittings and transition probabilities with $\beta_2\approx$0.28 and γ = -22.5^o in ^{107}Tc and near maximum triaxiality, γ = -28^o for ^{111,113}Rh. A shape coexisting K=1/2 band built on the 1/2$^+$[431] intruder orbital is also seen in ^{105,107}Tc and ^{111,113}Rh.

2. ^{99,101}Y and ^{101,105}Nb

The levels of ^{101}Y and ^{101}Nb are shown in Fig. 1. Note the strong population of the bands assigned 5/2$^+$[422]$\pi g_{9/2}$ as seen in ^{99,101}Y and ^{101,105}Nb. The signature splittings of these bands in ^{99,101}Y are very small (see Fig. 2), the order of 0.02 to 0.04 to high spin where S(I) = $\frac{[E(I)-E(I-1)][I(I+1)-(I-2)(I-1)]}{[E(I)-E(I-2)][I(I+1)-(I-1)I]}$ - 1. The ^{99,101}Y nuclei are considered as a proton coupled to the super-deformed ^{98,100}Sr cores and exhibit properties of a well deformed prolate rotor. The ground state configuration in ^{101,105}Nb is the same as in ^{99,101}Y, $\pi 5/2^+$[422]. Their deformation and

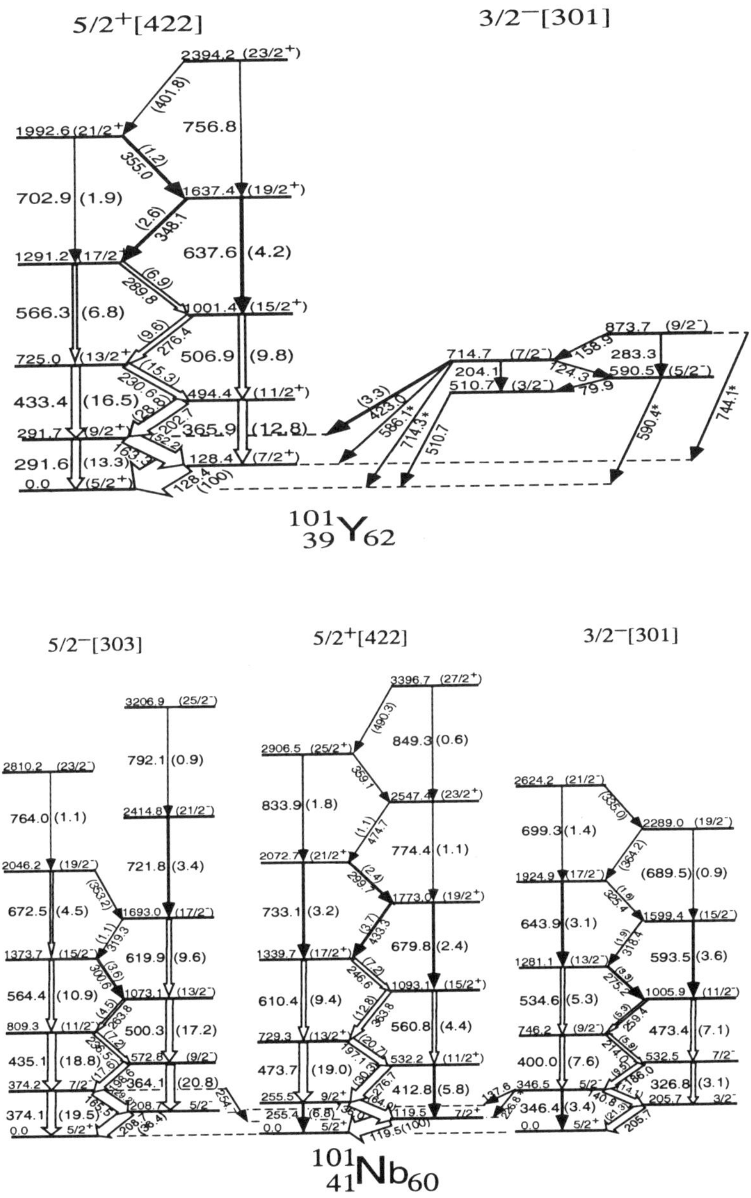

Figure 1. New levels in ^{101}Y and ^{101}Nb from Ref. [8]

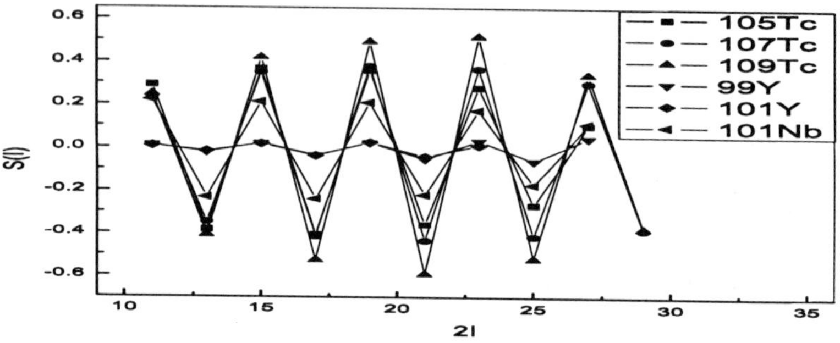

Figure 2. Signature splittings in ^{99,101}Yb [8] and ^{105,107,109}Tc[8,9]

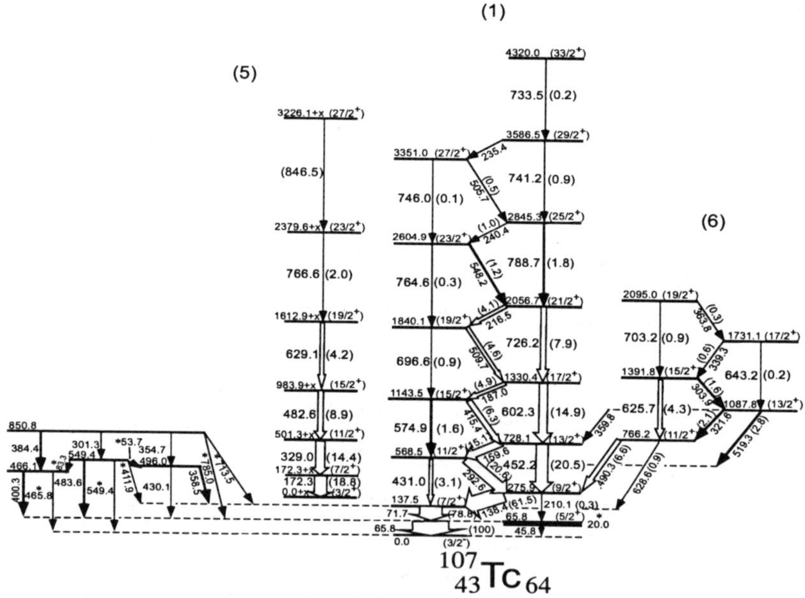

Figure 3. New levels in ^{107}Tc[8,9]

moments of inertia show a decrease with increasing Z. A shape transition from an axially-symmetric shape to one with a triaxial degree of freedom has been suggested between Zr(Z=40) and Mo(Z=42) [4]. S(I) increases significantly from ±(0.02 - 0.04) in ^{99,101}Y to ±(0.20) in ^{101,105}Nb to ±0.50 in ^{105,107,109}Tc (see Fig. 2). This marked increase in splitting, up to ±0.6 in the Tc and Rh isotopes where rigid triaxial rotor plus one particle calculations suggest triaxiality, indicates a transition from axially symmetric deformed shapes in Y to a triaxial configuration in Tc and Rh isotopes, with a transitional chracter in Nb isotopes with regard to triaxiality.

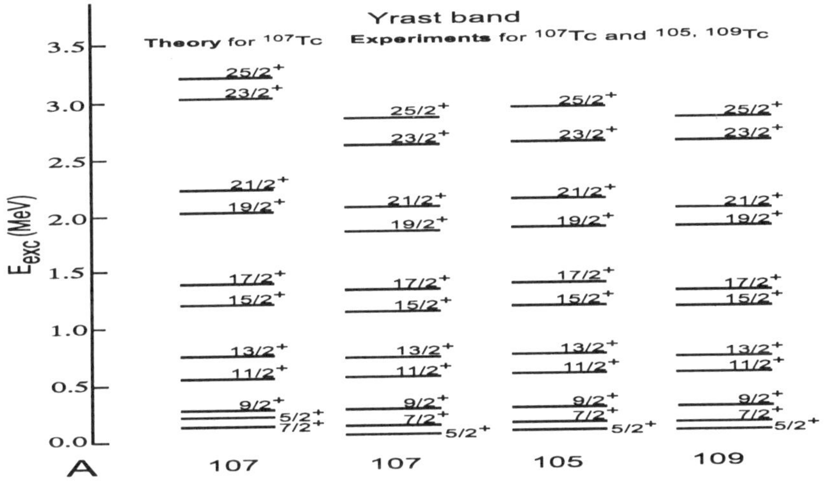

Figure 4. Comparison of rigid triaxial rotor plus particle calculations for ^{107}Tc with ^{105,107,109}Tc

3. ^{105,107,109}Tc, ^{111,113}Rh and ^{115,117}Ag

The levels in Fig. 3 for ^{107}Tc illustrate the new information on these nuclei. The band structures above the 137.5 keV ($7/2^+$) level are from refs. [8,9]. The dominate band in these nuclei is assigned $\pi 7/2^+[413]$. The Nilsson quantum numbers are assigned mainly for labeling since the triaxility will bring about considerable mixing. The signature splitting in these three Tc nuclei, as shown in Fig. 2, is two to three times greater than in the Nb nuclei. An excited $11/2^+$ state ($11/2^+_{exc}$) is observed with a band built on it. The $11/2^+_{exc}$ state has strong E2 strength to the $9/2^+$ member of the $\pi 7/2^+$ band and very small strength to the $7/2^+$ member. This is found in all three Tc nuclei and in ^{111,113}Rh where the quenching of the $11/2^+_{exc} \rightarrow$

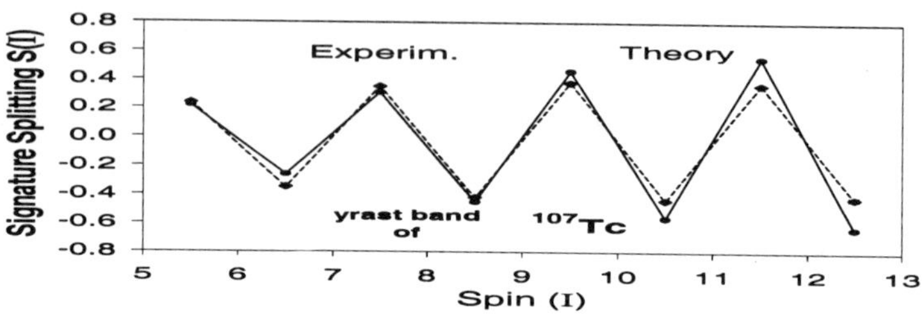

Figure 5. Comparison of theoretical (solid line) and experimental (dashed line) signature splittings in ^{107}Tc

$7/2_1$ transition was explained by examining the wave functions. The main core components of the initial and final states are the first 2^+ core state so the E2 strength is mainly dictated by the diagonal E2 reduced matrix element which vanishes for γ = -30°, while the main core component for the $9/2_1$ state is the 0^+ core state to give a large B(E2: $11/2^+_{exc} \rightarrow 9/2_1$). This quenching in Tc and Rh provides strong evidence for triaxially.

Rigid triaxial rotor plus one particle calculations were carried out for ^{107}Tc and are compared with experiment in Fig. 4. The best fit to the excitation energies, signature splittings and branching ratios is for $\epsilon(\approx \beta_2)\approx 0.32$ and γ = -22.5^o on the prolate side of maximum triaxiality. There is backbending in these nuclei above 1.8 MeV so the comparison of the theory that includes only one particle is not expected to be good above that energy. The calculations reproduce nicely the energies of the $11/2^+_{exc}$ bands and the strong signature splitings as seen in Fig. 5.

A band assigned as a $1/2^+$[431] intruder band from the $\pi(g_{7/2}, d_{5/2})$ subshell is seen in ^{105,107}Tc. The calculations do not reproduce this band as seen in Fig. 6. These orbitals have a strong prolate-deformation driving effect. The "anomalous"level spacings where the 1/2, 5/2 ... are above the 3/2, 7/2, ... levels are characteristic of K=1/2 bands. This irregular sequence is seen in similar bands in ^{111,113}Rb and is explained by a decoupling parameter a of -1 to -2. These strong prolate $1/2^+$[431] bands provide an example of triaxial-asymmetric and symmetric shape coexistence.

Examples of our new ^{111,113}Rh and ^{115,117}Ag data are shown in Fig. 7. Note the presence of the $\pi 7/2^+$ ground band with strong signature splittings, a $11/2^+_{exc}$ band with very weak $11/2^+_{exc} \rightarrow 7/2^+_1$ strength and a

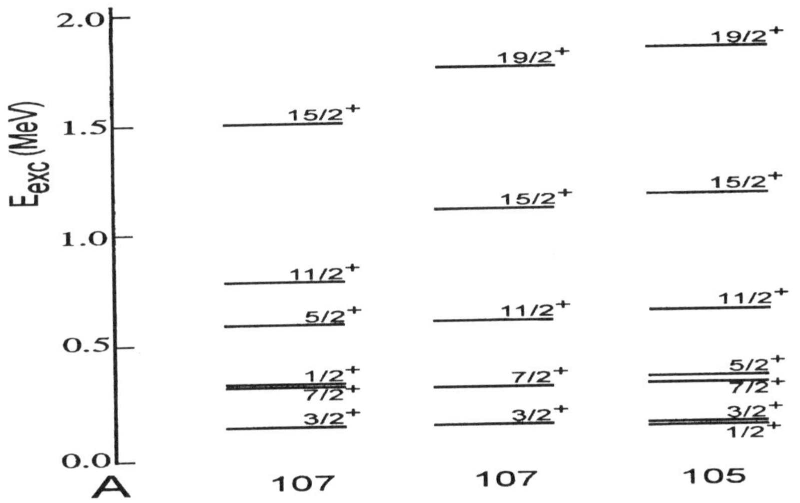

Figure 6. $1/2^+[431]$ bands in $^{105,107}Tc$ and theoretical levels on left

$1/2^+[431]$ band with level inversion. As in Tc, the $^{109,111,113}Rh$ all exhibit backbending (see Fig. 8). The magnitude of the spin alignment in $^{111,113}Rh$ and the absence of backbending in odd-odd ^{112}Rh indicate the Tc and Rh ground band backbendings are associated with an $h_{11/2}$ neutron pair.

Rigid triaxial rotor plus one particle calculations were also done for $^{111,113}Rh$. The best fits to the level energies, splittings and transition probabilities were for β_2 = 0.28 and γ = -28° as shown for ^{111}Rh in Fig. 9 and for β_2 = 0.27 and γ = -28° for ^{113}Rh. Here γ is very near maximum triaxiality. A comparison of theory and experiment for the signature splitting is shown in Fig. 10. The levels in ^{117}Ag are shown in Fig. 7. The $\pi 7/2^+$ ground bands show in $^{115,117}Ag$ [11] the same very strong signature splitting.

The rapidly changing magnitude of the signature splitting for the N = 60 ^{99}Y, ^{101}Nb, ^{103}Tc, and ^{105}Rh are seen in Fig. 11. Similar results are found for the N = 62 and 64 isotones as can be seen by comparing the level systematics in the Y to Rh nuclei shown in Fig. 12. The good agreement of the experimental data for ^{107}Tc and $^{111,113}Rh$ with the rigid triaxial rotor plus one particle calculation indicate an evolution of the nuclear structure from symmetric, strongly deformed shapes in $^{99,101}Y$ to near maximum triaxiality in $^{111,113}Rh$. Symmetric-asymmetric shape coexistence is seen

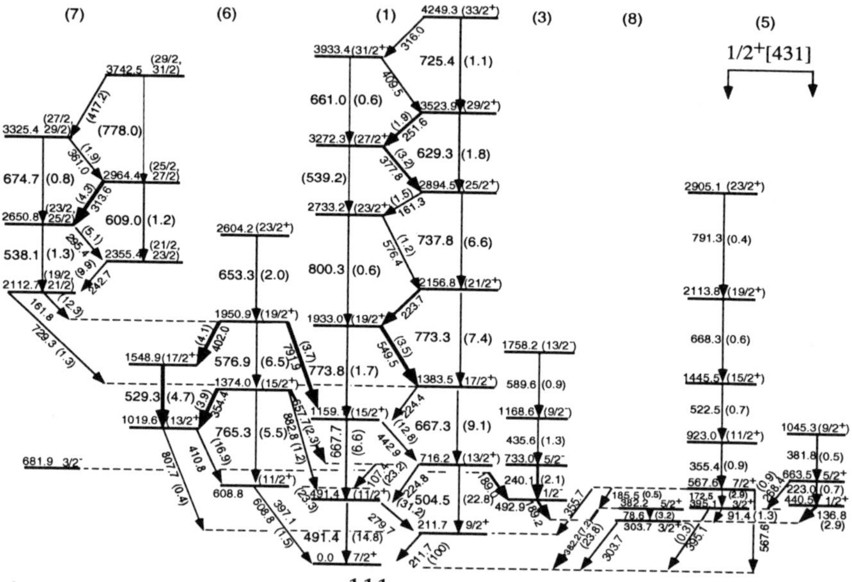

$^{111}_{45}Rh_{66}$

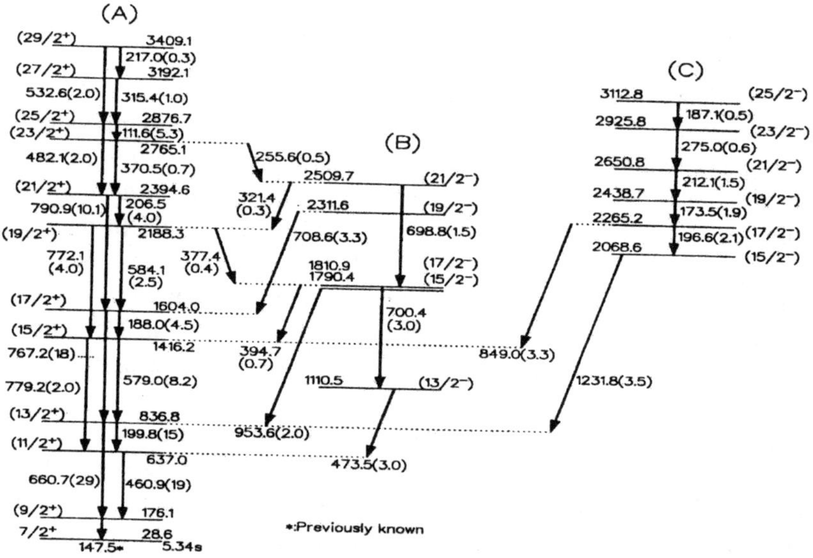

^{117}Ag

Figure 7. Levels in ^{111}Rh[10] and ^{117}Ag[11]

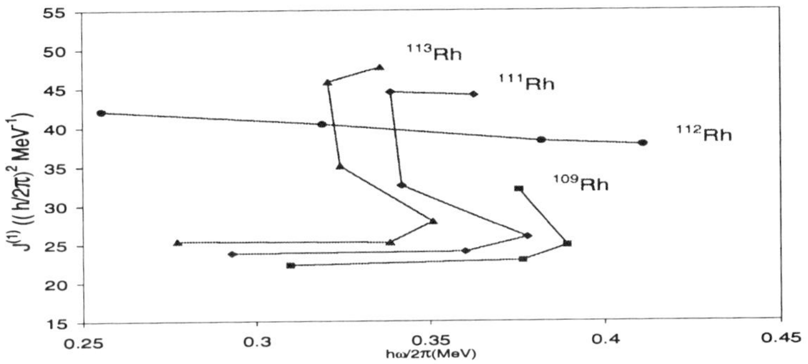

Figure 8. Kinematic moments of inertia vs. rotational frequency for the ground bands in ^{109,111,113}Rh [10]

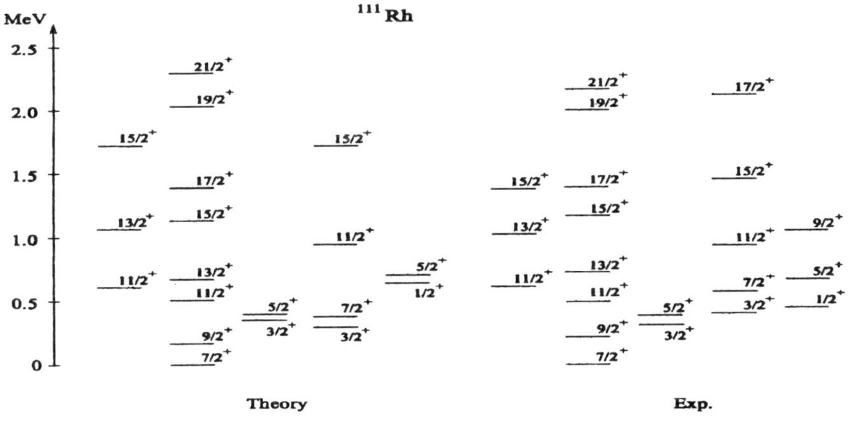

Figure 9. Comparison of rigid triaxial rotor plus particle calculations and experimental levels for ^{111}Rh

in the Tc and Rh nuclei by the presence of K = 1/2, $1/2^+[431]$ intruder band.

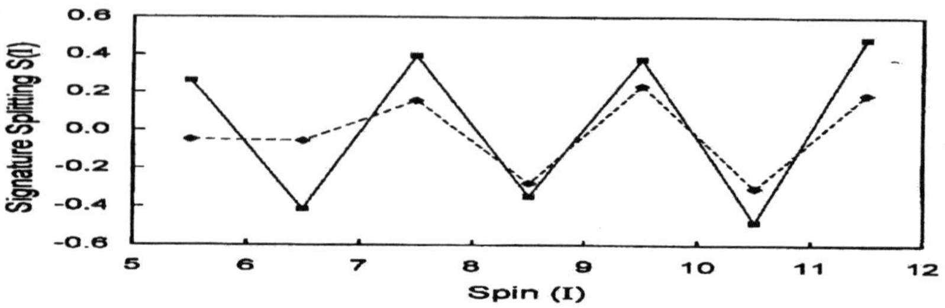

Figure 10. Comparison of theory(solid line) and experiment(dashed line) for signature splittings in ^{111}Rh

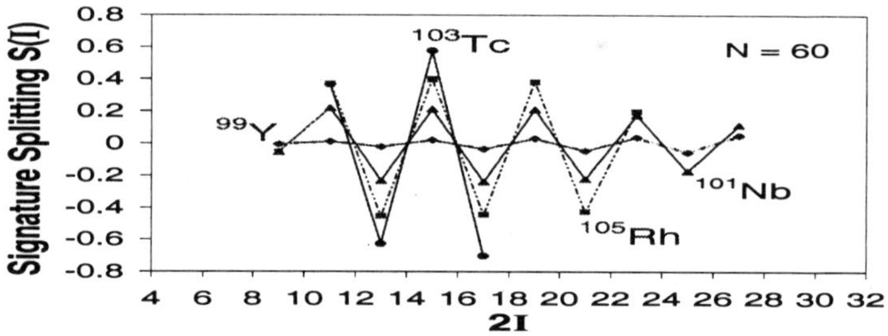

Figure 11. Signature splittings in N=60 isotopes with Z=39-45.

4. Acknowledgements

Work at Vanderbilt University, Lawrence Berkeley and Lawrence Livermore National Laboratories was supported by U.S. Department of Energy Grant DE-FG05-88ER40407 and Contract W-7405ENG48 and DE-AC03-76SF00098. Work at JINR, Dubna, Russia, was supported by the U.S. Department of Energy Contract DE-AC011-00NN4125, BBW1 Grant 3498 (CRDF Grant RPO-10301-INEEL) and by the joint RFBR-DFG Grant [RFBR Grant p2-02-04004, DFG Grant 436RUS 113/673/0-1(R)]. Work at Tsinghua was supported by the Major State Basic Research Development Program Contract G2000077405, the National Natural Science Foundation of China Grant 10375032, and the Special Program of Higher Education

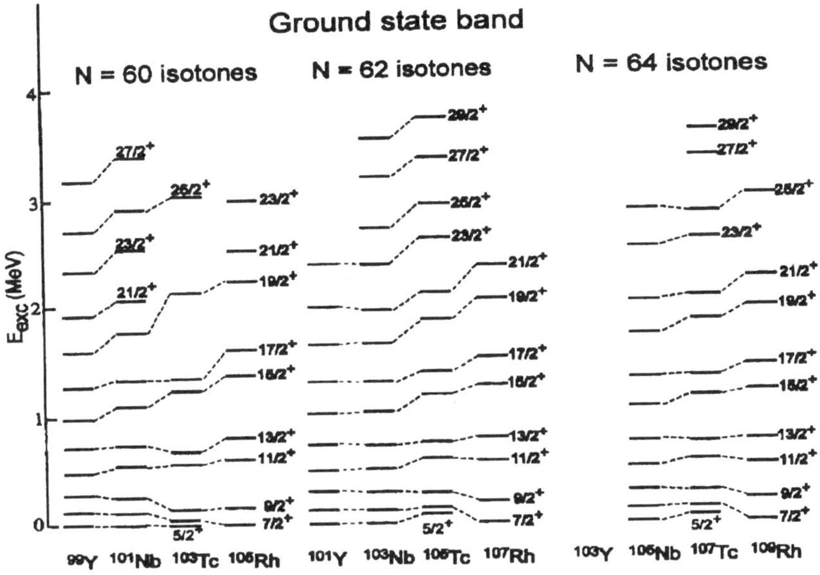

Figure 12. Ground state bands in N=60,62,64 isotones

Science Foundation, Grant 200300090.

References

1. J. H. Hamilton, in *Treatise on Heavy-Ion Science*, edited by D. A. Bromley (Plenum Press, New York 1989) Vol 8, p. 2.
2. J. H. Hamilton et al., Prog. Part. Nucl. Phys. **35**, 635 (1995).
3. D. Troltenier et al., Nucl. Phys. **A601**, 56 (1996).
4. H. Hua et al., Phys. Rev. **C35**, 014317 (2004).
5. J. H. Hamilton et al., J. Phys. **G10**, L87 (1984).
6. J. H. Hamilton, it Proc. Int. Symp. on Nuclear Shell Models, edited by M. Vallieres and B. H. Wildenthal (World Scientific Pub., Singapore 1985) p. 31. and J. H. Hamilton, Prog. Part. Nucl. Phys., **15**, 107 (1985).
7. J. H. Hamilton, *Int. Conf. Shells-50*, edited by Yu. Ts. Oganessian and R. Kalpakchieva (World Scientific Pub., Singapore 2000) p. 88.
8. Y. X. Luo et al., to be published.
9. J. K. Hwang et al., Phys. Rev. **C57**, 2250 (1998).
10. Y, X. Luo et al., Phys. Rev. **C69**, 024315 (2004).
11. J. K. Hwang et al., Phys. Rev. **C65**, 054314 (2002).

KEY TOPICS IN NUCLEAR STRUCTURE
Proceedings of the 8th Int. Spring Seminar on Nuclear Physics
edited by Aldo Covello
© 2005 World Scientific Publishing Co.

NUCLEAR BAND STRUCTURES IN ^{93,95}Sr AND HALF-LIFE MEASUREMENTS

J.K. HWANG[1], A.V. RAMAYYA[1], J.H. HAMILTON[1], J.O. RASMUSSEN[2], Y.X. LUO[1,2], P.M. GORE[1], E.F. JONES[1], K. LI[1], D. FONG[1], I.Y. LEE[2], P. FALLON[2], A. COVELLO[3], L. CORAGGIO[3], A. GARGANO[3], N. ITACO[3], AND S.J. ZHU[4]

[1] *Department of Physics, Vanderbilt University, Nashville, TN 37235 USA*
[2] *Lawrence Berkeley National Laboratory, Berkeley, CA 94720 USA*
[3] *Università di Napoli Federico II and INFN, Complesso Universitario di Monte S. Angelo Via Cintia, 80126 Napoli, Italy*
[4] *Department of Physics, Tsinghua University, Beijing 100084, People's Republic of China*

Many neutron hole and particle states were identified in ^{93,95}Sr by using the spontaneous fission of ^{252}Cf and Gammasphere array. 21 and 20 new gamma transitions were observed in ^{93}Sr and ^{95}Sr, respectively. The level schemes of ^{93}Sr and ^{95}Sr are interpreted in part as the weak coupling of the $2d_{5/2}$ neutron hole and $1g_{7/2}$ neutron particle, respectively, to the excited configurations of ^{94}Sr core. Also, the shell model calculations have been done for comparison with the levels of ^{93}Sr. Half-lives ($T_{1/2}$) of several states which decay by delayed γ transitions were determined from time-gated triple γ coincidence method. We determined, for the first time, the half-life of 330.6+x state in ^{108}Tc and the half-life of $19/2^-$ state in ^{133}Te based on the new level schemes. Five half-lives of ^{95,97}Sr, ^{99}Zr, ^{134}Te and ^{137}Xe are consistent with the previously reported ones. These results indicate that this new method is useful for measuring the half-lives.

1. Introduction

In the weakly deformed Zr and Sr isotopes between the two shell gaps of N=50 and 56, a comparison of the excitation patterns is necessary to extract information on the hole and particle residual interactions between the valence neutrons and protons. The high spin states in $^{89-97}$Sr and ^{92,93,94,95}Zr have been so far interpreted from this viewpoint [1,2,3]. An abrupt change from a spherical to a deformed shape takes place as the neutron number increases from N=56 to N=64. Also, the coexistence of spherical and deformed shapes has been observed in ^{96,97,98}Sr, and in ^{98,99,100}Zr. The nucleus ^{96}Sr has a spherical ground state with deformed excited states while in ^{93,94}Sr only the spherical ground state has been observed. The ground and first excited states in ^{97}Sr have spins and parities of $1/2^+$ and $3/2^+$, respectively. The $1/2^+$ ground state in ^{97}Sr has a spherical shape and high spin states in ^{97}Sr with a very deformed shape of $\beta_2 \approx 0.4$ were reported by us, recently [3]. For the ground and first excited states in ^{95}Sr, $1/2^+$ and ($1/2^+$,$3/2^+$), respectively, were assigned without any information on the deformation. We identified the high spin states in ^{93,95}Sr in order to look for neutron particle and hole states [1,2]. Weakly coupled bands based on $2d_{5/2}$ and $1g_{7/2}$ neutron configurations are reported in ^{93,95}Sr, for the first time, in the Sr and Zr region [1,2]. We have tried to give a qualitative description of the properties of ^{93}Sr by performing a shell-model study in which we assume that ^{88}Sr is a closed core. The calculated spectra have been compared with the experimental data [1].

Previously, half-lives of several states in neutron-rich nuclei have been determined by single γ or $\gamma-\gamma$ coincidence relations for the delayed γ transitions emitted from the isotopes produced in the fission of ^{235}U, ^{239}Pu, ^{248}Cm, and ^{252}Cf [5]. Most of the previous results were obtained from the coincidence measurement between the γ transition and the fission fragment after fission. And some of them were obtained from the delayed time measurement of the γ transition following the β decay after fission.

Usually, more than 100 isotopes are produced in the fission of these heavy nuclei, with each isotope emitting many γ rays. With such complex spectra, it is very difficult to isolate a single γ ray peak. Coincidences from other transitions with energies essentially equal to that of the transition of interest can lead to significant errors in the half-life values. The triple coincidence method can reduce the error associated with complexity of the γ ray spectra in spontaneous fission. Because several new nuclei and many new levels in the known nuclei have been identified in the spontaneous fission (SF) of ^{252}Cf, the present time-gated triple γ coincidence method is very useful for the half-life measurements of nuclear states in neutron-rich nuclei. We applied this method, for the first time, to extract the half-lives of two states in ^{95,97}Sr [3]. Also, in the present work, five other cases namely ^{99}Zr, ^{133,134}Te, ^{137}Xe and ^{108}Tc are investigated. Recently, the new level schemes of ^{133}Te and ^{108}Tc have been reported from the SF work of ^{252}Cf. Based on these new level schemes, the half-lives of 1610.4 keV state in ^{133}Te and 330.6+x keV state in ^{108}Tc are reported [4].

2. Experimental procedures

In the present work, the measurements were carried out at the Lawrence Berkeley National Laboratory by using a spontaneously fissioning ^{252}Cf source inside the Gammasphere array. A ^{252}Cf source of strength $\approx$62μCi was sandwiched between two Fe foils of thickness 10 mg/cm^2 and was mounted in a 7.62 cm diameter plastic (CH) ball to absorb β rays and conversion electrons. The source was placed at the center of the Gammasphere array which, for this experiment, consisted of 102 Compton suppressed Ge detectors. A total of 5.7$\times$10^{11} triple and higher fold coincidence events were collected. The coincidence data were analyzed with the RADWARE software package [5]. The width of the coincidence time window was about 1 μs. The ordering of transitions in bands is based on relative intensities, coincidence relationships, and the feeding and decaying intensity balances for levels.

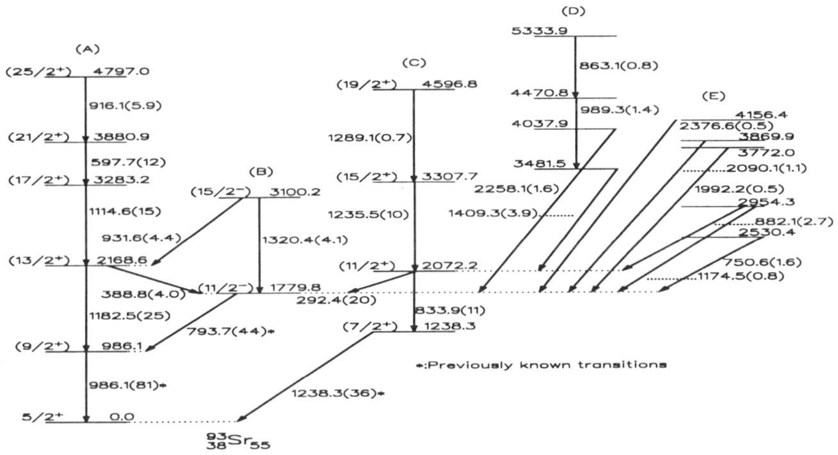

Figure 1. High spin states in ^{93}Sr [1]. The intensities of the transitions are shown in parentheses.

Also, the $\gamma-\gamma-\gamma$ coincidence measurements were done by using the Gammasphere facility with 72 Ge detectors and a ^{252}Cf SF source of strength $\sim$28 μCi at LBNL. Several $\gamma-\gamma-\gamma$ coincidence cubes with different time windows, t_w, were built for the three- and higher-fold data by using the Radware format. That is, a time-gated cube will contain all triple-coincidence events for which all these time differences are less than the specified time value. These time gated cubes have been used for the

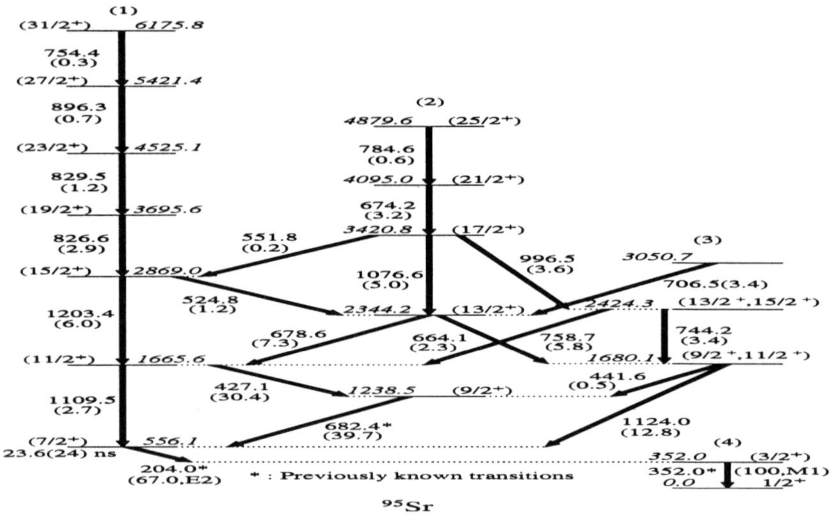

Figure 2. High spin states in ^{95}Sr [2]. The intensities of the transitions are shown in parentheses.

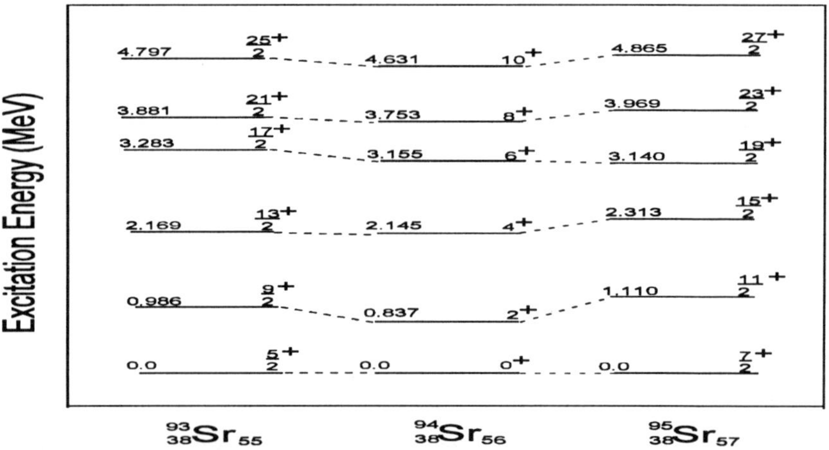

Figure 3. Comparison between the experimental excited energies in ^{95}Sr, ^{93}Sr[2] and ^{94}Sr [1,2]. A value of 0.5561 MeV is subtratced from each level energy in band -1 in ^{95}Sr to normalize it with the values in ^{93,94}Sr.

measurement of half-lives of several states in neutron rich nuclei.

3. High spin states in ^{93,95}Sr

The partner fragments of ^{93,95}Sr in spontaneous fission of ^{252}Cf are ^{156}Nd, ^{154}Nd, ^{153}Nd and ^{152}Nd. When we set double gates on two known transitions belonging to one of Nd isotopes, the previously known transitions in ^{93,95}Sr isotpes from the β decay works are clearly seen in our spectra. By double-gating on these known transitions we observed several new transitions belonging to ^{93,95}Sr. Also, by comparing coincidence spectra with double gates on one transition in ^{93,95}Sr and another on a transition in one of Nd partner isotopes, 21 and 20 new transitions in ^{93}Sr and ^{95}Sr, respectively, were identified. High spin states in ^{93,95}Sr are shown in Figs. 1 and 2.

In this mass region, according to the Nilsson deformed shell model, the $2d_{5/2}$, and $1g_{7/2}$ orbitals are below and above the N=56 spherical sub-shell gap, respectively. In ^{93}Sr, the bands were interpreted as originating from the weak coupling of the $2d_{5/2}$ neutron hole to the levels of the N=56 ^{94}Sr core. Since ^{95}Sr has one neutron in the $1g_{7/2}$ near spherical orbital, the levels in ^{95}Sr can be thought of as arising from the weak coupling of the $g_{7/2}$ neutron to the ^{94}Sr core. On this basis, the spins and parities to bands in ^{93,95}Sr are assigned. The E2 assignments to the transitions in the bands are based on the fact that the transition energies in bands in ^{93,95}Sr are similar to those in ^{94}Sr. A comparison of the level energies in ^{93,94,95}Sr is shown in Fig. 3. Since the band head energy of band -1 in ^{95}Sr is 556.1 keV above the ground state energies of ^{93,94}Sr, a value of 556.1 keV is subtracted from each level energy in band -1 for comparison. Then bands in ^{93,95}Sr will have a spherical shape similar to that of a ground state band of ^{94}Sr.

We have also performed realistic shell-model calculations for ^{93}Sr. The calculated level energies are in quantitative agreement with the experimental ones only for band -C. So, we need to consider the proton degrees of freedom but only large-scale calculations may shed light on this point [1].

4. Half-lives of several states in neutron rich nuclei from SF of ^{252}Cf

Let's consider a downward cascade consisting of $\gamma_3 - \gamma_2 - \gamma_1 - \gamma_0$ transitions where γ_0 is the outgoing transition from a state with long half-life and γ_1 is the incoming transition into the same state. Other higher states in this cascade are assumed to have very short lifetimes. We set a double gate on E_{γ_3} and E_{γ_1} and compare the intensities of transitions, γ_0 and γ_2, $N(\gamma_0)$ and $N(\gamma_2)$ in the spectra. In the present work, γ_1, γ_2, and γ_3, are in prompt coincidence. Therefore, the delay-time between γ_1 and γ_3 will be negligible. Since γ_0 is the ending transition in this cascade, the coincidence time window (t_w) limits the TDC time difference, t_{10}, between the γ_1 and γ_0 transitions, and the intensity N(γ_0) observed from the state with the long lifetime. The N(γ_0) intensity determines the fraction of N(γ_2) intensity observed from the state with the long half-life with decay constant, λ. Therefore, $N(\gamma_0)/ N(\gamma_2) = C(1 - e^{-\lambda t_w})$ can be applied in this case, where C is a constant [4].

In Fig. 4, the level scheme of ^{97}Sr is shown. For ^{97}Sr, we double-gated on the γ transitions of 239.6 and 272.5 keV to compare the 205.9 and 522.7 keV transition intensities with the coincidence time windows(t_w) of t_w=100, 300 and 500 nsec as shown in Fig. 5. We can decide the half-life by using the graph shown in Fig. 6. The measured half-life of 830.8 keV state is 265(27) nsec which is much less than the values of 382(11) nsec (the delayed 522.7 keV transition measurement in the ^{235}U(n,f) and ^{239}Pu(n,f) neutron induced fission) and 515(15) nsec(the delayed 522.7 keV transition measurement from the ^{252}Cf(SF)). But recently, Pfeiffer corrected their value of 515(15) nsec to 255(10) nsec [4] because of calibration error in data analysis. This corrected half life is consistent with our value.

In the present work, we report half-lives of five states in ^{99}Zr, ^{108}Tc, ^{133}Te, ^{134}Te, and ^{137}Xe by using the new time-gated triple coincidence method. We determined, for the first time, half-lives of ^{108}Tc and ^{133}Te based on the new level schemes. The half-lives of states in ^{99}Zr, ^{134}Te and ^{137}Xe are compared with the previously reported ones. The measured half-lives are consistent with the previously reported ones. These results indicate that this new method is useful for the half-life measurements. The measured half-lives are shown in Table 1.

Table 1. Half-lives ($T_{1/2}$ nsec) of several states (E_{IS}, keV) [4]. $E(\gamma_1)/E(\gamma_3)$ are the double-gated transition energies. For ^{97}Sr, $E(\gamma_2)/E(\gamma_3)$ and $E(\gamma_1)$ are used instead.

Nuclei	E_{IS}	$E(\gamma_1)/E(\gamma_3)$	$E(\gamma_2)$	$E(\gamma_0)$	Present $T_{1/2}$
^{95}Sr	556.1	682.4/678.6	427.1	204.0	23.6(24)
^{97}Sr	830.8	239.6/272.5	205.9	522.0	265(27)
^{99}Zr	252.0	426.4/415.2	142.5	130.2	316(48)
^{108}Tc	330.6+x	123.4/341.6	125.7	154.0	94(10)
^{133}Te	1610.4	721.1/933.4	738.6	125.5	102(15)
^{134}Te	1692.0	2322.0/516.0	549.7	115.2	197(20)
^{137}Xe	1935.2	311.3/304.1	1046.4	314.1	10.1(9)

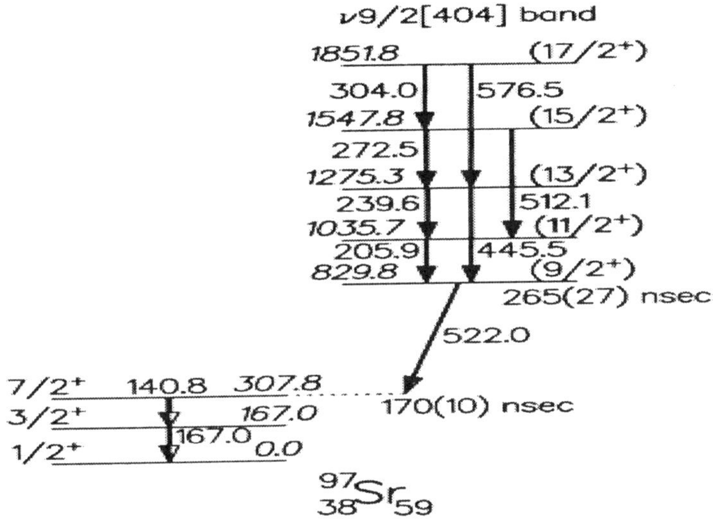

Figure 4. ^{97}Sr level scheme [3]

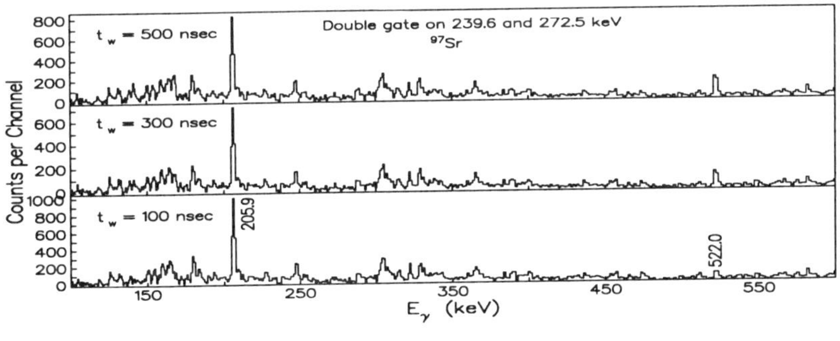

Figure 5. Coincidence spectra with double gates set on 239.6- and 272.5- keV transitions in ^{97}Sr with coincidence time windows (t_w) of 100, 300 and 500 nsec.

5. Conclusions

21 and 20 new γ transitions in ^{93}Sr and ^{95}Sr, respectively, were identified. States in ^{95}Sr, based on the These bands show a close similarity in transition

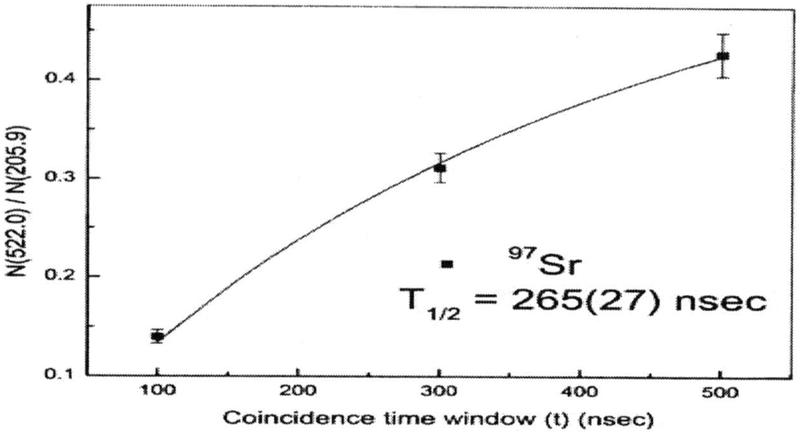

Figure 6. N(205.9)/N(522.7) versus coincidence time window (t_w) plot for ^{97}Sr.

energies to the ground state band of the neighboring ^{94}Sr which has a spherical shape. This similarity leads to an interpretation of the ^{93,95}Sr levels as arising from the weak coupling of the $2d_{5/2}$ neutron hole and $1g_{7/2}$ neutron particle, respectively, to the levels of the $^{94}_{38}Sr_{56}$ core. Also, the shell model calculations have been done for comparison with the levels of ^{93}Sr.

Half-lives ($T_{1/2}$) of several states which decay by delayed γ transitions were determined from time-gated triple γ coincidence method. We determined, for the first time, the half-life of 330.6+x state in ^{108}Tc and the half-life of $19/2^-$ state in ^{133}Te based on the new level schemes. Five half-lives of ^{95,97}Sr, ^{99}Zr, ^{134}Te and ^{137}Xe are consistent with the previously reported ones. These results indicate that this new method is useful for measuring the half-lives.

References

1. J.K. Hwang et al., *Phys. Rev.* **C67**, 014317 (2003) and References therein.
2. J.K. Hwang et al., *Phys. Rev.* **C69**, 67302 (2004) and References therein.
3. J.K. Hwang et al., *Phys. Rev* **C67**, 054304 (2003) and References therein.
4. J.K. Hwang et al., *Phys. Rev* **C69**, 57301 (2004) and References therein.
5. D.C. Radford, *Nucl. Instrum. Methods Phys. Res.* **A361**, 297 (1995).

KEY TOPICS IN NUCLEAR STRUCTURE
Proceedings of the 8th Int. Spring Seminar on Nuclear Physics
edited by Aldo Covello
© 2005 World Scientific Publishing Co.

SHIFTED IDENTICAL BANDS FROM Pt TO Pb

P. M. GORE, E. F. JONES, J. H. HAMILTON, AND A. V. RAMAYYA
Department of Physics, Vanderbilt University, Nashville, TN 37235, USA

From the Brookhaven and Lawrence Berkeley National Laboratory Data Retrieval websites, we present new insights into shape coexistence in the Pt to Pb region. The phenomenon of shifted identical bands is now found in proton-rich nuclei in the Pt to Pb region. Shifted identical bands are bands of two neighboring nuclei differing by 2n, 2p, α, n, p, and other combinations, where the transition energies in the bands differ by the same constant fraction over a range of spins. Identical bands and SIBs are found above spins of 6^+ in even-even nuclei or equivalent in odd-A nuclei, where the well-deformed structures become yrast with little mixing between the shape-coexisting structures. As an example, $E_\gamma(^{180}Hg)$ = 1-0.022($^{+3}{}_{-1}$)$E_\gamma(^{182}Hg)$ from 6^+ to 16^+.

1. Introduction

The new phemonenon we call shifted identical bands, SIBs, has been found to occur in the most neutron-rich stable to neutron-rich radioactive nuclei from Ce to W. The SIBs occur when yrast bands in two even-even nuclei separated by 2n, 2p, α, α+2n, etc. are found to have identical transition energies when one set of energies is shifted the same constant amount [1]. The equations which we use to characterize SIBs are:

$$\frac{\Delta E_\gamma}{E_\gamma} = \frac{(E_{\gamma nuclide1} - E_{\gamma nuclide2})}{E_{\gamma nuclide2}} = \kappa$$

$$= -\frac{\Delta J_1}{J_1} = -\frac{(J_{1nuclide1} - J_{1nuclide2})}{J_{1nuclide1}}$$

It follows that:

$$E_i = (1+\kappa)E_j$$

and

$$(1+\kappa)J_{1i} = J_{1j}$$

We define identical bands - IBs - as those in which the absolute value of average $\kappa < 1\%$ and the total spread in κ is within $\pm1\%$. We define

SIBs as those in which the absolute value of average $\kappa > 1\%$ and the total spread in κ is within $\pm 1\%$. In SIBs, the energies of at least 4 transitions in two neighboring nuclei (separated by 2 neutrons, 2 protons, or other combinations) differ by the same constant fraction over a range of spins, e.g. the transition energies $E_\gamma(^{158}Sm) = 1.033(\frac{3}{2})E_\gamma(^{160}Sm)$ from the $2^+ \rightarrow 0^+$ transition up to the $12^+ \rightarrow 10^+$ transition. Note the spread in the energies is remarkably constant, only 2 to 3 parts in 1000. We examined yrast bands in 1,448 Xe to Os nuclei from proton-rich to neutron-rich for shifted identical bands and found 107 cases concentrated in Sm to Dy with none in Xe, Ba, or Os for various separations. In none of these do the moments of inertia follow the expected $A^{5/3}$ dependence. These SIBs occur in the most neutron-rich stable to neutron-rich radioactive rare-earth nuclei. In only eleven cases are IBs observed. SIBs are not observed in rare earth proton-rich nuclei. However, in the A = 180 region of Pt through Pb, where shape coexisting bands occur near the ground state [2], IBs and a new type of SIB are found.

2. Results and Analysis

As reported here, comparing transitions in well-deformed bands in neighboring even-even, even-odd, odd-even, and odd-odd Pt to Pb nuclei, we found SIBs and IBs. The SIBs and IBs in even-even nuclei of this region start around 6^+. Below 6^+, the energy levels built on the two different shapes interact to shift their energies. Figure 1 gives an example of the shape coexistence of spherical and well-deformed bands in light-mass isotopes of Hg. In each level scheme, the left ground band is near spherical and the right excited band is well deformed. As we go up the deformed band, the interaction with the spherical band decreases significantly above 6^+ as the same spin states in these bands separate in energy. It is in this region above 6^+ the deformed bands in neighboring nuclei form IBs and SIBs for several 2n, 2p, and α separations as well as for several 1n and 1p separations, for example, $^{180-182}$Hg, ^{178}Pt-^{180}Hg, $^{182-183}$Hg, ^{182}Hg-^{183}Tl, and $^{185-187}$Tl.

The results for the even-Z even-N Hg isotopes are shown in Figure 2 and given in Table 1. Only one SIB in $^{180-182}$Hg and one nearly identical band in $^{182-184}$Hg are found. These occur for N = 100 to 104 around midshell at N = 104 where the deformation is the largest but they are not found for higher or lower N. In the Pt nuclei, the ground bands are the well-deformed bands that are the yrast states. Thus there is one SIB starting at 2^+ in $^{180-182}$Pt with N = 102-104 as seen in Table 1. Also in Table 1 are given

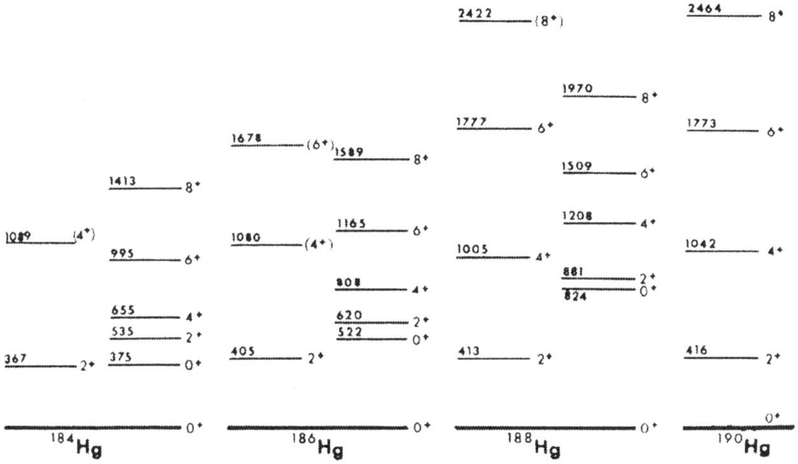

Figure 1. Spherical and well-deformed bands in light-mass isotopes of Hg.

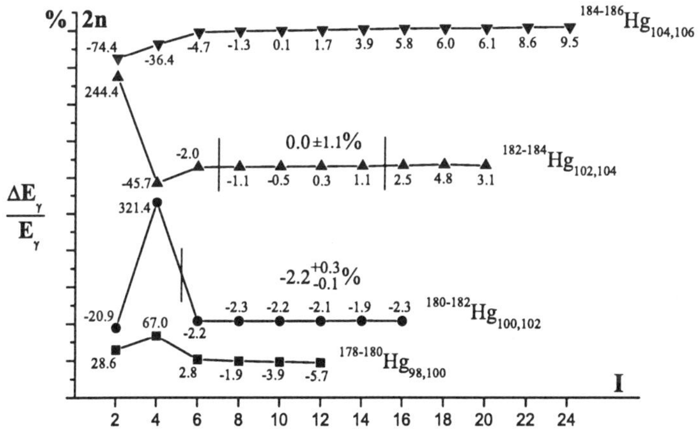

Figure 2. 2n-comparisons in even-Z even-N isotopes of Hg.

the results for $^{183-185}$Tl and $^{185-187}$Tl, both of which form IBs or small κ SIBs starting at $29/2^+$ for the bands built on the odd $i_{13/2}$ proton coupled to the well-deformed excited bands in the even-even Hg cores.

Now look at Pt-Hg 2p comparisons for the same N as shown in Figure 3 for the Pt-Hg and in Figure 4 for Hg-Pb and given in Table 2. Five 2p

Table 1. 2n-comparison IBs and SIBs in Pt, Au, Hg, and Tl isotopes.

2n Separation				
Z / N	Pair	Range	$\Delta E_\gamma/E_\gamma$ (%)	†
78/100-102	$^{178-180}Pt$	6 - 16	$-2.1^{+0.8}_{-0.6}$	PP
/102-104	$^{180-182}Pt$	2 - 8	$-1.9^{+0.9}_{-0.6}$	PP
79/100-102	$^{179-181}Au$	17/2-29/2	$-3.2^{+1.3}_{-0.9}$	PP
80/100-102	$^{180-182}Hg$	6 - 16	$-2.2^{+0.3}_{-0.1}$	PP
/102-104	$^{182-184}Hg$	8 - 14	0.0±1.1 ‡	PP
81/102-104	$^{183-185}Tl$	29/2-41/2	$-0.3^{+1.1}_{-1.0}$ ‡	PP
/104-106	$^{185-187}Tl$	29/2-43/2	$1.0^{+1.0}_{-0.8}$ ‡	PP
† P, π rich: N, ν rich: S, stable			‡ Identical Bands	

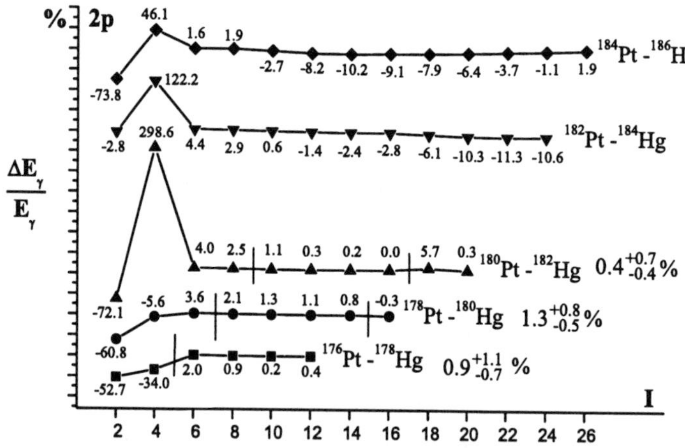

Figure 3. 2p-comparisons of Pt-Hg isotopes.

cases are IBs and only one SIB. The IBs and SIBs occur for N = 98, 100, and 102, but not for 104 and 106. Recall in Pt nuclei the ground-state band is well-deformed, but in the Hg nuclei it is the excited band. Here above 6^+ we are comparing the well-deformed ground bands in Pt with the well-deformed excited bands in Hg nuclei to show their very remarkably similar energies and moments of inertia.

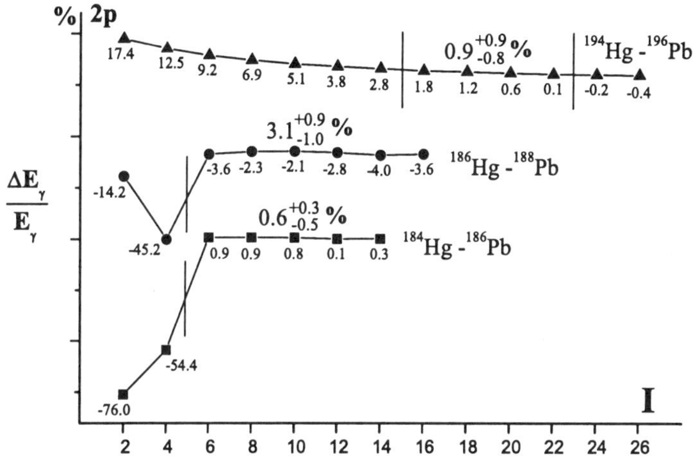

Figure 4. 2p-comparisons of Hg-Pb isotopes.

Table 2. 2p-comparison IBs and SIBs in Pt-Hg and Hg-Pb isotopes.

Z / N	Pair	Range	$\Delta E_\gamma/E_\gamma$(%)	†
78-80/98	^{176}Pt -^{178}Hg	6 - 12	$0.9^{+1.1}_{-0.7}$ ‡	PP
/100	^{178}Pt -^{180}Hg	8 - 14	$1.3^{+0.8}_{-0.5}$	PP
		10 - 16	$0.7^{+0.6}_{-1.0}$ ‡	
/102	^{180}Pt -^{182}Hg	10 - 16	$0.4^{+0.7}_{-0.4}$ ‡	PP
80-82/104	^{184}Hg-^{186}Pb	6 - 14	$0.6^{+0.3}_{-0.5}$ ‡	PP
/106	^{186}Hg-^{188}Pb	6 - 16	$3.1^{+0.9}_{-1.0}$	PP
/114	^{194}Hg-^{196}Pb	16 - 22	$0.9^{+0.9}_{-0.8}$ ‡	PP
† P, π rich: N, ν rich: S, stable			‡ Identical Bands	

Clearly the deformed structures in the ground states of Pt and excited bands in Hg and Pb around N = 98-106 are remarkably similar. In the 1p Hg-Tl comparisons as for the 2n Hg cases, the SIBs are for N =102 and 104 around midshell where deformation is the largest, see Figure 5 and Table 3. For the Pt-Au cases, N = 100 and 102, and for the one SIB case of Au-Hg, N = 98. There is only one 1n separation, $^{182-183}$Hg, that shows an SIB.

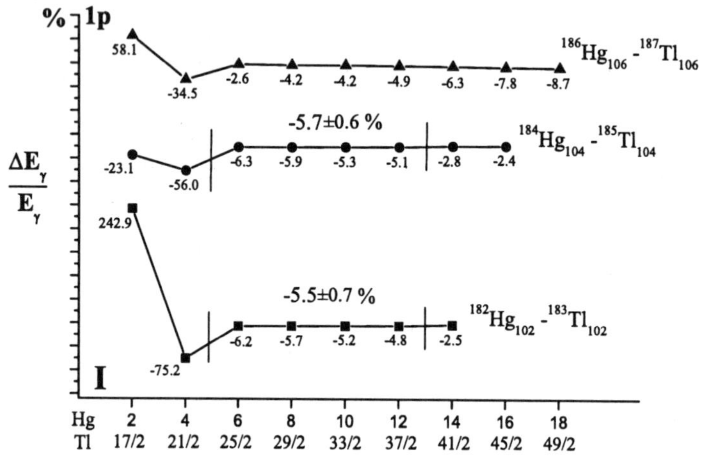

Figure 5. 1p-comparisons of Hg-Tl isotopes.

Table 3. 1p- and 1n-separation SIBs in Pt-Au, Au-Hg, Hg-Tl, and Hg-Hg comparisons.

		1p Separation		
Z / N	**Pair**	**Range**	**$\Delta E_\gamma/E_\gamma$ (%)**	**†**
78-79/100	^{178}Pt-^{179}Au	6 - 12	-4.6±0.2	PP
		(25/2 - 37/2)		
/102	^{180}Pt-^{181}Au	4 - 10	$-5.2^{+0.9}_{-0.6}$	PP
		(21/2 - 33/2)		
79-80/98	^{177}Au-^{178}Hg	29/2 - 37/2	$2.9^{+0.2}_{-0.1}$	PP
		(8 - 12)		
80-81/102	^{182}Hg-^{183}Tl	6 - 12	-5.5±0.7	PP
/104	^{184}Hg-^{185}Tl	6 - 12	-5.7±0.6	PP
		(25/2 - 37/2)		
		1n Separation		
80/102-103	$^{182\text{-}183}$Hg	6 - 16	$-4.6^{+0.8}_{-0.4}$	PP
		(19/2 - 39/2)		
† P, π rich: N, ν rich: S, stable				

The odd particle in ^{183}Hg acts as a spectator to the well-deformed excited band ^{182}Hg core. If we delete the 6^+ level, the value is $-4.7^{+0.4}_{-0.3}$ with significantly smaller spread.

Table 4. 1α- and 2α-separation SIBs in Pt-Hg, Au-Tl, and Hg-Pb comparisons.

1α Separation				
Z / N	Pair	Range	$\Delta E_\gamma/E_\gamma$(%)	†
78-80/100-102	^{178}Pt-^{182}Hg	8 - 14	$-0.8^{+0.6}_{-0.3}$ ‡	PP
/102-104	^{180}Pt-^{184}Hg	6 - 14	$1.2^{+0.7}_{-0.6}$	PP
/104-106	^{182}Pt-^{186}Hg	8 - 14	$1.1^{+0.5}_{-0.8}$	PP
79-81/100-102	^{179}Au-^{183}Tl	25/2-37/2	$-1.2^{+0.6}_{-0.2}$	PP
/102-104	^{181}Au-^{185}Tl	21/2-37/2	$0.4^{+1.0}_{-1.1}$ ‡	PP
		21/2-33/2	$0.6^{+0.8}_{-1.0}$ ‡	
80-82/102-104	^{182}Hg-^{186}Pb	6 - 14	$0.2^{+1.3}_{-1.4}$ ‡	PP
		6 - 12	$-0.2^{+0.7}_{-1.0}$ ‡	
2α Separation in + Parity Yrast Bands				
Z / N	Pair	Range	$\Delta E_\gamma/E_\gamma$ (%)	†
78-82/100-104	^{178}Pt-^{186}Pb	4 - 14	$-0.5^{+0.8}_{-0.9}$ ‡	PP
		6 - 14	$-0.3^{+0.6}_{-0.3}$ ‡	
† P, π rich: N, ν rich: S, stable			‡ Identical Bands	

Now look at the 1α separation IBs and SIBs found in Pt-Hg, Au-Tl, and Hg-Pb shown in Table 4. Five α-cases are IBs and the 3 SIBs are almost IBs with differences of only 1.1-1.2 %. A 2α-separation IB is found in ^{178}Pt-^{196}Pb, where the two 1α-separations that form this 2α IB are themselves IBs.

3. Summary

About half (or more) of the SIBs and IBs in the Pt to Pb region are identical bands (with an additional four cases being much smaller SIBs, so the spreads overlap IBs) which is quite different from the neutron-rich Ce to W region where only 9% are IBs. For the 2n separation IBs and SIBs in Pt, Au, Hg, and Tl, of the four SIBs, three with N = 100-102, and one N = 102-104, surprisingly the energies are larger and moments of inertia smaller for the heavier-mass nucleus which is nearer midshell at N = 104, where maximum deformation is expected. These data indicate the deformation is larger for N = 100 than N = 102 and for N = 102 compared to N = 104 in every case. For the 2p Pt-Hg and Hg-Pb separations, five are IBs with one SIB. The 1p separation and 1n separation comparisons showed SIBs in Pt-Au, Au-Hg, Hg-Tl, and Hg-Hg comparisons. Note when the $i_{13/2}$ proton is added to ^{178,180}Pt and to ^{182,184}Hg to form ^{179,181}Au and ^{183,185}Tl respectively, the moment of inertia goes down an average of more than 5% in each case, which is surprising. One could have expected this high-j particle to have increased, not decreased, the moment of inertia. Adding a second high-j particle to ^{177}Au to form ^{178}Hg does increase the moment of inertia. When the odd $i_{13/2}$ proton is added to Au cores to form a pair in ^{178}Hg, the moment of inertia goes up in ^{178}Hg. For ^{183,185}Tl isotopes, almost the same energy is predicted for prolate and oblate $i_{13/2}$ states. If we add a high-j proton in an oblate state, then we could decrease the MOI.

These well-deformed ground-state Pt and excited Au to Pb bands are the only region of SIBs in proton-rich nuclei from Xe to Pb. About half of the cases (15 of 32) form IBs with 4 more cases having small shifts, 1.1-1.3%, that overlap IBs. Nearly half the SIBs (7 of 17) are 1p, 1n, or 1n-1p separations, where the odd particle is a spectator on the well-deformed shape. These SIBs and IBs illustrate the remarkable stability and similarity of these well-deformed intruder structures in the Pt-Pb region and point up the unusual spectator role of the high-j odd-proton and odd-neutron orbitals in this region. They provide new challenges for microscopic theories.

The work at Vanderbilt Univ. was supported by the Dept. of Energy through Grant DE-FG05-88ER40407.

References

1. E. F. Jones, P. M. Gore, J. H. Hamilton, A. V. Ramayya, A. P. deLima, R. S. Dodder, J. Kormicki, J. K. Hwang, C. J. Beyer, X. Q. Zhang, S. J. Zhu, G. M. Ter-Akopian, Yu. Ts. Oganessian, A.V. Daniel, J.O. Rasmussen, I.Y. Lee, J. D. Cole, M. W. Drigert, W.-C. Ma, and GANDS95 Collaboration, *Nucl. Phys.* **A682**, 79c (2001).
2. J. H. Hamilton, in *Treatise on Heavy Ion Science*, Allen Bromley, ed., New York: Plenum Press (1989) Vol. 8, p. 2.

KEY TOPICS IN NUCLEAR STRUCTURE
Proceedings of the 8th Int. Spring Seminar on Nuclear Physics
edited by Aldo Covello
© 2005 World Scientific Publishing Co.

UNEXPECTED DECREASE IN MOMENT OF INERTIA BETWEEN N = 98 - 100 IN ^{162,164}GD

E. F. JONES[1,2], J. H. HAMILTON[1], P. M. GORE[1], A. V. RAMAYYA[1], J. K. HWANG[1], A. P. DELIMA[1,2], S. J. ZHU[1,3,4], Y. X. LUO[1,3,5], C. J. BEYER[1], J. KORMICKI[1], X. Q. ZHANG[1], W. C. MA[6], I. Y. LEE[5], J. O. RASMUSSEN[5], S. C. WU[5], T. N. GINTER[5,7], P. FALLON[5], M. STOYER[8], J. D. COLE[9], A. V. DANIEL[10], G. M. TER-AKOPIAN[10], AND R. DONANGELO[11]

[1] *Department of Physics and Astronomy, Vanderbilt University, Nashville, Tennessee 37235, USA*
[2] *Department of Physics, University of Coimbra, 3000 Coimbra, Portugal*
[3] *Joint Institute for Heavy Ion Research, Oak Ridge, Tennessee 37830, USA*
[4] *Department of Physics, Tsinghua University, Beijing 100084, Peoples Republic of China*
[5] *Lawrence Berkeley National Laboratory, Berkeley, California 94720, USA*
[6] *Department of Physics, Mississippi State University, Mississippi 39762, USA*
[7] *National Superconducting Cyclotron Laboratory, Michigan State University, East Lansing, MI 48824, USA*
[8] *Lawrence Livermore National Laboratory, Livermore, California 94550, USA*
[9] *Idaho National Engineering and Environmental Laboratory, Idaho Falls, Idaho 83415, USA*
[10] *Flerov Laboratory for Nuclear Reactions, JINR, Dubna, Russia*
[11] *Universidade Federal do Rio de Janeiro, CP 68528, RG Brazil*

From prompt $\gamma - \gamma - \gamma$ coincidence studies with a ^{252}Cf source, the yrast levels were identified from 2^+ to 16^+ and 14^+ in neutron-rich ^{162,164}Gd, respectively. Transition energies between the same spin states are higher and moments of inertia lower at every level in N = 100 ^{164}Gd than in N = 98 ^{162}Gd. The same trend is seen in ^{164,166}Dy. These observations are in contrast to the continuous decrease in the 2^+ energy to a minimum at neutron midshell (N = 104) in Er, Yb, and Hf nuclei. The lowest known 2^+ energy in this region now occurs for N = 96 ^{156}Nd and next for N = 98 ^{160}Sm, which are well removed from midshell for both protons and neutrons.

1. Introduction

The deformation of a nucleus increases as the proton and neutron orbitals fill beyond the spherical magic closed shells. Theoretically, one expects that the minimum in the 2^+ energies and maximum in deformation in nuclei should occur when the neutrons and protons are near midshell between two spherical magic numbers [1]. Thus, ^{170}Dy with Z = 66 and N = 104 should be the best rotational nucleus possible since it has both protons and neutrons exactly at midshell between the spherical Z = 50, 82 and N = 82, 126 closed shells. Indeed, a minimum in the 2^+ energies and maximum in deformation at N = 104 are found in Er, Yb, and Hf nuclei [2], see Figure 1.

From $\gamma-\gamma-\gamma$ coincidence studies of the SF of ^{252}Cf, levels in ^{162,164}Gd were identified for the first time through their coincidences with their ^{84,86}Se partners. Our $\gamma-\gamma-\gamma$ coincidence study of the SF of ^{252}Cf with 102 detectors and a 62 μCi ^{252}Cf source in Gammasphere in the year 2000 yielded much higher statistical data than previous data sets. Both internal and external energy standards were used to obtain more accurate energies, particularly in the low energy region. Further experimental details are found in Luo et al. [3].

2. Results and Analysis

The transitions in ^{164}Gd were identified in a double coincidence gate on the known 1455.1-keV, $2^+ \rightarrow 0^+$ ^{84}Se transition and the transition identified as the 261.3-keV $6^+ \rightarrow 4^+$ transition in ^{164}Gd, as shown in Figure 1(a.). The more intense higher energy peaks belong to ^{164}Gd, based on the more intense 4n channel, as expected from the yields [4]. In addition, we see the previously known 667.5-keV, $4^+ \rightarrow 2^+$ transition in ^{84}Se in Figure 1(a.).

When we gate on the $6^+ \rightarrow 4^+$ transitions in the band assigned to ^{162}Gd, and the $2^+ \rightarrow 0^+$ transition of ^{84}Se, we see the other transitions in ^{162}Gd as shown in Figure 1(b.) up to the 16^+ state, together with the 667.1-keV $4^+ \rightarrow 2^+$ transition in ^{84}Se. A 165-keV $6^+ \rightarrow (5^+)$ transition in ^{84}Se is also found. In a double gate on the 253.6-keV $6^+ \rightarrow 4^+$ and 336.2-keV $8^+ \rightarrow 6^+$ transitions in ^{162}Gd as shown in the clear spectrum of Figure 1(c.), we can see the transitions of ^{162}Gd up to the 16^+ state. The level schemes of ^{162}Gd and ^{164}Gd based on our coincidence analysis studies are given in Figure 2.

In Figure 3 are shown plots of moment of inertia (MOI) J_1 for ^{156,158,160}Sm, ^{160,162,164}Gd, and ^{162,164,166,168}Dy. The energies from $2^+ \rightarrow 0^+$ to $14^+ \rightarrow 12^+$ all decrease from N = 94 to 98 in ^{156,158,160}Sm, and their J_1 and J_2 MOIs increase in a systematic pattern. Unfortunately, the levels of N = 100 ^{162}Sm are not yet known. The J_1 and J_2 values for ^{164}Gd fall

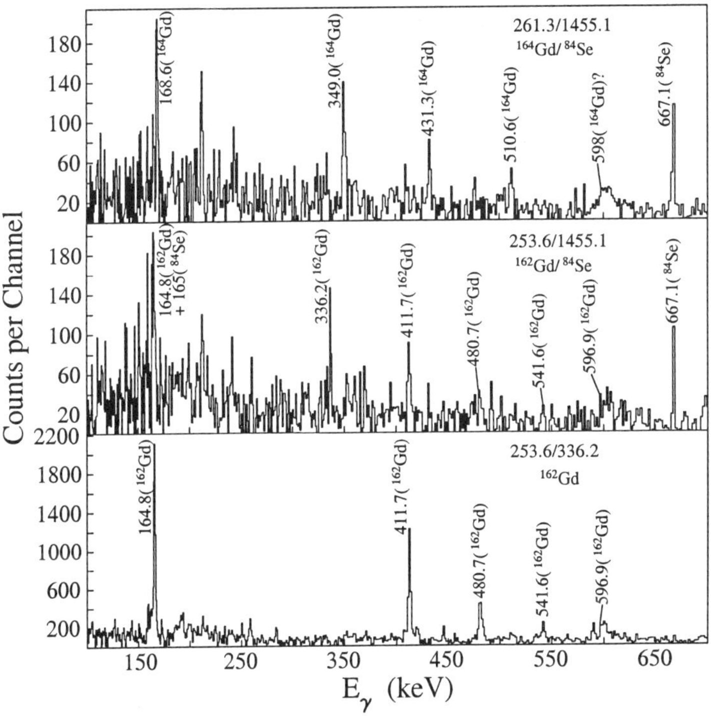

Figure 1. (a.) Double gate on ^{164}Gd 261.3 keV and ^{84}Se 1455.1 keV. (b.) Double gate on ^{162}Gd 253.6 keV and ^{84}Se 1455.1 keV. (c.) Double gate on 253.6 keV and 336.2 keV in ^{162}Gd. All gates have gate width = 0.33 keV.

between those of ^{160,162}Gd from $2^+ \rightarrow 0^+$ up to $10^+ \rightarrow 8^+$, then become less than those of even ^{160}Gd at 10^+. The J_1 MOI of ^{164}Gd from 2^+ to 14^+ does not change as much as those of ^{160}Gd and ^{162}Gd, thus ^{164}Gd has less stretching with increasing rotation. The N = 98, 100 ^{164,166}Dy [5] likewise have transition energies that increase in going from N = 98 to 100 as in ^{162,164}Gd, and the J_1 and J_2 values of ^{166}Dy similarly fall between those of ^{162,164}Dy from $2^+ \rightarrow 0^+$ up to $12^+ \rightarrow 10^+$, then become less than those of ^{162}Dy at 12^+. However, Asai et al. [6] found that the 2^+ and 4^+ energies in ^{168}Dy decrease again such that the J_1 values for N = 102 of ^{168}Dy are above the N = 100 values, but still below the N = 98 values.

In Figure 4, the Gd and Dy isotopes are seen to have a minimum in $E(2^+)$ energies, and presumably a maximum in β_2 quadrupole deformation,

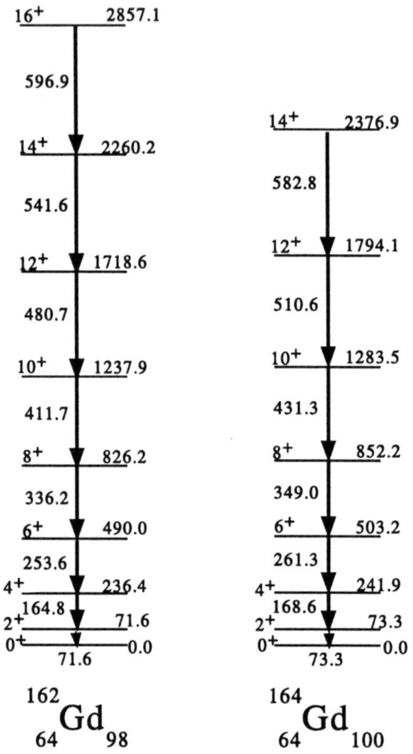

Figure 2. Level schemes of ^{162,164}Gd determined in this work.

at N = 98, while the Yb and Hf nuclei have minima in E(2^+) at N = 104. The E(2^+) values for the Er isotopes follow this latter trend, decreasing to N = 104. However, the lowest known E(2^+) energies in this region for N = 92 - 110 are for Z = 60 Nd, followed by Z = 62 Sm and then Z = 64 Gd, with Z = 58 Ce E(2^+) values curiously falling between the Gd and Dy values at N = 92 and 94.

The β_2 values do not absolutely track the 2^+ level energies in two ways. First, for Er, Yb, and Hf, the measured β_2 values reach a nearly constant maximum value for N = 96 - 102 (Er), 98 - 104 (Yb), and 102 - 106 (Hf), while their 2^+ energies continue to decrease to a minimum at 104. Next, the largest previously known experimental β_2 values [0.348(2), 0.353(2), and 0.348(2)] in the region occur for 2^+ energies of 79.5, 75.3, and 73.4 keV in ^{158,160}Gd and ^{164}Dy, respectively. Now that we know that both the Gd and Dy 2^+ energies start to increase from N = 98 to 100, the most neutron-

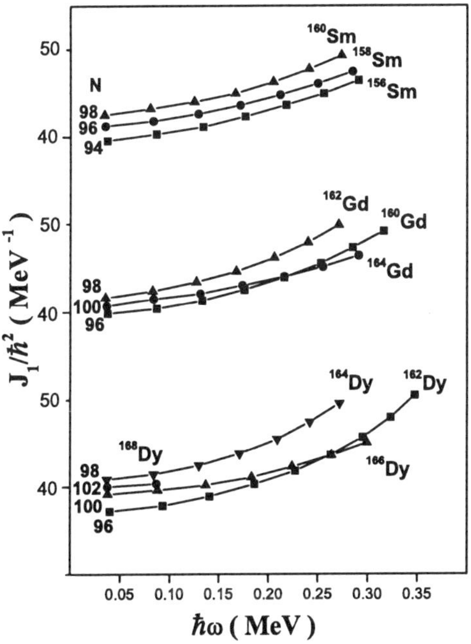

Figure 3. Values of $J_1/\ ^2(\mathrm{MeV}^{-1})$ vs. ω(MeV) for ^{156,158,160}Sm, ^{160,162,164}Gd, and ^{162,164,166,168}Dy.

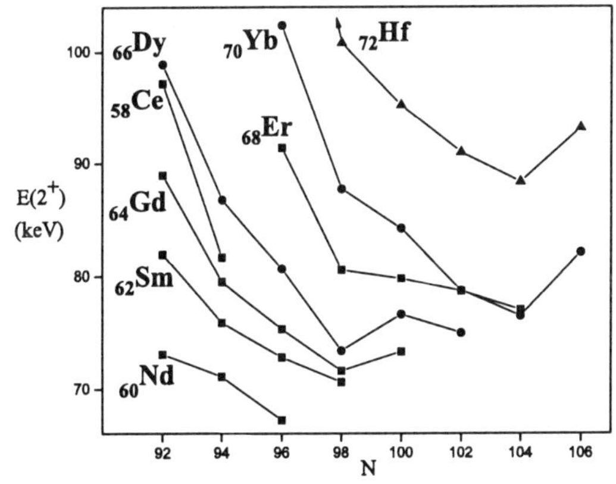

Figure 4. Plot of 2^+ level energies vs. neutron number.

rich known ^{156}Nd and ^{160}Sm have the lowest 2^+ energies of any known nucleus in this region. However, their theoretical β_2 values, 0.338, 0.344 [7], are somewhat lower than the largest theoretical values in this region,

0.38 (present work), 0.357 [7], which occur for ^{162,166}Gd, respectively. If we scale their 2^+ energies by $A^{-5/3}$ up to $A \approx 172$, then the ^{156}Nd, ^{160}Sm scaled energies would be even smaller. The Nd isotopes, with Z = 60, are well removed from the proton midshell at Z = 66, and the most neutron-rich N = 96 is 8 neutrons away from midshell. Very recently we determined $\beta_2 = 0.46(5)$ from a lifetime measurement of the 2^+ level in ^{158}Sm [8]. This result likewise supports the unexpected lowest 2^+ energies and presumably largest deformation being well below proton and neutron midshell. Thus, our ^{162,164}Gd data along with the ^{164,166,168}Dy data raise a new question of why is it that the most neutron-rich known Z = 60, 62 Nd, Sm isotopes have the lowest 2^+ energies and presumably the largest deformation in the deformed region bounded by Z = 50 - 82 and N = 82 - 126.

3. Summary

It is surprising that the Gd and Dy 2^+ energies increase from N = 98 to 100 since E(2^+) values of Er - Yb - Hf continuously decrease to a minimum at N = 104, as expected theoretically. These Gd, Dy results also show ^{156}Nd and ^{160}Sm unexpectedly have the lowest known 2^+ energies in this region. These data present a new challenge for microscopic theories.

Work at VU, INEEL, LBNL, LLNL, and MSU is supported by U.S. DOE grants and contracts DE-FG05-88ER40407, DE-AC07-99ID13727, W-7405-ENG48, DE-AC03-76SF00098, and DE-FG05-95ER40939; Tsinghua by State Basic Res. Dev. Prog. G2000077400 and Nat'l. Nat. Sci. Found. of China, 19775028. The JIHIR is supported by U.TN, VU, ORNL.

References

1. J. M. Eisenberg and W. Greiner, *Nuclear Theory, vol. 1: Nuclear Models*, (North-Holland, Amsterdam, 1987), Chapter 1.
2. *Table of Isotopes*, 8th ed., edited by R. B. Firestone and V. S. Shirley (Wiley, New York, 1996).
3. Y. X. Luo et al., Phys. Rev. C **64**, 054306 (2001).
4. G. M. Ter-Akopian et al., Phys. Rev. C **55**, 1146 (1997).
5. C. Y. Wu et al., Phys. Rev. C **57**, 3466 (1998).
6. M. Asai et al., Phys. Rev. C **59**, 3060 (1999).
7. G. A. Lalazissis, M. M. Sharma, and P. Ring, Nucl. Phys. **A 597**, 35 (1996).
8. J. K. Hwang et al., private communication, Vanderbilt University.

KEY TOPICS IN NUCLEAR STRUCTURE
Proceedings of the 8th Int. Spring Seminar on Nuclear Physics
edited by Aldo Covello
© 2005 World Scientific Publishing Co.

EXACTLY SOLVABLE PAIRING MODELS

J. P. DRAAYER, V. G. GUEORGUIEV, K. D. SVIRATCHEVA, C. BAHRI
Department of Physics and Astronomy, Louisiana State University, Baton Rouge, LA 70803, USA

FENG PAN
Department of Physics, Liaoning Normal University, Dalian, 116029, P. R. China

A. I. GEORGIEVA
Institute of Nuclear Research and Nuclear Energy, Bulgarian Academy of Sciences, Sofia, Bulgaria

Abstract

Some results for two distinct but complementary exactly solvable algebraic models for pairing in atomic nuclei are presented: 1) binding energy predictions for isotopic chains of nuclei based on an extended pairing model that includes multi-pair excitations; and 2) fine structure effects among excited 0^+ states in $N \approx Z$ nuclei that track with the proton-neutron (pn) and like-particle isovector pairing interactions as realized within an algebraic $sp(4)$ shell model. The results show that these models can be used to reproduce significant ranges of known experimental data, and in so doing, confirm their power to predict pairing-dominated phenomena in domains where data is unavailable.

1. Introduction

Pairing is an important interaction that is widely used in nuclear and other branches of physics. In this contribution we present some results that follow from exact algebraic solutions of an extended pairing model that includes multi-pair excitations and that is designed to reproduce binding energies of deformed nuclei,[1] and the $sp(4)$ pairing model that can be used to track fine structure effects in excited 0^+ states in medium mass nuclei.[2] The results show that these models can be used to reproduce significant ranges of known experimental data, and in so doing, confirm their power to predict pairing-like phenomena in domains where data is unavailable or simply not well understood, such as binding energies for proton or neutron rich nuclei far off the line of stability and the fine structure of proton-neutron systems that are critical to understanding the *rp*-process in nucleosynthesis.

The Bardeen-Cooper-Schrieffer (BCS)[3] and Hartree-Fock-Bogolyubov (HFB)[4] methods for finding approximate solutions when pairing plays an important role are well known. However, the limitations of BCS methods, when applied in nuclear physics, are also well understood. First of all, the number of valence particles ($n \sim 10$) that dominate the behavior of low-lying states is too few to support the underlying assumptions of the approximations, that is, particle number fluctuations are non-negligible. As a result, particle number-nonconservation effects can lead to serious problems such as spurious states, nonorthogonal solutions, and so on. In addition, an essential feature of pairing correlations are differences between neighboring even and odd mass nuclei, which are driven mainly by Pauli blocking effects. It is difficult to treat these even-odd differences with either the BCS or HFB theories because different quasi-particle bases must be introduced for different blocked levels. Another difficulty with approximate treatments of the pairing interaction is related to the fact that both the BCS and the HFB approximations break down for an important class of physical situations. A remedy that uses particle number projection techniques complicates these methods and does not help achieve a better description of higher-lying states.

2. Mean-field plus Extended Pairing Model

The importance of having exact solutions of the pairing Hamiltonian has driven a great deal of work in recent years. In particular, building on Richardson's early work[5] and extensions to it based on the Bethe ansatz, several authors have introduced novel approaches.[6,7] For the algebraic approaches based on the Bethe ansatz, the solutions are provided by a set of highly non-linear Bethe Ansatz Equations (BAE). Although these applications demonstrate that the pairing problem is exactly solvable, solutions are not easily obtained and normally require extensive numerical work, especially when the number of levels and valence pairs are large. This limits the applicability of the methodology to relatively small systems; in particular, it cannot be applied to large systems such as well-deformed nuclei.

2.1. *Algebraic Underpinnings of the Theory*

The standard pairing Hamiltonian for well-deformed nuclei is given by

$$\hat{H} = \sum_{j=1}^{p} \epsilon_j n_j - G \sum_{i,j=1}^{p} a_i^+ a_j, \tag{1}$$

where p is the total number of single-particle levels, $G > 0$ is the pairing strength, ϵ_j is single-particle energies taken for example from a Nilsson model, $n_j = c_{j\uparrow}^\dagger c_{j\uparrow} + c_{j\downarrow}^\dagger c_{j\downarrow}$ is the fermion number operator for the j-th single particle level, and $a_i^+ = c_{i\uparrow}^\dagger c_{i\downarrow}^\dagger$ ($a_i = (a_i^+)^\dagger = c_{i\downarrow} c_{i\uparrow}$) are pair creation (annihilation) operators. The up and down arrows in these expressions denote time-reversed states. Since each level can only be occupied by one pair due to the Pauli Exclusion Principle, the Hamiltonian (1) is also equivalent to a finite site hard-core Bose-Hubbard model with infinite range one-pair hopping and infinite on-site repulsion. Specifically, the operators a_i^+, a_i, and $n_i^a = n_i/2$ satisfy the following hard-core boson algebra:

$$(a_i^+)^2 = 0, \quad [a_i, a_j^+] = \delta_{ij}(1 - 2n_i^a), \quad [a_i^+, a_j^+] = [a_i, a_j] = 0. \tag{2}$$

The extended pairing Hamiltonian adds multiple-pair excitations to the standard pairing interaction (1):

$$\hat{H} = \sum_{j=1}^{p} \epsilon_j n_j - G \sum_{i,j=1}^{p} a_i^+ a_j - G \sum_{\mu=2}^{\infty} \frac{1}{(\mu!)^2} \sum_{i_1 \neq \cdots \neq i_{2\mu}} a_{i_1}^+ \cdots a_{i_\mu}^+ a_{i_{\mu+1}} \cdots a_{i_{2\mu}}, \tag{3}$$

where no pair of indices among the $\{i_1, i_2, \cdots, i_{2\mu}\}$ are the same for any μ. With this extension, the model is exactly solvable.[1] In particular, the k-pair excitation energies of (3) are given by the expression:

$$E_k^{(\zeta)} = \frac{2}{x^{(\zeta)}} - G(k-1), \tag{4}$$

where the undetermined variable $x^{(\zeta)}$ satisfies

$$\frac{2}{x^{(\zeta)}} + \sum_{1 \le i_1 < i_2 < \cdots < i_k \le p} \frac{G}{(1 - x^{(\zeta)} \sum_{\mu=1}^{k} \epsilon_{i_\mu})} = 0. \tag{5}$$

The additional quantum number ζ can be understood as the ζ-th solution of (5). Similar results can be shown to hold for even-odd systems except that the index j of the level occupied by the single nucleon should be excluded from the summation and the single-particle energy term ϵ_j contributing to the eigenenergy from the first term of (3) should be included. Extensions to many broken-pair cases are straightforward. If (5) is rewritten in terms of a new variable $z^{(\varsigma)} = 2/[Gx^{(\varsigma)}]$ and the dimensionless energy of a 'grand' boson $\tilde{E}_{i_1 i_2 \ldots i_k} = \sum_{\mu=1}^{k} \frac{2\epsilon_{i_\mu}}{G}$, (5) reduces to:

$$1 = \sum_{1 \leq i_1 < i_2 < \cdots < i_k \leq p} \frac{1}{(\tilde{E}_{i_1 i_2 \ldots i_k} - z^{(\varsigma)})}. \tag{6}$$

Since there is only a single variable $z^{(\varsigma)}$ in (6), the zero points of the function can be determined graphically in a manner that is similar to the one-pair solution of the TDA and RPA approximations with separable potentials.[4]

2.2. *Application to the $^{154-181}$Yb Isotopes*

A study of the binding energies of well-deformed nuclei within the framework of the extended pairing model is currently in progress.[9] Typically, the single-particle energies of each nucleus are calculated within the deformed Nilsson shell model with deformation parameters taken from Moller and Nix;[10] experimental binding energies are taken from Audi, et al;[8] and, theoretical binding energies are calculated relative to a particular core. For an even number of neutrons, only pairs of particles (bosons-like structures) are considered. For an odd number of neutrons, Pauli blocking of the Fermi level of the last unpaired fermion is envoked with the remaining fermions are considered to be an even A fermion system. Using (4) and (5), values of G are calculated so that the experimental and theoretical binding energy match exactly. Note that for a given set of single-particle energies there is an upper limit to the binding energy for which a physically meaningful exact solution can be constructed. This upper value on the binding energy is given by the energy of the lowest 'grand' boson, with energy given by $\sum_{\mu=1}^{k} 2\epsilon_\mu$.

As a first application of the theory, we calculated the binding energies for the $^{154-181}$Yb isotopes and extracted the corresponding $\log(G)$ values for the extended pairing model. The binding energy of the closed neutron shell nucleus ^{152}Yb was taken to be the zero-energy reference point. Its odd-A ^{153}Yb neighbor was assumed to be well described by the independent particle model with Nilsson single-particle energies; this means that the pairing interaction terms have no affect on ^{153}Yb. The energy scale applied to the Nilsson single-particle energies, which is 3/4 for pure harmonic oscillator interaction, was set so that the binding energy of ^{153}Yb is reproduced by the independent particle model.[4] For all the other nuclei we solved for the pairing strength $G(A)$ that reproduces the experimental binding energies exactly within the selected model space, the latter con-

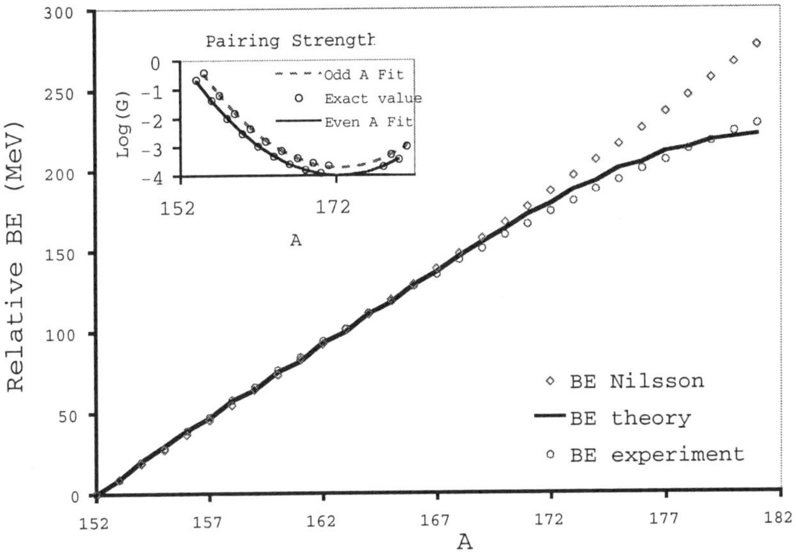

Figure 1. The solid line gives the theoretical binding energies of the Yb isotopes relative to that of the ^{152}Yb core. The single-particle energy scale is set from the binding energy of ^{153}Yb. The inset shows the fit to values of G that reproduce exactly the experimental data. The two fitting functions are: $\log(G(A)) = 662.2247 - 7.7912A + 0.0226A^2$ for even values of A and $\log(G(A)) = 716.3279 - 8.4049A + 0.0244A^2$ for odd values of A. The Nilsson BE energy is the lowest configuration energy of the non-interacting system.

sisting of the neutron single-particle levels between the closed shells with magic numbers 50 and 82. The structure of the model space is reflected in the values of $G(A)$. In particular, $\log(G(A))$ has a smooth quadratic behavior for even- and odd-A values with a minimum in the middle of the model space where the size of the space is a maximal. As shown in Figure 1, although the even- and odd-A curves are very similar, they are shifted from one another due to the even-odd mass difference.

To summarize, in this section we reviewed the extended pairing model and tested its predictive power using the $^{172-177}$Yb isotopic chain as an example. In particular, calculations of the pairing strength G were carried out for the $^{154-171}$Yb and $^{178-181}$Yb isotopes but not for the $^{172-177}$Yb isotopes that are in the middle of the model space where the computations are more involved. The even- and odd-A $\log(G)$ curves, which were assumed to have a quadratic polynomial form and therefore determined by three parameters, were fit to the two date sets which consist of 11 data points each, one for the even-A isotopes and another for odd A. From the quadratic polynomial fit to the $\log(G)$ values, we then calculate the theoretical values of the binding energy for all the nuclei shown in Figure 1. The prediction is very good when compared to the experimental numbers. Thus, based on experimental data of the nuclei in the upper and lower parts of the shell and an assumed quadratic from for $\log(G)$ that was fit to this data, we were able to make reasonable estimates for the binding energies of mid-shell nuclei. Based on this simple exercise, we conclude that the extended pairing model has good predictive power for binding energies. Indeed, this early success suggests that the extended pairing model may have broader applicability to other well-deformed nuclei as well as other physical systems where pairing plays an important role.

3. Algebraic $sp(4)$ Pairing Model

The recent renaissance of studies on pairing is related to the search of a reliable microscopic theory for a description of medium nuclei around the $N = Z$ line, where like-particle pairing comprises only a part of the complicated nuclear interaction in this region. This is because for such nuclei protons and neutrons occupy the same major shells and their mutual interactions are expected to influence significantly the structure and decay of these nuclei. Such a microscopic framework is as well essential for astrophysical applications, for example the description of the rp-process in nucleosynthesis, which runs close to the proton-rich side of the valley of stability through reaction sequences of proton captures and competing β decays.[11] The revival of interest in pairing correlations is also prompted by the initiation of radioactive beam experiments, which advance towards exploration of 'exotic' nuclei, such as neutron-deficient or $N \approx Z$ nuclei far off the valley of stability.

In our search for a microscopic description of pairing in the broad range of nuclei with mass numbers $32 \leq A \leq 100$ with protons and neutrons filling the same major shell, we employ an $sp(4)$ algebraic model that accounts for proton-neutron and like-particle pairing correlations and higher-J proton-neutron interactions, including the so-called symmetry and Wigner energies.[2] The nuclei classified within a major shell possess a clear $Sp(4)$ dynamical symmetry. The basis operators of the $sp(4)$ algebra ($\sim so(5)$[12,13]) have a distinct physical meaning: $N_{\pm 1}$ counts the total number of protons (neutrons) (and hence $\hat{N} = N_{+1} + N_{-1}$ is the total number operator), the operators $T_{0,\pm}$ are related to isospin (where

$T_0 = (N_{+1} - N_{-1})/2$ is the third projection of isospin), while the six operators $A^\dagger_{-1,0,1}$ ($A_{-1,0,1}$) create (annihilate) a pair of total angular momentum $J^\pi = 0^+$ and isospin $T = 1$. The model Hamiltonian with an $Sp(4)$ dynamical symmetry,

$$H = -G \sum_{i=-1}^{1} A_i^\dagger A_i - F A_0^\dagger A_0 - \frac{E}{2\Omega}(T^2 - \frac{3\hat{N}}{4}) \\ -D(T_0^2 - \frac{\hat{N}}{4}) - C\frac{\hat{N}(\hat{N}-1)}{2} - \epsilon \hat{N}, \quad (7)$$

includes a two-body isovector ($T = 1$) pairing interaction and a diagonal isoscalar ($T = 0$) force, which is proportional to a symmetry and Wigner term ($T(T+1)$-like dependence). In addition, the D-term introduces isospin symmetry breaking and the F-term accounts for a plausible, still extremely weak, isospin mixing. This Hamiltonian conserves the number of particles (N), the third projection of isospin (T_0) and angular momentum, and changes the like-particle seniority quantum number by zero or ± 2, the latter implies scattering of a pp pair and a nn pair into two pn pairs and vice versa. The interaction strength parameters in (7) are estimated in optimum fits to the lowest isobaric analog 0^+ state experimental energies of total of 149 nuclei[2] and are found to have a smooth dependence on the nuclear mass A,

$$\frac{G}{\Omega} = \frac{23.9 \pm 1.1}{A}, \quad \frac{E}{2\Omega} = \frac{-52 \pm 5}{A}, \\ D = \frac{-37 \pm 5}{A} + (-0.24 \pm 0.09), \quad C = \left(\frac{32 \pm 1}{A}\right)^{1.7 \pm 0.2}, \quad (8)$$

where $2\Omega = \Sigma_j (2j + 1)$ is the shell dimension.

The basis states are constructed as ($T = 1$)-paired fermions, $|n_1, n_0, n_{-1}) = \left(A_1^\dagger\right)^{n_1} \left(A_0^\dagger\right)^{n_0} \left(A_{-1}^\dagger\right)^{n_{-1}} |0\rangle$, and model the 0^+ ground state for even-even and some odd-odd nuclei and the corresponding isobaric analog excited 0^+ state for even-A nuclei in a significant range of nuclei, $32 \leq A \leq 100$. The properties of these states are described well by the $Sp(4)$ dynamical symmetry model, including quite good agreement of the isobaric analog 0^+ state energy spectra with experiment, and in addition the remarkable reproduction of their detailed structure properties.

3.1. *Energy Spectra of Isobaric Analog* 0^+ *States*

The $Sp(4)$ model leads to a very good reproduction of the experimental energies of the lowest isobaric analog 0^+ state for even-A nuclei (that is, binding energies for even-even and some odd-odd nuclei) with nuclear masses

$32 \leq A \leq 100$.[2] This result follows from the very small deviation (estimated by the χ-statistics) between experimental energies and the corresponding theoretical energies predicted in optimization procedures, namely $\chi = 0.496$ in the $1d_{3/2}$ shell, $\chi = 0.732$ in the $1f_{7/2}$ shell and $\chi = 1.787$ in the $1f_{5/2}2p_{1/2}2p_{3/2}1g_{9/2}$ major shell. Without varying the values of the interaction strength parameters, the energy of the higher-lying isobaric analog 0^+ states can be theoretically calculated and they agree remarkably well with the available experimental values for the single-j $1d_{3/2}$ and $1f_{7/2}$ orbits (Figure 2). However, such a comparison to experiment is impossible for the nuclei in the region with nuclear masses $56 < A < 100$, since their energy spectra are not yet completely measured, especially the higher-lying 0^+ states.

The agreement, which is observed throughout both single-j shells, represents an important result. This is because the higher-lying isobaric analog 0^+ states constitute an experimental set independent of the data that determines the interaction strength parameters in (7). Therefore, such a result is, first, an independent test of the physical validity of the strength parameters, and, second, an indication that the interactions interpreted by the model Hamiltonian are the main driving force that defines the properties of these states. In this way, the simple $Sp(4)$ model provides for a reasonable prediction of the isobaric analog (ground and/or excited) 0^+ states in proton-rich nuclei with energy spectra not yet experimentally fully explored. For example, in the case of the $1f_{7/2}$ level the binding energy of the proton-rich ^{48}Ni nucleus is estimated to be $E_0 = 348.19$ MeV, which is 0.07% greater than the sophisticated semi-empirical estimate of Moller and Nix.[10] Likewise, for the odd-odd nuclei that do not have measured energy spectra the theory can predict the energy of their lowest 0^+ isobaric analog state: 358.62 MeV (^{44}V), 359.34 MeV (^{46}Mn), 357.49 MeV (^{48}Co), 394.20 MeV (^{50}Co). The $Sp(4)$ model predicts the relevant 0^+ state energies for additional 165 even-A nuclei in the medium mass region of the $1f_{5/2}2p_{1/2}2p_{3/2}1g_{9/2}$ major shell. The binding energies for 25 of them are also calculated in Moller and Nix.[10] For these even-even nuclei, we predict binding energies that on average are by 0.05% less than the semi-empirical approximation.[10]

3.2. $N = Z$ *Irregularities, Staggering and the Pairing Gap*

The theoretical $Sp(4)$ model can be further tested through second- and higher-order discrete derivatives of the energies of the lowest isobaric ana-

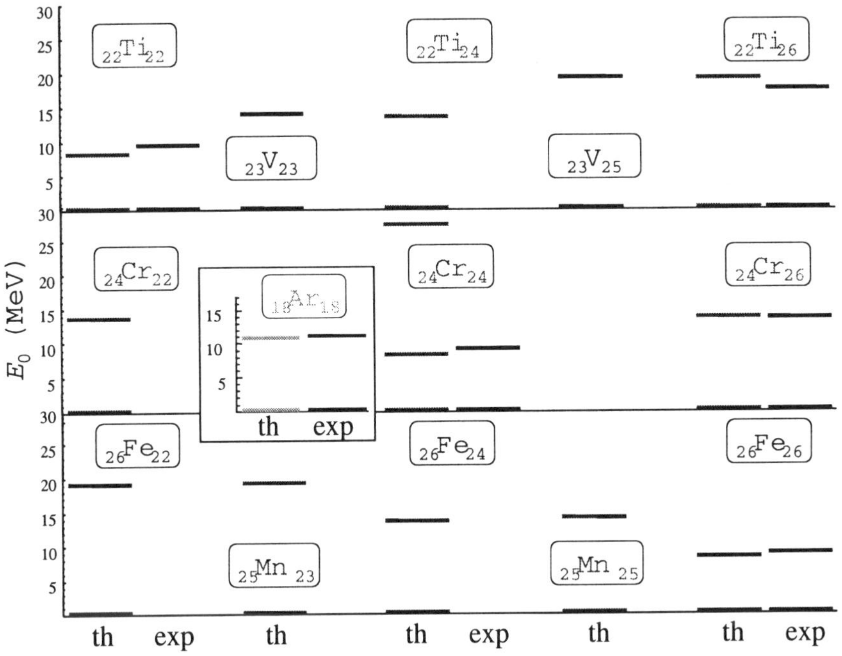

Figure 2. Theoretical ('th') and experimental ('exp') energy spectra of the higher-lying isobaric analog 0^+ states for isotopes in $1f_{7/2}$ (in $1d_{3/2}$ (insert)).

log 0^+ states in the $Sp(4)$ systematics, without any parameter variation. The theoretical discrete derivatives under investigation not only follow the experimental patterns but their magnitude was found to be in a remarkable agreement with the data. The proposed model has been used to successfully interpret: the two-proton (two-neutron) separation energy $S_{2p(2n)}$ for even-even nuclei (hence determined the two-proton drip line), the S_{pn} energy difference when a pn $T = 1$ pair is added, the observed[14] irregularities around $N = Z$ (Figure 3), the like-particle and pn isovector pairing gaps, and the prominent "*ee-oo*" staggering between even-even and odd-odd nuclides. We suggest that the oscillating "*ee-oo*" effects correlate with the alternating of the seniority numbers related to the pn and like-particle isovector pairing, which is in addition to the larger contribution due to the discontinuous change in isospin values associated with the symmetry energy.[15]

The present study brings forward a very useful result. We find a finite

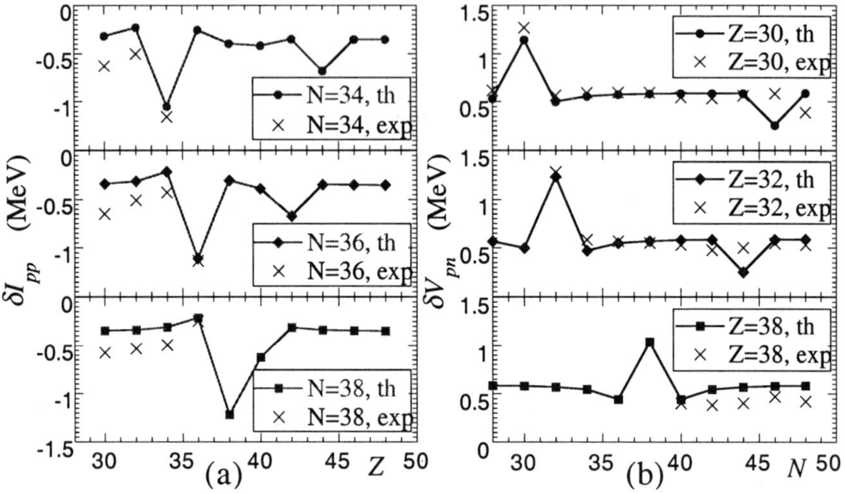

Figure 3. Second discrete derivatives of the energy function E_0: (a) $\delta I_{pp(nn)}(N_{\pm1}) = \frac{E_0(N_{\pm1}+2)-2E_0(N_{\pm1})+E_0(N_{\pm1}-2)}{4}$ versus $N_{\pm1}$, as an estimate for the non-pairing like-particle nuclear interaction in MeV for the $N(Z) = 34, 36, 38$-multiplets; (b) $\delta V_{pn}(N_{+1}, N_{-1}) = \frac{E_0(N_{+1}+2,N_{-1}+2)-E_0(N_{+1}+2,N_{-1})-E_0(N_{+1},N_{-1}+2)+E_0(N_{+1},N_{-1})}{4}$ versus N_{+1} and N_{-1}, as an estimate for the residual interaction between the last proton and the last neutron in MeV for Zn, Ge, Sr isotopes.

energy difference of the energy function E_0,

$$E_0(N, T_0 + 1) - 2E_0(N, T_0) + E_0(N, T_0 - 1) =$$
$$= E_0(N_{+1} + 1, N_{-1} - 1) - 2E_0(N_{+1}, N_{-1}) + E_0(N_{+1} - 1, N_{-1} + 1), \quad (9)$$

that, for the specific case $T_0 = 0$ (or $N = Z$), can be interpreted as an isovector pairing gap, $\tilde{\Delta} = \Delta_{pp} + \Delta_{nn} - 2\Delta_{pn}$, which is related to the like-particle and pn isovector pairing gaps. Indeed, they correspond to the $T = 1$ pairing mode because we do not consider the binding energies for all the nuclei but the respective isobaric analog 0^+ states for the odd-odd nuclei with a $J \neq 0^+$ ground state. This investigation is the first of its kind. Moreover, the relevant energies are corrected for the Coulomb interaction and therefore the isolated effects reflect solely the nature of the nuclear interaction. In addition, the discrete derivative filter (9) can be used to estimate the pairing gaps for all the nuclei within a major shell when only the contribution of the pairing energy is considered in the E_0 energy function. In this way, the like-particle pairing gap is found to be in a very good agreement with the $12/\sqrt{A}$ experimental approximation.[16]

Small deviations from the experimental data are attributed to other two-body interactions or higher-order correlations that are not included in the theoretical model.

In summary, the symplectic $Sp(4)$ scheme allows not only for an extensive systematic study of various experimental patterns of the even-A nuclei, it also offers a simple $sp(4)$ algebraic model for interpreting the results and predicting properties of nuclei that are not yet experimentally explored. The outcome of the present investigation shows that, in comparison to experiment, the $sp(4)$ algebraic approach reproduces not only overall trends of the relevant energies but as well the smaller fine features driven by isovector pairing correlations and higher-J pn and like-particle nuclear interactions.

4. Conclusion

Results for two distinct but complementary exactly solvable algebraic models for pairing in atomic nuclei have been presented: 1) binding energy predictions for isotopic chains of nuclei based on an extended pairing model that includes multi-pair excitations; and 2) fine structure effects among excited 0^+ states in $N \approx Z$ nuclei that track with the proton-neutron (pn) and like-particle isovector pairing interactions as realized within an algebraic $sp(4)$ shell model. The results show that both models can be used to reproduce significant ranges of known experimental data, and in so doing confirm their power to predict pairing-dominated phenomena in domains where data is either not, or only partially available or simply not well understood in terms of applicable models.

In addition, it is important to reiterate that both approaches, the extended pairing model and the algebraic $sp(4)$ model, yield exact analytic solutions to their respective pairing problems. As the examples show, this is important for applications, but it is also important for theory as having exact solutions available gives one an opportunity to test approximate and perhaps simpler to apply approaches, such as the BCS scheme. Other limits as well as extensions of these theories are under investigation.

Acknowledgments

Support from the U.S. National Science Foundation (0140300), the Natural Science Foundation of China (10175031), and the Education Department of Liaoning Province (202122024) is acknowledged.

References

1. Feng Pan, V. G. Gueorguiev and J. P. Draayer, Phys. Rev. Lett. **92**, 112503 (2004).
2. K. D. Sviratcheva, A. I. Georgieva and J. P. Draayer, J. Phys. G: Nucl. Part. Phys. **29**, 1281 (2003).
3. J. Bardeen, L. N. Cooper and J. R. Schrieffer, Phys. Rev. **108**, 1175 (1957).
4. P. Ring and P. Schuck, *The Nuclear Many-Body Problem* (Springer Verleg, Berlin, 1980).
5. R. W. Richardson, Phys. Lett. **3**, 277 (1963); Phys. Lett. **5**, 82 (1963); R. W. Richardson and N. Sherman, Nucl. Phys. **52**, 221 (1964).
6. Feng Pan, J. P. Draayer and W. E. Ormand, Phys. Lett. **B422**, 1 (1998); Feng Pan and J. P. Draayer, Phys. Lett. **B442**, 7 (1998); Feng Pan, J. P. Draayer and Lu Guo, J. Phys. A: Math. Gen. **33**, 1597 (2000); J. Dukelsky, C. Esebbag and P. Schuck, Phys. Rev. Lett. **87**, 066403 (2001); J. Dukelsky, C. Esebbag and S. Pittel, Phys. Rev. Lett. **88**, 062501 (2002); H.-Q. Zhou, J. Links, R. H. McKenzie and M. D. Gould, Phys. Rev. B **65**, 060502(R) (2002).
7. Feng Pan and J. P. Draayer, Ann. Phys. (NY) **271**, 120 (1999).
8. G. Audi and A. H. Wapstra Nucl. Phys. **A595**, 409 (1995); G. Audi, O. Bersillon, J. Blachot and A. H. Wapstra, Nucl. Phys. **A624**, 1 (1997); (http://csnwww.in2p3.fr/AMDC/web/amdcw_en.html).
9. V. G. Gueorguiev, Feng Pan and J. P. Draayer, nucl-th/0403055.
10. P. Möller, J. R. Nix and K. L. Kratz, Atomic Data Nucl. Data Tables **66**, 131 (1997); P. Möller, J. R. Nix, W. D. Myers and W. J. Swiatecki, Atomic Data Nucl. Data Tables **59**, 185-381 (1995); (http://t2.lanl.gov/data/astro/molnix96/massd.html).
11. K. Langanke, Nucl. Phys. **A630**, 368c (1998); H. Schatz *et. al*, Phys. Rep. **294**, 167 (1998).
12. K. Helmers, Nucl. Phys. **23**, 594 (1961); K. T. Hecht, Nucl. Phys. **63**, 177(1965); Phys. Rev. **139**, B794 (1965); Nucl. Phys. **A102**, 11 (1967); J. N. Ginocchio, Nucl. Phys. **74**, 321 (1965).
13. J. Engel, K. Langanke and P. Vogel, Phys. Lett.**B389**, 211 (1996).
14. D. S. Brenner, C. Wesselborg, R. F. Casten, D. D. Warner and J.-Y. Zhang, Phys. Lett. **B243**, 1 (1990); N. V. Zamfir and R. F. Casten, Phys. Rev. **C43**, 2879 (1991).
15. K. D. Sviratcheva, A. I. Georgieva and J. P. Draayer, Phys. Rev. **C69** (2004) 024313.
16. A. Bohr and B. R. Mottelson, *Nuclear Structure* (Benjamin, New York), Vol. I (1969); Vol. II (1975).

KEY TOPICS IN NUCLEAR STRUCTURE
Proceedings of the 8th Int. Spring Seminar on Nuclear Physics
edited by Aldo Covello
© 2005 World Scientific Publishing Co.

MANY-BODY EFFECTS AND PAIRING CORRELATIONS IN FINITE NUCLEI

E. VIGEZZI, P.F. BORTIGNON, G. COLÒ, G. GORI AND F. RAMPONI
INFN Sez. Milano and Dipartimento di Fisica, Università di Milano, Via Celoria 16, 20123 Milano, Italy

F. BARRANCO
Departamento de Fisica Aplicada III; Escuela Superior de Ingenieros, Camino de los Descubrimientos s/n, 41092 Sevilla, Spain

R.A. BROGLIA
INFN Sez. Milano and Dipartimento di Fisica, Università di Milano, Via Celoria 16, 20123 Milano, Italy and The Niels Bohr Institute, University of Copenhagen, Blegdamsvej 17, 2100 Copenhagen Ø, Denmark

Many-body effects associated with the coupling of particles and vibrations contribute in an important way to pairing correlations in finite nuclei.

1. Introduction

Single-particle motion in nuclei is strongly renormalized by the coupling to collective vibrations, leading in particular to an increase of the effective mass around the Fermi energy, and to the fragmentation of the single-particle levels, which affect the level density and the spectroscopic factors[1]. Particle-vibration coupling also plays a crucial role in the quantitative description of Cooper pairs and pairing correlations in metals as well as in superfluid Fermi systems. Concerning nuclear systems, it has been found that in neutron matter the bare interaction acting in the $^1S_0$ channel between pairs of nucleons moving in time-reversal states is renormalized by the exchange of spin and density modes. As a consequence, the pairing gap is strongly reduced, as compared to the values obtained in BCS theory ignoring this renormalization [2]. In the last years, a specific investigation of these effects in finite nuclei has been carried out [3,4,5]. It has been found that the exchange of collective vibrations (mainly of surface character), enhances in this case the pairing gap. In the following we shall present a unified description of these phenomena. We first focus on the effects the exchange of spin and density vibrations has on pairing correlations in nuclei and in neutron stars, using a simple approach in which self-energy and fragmentation effects are neglected, and only the induced interaction is considered. We then report on the results obtained within the framework of a more comprehensive model, namely the Nambu-Gor'kov description of superconductivity [6], adding the bare and the induced interaction as well as vertex corrections and self-energy effects in a consistent fashion.

2. Particle-vibration coupling and the induced interaction

We want to calculate the state-dependent neutron pairing gap Δ_ν arising from the interaction v^{ind} induced by the exchange of phonons in the case of ^{120}Sn, taken as a typical example of spherical, superfluid nucleus. We shall first evaluate the matrix elements $v^{ind}_{\nu,\nu'}$ between pairs of neutrons in time-reversal orbits, characterized by the quantum numbers $\nu(\equiv nlj)$ and m :

$$v^{ind}_{\nu,\nu'} = \sum_{mm'} < \nu m \nu \bar{m} | v^{ind} | \nu' m' \nu' \bar{m}' >, \tag{1}$$

and then we shall insert them in the BCS gap equation

$$\Delta_\nu = -\frac{1}{2j+1} \sum_{\nu'} \frac{\Delta_{\nu'}}{2E_{\nu'}} v^{ind}_{\nu,\nu'} \tag{2}$$

and in the associated number equation, from which we shall obtain the gap produced by the induced interaction. In order to calculate $v^{ind}_{\nu,\nu'}$, we need the single-particle energies ϵ_ν and the associated wavefunctions ϕ_ν, which are obtained from a Hartree-Fock calculation employing the Skyrme SkM* interaction. We also need the energies $\hbar\omega_i$ of the vibrations (phonons), classified according to their angular momentum J and parity π, as well as the associated transition densities $\delta\rho^i_{J\pi}$ (we denote by $\delta\rho^i_{J\pi n}$ and $\delta\rho^i_{J\pi p}$ the neutron and proton contributions). The phonons are obtained by a self-consistent QRPA calculation, employing the particle-hole interaction v_{ph} associated with the SkM* force. The particle-phonon coupling matrix elements (cf. Fig. 1a) can be evaluated in terms of the generalized Landau-Migdal parameters $F_0(r), F_0'(r), G_0(r)$ and $G_0'(r)$ of the SkM* interaction (we neglect the contributions of the momentum-dependent part of the interaction to the coupling) : those associated with the spin-independent part

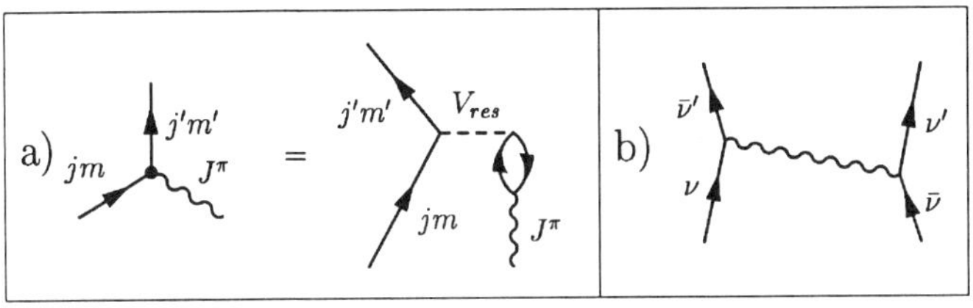

Figure 1. Diagrams depicting (a) the particle-vibration coupling vertex and (b) the pairing interaction induced by the exchange of phonons.

of v_{ph} are given by [7]

$$f^{\nu' m'}_{\nu m; J^\pi M i} = i^{l-l'} \langle j'm'|(i)^J Y_{JM}|jm\rangle \\ \times \int dr \varphi_{\nu'} [(F_0 + F_0')\delta\rho^i_{J^\pi n} + (F_0 - F_0')\delta\rho^i_{J^\pi p}]\varphi_\nu, \quad (3)$$

and those associated with the spin-dependent part by

$$g^{\nu' m'}_{\nu m J^\pi M i} = \sum_{L=J-1}^{J+1} i^{l-l'} \langle j'm'|(i)^L [Y_L \times \sigma]_{JM}|jm\rangle \times \\ \int dr \varphi_{\nu'} [(G_0 + G_0')\delta\rho^i_{J^\pi Ln} + (G_0 - G_0')\delta\rho^i_{J^\pi Lp}]\varphi_\nu. \quad (4)$$

The matrix elements of the induced interaction are then given by

$$< \nu' m' \nu' \bar{m}' | v_{ind} | \nu m \nu \bar{m} > = \sum_{J^\pi M i} \frac{2(f+g)^{\nu' m'}_{\nu m; J^\pi M i} (f-g)^{\nu' m'}_{\nu m; J^\pi M i}}{E_0 - E_{int}} \quad (5)$$

in which E_{int} is the energy of the intermediate state given by two particles and one vibration (cf. Fig. 1b), and E_0 is the energy of the correlated two-particle state, which must be obtained self-consistently with the solution of the gap equation. The denominator is always negative, and the sign of the matrix element then depends on the relative magnitude of the attractive contributions f^2 arising from the spin-independent terms, and of the repulsive contributions $-g^2$ arising from the spin-dependent term.

The diagonal induced pairing matrix elements (i.e., Eq. 5 with $\nu = \nu'$) are shown in Fig. 2(a). In order to make a comparison with the case of neutron matter, it is useful to distinguish between the contributions arising from phonons with natural and non-natural parity, which turn out to be negative and positive, respectively. In fact one can show [7] that for non-natural parity modes only the spin dependent vertices (4) contribute, while for natural parity the spin independent vertices (3) are the dominant ones. The attractive contributions dominate, and solving the BCS gap equation one obtains the state-dependent pairing gap shown in Fig. 2(b), which is of the order of 0.7 MeV close to the Fermi energy. Neglecting the

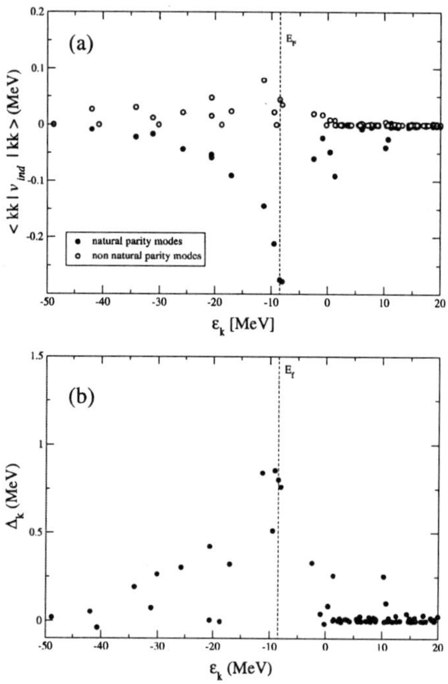

Figure 2. (a) Diagonal induced pairing matrix elements resulting from the exchange of phonons with natural parity (filled circles) and those resulting from the exchange of phonons with non-natural parity vibrations (empty circles), displayed as a function of the energy of the single-particle state ϵ_ν. (b) State dependent pairing gap obtained solving the BCS equations associated with the matrix elements of the induced interaction.

spin-dependent contribution increases the gap by about 30%. From a detailed analysis of the matrix elements [7], it turns out that, as expected, the main attractive contributions arise from the coupling to low-lying vibrations characterized by a transition density localized at the nuclear surface, the main contribution to the matrix element being associated with the neutron-proton interaction (i.e., with the $F_0 - F_0'$ term in Eq. (3)). These two elements (presence of the surface and of the neutron-proton interaction) explain qualitatively the different behavior respect to neutron matter, where instead the repulsive contribution associated with spin modes dominates. Quantitatively, the results may vary for other interactions, depending on the values of the Landau parameters; one should take care, that the experimental properties of the low-lying vibrations are reasonably well reproduced.

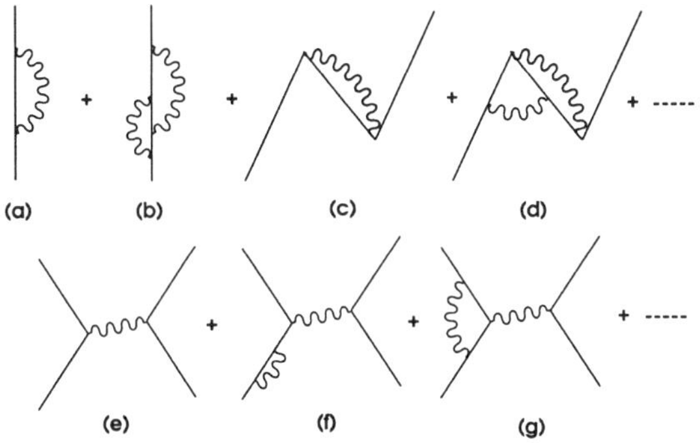

Figure 3. Renormalization processes arising from the particle-vibration coupling phenomenon.

3. Solution of the Dyson equation

The calculations presented in the previous section show that the induced interaction in ^{120}Sn is attractive, that it is mostly due to the coupling with surface vibrational modes and that it can give rise to a sizeable pairing gap. However, they are too schematic at least in three important respects:

1) They take into account only the induced interaction. Of course, to make a quantitative calculation of the pairing gap one has to start with the bare interaction and add the induced interaction to it in a consistent fashion.

2) The same particle-vibration coupling which gives rise to the induced interaction, is at the origin of other renormalization processes (self-energy effects and vertex corrections, cf. Fig. 3). They affect the level density and fragment the single-particle strength, and are ignored in the previous approach, which is based on the quasiparticle approximation and on the BCS gap equation.

3) Also the energy and transition strength of the phonons are renormalized and should be calculated self-consistently, together with the renormalization of the quasiparticles and of the pairing gap.

In the following we shall present calculations which overcome the first two shortcomings listed above [5,8]. We shall follow a simpler phenomenological approach concerning point 3) [9], and calculate the phonons using a separable multipole-multipole interaction [10], with coupling constants chosen so as to reproduce the measured energies and transition strengths of the low-lying states. Spin modes will be neglected, in keeping with the finding reported in Section 2, that the induced interaction is mostly a consequence

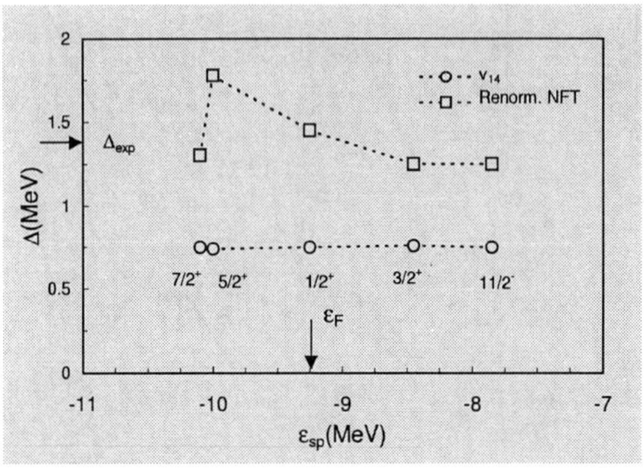

Figure 4. State-dependent pairing gaps for the levels close to the Fermi energy ϵ_F, obtained using BCS theory and the v_{14} Argonne potential (circles), or using the Dyson equation including renormalization effects (squares)

of the coupling to surface vibrations.

We first calculate the Hartree-Fock mean field using a Skyrme interaction, similarly to what has been discussed in Section 2. We then take into account the bare interaction (point 1) above), by performing an extended BCS calculation (i.e., including pairs of particles with different number of nodes in the Cooper wavefunction) with the Argonne v_{14} potential. This requires a high energy cutoff (of the order of 800 MeV), in order to take into account the scattering to high-energy states, caused by the presence of a strong repulsive core [11]. The resulting pairing gap for states around the Fermi energy is shown in Fig. 4, and is equal to about 0.7 MeV. We then calculate the renormalization of the quasiparticles, taking into account the processes arising from the particle-vibration coupling (cf. Fig. 3). This is obtained solving the Dyson (or Nambu-Gor'kov) equation for superfluid systems [12]. Denoting by E_a the energy of a quasiparticle resulting from the previous calculation with the Argonne potential, the Dyson equation reads

$$\left[\begin{pmatrix} E_a & 0 \\ 0 & -E_a \end{pmatrix} + \begin{pmatrix} \Sigma_{11}(\tilde{E}_a) & \Sigma_{12}(\tilde{E}_a) \\ \Sigma_{12}(\tilde{E}_a) & \Sigma_{22}(\tilde{E}_a) \end{pmatrix}\right] \begin{pmatrix} \tilde{x}_a \\ \tilde{y}_a \end{pmatrix} = \tilde{E}_a \begin{pmatrix} \tilde{x}_a \\ \tilde{y}_a \end{pmatrix}, \qquad (6)$$

where Σ_{ii} and $\Sigma_{ij}, (i \neq j)$ are the normal and abnormal self-energies. Eq. (6) is to be solved iteratively, and simultaneously for all the involved quasiparticle states [5,13]. At each iteration step, the original quasiparticle states a with occupation numbers u_a and v_a, become fragmented over the different eigenstates $\tilde{a}$ with probability $\tilde{u}_a^2 + \tilde{v}_a^2$, where the renormalized occupation factors are obtained from the components of the eigenvectors, $\tilde{x}_a$ and

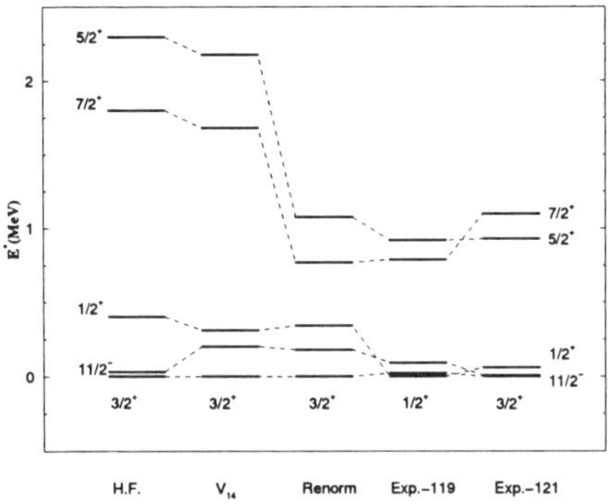

Figure 5. The spectra of the lowest quasiparticle states in ^{120}Sn calculated using Hartree-Fock theory, BCS with the Argonne v_{14} potential, and after renormalization, are compared to the experimental levels in the odd neighbouring nuclei ^{119}Sn and ^{121}Sn.

$\tilde{y}_a$, according to the relations $\tilde{u}_a = \tilde{x}_a u_a + \tilde{y}_a v_a \tilde{v}_a = -\tilde{y}_a u_a + \tilde{x}_a v_a$. The quantities $\tilde{u}_a$ and $\tilde{v}_a$ are related to the spectroscopic factors measured in one-nucleon stripping and pick-up reactions, respectively. One can also define [5,12] a renormalized state-dependent pairing gap, through the relation $\tilde{\Delta}_a = 2\tilde{E}_a \tilde{u}_a \tilde{v}_a / (\tilde{u}_a^2 + \tilde{v}_a^2)$, which in the limit of no fragmentation reduces to the usual BCS expression. The resulting value (cf. Fig. 4) turns out to be fairly close to the value Δ_{exp} obtained from the experimental odd-even mass differences. Also the energy of the lowest quasiparticles is in reasonable agreement with experiment, as shown in Fig.5. As an example of the fragmentation of single-particle levels, we show in Fig. 6 the calculated spectroscopic factors associated with the quantum numbers 5/2+. Unfortunately, it is difficult to make a detailed comparison of these results with the experiment, because only low resolution experiments are available [14].

4. Conclusions

We can conclude that the exchange of collective vibrations gives rise to an attractive interaction between pairs of nucleons in time reversal states, which plays an important role in determining the superfluid properties of open shell nuclei. Because of the key role played by the neutron-proton interaction and by the nuclear surface, this interaction increases the pairing gap, at variance from neutron matter.

Pairing correlations in finite nuclei are often calculated in mean field theory by means of phenomenological effective forces. However, in this way one cannot obtain neither correct level densities nor spectroscopic factors.

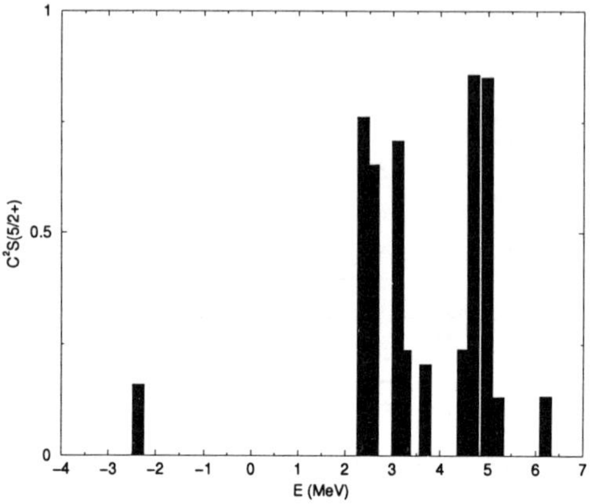

Figure 6. Calculated spectroscopic factors associated to the 5/2+ (left) states. Peaks lying at positive (negative) energy correspond to states in ^{119}Sn (^{121}Sn).

Solving the Dyson equation, one can instead simultaneously obtain pairing gaps of the correct magnitude and account for the level density and the fragmentation of single-particle strength.

References

1. C.Mahaux et al., *Phys. Rep.* **120** (1985) 1.
2. U. Lombardo and H.J. Schultze, in *Physics of neutron star interiors*, D. Blaschke, N.K. Glendenning and A. Sedrakian eds., Springer-Verlag (2001), p.30, and refs. therein.
3. A. V. Avdeenkov and S. P. Kamerdzhiev, *Phys. Lett.* **B 459**, 423 (1999).
4. F. Barranco et al., *Phys. Rev. Lett.* **83**, 2147 (1999).
5. J. Terasaki et al., *Nucl. Phys.* **A697**, 126 (2002); *Progr. Theor. Phys.* **108**, 495 (2002).
6. Y. Nambu, *Phys. Rev.* **117**, 648 (1960); L.P. Gor'kov, *Sov. Phys. JETP* **7**, 505 (1958).
7. G. Gori et al., to be published.
8. F. Barranco et al., *Eur. Phys. J.* **A21**, 57 (2004).
9. In ref. 8 we have also discussed the renormalization of the properties of quadrupole vibrations.
10. A. Bohr and B.R. Mottelson, *Nuclear Structure*, Vol. II, Benjamin (1975).
11. F. Barranco et al., *Phys. Lett.* **B390**, 13 (1997).
12. J.R. Schrieffer, *Theory of superconductivity*, Addison-Wesley (1964).
13. V. Van der Sluys et al., *Nucl. Phys.* **A551**, 210 (1993).
14. M.J. Bechara and O. Dietzch, *Phys. Rev.* **C12**, 90 (1975); E. Gerlic et al., *Phys. Rev.* **C21**, 124 (1980).

KEY TOPICS IN NUCLEAR STRUCTURE
Proceedings of the 8th Int. Spring Seminar on Nuclear Physics
edited by Aldo Covello
© 2005 World Scientific Publishing Co.

MICROSCOPIC STUDY OF LOW-LYING 0^+ STATES IN DEFORMED NUCLEI

N. LO IUDICE
Dipartimento di Scienze Fisiche, Universitá di Napoli "Federico II" and Istituto Nazionale di Fisica Nucleare, Monte S Angelo, Via Cinthia I-80126 Napoli, Italy
E-mail: loiudice@na.infn.it

A. V. SUSHKOV, AND N. YU. SHIRIKOVA
Bogoliubov Laboratory of Theoretical Physics, Joint Institute for Nuclear Research, 141980 Dubna, Russia
E-mail: Sushkov@jinr.thsun1.ru

The microscopic structure of the 14 low-lying excited 0^+ states, observed recently in ^{158}Gd, is investigated within the quasiparticle-phonon model. The results are compared with the predictions by other approaches for a complete characterization of these states.

1. Introduction

A recent (p,t) experiment [1] has established the existence of 14 excited 0^+ states in ^{158}Gd, below an excitation energy of approximately 3.1 MeV. The appearance of so many 0^+ levels at low energy in a single nucleus has stirred up an immediate theoretical interest. In a study carried out within the interacting boson model (IBM) [2], it was found that the IBM, if based on *s* and *d* bosons, can account for only five states below $\sim$ 3.2 MeV. It can predict nearly all the states observed if octupole bosons are included. One should then forcefully conclude that, if the IBM predictions were correct, most of the detected states should be collective and many should be two-phonon octupole states. As pointed out by the same authors, however, it is unnatural to expect that all states detected are collective.

At the other extreme, a projected shell model (PSM) calculation yielded all the states observed[3]. The additional outcome of this calculation was that all 0^+ excitations are non collective, being dominated by either a single two quasi-particle or a single four quasi-particle configuration. The same authors, however, point out that the absence of collective states might be an artifact of their approach, which is non collective by its own nature.

A microscopic calculation within the quasiparticle-phonon model (QPM) [4] was carried out recently [5]. This generates microscopic phonons in random-phase approximation (RPA) and then diagonalizes a Hamiltonian composed of a sum of separable two-body potentials with different multipolarity in a basis of multiphonon states. Since both collective and non collective RPA phonons are included in the multiphonon basis, the approach allows to determine the extent of collectivity of these 0^+ states.

2. Algebraic and shell model model predictions

2.1. *IBM predictions*

Calculations carried out within standard IBM, employing s and d bosons only, can account for very few, only five, 0^+ states. In order to increment the number of these states, Zamfir *et al.* [2] have enlarged the IBM boson space by adding dipole (p) and octupole (f) bosons. In this enlarged *spdf* boson space, they adopted the simple Hamiltonian

$$H = \epsilon_d \hat{n}_d + \epsilon_p \hat{n}_p + \epsilon_f \hat{n}_f - \kappa \hat{Q}_{spdf} \cdot \hat{Q}_{spdf}. \tag{1}$$

where ϵ_p, ϵ_d, and ϵ_f are the p, d, and f boson energies, while $\hat{n}_p$, $\hat{n}_d$, and $\hat{n}_f$ are the corresonding boson number operators. The quadrupole operator is

$$\hat{Q}_{spdf} = \hat{Q}_{sd} + \hat{Q}_{pf} \tag{2}$$

where $\hat{Q}_{sd}$ is given by

$$\hat{Q}_{sd} = (s^\dagger \tilde{d} + d^\dagger s) + \chi (d^\dagger \tilde{d})^2 \tag{3}$$

and $\hat{Q}_{pf}$, of similar structure, is defined as the generator of SU(3) in the pf boson sector.

By a proper choice of the Hamiltonian parameters, it was possible to predict about nine 0^+ levels below $\sim$ 3.2 MeV and 14 of them below 4.0 MeV. The agreement between IBM and experimental spectra is fair (Fig. 1). The remarkable conclusion to be drawn from this calculation is that, if the IBM predictions are correct, most of the 0^+ states should be collective and an appreciable number of them should be two-phonon states built out of octupole bosons.

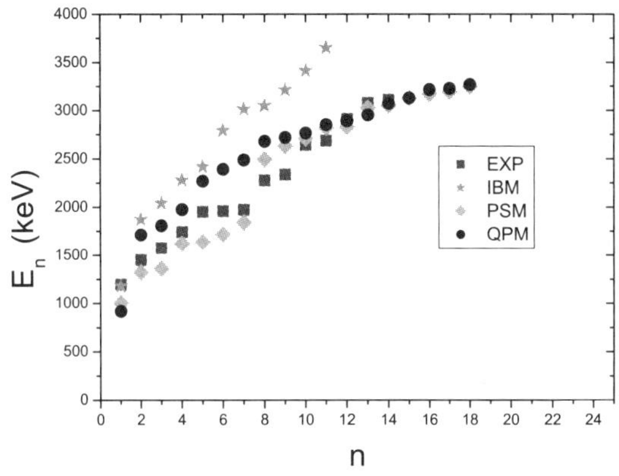

Figure 1. (Color online) Energies of 0^+ states calculated in IBM, PSM, and QPM compared to the experimental data.

2.2. *The PSM predictions*

In the PSM,one starts with a deformed Nilsson basis to build a shell model space spanned by 0-, 2-, and 4-quasiparticle states

$$| \phi_\alpha\rangle = \{| 0\rangle, | p_1p_2\rangle, | n_1n_2\rangle, | p_1p_2n_1n_2\rangle \tag{4}$$

where p_i and n_i label quasi-proton and quas-ineutron states with Nilsson quantum numbers. Out of these multi-quasiparticle deformed states, one generates by projection the spherical shell model basis

$$\Psi^I_M\rangle = \sum_\alpha c^I_\alpha P^I_{MK_\alpha} \mid \phi_\alpha\rangle \tag{5}$$

In the space spanned by these states, one diagonalizes the Hamiltonian

$$H = H_0 - G_0 P_0^\dagger P_0 - G_2 P_2^\dagger P_2 - \frac{1}{2}\kappa Q^\dagger(0)\cdot Q(0) \tag{6}$$

where H_0 is the spherical single particle Hamiltonian and the two-body potential is composed of monopole and quadrupole pairing interactions with respective strengths G_0 and G_2 plus a separable isoscalar quadrupole-quadrupole interaction.

The PSM calculation predicts about 18 0^+ levels below 3.2 MeV in good agreement with experiments (Fig. 1). The corresponding eigenvectors are non collective states dominated by a single two- or four-quasiparticle configuration, in contradiction with the IBM predictions. The strengths of the E2 transitions to the 2^+ member of the ground band are either negligible or of the order of one W.u. (Fig. 2). Both PSM and experimental values are much smaller than the E2 strength typical of a β vibrational mode.

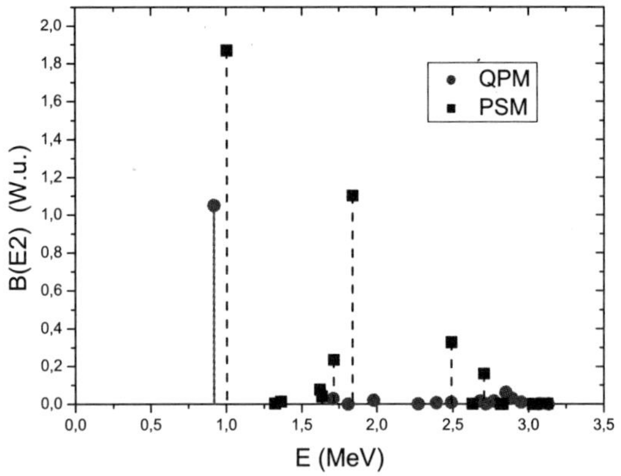

Figure 2. (Color online) Strengths of the E2 transitions from the 0^+ states to the ground 2^+, calculated in PSM and QPM.

3. QPM study

In the QPM, one basically diagonalizes a Hamiltonian of general separable form in a space spanned by one- and two-phonon states, the phonons being generated in RPA.

The Hamiltonian is considerably more complex than the one adopted in the PSM. It has, in fact, the following structure

$$H = H_0 + V^{(P)} + V^{(M)}. \tag{7}$$

The one-body piece H_0 is composed of a kinetic term plus an axially deformed Woods-Saxon potential $V_{WS}(r, \beta_2, \beta_4)$, where β_2 and β_4 are the quadrupole and hexadecapole deformation parameters.

The two-body potential consists of two main pieces acting in the particle-particle and particle-hole channels respectively The particle-hole piece is the sum of several separable potentials of different multipolarity

$$V^{(M)} = -\sum_{\tau_1\tau_2}\sum_{\lambda} \kappa_{\lambda\mu}^{(\tau_1\tau_2)} M_{\lambda\mu}^{\dagger}(\tau_1) M_{\lambda\mu}(\tau_2). \tag{8}$$

where $\tau = p, n$ and

$$M_{\lambda\mu} = R_{\lambda}(r)\, Y_{\lambda\mu}(\theta, \varphi) \tag{9}$$

is the λ-multipole field and $R_{\lambda}(r)$ is the derivative of the W-S potential at zero deformation. The particle-hole interaction includes quadrupole-quadrupole as well as octupole-octupole potentials.

The particle-particle term consists of a monopole plus a sum of multipole pairing potentials

$$V^{(M)} = -\sum_{\tau=p,n} G_0^{(\tau)} P_0^{\dagger} P_0 - \frac{1}{2} \sum_{\tau=p,n} G_{\lambda}^{(\tau)} \sum_{\lambda \neq 0 \mu} P_{\lambda\mu}^{\dagger} P_{\lambda\mu}, \tag{10}$$

where

$$P_0^{\dagger} = \sum_q a_q^{\dagger} a_{\bar{q}}^{\dagger}, \qquad P_{\lambda\mu}^{\dagger} = \sum_{q_1 q_2} \langle q_1 \mid M_{\lambda\mu} \mid q_2 \rangle a_{q_1}^{\dagger} a_{\bar{q}_2}^{\dagger} \tag{11}$$

are, respectively, the monopole and λ multipole pairing operators.

In the QPM one expresses the Hamiltonian in terms of quasi-particle creation ($\alpha_q^{\dagger}$) and annihilation (α_q) operators by means of the Bogolyubov canonical transformation and, then, in terms of the RPA phonon operators

$$Q_{i\nu}^{\dagger} = \frac{1}{2} \sum_{q_1 q_2} \left(\psi_{q_1 q_2}^{i\nu} A_{q_1 q_2}^{\dagger} - \phi_{q_1 q_2}^{i\nu} A_{q_1 q_2} \right) \tag{12}$$

where $A_{q_1 q_2}^{\dagger} = \alpha_{q_1}^{\dagger} \alpha_{\bar{q}_2}^{\dagger}$. This procedure yields the interacting phonon Hamiltonian

$$H_{QPM} = \sum_{v_i} \omega_{v_i} Q_{v_i}^{\dagger} Q_{v_i} + H_{vq}, \tag{13}$$

where $v_i = \{i\, \lambda_i \mu_i\}$. The first term is an unperturbed Boson Hamiltonian diagonal in the basis of the RPA phonon states $\mid v_i \rangle = Q_{v_i}^{\dagger} \mid 0 \rangle$ of energies ω_{v_i}. These are coupled by the term H_{vq}.

The interacting phonon Hamiltonian is put in diagonal form through the variational principle with the trial wave function

$$\Psi_{nK^{\pi}=0^+} = \sum_i C_i(n) \mid i, K^{\pi} = 0^+ \rangle + \sum_{v_1 v_2} C_{v_1 v_2}(n) \mid \left[v_1 \otimes v_2 \right]_0^+ \rangle, \tag{14}$$

where $\mid i, K^{\pi} = 0^+ \rangle = Q_{i0^+}^{\dagger} \mid 0 \rangle$ and $\mid \left[v_1 \otimes v_2 \right]_0^+ \rangle = \left[Q_{v_1}^{\dagger} \otimes Q_{v_2}^{\dagger} \right]_0^+ \mid 0 \rangle$. The two-phonon basis contains phonons of different multipolarity, including the octupole ones.

3.1. *Numerical results*

To construct our phonon basis, we included twenty $K^\pi = 0^+$ RPA phonons and ten phonons of different multi multipolarities $\lambda\mu$. For $\nu = K^\pi = 0^+$, we have eliminated the spurious admixtures, induced by the violation of the proton and neutron number conservation, by imposing the vanishing of the lowest 0^+ RPA root.

The QPM calculation generates about 14 levels below ~ 3.2 MeV, in accordance with experiments, and about 18 below ~ 3.5 MeV in agreement with the PSM results. As shown in Fig. 1, the calculation overstimates the energy of the nearly degenerate levels located around 2 MeV. The latter are, instead, slightly underestimated by the PSM calculation.

As shown in Fig. 2, the strengths of the E2 transitions to the 2^+_g state are even smaller than the ones obtained in the PSM. Apparently, the correlations present in the QPM 0^+ states do not induce any E2 collectivity. The lack of quadrupole collectivity in these states is supported by the analysis of the monopole transition strengths. These resulted to be quite smaller than the typical β vibrational values $\rho^2(E0; 0^+_\beta \to 0^+_g) \sim 100$.

To try to undertstand the physics underlying these puzzling results, we analyse the phonon compositon of the QPM states. The structure of few representative states is shown in Table 1. About six of the QPM states have a single dominant component, namely a RPA quadrupole $\lambda = 2, \mu = 0$ phonon. Very few are linear combinations of one-phonon states. A large number is instead characterized by a two-phonon component with a sizeable, huge in some cases, amplitude. It is remarkable that most of these *two-phonon* states are built out of octupole phonons, consistently with the IBM calculation [2]. There are, however, two-phonon states dominated by $\lambda = 4$ and $\lambda = 5$ phonons which suggest that the g boson should also be included in the IBM scheme.

Further insight can be gained by inspecting the shell structure of the dominant phonons. Except for one, all the states are linear combinations of several two quasi-particle configurations. Few typical examples are given in Table 2. This implies that the QPM states are collective, at variance with the PSM findings. The collectivity, however, is not of quadrupole nature. Most of the two quasi-particle components are in fact paired correlated $q\bar{q}$ states. Such a collectivity is to be probed by other means. To this purpose we have computed in RPA the (p,t) two-nucleon transfer spectroscopic amplitudes. As shown in Fig. 3, some of them are sizeable, indicating that some of these states are characterized by strong neutron

Table 1. Energies, E2 decay strengths and phonon structure of few representative 0^+ states computed in QPM. The symbol $[v]_i$ denotes either the one phonon ($[\lambda\mu]_i$) or the two-phonon ($\left[(\lambda\mu)_i \otimes (\lambda\mu)_j\right]$) components, C_i their corresponding amplitudes.

n	E_n (MeV)	$B_n(E2)$ (W.u.)	$\sum_i C_i^2\,(n)[v]_i$
1	0.93	0.658	$0.92\ [20]_1$
7	2.62	0.002	$0.25\ [20]_6 + 0.55\ \left[(31)_1 \otimes (31)_1\right]$
11	3.04	0.0007	$0.97\ \left[(44)_1 \otimes (44)_1\right]$
14	3.24	0.006	$0.14\ [20]_8 + 0.69\ \left[(30)_1 \otimes (30)_1\right]$
15	3.30	0.045	$0.14\ [20]_7 + 0.25\ [20]_{12} + 0.25\ \left[(22)_1 \otimes (22)_1\right]$

Table 2. Two quasi-particle structure of lowest two $K^\pi = 0^+$ RPA states.

n	E_n (MeV)	$\sum_i c_i^2\,(n)[(q_1) \otimes (q_2)]_\tau$
1	1.135	$0.195[(521\uparrow) \otimes (521\uparrow)]_n + 0.184\ [(505\uparrow) \otimes (505\uparrow)]_n$ $+ 0.157\ [(523\downarrow) \otimes (523\downarrow)]_n + 0.105\ [(402\downarrow) \otimes (402\downarrow)]_n$ $+ 0.063\ [(400\uparrow) \otimes (400\uparrow)]_n + 0.044\ [(520\uparrow) \otimes (520\uparrow)]_n$ $+ 0.152\ [(411\uparrow) \otimes (411\uparrow)]_p + 0.032\ [(413\downarrow) \otimes (413\downarrow)]_p$ $+ 0.015\ [(532\uparrow) \otimes (532\uparrow)]_p$
2	1.767	$0.0623[(505\uparrow) \otimes (505\uparrow)]_n + 0.026\ [(402\downarrow) \otimes (402\downarrow)]_n$ $+ 0.014\ [(400\uparrow) \otimes (400\uparrow)]_n + 0.609\ [(411\uparrow) \otimes (411\uparrow)]_p$ $+ 0.100\ [(413\downarrow) \otimes (413\downarrow)]_p + 0.046\ [(532\uparrow) \otimes (532\uparrow)]_p$

pairing correlations. This is supported also by the experimental data, although the agreement between theory and experiments is only qualitative and not conclusive. We have in fact to include the anharmonic effect due to the coupling with two-phonon states. In Fig. 3 we give the spectroscopic factor normalized to the ground state.

4. Conclusion

Our analysis suggests that the appearance of so many low-lying 0^+ states in the deformed ^{158}Gd is a manifestation of the complex shell structure of this nucleus. Enveiling the true nature of all these states represents a quite difficult task. According to thee QPM calculation, none of the QPM 0^+ states is quadrupole collective. This prediction needs to be tested by more reliable measurements of the E2 transition strengths. The only collectivity we found is the one induced by pairing, in qualitative agreement with the

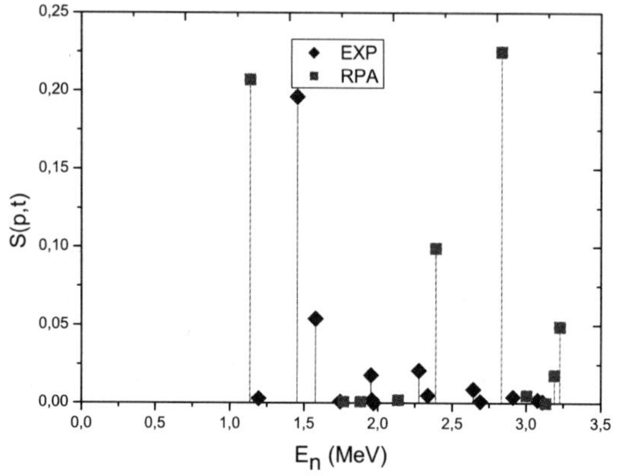

Figure 3. (Color online) QPM two-neutron transfer spectroscopic factor for the different 0^+ states compared to experiments normalized to the ground to ground state spectroscopic factor.

experimental two-neutron transfer spectroscopic factors. It is remarkable that octupole phonons account for a fair number of QPM 0^+ states in agreement with the IBM prediction. A comparable number of 0^+ states is expected in several other nuclei. Experimental as well as theoretical work devoted to the study of these new states is under way.

Acknowledgments

This work is partly supported by the Ministero dell' Istruzione, Università e Ricerca (MIUR).

References

1. S. R. Lesher, A. Aprahamian, L. Trache, A. Oros-Peusquens, S. Deyliz, A. Gollwitzer, R. Hertenberger, B. D. Valnion, and G. Graw, Phys. Rev. C **65**, 031301(R) (2002).
2. N. V. Zamfir, Jing-ye Zhang, and R. F. Casten, Phys. Rev. C **66**, 057303 (2002)
3. Y. Sun, A. Aprahamian, J. Zhang, C. Lee, Phys. Rev. C **68**, 061301 (2003).
4. V. G. Soloviev, *Theory of Atomic Nuclei: Quasiparticles and Phonons* (Institute of Physics Publishing, Bristol, 1992).
5. N. Lo Iudice, A. V. Sushkov, and N. Yu. Shirikova, to be submitted for publication.
6. V.G. Soloviev, Z. Phys. A - Atomic Nuclei **334**, 143 (1989).

7. F. A. Gareev, S. P. Ivanova, V. G. Soloviev, and S. I. Fedotov, Sov. J. Part. Nucl. **4**, 357 (1973).
8. V.G. Soloviev, A.V. Sushkov, N. Yu. Shirikova, and N. Lo Iudice, Nucl. Phys. A **600**, 155 (1996).
9. V.G. Soloviev, A.V. Sushkov, and N. Yu. Shirikova, Z. Phys. A **358**, 287 (1997).
10. V.G. Soloviev, A.V. Sushkov, and N. Yu. Shirikova, Phys. Rev. C **56**, 2528 (1997).
11. L. A. Malov and V. G. Soloviev, Phys. Part. Nucl. **11**, 301 (1980).

KEY TOPICS IN NUCLEAR STRUCTURE
Proceedings of the 8th Int. Spring Seminar on Nuclear Physics
edited by Aldo Covello
© 2005 World Scientific Publishing Co.

MICROSCOPIC STUDY OF LOW-LYING STATES IN ^{92}Zr

CH. STOYANOV
Institute for Nuclear Research and Nuclear Energy,
1784 Sofia, Bulgaria
E-mail:stoyanov@inrne.bas.bg

N. LO IUDICE
Dipartimento di Scienze Fisiche, Universitá di Napoli "Federico II"
and Istituto Nazionale di Fisica Nucleare,
Monte S Angelo, Via Cinthia I-80126 Napoli, Italy
E-mail: loiudice@na.infn.it

We studied in a microscopic multiphonon approach the proton-neutron symmetry and phonon structure of some low-lying states recently discovered in ^{92}Zr. We confirm the breaking of F-spin symmetry, but argue that the breaking mechanism is more complex than the one suggested in the original shell model analysis of the data. We found other new intriguing features of the spectrum, like a pronounced multiphonon fragmentation of the states and a tentative evidence of a three-phonon mixed symmetry state.

1. Introduction

The experiments on ^{94}Mo [1,2,3,4], ^{96}Ru [5,6] and other nuclei [7] have provided a conclusive evidence in favor of the existence of F-spin mixed symmetry states in nearly spherical nuclei, thereby confirming a major prediction of the proton-neutron interacting boson model (IBM2)[8]. The mixed-symmetry states are qualified as as scissors-like excitations [9,10] built on quadrupole vibrational states.

The phonon structure as well as the proton-neutron (p-n) symmetry were confirmed by a microscopic calculation using a multiphonon basis, whose phonons were generated in random-phase approximation (RPA) [11,12]. This approach, known as quasiparticle-phonon model (QPM) [13], accounted with good accuracy for the energies, transition probabilities and selection rules. A good description was also given by a truncated shell model (SM) calculation [14].

In the attempt of establishing how far the F-spin symmetry holds valid, the properties of low-lying states of ^{92}Zr were measured [15]. The most meaningful outcome of the analysis was that the two lowest 2^+ states are one-phonon excitations, at variance with the other nuclei of the region where the second 2_2^+ is a two-phonon state.

This latter point, implying a severe breaking of F-spin, gave us the main motivation for the present QPM study.

2. Calculation and results

We used the same Woods-Saxon potential and the same separable two-body Hamiltonian adopted in [11,12]. We used also the same single particle basis, which encompasses all bound states from the bottom of the well up to the quasi-bound states embedded into the continuum. Following the same strategy, we fit the strength κ_2 of the quadrupole-quadrupole (Q-Q) interaction on the energy and E2 decay strength of the 2_1^+ and the coupling constant G_2 of the quadrupole pairing on the overall properties of the low-lying 2^+ isovector state. The other Hamiltonian parameters remained unchanged. Because of the large model space, we used effective charges very close to the bare values, namely $e_p = 1.1$ for protons and $e_n = 0.1$ for neutrons. We also used the spin-gyromagnetic quenching factor $g_s = 0.7$.

Following the QPM procedure the RPA phonon operators is defined as follows :

$$Q^\dagger_{i\lambda\mu} = \frac{1}{2}\sum_{jj'}\left\{\psi^{i\lambda}_{jj'}[\alpha^\dagger_j\alpha^\dagger_{j'}]_{\lambda\mu} - (-1)^{\lambda-\mu}\varphi^{i\lambda}_{jj'}[\alpha_{j'}\alpha_j]_{\lambda-\mu}\right\}. \tag{1}$$

The phonons have the multipolarity $\lambda\mu$ and the energy $\omega_{i\lambda}$; $\alpha^\dagger_{jm}(\alpha_{jm})$ are quasiparticle operators obtained from the corresponding particle operators through a Bogoliubov transformation. The phonon operators fulfil the normalization conditions

$$\frac{1}{2}\sum_{jj'}\left[\psi^{i\lambda}_{jj'}\psi^{i'\lambda}_{jj'} - \varphi^{i\lambda}_{jj'}\varphi^{i'\lambda'}_{jj'}\right] = \delta_{ii'}\delta_{\lambda\lambda'}. \tag{2}$$

It is worth to point out that, among the RPA phonons, only few are collective, composed of a coherent linear combination of two-quasiparticle configurations. The QPM Hamiltonian is then diagonalized in a space spanned by RPA multiphonon states. As in Ref. [11,12], we included up to three-phonon states.

2.1. *RPA analysis*

To test the proton-neutron symmetry of the lowest RPA states, denoted as $[2_1^+]_{RPA}$ and $[2_2^+]_{RPA}$, we compute the ratios

$$R_{i=1,2}(IV/IS) = \frac{M_i^{(n)} - M_i^{(p)}}{M_i^{(n)} + M_i^{(p)}}, \tag{3}$$

$$R_i^2(IV/IS) = |R_i(IV/IS)|^2 \tag{4}$$

where

$$M_i^{(n/p)} = \langle 2_i^+ \parallel \sum_k^{(p/n)} r2_k Y_2(\Omega_k) \parallel g.s.\rangle. \tag{5}$$

If F-spin is preserved to a good extent, we must have $R_1^2(IV/IS) < 1$ and $R_2^2(IV/IS) > 1$.

Table 1. Neutron and proton quadrupole transition amplitudes and IV/IS ratios for the lowest two $[2^+]_{RPA}$ states.

$g.s. \to [2_i^+]_{RPA}$	G_2	$M_i^{(n)}(fm^2)$	$M_i^{(p)}(fm^2)$	$R_i(IV/IS)$	$R_i^2(IV/IS)$
$g.s. \to [2_1^+]_{RPA}$	0.0	67.67	41.98	0.234	0.055
	1.0	72.2	49.6	0.185	0.034
$g.s. \to [2_2^+]_{RPA}$	0.0	52.67	46.40	0.063	0.004
	1.0	14.4	28.4	- 0.327	0.107

In order to meet all the above requirements we have only one parameter at our disposal, the quadrupole pairing constant. For $G_2 = 0$, the ratios shown in Table 1 qualify not only the first but also the second $[2^+]_{RPA}$ as p-n symmetric. The second is, actually, even more symmetric and collective than the first. As we increase G_2, the transition amplitudes, specially the neutron one, decrease. For $G_2/\kappa = 1$ the neutron amplitude is small but positive, so that the corresponding IV/IS ratio is much larger than in the case of vanishing G_2, but, still, appreciably smaller than one. Only for values of G_2 considerably larger than κ, the $[2_2^+]_{RPA}$ becomes p-n non symmetric ($R(IV/IS) > 1$), but looses completely its collectivity. The corresponding E2 strength is negligible, at variance with experiments. We therefore chose $G_2 = \kappa$ which allows to fulfils more closely the experimental requirements.

For such a value, the lowest RPA 2_1^+ is collective, though to a less extent than in other nuclei of the same region. Its RPA E2 decay strength (Table 2)

Table 2. Structure of the lowest RPA phonons in ^{92}Zr.

λ_i^π	$\omega_{\lambda_i^\pi}$ (MeV)	$B(E2)\downarrow$ ($w.u.$)	Structure	
2_1^+	1.18	7.6	$0.99(2d_{5/2}\otimes 2d_{5/2})_n$	48.4%
			$+0.23(2d_{5/2}\otimes 3s_{1/2})_n$	4.8%
			$+0.17(2d_{5/2}\otimes 1g_{9/2})_n$	2.5%
			total	60%
			$+0.64(1g_{9/2}\otimes 1g_{9/2})_p$	20.5%
			$+0.23(1f_{5/2}\otimes 2p_{1/2})_p$	5.5%
			$+0.23(2p_{3/2}\otimes 2p_{1/2})_p$	5.2%
			$+0.15(1f_{5/2}\otimes 1f_{5/2})_p$	1.1%
			$+0.14(1g_{9/2}\otimes 2d_{5/2})_p$	2.1%
			total	40%
2_2^+	2.07	2.2	$-0.99(2d_{5/2}\otimes 2d_{5/2})_n$	48.8%
			$+0.09(2d_{5/2}\otimes 3s_{1/2})_n$	0.72%
			$+0.11(2d_{5/2}\otimes 1g_{9/2})_n$	0.82%
			total	51%
			$+0.79(1g_{9/2}\otimes 1g_{9/2})_p$	30.7%
			$+0.25(1f_{5/2}\otimes 2p_{1/2})_p$	5.7%
			$+0.25(2p_{3/2}\otimes 2p_{1/2})_p$	5.7%
			$+0.18(1f_{5/2}\otimes 1f_{5/2})_p$	1.3%
			$+0.15(1g_{9/2}\otimes 2d_{5/2})_p$	1.9%
			total	49%

is smaller than the corresponding one in ^{94}Mo [12] by more than a factor two. Such a quenching reflects the diminished role of the proton with respect to the neutron component in ^{92}Zr. The neutron dominance, however, is far less pronounced than in SM [15] and does not alter dramatically the symmetry of the state. In fact, not only all proton and neutron components are in phase, but also the ratio of the isovector to the isoscalar quadrupole transition amplitudes is small, though not negligible. Such a test qualifies the 2_1^+ as a $\Delta T = 0$ p-n symmetric state with a small, though non negligible, admixture of non symmetric pieces.

The F-spin breaking is more substantial in the second $[2_2^+]_{RPA}$. Indeed, its isoscalar to isovector ratio $R_2(IV/IS)$ is considerably smaller than unity, indicating that the transition is promoted with comparable strengths by both isoscalar and isovector quadrupole operators.

2.2. *QPM results*

Let us now investigate the QPM states (Table 3) and how their phonon composition affects the E2 (Tables 4) as well as the E1 and M1 transitions (Table 5).

The first 2_1^+ is mostly accounted for by the lowest RPA one-phonon component. The appreciable neutron dominance is therefore confirmed and is consistent with the magnetic properties of the state. The QPM yields for the gyromagnetic factor $g(2_1^+) = -0.20$, very close to the experimental value $g_{exp}(2_1^+) = -0.18(1)$.

The second 2_2^+ is a one-phonon state, dominated by the second $[2_2^+]_{RPA}$. As pointed out already, this is peculiar of ^{92}Zr, since the 2_2^+ in the nearby nuclei ^{94}Mo and ^{136}Ba was found to be a two-phonon symmetric state [12]. This one-phonon 2_2^+ undergoes an E2 decay to the ground state (Table 4) and a M1 transition to the symmetric 2_1^+ (Table 5) . The computed M1 strength is smaller than in ^{94}Mo, while the E2 strength is unusually large. These two correlated facts, already pointed out in the experimental analysis [15], indicate that the 2_2^+ is not a pure "mixed symmetry" state but has an appreciable F-spin symmetric component. The latter is responsible for the enhancement of the E2 and quenching of the M1 strengths, respectively, with respect to ^{94}Mo. This overestimates the experimental M1 strength and underestimates the E2 transition probability roughly by the same factor ~ 1.5. Our QPM result is therefore at variance with the SM findings [15]. This discrepancy emerges clearly from the analysis of the magnetic moments. The QPM g-factor for the 2_2^+ state is $g(2_2^+) = -0.31$, quite different in sign and magnitude from the SM value $g_{SM}(2_2^+) = 0.9$. Clearly, a measure of this quantity would discriminate between the two descriptions and, therefore, would shed light on the p-n symmetry of this state.

Table 3. Energy and phonon structure of selected low-lying excited states in ^{92}Zr. Only the dominant components are shown.

State J^{π}	E (keV) EXP	QPM	Structure,%
2_1^+	934	1017	$92.6\%[2_1^+]_{RPA}$
2_2^+	1847	1945	$90\%[2_2^+]_{RPA}$
2_3^+	2067	2008	$19\%[2_3^+]_{RPA} + 65\%[2_1^+ \otimes 2_1^+]_{RPA}$
2_7^+		3386	$32\%[2_5^+]_{RPA} + 51\%[2_1^+ \otimes 2_2^+]_{RPA}$
1_1^+	3472	3235	$93\%[2_1^+ \otimes 2_2^+]_{RPA} + 6\%[1_1^+]_{RPA}$
1_2^+	3638	3781	$91\%[1_1^+]_{RPA}$
3_2^+		3180	$74\%[2_1^+ \otimes 2_2^+]_{RPA}$

Another distinguishing feature of ^{92}Zr with respect to the nearby nuclei is the fragmentation of the QPM states into several multiphonon components. The symmetric two-phonon $[2_1^+ \otimes 2_1^+]_{RPA}$ accounts only for 65% of the 2_3^+. For the sake of comparison, the two-phonon counterpart in ^{94}Mo represents the 82% of the 2_2^+ state. The non symmetric $[2_1^+ \otimes 2_2^+]_{RPA}$ is spread over several 2^+ states. It is dominant in the 2_7^+ and sizeable in 2_8^+. These two states are therefore predicted to have strong E2 decays to the p-n non symmetric 2_2^+ state. The few available data are closely reproduced by the calculation.

The 1^+ and 3^+ states are also of importance for testing the p-n symmetry. As shown in Table 3, only the first 1_1^+ is predominantly a two-phonon non symmetric state and, therefore, would be the analogue of the IBM mixed-symmetry state, if F-spin were conserved. The other has a dominant spin excitation component. Spin indeed contributes mainly to the strength of the $M1$ decay of the second 1_2^+. It gives also a small but non negligible contribution to the decay of the 1_1^+. Such a contribution is crucial for attaining a good agreement with experiments. The strong E2 decay of the 1_1^+ to the symmetric 2_1^+ is also consistent with the experiments and mirrors the unusually strong E2 decay of the 2_2^+ to the ground state. It represents, therefore, an additional signature of F-spin breaking. A further confirm may be provided by the E2 decay of the 3_2^+. This contains a very large $\left[2_1^+ \otimes 2_2^+\right]_{RPA}$ component and is predicted to decay to the 2_1^+ with a strong E2 transition. An experimental test would be desirable. Very

Table 4. $E2$ transitions connecting some excited states in ^{92}Zr calculated in QPM.

$B(E2; J_i \rightarrow J_f)(w.u.)$	EXP	QPM
$B(E2; 2_1^+ \rightarrow g.s.)$	6.4(6)	6.5
$B(E2; 2_2^+ \rightarrow g.s.)$	3.7(8)	2.1
$B(E2; 2_2^+ \rightarrow 2_1^+)$	0.3(1)	0.39
$B(E2; 2_3^+ \rightarrow 2_1^+)$		6.8
$B(E2; 1_1^+ \rightarrow 2_1^+)$	<8.0(6)	2.0
$B(E2; 3_2^+ \rightarrow 2_1^+)$		1.6
$B(E2; 3_2^+ \rightarrow 2_2^+)$		4.4

intriguing is the case of the $2^+_{3.263}$ level observed at $E = 3.263$ MeV. This state decays to the first 2_1^+ with an appreciable M1 strength and an E2 strength of the order of one single particle unit. This level is close to the energy $E \simeq 3.7$ MeV of the three phonon state $\mid 2^+_{3ph}\rangle = \left[2_1^+ \otimes 2_1^+ \otimes 2_2^+\right]_{2^+}$. Moreover, the strength of the M1 transition of this three phonon state to the first 2_1^+ has the same structure of and is comparable in magnitude to the strength of the M1 decay of the non symmetric 1_1^+ to the ground state. For a pure, properly antisymmetrized, three-phonon state, we get $B(M1, 2^+_{3ph} \rightarrow 2_1^+) = 0.06\mu_N 2$, close to $B(M1, 1_1^+ \rightarrow 0_{gr}) = 0.07\mu_N 2$ and smaller than the measured strength by a factor two. Thus, it is tempting to consider this $2^+_{3.263}$ as a good candidate for being a three-phonon excitation with small admixture of two-phonon components. If confirmed by more complete calculations, this level would provide the first evidence of a three-phonon non symmetric 2^+ state.

Table 5. QPM versus experimental $M1$ transitions between some excited states in ^{92}Zr. $^{(a)}$ Computed under the assumption that the state is a pure, properly antisymmetrized, three-phonon state.

$B(M1; J_i \rightarrow J_f)(\mu_N^2)$	EXP	QPM	QPM ($g_s = 0$)
$B(M1; 2^+_2 \rightarrow 2^+_1)$	0.46(15)	0.68	0.22
$B(M1; 2^+_{3.263} \rightarrow 2^+_1)$	0.16(2)	0.06[a]	
$B(M1; 1^+_1 \rightarrow g.s.)$	0.094(4)	0.069	0.031
$B(M1; 1^+_2 \rightarrow g.s.)$		0.081	0.018
$B(M1; 1^+_1 \rightarrow 2^+_1)$	$< 0.089(6)$	1×10^{-4}	5×10^{-5}

3. Concluding remarks

On the ground of the present study, we may draw the conclusion that, consistently with the experimental analysis [15], the lowest two 2^+ are RPA one-phonon states. At variance with the conclusion drawn in [15], based on a calculation carried out within a too severely truncated SM space, we found that the 2^+_1 state has appreciable but not huge neutron dominance which does not destroy its p-n symmetric character. The F-spin, instead, is broken more substantially in the second 2^+ state.

The present study offers also the arguments in favor of the first experimental evidence of a three-phonon non symmetric 2^+ state.

More details about the calculations of ^{92}Zr within QPN can be found in Ref. [18].

Acknowledgements

Work partially supported by the Italian Ministero dell'Istruzione, Università e Ricerca (MIUR) and by the grant Ph 1311 of the Bulgarian Science Foundation.

References

1. N. Pietralla, C. Fransen, D. Belic, P. von Brentano, C. Friessner, U. Kneissl, A. Linnemann, A. Nord, H. H. Pitz, T. Otsuka, I. Schneider, V. Werner, and I. Wiedenhover, Phys. Rev. Lett. **83**, 1303 (1999).

2. N. Pietralla, C. Fransen, P. von Brentano, A. Dewald, A. Fitzler, C. Friessner, J. Gableske, Phys. Rev. Lett. **84**, 3775 (2000).
3. C. Fransen, N. Pietralla, P. von Brentano, A. Dewald, J. Gableske, A. Gade, A. F. Lisetskiy, V. Werner, Phys. Lett. **B 508**, 219 (2001).
4. C. Fransen, N. Pietralla, Z. Ammar, D. Bandyopadhyay, N. Boukharouba, P. von Brentano, A. Dewald, J. Gableske, A. Gade, J. Jolie, U. Kneissl, S. R. Lesher, A. F. Lisetskiy, M. T. McEllistrem, M. Merrick, H. H. Pitz, N. Warr, V. Werner, and S. W. Yates, Phys. Rev. C **67**, 024307 (2003).
5. N. Pietralla, C. J. Barton III, R. Krücken, C. W. Beausang, M. A. Caprio, R. F. Casten, J. R. Cooper, A. A. Hecht, H. Newman, J. R. Novak, and N. V. Zamfir, Phys. Rev. C **64**, 031301(R) (2001).
6. H. Klein, A. F. Lisetskiy, N. Pietralla, C. Fransen, A. Gade, and P. von Brentano, Phys. Rev. C **65**, 044315 (2002).
7. A. Gade, H. Klein, N. Pietralla, and P. von Brentano, Phys. Rev. C **65**, 054311 (2002).
8. A. Arima, T. Otsuka, F. Iachello, and I. Talmi, Phys. Lett. **B 508**, 219 (2001).
9. N. Lo Iudice and F. Palumbo, Phys. Rev. Lett. **41**, 1532 (1978).
10. D. Bohle, A. Richter, W. Steffen, A. E. L. Dieperink, N. Lo Iudice, F. Palumbo, and O. Scholten, Phys. Lett. **B137**, 27 (1984).
11. N. Lo Iudice and Ch. Stoyanov, Phys. Rev. C **62**, 047302 (2000).
12. N. Lo Iudice and Ch. Stoyanov, Phys. Rev. C **65**, 064304 (2002).
13. V. G. Soloviev, *Theory of atomic nuclei : Quasiparticles and Phonons* (Institute of Physics Publishing, Bristol and Philadelphia, 1992).
14. A. F. Lisetskiy, N. Pietralla, C. Fransen, R.V. Jolos, P. von Brentano, Nucl. Phys. **A 677**, 1000 (2000).
15. V. Werner, D. Belic, P. von Brentano, C. Fransen, A. Gade, H. von Garrel, J. Jolie, U. Kneissl, C. Kostall, A. Linnemann, A. F. Lisetskiy, N. Pietralla, H. H. Pitz, M. Scheck, K.-H. Speidel, F. Stedile, S. W. Yates, Phys. Lett. **B 550**, 140 (2002).
16. V. Yu. Ponomarev, Ch. Stoyanov, N. Tsoneva, M. Grinberg, Nucl. Phys. **A 635**, 470 (1998).
17. N. Pietralla, C. Fransen, A. Gade, N. A. Smirnova, P. von Brentano, V. Werner, and S. W. Yates, Phys. Rev. C. **68**, 031305(R) (2003).
18. N. Lo Iudice and Ch. Stoyanov, Phys. Rev. C **69**, 044312 (2004).

KEY TOPICS IN NUCLEAR STRUCTURE
Proceedings of the 8th Int. Spring Seminar on Nuclear Physics
edited by Aldo Covello
© 2005 World Scientific Publishing Co.

FINITE RANK APPROXIMATION FOR NUCLEAR STRUCTURE CALCULATIONS WITH SKYRME INTERACTIONS

A.P. SEVERYUKHIN *

Service de Physique Nucléaire Théorique, Université Libre de Bruxelles, Case Postale 229, B-1050 Bruxelles, Belgium

V.V. VORONOV

Bogoliubov Laboratory of Theoretical Physics, JINR, 141980 Dubna, Moscow Region, Russia
E-mail: voronov@thsun1.jinr.ru

NGUYEN VAN GIAI

Institut de Physique Nucleaire, Universite Paris-Sud, F-91406 Orsay Cedex, France

Starting from an effective Skyrme interaction we present a method to take into account the coupling between one- and two-phonon terms in the wave functions of excited states. The approach is a development of a finite rank separable approximation for the quasiparticle RPA calculations proposed in our previous work. The influence of the phonon-phonon coupling on energies and transition probabilities for the low-lying quadrupole in the neutron-rich Sn isotopes is studied.

1. Introduction

The experimental and theoretical studies of properties of the excited states in nuclei far from the β-stability line are presently the object of very intensive activity. The random phase approximation (RPA) [1,2] is a well-known and successful way to treat nuclear vibrational excitations. Using different effective nucleon-nucleon interactions[3,4] the most consistent models can describe the ground states within the Hartree-Fock (HF) or Hartree-Fock-Bogoliubov (HFB) approximations and the excited states within the RPA

*on leave from Bogoliubov Laboratory of Theoretical Physics, JINR, Dubna

and quasiparticle RPA (QRPA). Such models are quite successful to reproduce the nuclear ground state properties [5,6] and the main features of nuclear excitations in nuclei[7,8,9]. For the open-shell nuclei the pairing correlations are very important.

Due to the anharmonicity of vibrations there is a coupling between one-phonon and more complex states [1,2] and the complexity of calculations beyond standard RPA or QRPA increases rapidly with the size of the configuration space, so one has to work within limited spaces. Making use of separable forces one can perform calculations of nuclear characteristics in very large configuration spaces since there is no need to diagonalize matrices whose dimensions grow with the size of configuration space [2], but it is very difficult to extrapolate the phenomenalogical parameters of the nuclear hamiltonian to new regions of nuclei.

That is why a finite rank approximation for the particle–hole (p-h) interaction resulting from the Skyrme forces has been suggested in our previous work[10]. Thus, the self-consistent mean field can be calculated with the original Skyrme interaction whereas the RPA solutions would be obtained with the finite rank approximation to the p-h matrix elements.

Recently, this approach has been generalized to take into account the pairing correlations [11]. The QRPA was used to describe characteristics of the low-lying 2^+ and 3^- states and giant resonances in nuclei with very different mass numbers [11]. It was found that there is room for the phonon-phonon coupling effects in many cases [12,13].

As an application of our method we present results for low-lying 2^+ states in neutron-rich Sn isotopes and compare them with recent experimental data [14] and other calculations [15,16,17].

2. Method of calculations

We start from the effective Skyrme interaction[4] and use the notation of Ref.[18] containing explicit density dependence and all spin-exchange terms. The single-particle spectrum is calculated within the HF method. The continuous part of the single-particle spectrum is discretized by diagonalizing the HF hamiltonian on a harmonic oscillator basis[19]. The p-h residual interaction $\tilde{V}_{res}$ corresponding to the Skyrme force and including both direct and exchange terms can be obtained as the second derivative of the energy density functional with respect to the density[20]. Following our previous papers[10] we simplify $\tilde{V}_{res}$ by approximating it by its Landau-Migdal form. Here, we keep only the $l = 0$ terms in V_{res} and in the coordinate representation one can write it in the following form:

$$V_{res}(\mathbf{r}_1,\mathbf{r}_2) = N_0^{-1}[F_0(r_1) + G_0(r_1)(\sigma_1\cdot\sigma_2) \\ +(F_0^{'}(r_1) + G_0^{'}(r_1)(\sigma_1\cdot\sigma_2))(\tau_1\cdot\tau_2)]\delta(\mathbf{r}_1-\mathbf{r}_2) \quad (1)$$

where σ_i and τ_i are the spin and isospin operators, and $N_0 = 2k_F m^*/\pi^2\hbar^2$ with k_F and m^* standing for the Fermi momentum and nucleon effective mass. The expressions for $F_0, G_0, F_0^{'}, G_0^{'}$ in terms of the Skyrme force parameters can be found in Ref.[18].

In what follows we use the second quantized representation and V_{res} can be written as:

$$\hat{V}_{res} = \frac{1}{2}\sum_{1234} V_{1234} : a_1^+ a_2^+ a_4 a_3 : \quad (2)$$

where a_1^+ (a_1) is the particle creation (annihilation) operator and 1 denotes the quantum numbers $(n_1 l_1 j_1 m_1)$,

$$V_{1234} = \int \phi_1^*(\mathbf{r}_1)\phi_2^*(\mathbf{r}_2)V_{res}(\mathbf{r}_1,\mathbf{r}_2)\phi_3(\mathbf{r}_1)\phi_4(\mathbf{r}_2)d\mathbf{r}_1 d\mathbf{r}_2. \quad (3)$$

After integrating over the angular variables one needs to calculate the radial integrals. As it is shown in [10,11] the radial integrals can be calculated accurately by choosing a large enough cutoff radius R and using a N-point integration Gauss formula with abscissas r_k and weights w_k. Thus, the two-body matrix element is a sum of N separable terms, i.e., the residual interaction takes the form of a rank N separable interaction.

We employ a hamiltonian including an average HF field, pairing interactions, the isoscalar and isovector particle–hole (p–h) residual forces in a finite rank separable form [11]:

$$H = \sum_\tau \left(\sum_{jm}^{\tau} (E_j - \lambda_\tau) a_{jm}^\dagger a_{jm} - \frac{1}{4} V_\tau^{(0)} : P_0^\dagger(\tau) P_0(\tau) : \right) + \hat{V}_{res}, \quad (4)$$

where

$$P_0^+(\tau) = \sum_{jm}^{\tau} (-1)^{j-m} a_{jm}^+ a_{j-m}^+. \quad (5)$$

We sum over the proton(p) and neutron(n) indexes and the notation $\{\tau = (n,p)\}$ is used. A change $\tau \leftrightarrow -\tau$ means a change $p \leftrightarrow n$. The single-particle states are specified by the quantum numbers (jm), E_j are the single-particle energies, λ_τ the chemical potentials. $V_\tau^{(0)}$ is the interaction strength in the particle-particle channel. The hamiltonian (4) has the same form as the QPM hamiltonian with N separable terms [2], but the single-particle spectrum and parameters of the p-h residual interaction are calculated making use of the Skyrme forces.

In what follows we work in the quasiparticle representation defined by the canonical Bogoliubov transformation:

$$a^+_{jm} = u_j \alpha^+_{jm} + (-1)^{j-m} v_j \alpha_{j-m}. \tag{6}$$

The hamiltonian (4) can be represented in terms of bifermion quasiparticle operators and their conjugates [2]:

$$B(jj^{'};\lambda\mu) = \sum_{mm'} (-1)^{j^{'}+m'} \langle jmj^{'}m^{'} \mid \lambda\mu\rangle \alpha^+_{jm}\alpha_{j'-m'}, \tag{7}$$

$$A^+(jj^{'};\lambda\mu) = \sum_{mm'} \langle jmj^{'}m^{'} \mid \lambda\mu\rangle \alpha^+_{jm}\alpha^+_{j'm'}. \tag{8}$$

We introduce the phonon creation operators

$$Q^+_{\lambda\mu i} = \frac{1}{2}\sum_{jj'} \left(X^{\lambda i}_{jj'} A^+(jj^{'};\lambda\mu) - (-1)^{\lambda-\mu} Y^{\lambda i}_{jj'} A(jj^{'};\lambda-\mu) \right). \tag{9}$$

where the index λ denotes total angular momentum and μ is its z-projection in the laboratory system. One assumes that the ground state is the QRPA phonon vacuum $\mid 0\rangle$. We define the excited states for this approximation by $Q^+_{\lambda\mu i} \mid 0\rangle$. The quasiparticle energies (ε_j), the chemical potentials (λ_τ), the energy gap and the coefficients u,v of the Bogoliubov transformations (6) are determined from the BCS equations with the single-particle spectrum that is calculated within the HF method with the effective Skyrme interaction. Making use of the linearized equation-of-motion approach one can get the QRPA equations.

In QRPA problems there appear two types of interaction matrix elements, the $A^{(\lambda)}_{(j_1j_1')(j_2j_2')}$ matrix related to forward-going graphs and the $B^{(\lambda)}_{(j_1j_1')_\tau(j_2j_2')_{q\tau}}$ matrix related to backward-going graphs. Solutions of this set of linear equations yield the eigen-energies and the amplitudes X, Y of the excited states. The dimension of the matrices A, B is the space size of the two-quasiparticle configurations. For our case expressions for A, B and X, Y are given in [11]. Using the finite rank approximation we need to invert a matrix of dimension $4N \times 4N$ independently of the configuration space size [10,11]. Therefore, this approach enables one to reduce remarkably the dimensions of the matrices that must be inverted to perform structure calculations in very large configuration spaces.

Our calculations [11] show that, for the normal parity states one can neglect the spin-multipole terms of the p-h residual interaction (1). Using the completeness and orthogonality conditions for the phonon operators one can express bifermion operators $A^+(jj^{'};\lambda\mu)$ and $A(jj^{'};\lambda\mu)$ through the phonon ones and the initial hamiltonian (4) can be rewritten in terms of quasiparticle and phonon operators in the following form:

$$H = h_0 + h_{QQ} + h_{QB} \tag{10}$$

$$h_0 = \sum_{jm} \varepsilon_j \, \alpha_{jm}^{+} \alpha_{jm} \tag{11}$$

$$h_{QQ} = -\frac{1}{4} \sum_{\lambda\mu i i' \tau} W^{\lambda i i'}(\tau) \, Q_{\lambda\mu i}^{+} Q_{\lambda\mu i'} \tag{12}$$

$$h_{QB} = -\frac{1}{2} \sum_{\lambda\mu i\tau} \sum_{jj'}{}^{\tau} \Gamma_{jj'}^{\lambda i}(\tau) \left((-)^{\lambda-\mu} Q_{\lambda\mu i}^{+} + Q_{\lambda-\mu i} \right) B(jj^{'};\lambda-\mu) \tag{13}$$

The coefficients W, Γ of the hamiltonian (10) are sums of N combinations of phonon amplitudes, the Landau parameters, the reduced matrix element of the spherical harmonics and radial parts of the HF single-particle wave function (see [13]).

To take into account the mixing of the configurations in the simplest case one can write the wave functions of excited states as:

$$\Psi_\nu(\lambda\mu) = \{\sum_i R_i(\lambda\nu) Q_{\lambda\mu i}^{+} + \sum_{\lambda_1 i_1 \lambda_2 i_2} P_{\lambda_2 i_2}^{\lambda_1 i_1}(\lambda\nu) \left[Q_{\lambda_1\mu_1 i_1}^{+} Q_{\lambda_2\mu_2 i_2}^{+} \right]_{\lambda\mu} \}|0\rangle \tag{14}$$

with the normalization condition:

$$\sum_i R_i^2(J\nu) + 2 \sum_{\lambda_1 i_1 \lambda_2 i_2} (P_{\lambda_2 i_2}^{\lambda_1 i_1}(J\nu))^2 = 1 \tag{15}$$

Using the variational principle in the form:

$$\delta\ (\langle\,\Psi_\nu(\lambda\mu)\,|H\mid\Psi_\nu(\lambda\mu)\rangle\ -\ E_\nu\,(\langle\,\Psi_\nu(\lambda\mu)\,|\,\Psi_\nu(\lambda\mu)\rangle\ -\ 1))=0, \tag{16}$$

one obtains a set of linear equations for the unknown amplitudes $R_i(J\nu)$ and $P_{\lambda_2 i_2}^{\lambda_1 i_1}(J\nu)$. The number of linear equations that have the same form as the basic QPM equations[2] equals the number of one- and two-phonon configurations included in the wave function (14).

3. Details and results of calculations

We apply the present approach to study characteristics of the low-lying vibrational states in the neutron-rich Sn isotopes. In this paper we use the parametrization SLy4 [21] of the Skyrme interaction. Spherical symmetry is assumed for the HF ground states. The pairing constants V_τ^0 are fixed to reproduce the odd-even mass difference of neighboring nuclei. It is well known [8] that the constant gap approximation leads to an overestimating of occupation probabilities for subshells that are far from the Fermi level and it is necessary to introduce a cut-off in the single-particle space. Above this cut-off subshells don't participate in the pairing effect. In our calculations we choose the BCS subspace to include all subshells lying below 5 MeV.

In order to perform QRPA calculations, the single-particle continuum is discretized [19] by diagonalizing the HF hamiltonian on a basis of twelve harmonic oscillator shells and cutting off the single-particle spectra at the energy of 100 MeV. This is sufficient to exhaust practically all the energy-weighted sum rule.

Our previous investigations [11] enable us to conclude that N=45 for the rank of our separable approximation is enough for multipolarities $\lambda \leq 3$ in nuclei with $A \leq 208$.

The two-phonon configurations of the wave function (14) are constructed from natural parity phonons with multipolarities $\lambda = 2, 3, 4, 5$. All one-phonon configurations with energies below 8 MeV are included in the the wave function (14). The cut-off in the space of the two-phonon configurations is 21 MeV. An extension of the space for one- and two-phonon configurations does not change results for energies and transition probabilities practically.

As an application of the method we investigate effects of the phonon-phonon coupling on energies and transition probabilities to 2_1^+ in $^{124-134}$Sn.

Results of our calculations for the 2_1^+ energies and transition probabilities $B(E2)$ are compared with experimental data [14,22] in Figure 1.

As it is seen from Figure 1 there is a remarkable increase of the 2_1^+ energy and $B(E2\uparrow)$ in ^{132}Sn in comparison with those in ^{130,134}Sn. Such a behaviour of $B(E2\uparrow)$ is related with the proportion between the QRPA amplitudes for neutrons and protons in Sn isotopes. The neutron amplitudes are dominant in all Sn isotopes and the contribution of the main neutron configuration $\{1h_{11/2}, 1h_{11/2}\}$ increases from 81.2% in ^{124}Sn to 92.8% in ^{130}Sn when neutrons fill the subshell $1h_{11/2}$. At the same time the contribution of the main proton configuration $\{2d_{5/2}, 1g_{9/2}\}$ is decreasing from 9.3% in ^{124}Sn to 3.9% in ^{130}Sn. The closure of the neutron subshell $1h_{11/2}$ in ^{132}Sn leads to the vanishing of the neutron paring. The energy of the first neutron two-quasiparticle pole $\{2f_{7/2}, 1h_{11/2}\}$ in ^{132}Sn is greater than energies of the first poles in ^{130,134}Sn and the contribution of the $\{2f_{7/2}, 1h_{11/2}\}$ configuration in the doubly magic ^{132}Sn is about 61%. Furthermore, the first pole in ^{132}Sn is closer to the proton poles. This means that the contribution of the proton two-quasiparticle configurations is greater than those in the neighbouring isotopes and as a result the main proton configuration $\{2d_{5/2}, 1g_{9/2}\}$ in ^{132}Sn exhausts about 33%. In ^{134}Sn the leading contribution (about 99%) comes from the neutron configuration $\{2f_{7/2}, 2f_{7/2}\}$ and as a result the $B(E2)$ value is reduced. Such a behaviour of the 2_1^+ energies and $B(E2)$ values in the neutron-rich Sn isotopes reflects the shell structure in this region . It is worth to mention that the first prediction of the anomalous behaviour of 2^+ excitations around ^{132}Sn based on the QRPA calculations with a separable quadrupole-plus-pairing hamiltonian has been done in [15]. Other QRPA calculations with Skyrme [16] and Gogny [17] forces give similar results for Sn isotopes.

One can see from Figure 1 that the inclusion of the two-phonon terms results in a decrease of the energies and a reduction of transition probabilities. Note that the effect of the two-phonon configurations is important for the energies and this effect becomes weak in ^{132}Sn. There is some overestimate of the energies for the QRPA calculations and taking into account of the two-phonon terms improves the description of the 2_1^+ energies . The reduction of the $B(E2)$ values is small in most cases due to the crucial contribution of the one-phonon configuration in the wave function structure.

4. Conclusions

A finite rank separable approximation for the QRPA calculations with Skyrme interactions that was proposed in our previous work is extended

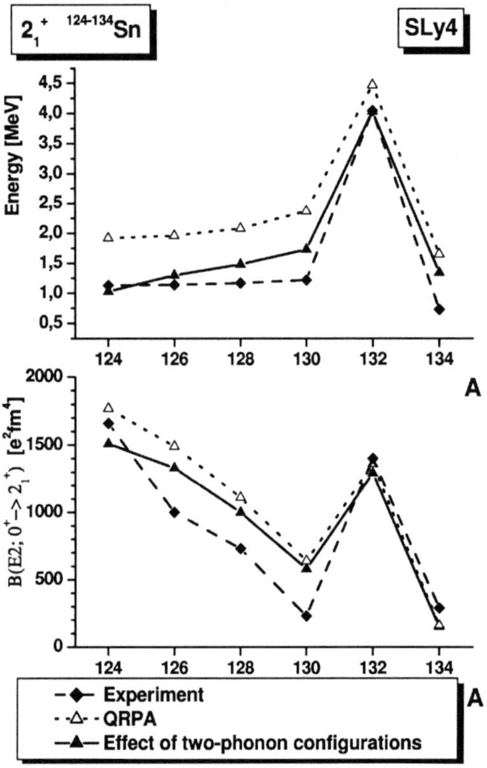

Figure 1. Energies and transition probabilities for Sn isotopes

to take into account the coupling between one- and two-phonon terms in the wave functions of excited states. The suggested approach enables one to reduce remarkably the dimensions of the matrices that must be diagonalized to perform structure calculations in very large configuration spaces. As an application of the method we have studied the behavior of the energies and transition probabilities to 2_1^+ states in $^{124-134}$Sn. The inclusion of the two-phonon configurations results in a decrease of the energies and a reduction of transition probabilities. It is shown that the effect of the two-phonon configurations is important, but this effect decreases in ^{132}Sn.

5. Acknowledgments

We are grateful to Prof. Ch.Stoyanov for valuable discussions and help. This work is partly supported by the IN2P3-JINR agreement.

References

1. A. Bohr and B. Mottelson, *Nuclear Structure* **vol.2** (Benjamin, New York, 1975).
2. V.G. Soloviev, *Theory of Atomic Nuclei: Quasiparticles and Phonons* (Institute of Physics, Bristol and Philadelphia, 1992).
3. D. Gogny, in *Nuclear Self-consistent Fields*, eds. G. Ripka and M. Porneuf (North-Holland, Amsterdam, 1975).
4. D. Vautherin and D.M. Brink, Phys. Rev. **C 5**, 626 (1972).
5. H. Flocard and P. Quentin, Ann. Rev. Nucl. Part. Sci. **28**, 523 (1978).
6. J. Dobaczewski, W. Nazarewicz, T.R. Werner, J.F. Berger, C.R. Chinn and J. Dechargé, Phys. Rev. **C 53**, 2809 (1996).
7. G. Colò, N. Van Giai, P.F. Bortignon and R.A. Broglia, Phys. Rev. **C50**, 1496 (1994).
8. G. Colò, N. Van Giai, P.F. Bortignon and M.R. Quaglia, Phys. Lett. **B485**, 362 (2000).
9. E. Khan, N. Sandulescu, M. Grasso and Nguyen Van Giai, Phys. Rev. **C66**,024309 (2002).
10. Nguyen Van Giai, Ch. Stoyanov and V.V. Voronov, Phys. Rev. **C57**,1204 (1998).
11. A.P. Severyukhin, Ch. Stoyanov, V.V. Voronov and Nguyen Van Giai, Phys. Rev. **C66**, 034304 (2002).
12. A.P. Severyukhin, V.V. Voronov, Ch. Stoyanov and Nguyen Van Giai, Nucl. Phys. **A722**, 123c, (2003).
13. A.P. Severyukhin, V.V. Voronov and Nguyen Van Giai, http://xxx.lanl.gov/nucl-th/0402096 (2004).
14. D.C. Radford et al., Phys. Rev. Lett. 88, 22501 (2002); Eur. Phys. J. **A 15**, 171 (2002); http:// www.phy.ornl.gov, HRIBF Newsletter, July 2003.
15. J. Terasaki, J. Engel, W. Nazarewicz and M. Stoitsov, Phys. Rev. **C66**, 054313 (2002).
16. G. Colò, P.F. Bortignon, D. Sarchi, D. T. Khoa, E. Khan and Nguyen Van Giai, Nucl. Phys. **A722**, 111c (2003).
17. G. Giambrone et al., Nucl. Phys. **A726**, 3 (2003).
18. Nguyen Van Giai and H. Sagawa, Phys. Lett. **106 B**, 379 (1981).
19. J.P. Blaizot and D. Gogny, Nucl. Phys. **A284**, 429 (1977).
20. G.F. Bertsch and S.F. Tsai, Phys. Reports **18 C**, 126 (1975).
21. E. Chabanat , P. Bonche, P. Haensel, J. Meyer and R. Schaeffer, Nucl. Phys. **A 635**, 231 (1998).
22. S. Raman, C.W. Nestor Jr. and P. Tikkanen, At. Data and Nucl. Data Tables **78**, 1 (2001).

KEY TOPICS IN NUCLEAR STRUCTURE
Proceedings of the 8th Int. Spring Seminar on Nuclear Physics
edited by Aldo Covello
© 2005 World Scientific Publishing Co.

GAMMA TRANSITIONS BETWEEN CONFIGURATIONS "QUASIPARTICLE ⊗ PHONON"

A. I. VDOVIN
Bogoliubov Laboratory of Theoretical Physics,
Joint Institute for Nuclear Research,
141980 Dubna, Russia E-mail: vdovin@thsun1.jinr.ru

N. YU. SHIRIKOVA
Laboratory of Information Technologies, Joint Institute for Nuclear Research,
141980 Dubna, Russia E-mail: shirik@thsun1.jinr.ru

A matrix element of the γ transition operator between the states of "quasiparticle ⊗phonon" type is evaluated and analyzed. A contribution of transitions of this type to the total $M2$ and $E3$ transition probabilities between states of different structure in several odd-mass spherical nuclei is computed. A role of these transitions appears to be somewhat stronger for magnetic transitions than for electric ones. Their effect is much more pronounced if states with dominating components of "quasiparticle⊗phonon" type are involved in the γ transition.

1. Motivation

A growing interest in structure of nuclear states at intermediate excitation energies is seen during last years. Excited states at intermediate energies are involved in processes of feeding and decay of isomeric states [1,2] as well as in two-step γ cascades following the thermal-neutron capture [3]. Structures of these states are already quite complicated. In odd-mass nuclei wave functions of many of them contain large components of a type "quasiparticle⊗phonon" or even more complicated ones. However, in calculations of γ excitation and decay of states at $E_x \geq 2 - 3$ MeV only single-particle and core polarization contributions are taken into account as a rule.

In the present report, we evaluate a matrix element of direct γ transition between "quasiparticle⊗phonon" components of nuclear excited states and estimate their contribution to γ transition rates in spherical odd-mass nuclei.

2. Transition matrix element

Our consideration is based on the quasiparticle-phonon model (QPM) which treats nuclear level structure in terms of interacting Bogoliubov quasiparticles and RPA phonons. A detailed description of the model can be found in the book by V. G. Soloviev [4] and the series of reviews [5,6,7].

We start with a standard one-body electromagnetic multipole operator $\mathfrak{M}(E/M, \lambda\mu)$ written in terms of the Bogoliubov quasiparticles α^+_{jm}, α_{jm} and RPA phonons $Q^+_{\lambda\mu i}$, $Q_{\lambda\mu i}$

$$\mathfrak{M}(E/M, \lambda\mu) = \frac{1}{\sqrt{2\lambda+1}} \sum_{j_1 j_2} F^{(\lambda)}_{j_1 j_2}(E/M) \left\{ v^{(\mp)}_{j_1 j_2} B(j_1, j_2; \lambda\mu) + \right.$$

$$\left. + \frac{1}{2} u^{(\pm)}_{j_1 j_2} \left(\psi^{\lambda i}_{j_1 j_2} \pm \varphi^{\lambda i}_{j_1 j_2}\right) \left[Q^+_{\lambda\mu i} \mp (-)^{\lambda-\mu} Q_{\lambda -\mu i} \right] \right\} \quad (1)$$

The RPA phonon operator is a linear superposition of forward- and backward going two-quasiparticle amplitudes

$$Q^+_{\lambda\mu i} = \frac{1}{2} \sum_{j_1 j_2} \left\{ \psi^{\lambda\mu}_{j_1 j_2} [\alpha^+_{j_1 m_1} \alpha^+_{j_2 m_2}]_{\lambda\mu} - (-)^{\lambda-\mu} \phi^{\lambda\mu}_{j_1 j_2} [\alpha_{j_2 m_2} \alpha_{j_1 m_1}]_{\lambda-\mu} \right\}.$$

In (1), $F^{(\lambda)}_{j_1 j_2}(E/M)$ is a reduced single-particle matrix element of multipole electric (E) or magnetic (M) operator; factors $v^{(\mp)}_{j_1 j_2}$ and $u^{(\pm)}_{j_1 j_2}$ are standard combinations of the Bogoliubov transformation coefficients. Square brackets $[\ \ ]_{\lambda\mu}$ stand for the coupling of single-particle angular momenta j_1, j_2 to the sum angular momentum λ. The operator $B(j_1, j_2; \lambda\mu)$ is defined as

$$B(j_1, j_2; \lambda\mu) = \sum_{m_1 m_2} (-)^{j_2+m_2} \langle j_1 m_1 j_2 m_2 | \lambda\mu \rangle \alpha^+_{j_1 m_1} \alpha_{j_2 -m_2}$$

The simplest trial wave function taking into account a coupling of quasiparticles and phonons in an odd-mass spherical nucleus reads

$$\Psi_\nu(J\mathcal{M}) = C_{J\nu} \{ \alpha^+_{J\mathcal{M}} + \sum_{\lambda_2 i_2 j_2} D^{\lambda_2 i_2}_{j_2}(J\nu) \left[\alpha^+_{j_2} Q^+_{\lambda_2 i_2} \right]_{J\mathcal{M}} \} |0\rangle, \quad (2)$$

where $|0\rangle$ is the ground state wave function of the even-even core-nucleus which is supposed to be a vacuum for phonon and odd-quasiparticle annihilation operators.

In principle, a wave function (2) is not appropriate for description of excited states which structures are dominated by "quasiparticle⊗phonon" ($|\alpha \otimes Q\rangle$) components. To this aim components of a type $|\alpha \otimes Q \otimes Q\rangle$ should be included in a trial wave function. Nevertheless, to get a qualitative estimation of the role of direct transitions $|\alpha_1 \otimes Q_1\rangle \to |\alpha_2 \otimes Q_2\rangle$ we shall use a wave function (2). In this way we suppose to get the upper limit for their role since a contribution of $|\alpha \otimes Q\rangle$ components to a trial wave function should decrease after adding more complicated components, i.e. after increasing of a phase space.

A general structure of matrix element of the operator (1) between an initial state $\Psi_\rho(IM)$ and a final state $\Psi_\nu(J\mathcal{M})$ is the following:

$$\langle \Psi_\nu(J\mathcal{M}) \| \; \mathfrak{M}(E/M,\lambda) \; \| \Psi_\rho(IM) \rangle = C_{J\nu} \; C_{I\rho} \; (M_{qp} + M_{ph} + M_{qph})$$

Above the term M_{qp} is a single-quasiparticle part of the total matrix element

$$M_{qp} = F_{JI}^{(\lambda)}(E/M) \; v_{JI}^{(\mp)}. \tag{3}$$

This part of the matrix element is due to the B-term of the operator (1).

In contrast with M_{qp}, the term M_{ph} is due to the second term (the Q-term) of the operator (1). It sums crossover transitions between single-quasiparticle $|\alpha\rangle$ and $|\alpha \otimes Q\rangle$ components of initial and final states. It reads

$$\begin{aligned} M_{ph} = \sum_i \Bigg[& \sqrt{\frac{2J+1}{2\lambda+1}} \; D_I^{\lambda i}(J\nu) \; \pm \; (-1)^{I-J+\lambda} \sqrt{\frac{2I+1}{2\lambda+1}} \; D_J^{\lambda i}(I\rho) \\ & + \sum_{\lambda_1 i_1 j_1} \sqrt{\frac{2J+1}{2\lambda+1}} \; D_{j_1}^{\lambda_1 i_1}(J\nu) \; \mathcal{L}_J^{\tau}(I\lambda i | j_1 \lambda_1 i_1) \\ & \pm (-1)^{I-J+\lambda_1} \sqrt{\frac{2I+1}{2\lambda+1}} \; D_{j_1}^{\lambda_1 i_1}(I\rho) \; \mathcal{L}_I^{\tau}(J\lambda i | j_1 \lambda_1 i_1) \Bigg] \mathfrak{R}^{\lambda i}, \end{aligned} \tag{4}$$

where

$$\mathfrak{R}^{\lambda i} = \sum_{j_1 j_2} u_{j_1 j_2}^{(\pm)} F_{j_1 j_2}^{(\lambda)}(E/M) \left(\psi_{j_1 j_2}^{\lambda i} \pm \varphi_{j_1 j_2}^{\lambda i} \right)$$

and

$$\mathcal{L}_I^{\tau}(j_1 \lambda_1 i_1 | j_2 \lambda_2 i_2) = \sqrt{2\lambda_1+1}\sqrt{2\lambda_2+1} \sum_{j_3}^{\tau} \begin{Bmatrix} j_2 & j_3 & \lambda_1 \\ j_1 & I & \lambda_2 \end{Bmatrix} \psi_{j_3 j_2}^{\lambda_1 i_1} \; \psi_{j_3 j_1}^{\lambda_2 i_2}.$$

Index τ at $\mathcal{L}^\tau$ and $\sum^\tau$ means that the summation is taken over only one of the two single-particle sets, either neutron, or proton, in correspondence with a type of the odd-quasiparticle. The matrix element M_{ph} (4) takes into account a core polarization or a contribution of virtual excitations of λ phonons of the even-even core. The function $\mathfrak{R}^{\lambda i}$ is just proportional to an excitation probability of a phonon $Q^+_{\lambda i}|0\rangle$ from the ground state $|0\rangle$ of the even-even core-nucleus.

We don't treat phonons and odd-quasiparticle as independent objects but take into account the Pauli principle corrections. It is done following the procedure from [8]. A function $\mathcal{L}^{\tau}_I(j_1\lambda_1 i_1|j_2\lambda_2 i_2)$ takes into account the Pauli principle corrections to overlapping of two different $|\alpha \otimes Q\rangle$ components.

The terms discussed so far, M_{qp} and M_{ph}, were usually included in calculations of γ transition rates whereas the third term, M_{qph}, was omitted. For the first time M_{qph} was analyzed in Ref. [9] where $M3$ transitions in near-spherical nuclei were studied.

For more detailed discussion it is convenient to divide the term M_{qph} into four parts as follows:

$$M_{qph} = \sum_{\substack{\lambda_1 i_1 j_1 \\ \lambda_2 i_2 j_2}} D^{\lambda_1 i_1}_{j_1}(I\rho) D^{\lambda_2 i_2}_{j_2}(J\nu)\ [(2I+1)(2J+1)]^{1/2} \times$$
$$\left[M^{\lambda_1 i_1 j_1}_{\lambda_1 i_1 j_2}(A) + M^{\lambda_1 i_1 j_1}_{\lambda_2 i_2 j_2}(B) + M^{\lambda_1 i_1 j_1}_{\lambda_2 i_2 j_1}(C) + M^{\lambda_1 i_1 j_1}_{\lambda_2 i_2 j_2}(D)\right] \quad (5)$$

The first term

$$M(A) = \delta_{\lambda_1 \lambda_2}\ \delta_{i_1 i_2}\ (-1)^{j_1 \pm I + \lambda_1} F^{(\lambda)}_{j_1 j_2}(E/M)\ v^{(\mp)}_{j_1 j_2} \begin{Bmatrix} I & \lambda & J \\ j_2 & \lambda_1 & j_1 \end{Bmatrix} \quad (6)$$

is rather like a single-quasiparticle term (3), the only difference is an additional geometric factor (6j-symbol). The term doesn't equal to zero only if both the components $|\alpha_1 \otimes Q_1\rangle$ and $|\alpha_2 \otimes Q_2\rangle$ contain the same phonon. Thus, the phonon resembles a "spectator" and the transition involves only quasiparticles.

The term $M(B)$ is the Pauli principle correction to the term $M(A)$

$$M(B) = \frac{1}{2}\sum_{j_3}{}^{\tau} (-1)^{j_3 + I - \lambda_1 - \lambda} F^{(\lambda)}_{j_3 j_1}(E/M) v^{(\mp)}_{j_3 j_1} \begin{Bmatrix} I & \lambda & J \\ j_3 & \lambda_1 & j_1 \end{Bmatrix} \mathcal{L}^{\tau}_J(j_3\lambda_1 i_1|j_2\lambda_2 i_2)$$
$$+ \begin{Bmatrix} \lambda_1 i_1 \Leftrightarrow \lambda_2 i_2 \\ j_1 \Leftrightarrow j_2 \end{Bmatrix} \quad (7)$$

In Eq.(7), the term in the second row is the same as the first term but with transposed groups of indexes j_1, λ_1, i_1 and j_2, λ_2, i_2, respectively.

In the term $M(C)$, a quasiparticle in the initial and final components should be the same and thus it is a "spectator". The transition is going between phonons

$$M(C) = \delta_{j_1 j_2}\ (-1)^{j_1-J+\lambda_2} \left\{ \begin{matrix} I & \lambda & J \\ \lambda_2 & j_1 & \lambda_1 \end{matrix} \right\} \sum_{j_3 j_4 \tau} F_{j_3 j_4}(E/M;\lambda) v^{(\mp)}_{j_3 j_4} \times \qquad (8)$$
$$V^{\lambda_1 i_1 \lambda_2 i_2}_{j_3 j_4}(E/M;\lambda).$$

Phonon structures appear through a function $V^{\lambda_1 i_1 \lambda_2 i_2}_{j_3 j_4}(E/M;\lambda)$, which reads

$$V^{\lambda_1 i_1 \lambda_2 i_2}_{j_3 j_4}(E/M;\lambda) = [(2\lambda_1+1)(2\lambda_2+1)]^{1/2} \sum_{j_5} \left\{ \begin{matrix} \lambda & \lambda_1 & \lambda_2 \\ j_5 & j_3 & j_4 \end{matrix} \right\} \times$$
$$\left[\psi^{\lambda_1 i_1}_{j_4 j_5} \psi^{\lambda_2 i_2}_{j_5 j_3} \pm \varphi^{\lambda_1 i_1}_{j_4 j_5} \varphi^{\lambda_2 i_2}_{j_5 j_3} \right].$$

The last term, $M(D)$, is the Pauli correction to $M(C)$. It contains both the functions $\mathcal{L}$ and V and is of the order $\sim \psi^4$

$$M(D) = \frac{1}{2} \sum_{\lambda_3 i_3} \left[(-1)^{j_1-J+\lambda_3} \left\{ \begin{matrix} I & \lambda & J \\ \lambda_3 & j_1 & \lambda_1 \end{matrix} \right\} \mathcal{L}_J(j_1 \lambda_3 i_3 | j_2 \lambda_2 i_2) \times \right.$$
$$\left. \sum_{j_3 j_4} F^{(\lambda)}_{j_3 j_4}(E/M)\ v^{(\mp)}_{j_3 j_4}\ V^{\lambda_1 i_1 \lambda_3 i_3}_{j_3 j_4}(E/M;\lambda) + \left\{ \begin{matrix} \lambda_1 i_1 \Leftrightarrow \lambda_2 i_2 \\ j_1 \Leftrightarrow j_2 \end{matrix} \right\} \right] \qquad (9)$$

3. Numerical results

To analyze a contribution of direct transitions $|\alpha_1 \otimes Q_1\rangle \to |\alpha_2 \otimes Q_2\rangle$ to total reduced probabilities of diffrent γ transitions as well as to compare the terms $M(A)$, $M(B)$, $M(C)$, $M(D)$ (6)–(9), we calculate reduced matrix elements of $M2$ and $E3$ transitions in several nuclei from region near closed shell $N = 82$. Although there are numerous experimental data on γ transition rates in these nuclei we avoid to compare them with the theory since this is not the aim of the present report. The $B(M2)$ values in odd-mass nuclei near closed shells Z=50 or N=82 were studied within the QPM in Refs. [10]. All the model parameters of the present study are taken from [10].

The results of calculation are presented in Table 1 (M2 transitions) and Table 2 (E3 transitions). Here, because of a limited size of the report, we present only a few typical examples although calculations are performed for a dozen of nuclides. A choice of initial and final states is stipulated by our wish to follow changes in M_{qph} with changing in structures of states involved in a γ transition. In accordance with the structure of states, $M2$

Table 1. The total $B(M2)$ factors and different parts (3)-(9) of the reduced matrix element for some $M2$ transitions in selected spherical odd-mass nuclei.

Nucleus	$B(M2)$	M_{qp}	M_{ph}	$M_{qph} = M(A) + M(B) + M(C) + M(D)$
		a) $M2$ transition: $11/2^-_1 \rightarrow 7/2^+_1$		
^{141}Pr	52.4	-48.0	20.2	0.33 = 0.42 - 0.13 + 0.07 - 0.03
^{143}Eu	42.1	-45.2	16.6	3.17 = 3.80 - 0.55 - 0.03 -0.05
^{149}Eu	36.2	-45.2	18.5	2.35 = 3.49 - 1.0 - 0.01 - 0.13
		b) $M2$ transition: $11/2^-_2 \rightarrow 7/2^+_1$		
^{141}Pr	$0.2{\cdot}10^{-2}$	-48.0	44.8	-3.3 = 6.2 + 7.0 - 22.0 + 5.5
^{143}Eu	$0.4{\cdot}10^{-5}$	-45.2	24.5	20.5 = -3.2 + 20.2 + 0.8 + 2.7
^{149}Eu	1.6	-45.2	-20.5	-6.8 = -18.6 + 9.7 + 0.7 + 1.4
		c) $M2$ transition: $11/2^-_2 \rightarrow 9/2^+_1$		
^{141}Pr	$0.2{\cdot}10^{-2}$	33.6	-27.7	33.0 = -9.9 - 56.6 + 0.7 + 98.8
^{143}Eu	6.9	43.2	-39.5	-1537 = -719 - 508 + 39 - 349
^{149}Eu	4.6	43.2	-36.4	-367 = -111 - 161 + 6 - 101

Note: $B(M2)$ – in units of μ_0^2 fm^2; matrix elements M_{qp}, M_{ph}, M_{qph} etc – in units of μ_0 fm. Effective gyromagnetic factors have the bare values.

and $E3$ transitions presented in the Tables are divided into three groups: a) $11/2^-_1 \rightarrow 7/2^+_1$ and $11/2^-_1 \rightarrow 5/2^+_1$; b) $11/2^-_2 \rightarrow 7/2^+_1$ and $11/2^-_2 \rightarrow 5/2^+_1$; c) $11/2^-_2 \rightarrow 9/2^+_1$. The structures of the states $11/2^-_1$, $7/2^+_1$ and $5/2^+_1$ are dominated by single-quasiparticle components, whereas the states $11/2^-_2$ and $9/2^+_1$ are predominantly of $|\alpha \otimes Q\rangle$ type[a].

In case when initial and final states have small admixtures of $|\alpha \otimes Q\rangle$ configurations (groups a) in Tables 1 and 2) a contribution of direct transitions between complex configurations to the total $B(M2)$ or $B(E3)$ values is also quite small. The changes of the reduced transition probabilities do not exceed 10-15%.

When one of the states coupled by γ transition appears to be of $|\alpha \otimes Q\rangle$ nature, a contribution of direct transitions between complex components increases as a rule. This is seen from the groups b) of the results in Tables 1 and 2. However, the resulting effect of increasing M_{qph} on $B(M2)$ and $B(E3)$ values is very different. The reason is that in $E3$ transitions $11/2^-_2 \rightarrow 5/2^+_1$ a dominating mechanism appears to be a core polarization,

[a]Of course, this component is different in different nuclei.

Table 2. The total $B(E3)$ factors and different parts (3)-(9) of the reduced matrix element for some $E3$ transitions in selected spherical odd-mass nuclei.

Nucleus	$B(E3)$	M_{qp}	M_{ph}	$M_{qph} = M(A) + M(B) + M(C) + M(D)$
		a) $E3$ transition: $11/2_1^- \to 5/2_1^+$		
^{141}Pr	$1.23 \cdot 10^4$	129	293	12.2 = 11.7 - 1.3 + 2.7 - 0.9
^{143}Eu	$4.3 \cdot 10^3$	66.4	178	9.1 = 8.5- 0.9 + 2.0 - 0.5
^{149}Eu	$1.2 \cdot 10^4$	66.4	359	18.0 = 16.3 - 2.2 + 5.4 - 1.5
		b) $E3$ transition: $11/2_2^- \to 5/2_1^+$		
^{141}Pr	$1.4 \cdot 10^2$	129	1853	15.0 = -48.9 + 2.4 + 12.8 + 48.7
^{143}Eu	$4.2 \cdot 10^4$	66.4	-12400	-90.1 = -138 + 5.0 + 18.0 + 24.5
^{149}Eu	$7.1 \cdot 10^4$	66.4	-6700	-41.9 = -87.9 + 8.3 + 12.3 + 25.4
		c) $E3$ transition: $11/2_2^- \to 9/2_1^+$		
^{141}Pr	4.5	-28.2	-137	-1604 = -374 - 579 - 843 + 192
^{143}Eu	0.5	-51.8	-122	603 = -99 + 75 + 614 + 13
^{149}Eu	3.0	-51.8	-332	90 = -103 + 44 + 168 - 19

Note: $B(E3)$ – in units of e$^2\cdot$fm^6; matrix elements M_{qp}, M_{ph}, M_{qph}, etc – in units of e$\cdot$fm^3.

i.e. the term M_{ph} is much larger than M_{qp} and M_{qph}. This is due to particular structures of the states $11/2_2^-$ and $5/2_1^+$ which contain a large component with admixture of the lowest collective octupole phonon $|\alpha \otimes 3_1^-\rangle$. At the same time changes of $B(M2)$ values are strong and irregular. Here, the M_{ph}, M_{qp} and M_{qph} terms appear to be of the same order, have different signs and hence their interference is quite unpredictable.

Transitions in groups c) connect almost pure $|\alpha \otimes Q\rangle$ states and M_{qph} is by one or two orders of magnitude larger than M_{qp} and M_{ph}. However, the total values of $B(M2)$ and especially $B(E3)$ are quite small.

Calculations don't reveal any special role of one of the four discussed mechanisms of transitions $|\alpha_1 \otimes Q_1\rangle \to |\alpha_2 \otimes Q_2\rangle$. In most cases, when M_{qph} is small or moderate, the first term $M(A)$ is the largest one although it is easy to find other examples in groups a) and b) of Tables 1 or 2. For transitions between pure $|\alpha \otimes Q\rangle$ states (groups c) several matrix elements are of the same order. So all the four matrix elements $M(A)$, $M(B)$, $M(C)$, $M(D)$ should be taken into account while calculating M_{qph}.

4. Conclusion

From the above consideration we conclude that analyzing a γ decay of excited nuclear states at energies $E_x \geq 2-3$ MeV one should take into account direct transitions between complex components of initial and final states. A corresponding contribution can be of importance for adequate understanding of properties of particular states. At the same time, the total contribution of such transitions to, e.g., intensities of γ cascades or other values "integrated" over more or less wide energy interval at intermediate excitation energies seems to be of minor importance because of small absolute values of corresponding transition rates. At least in γ transitions involving admixtures of low-lying collective vibrations, like $E2$ and/or $E3$, the core polarization mechanism should dominate. However, a situation with $E1$ and $M1$ transitions at intermediate excitation energies is still not clear and should be investigated.

Acknowledgments

The authors are grateful to A. Arefiev for collaboration and help in calculations.

References

1. V. Ponomarev, A. P. Dubenskiy, V. P. Dubenskiy, E. A. Boykova, *J. Phys.G: Part. Nucl. Phys.* **16**, 1727 (1990); A. P. Dubenskiy, V. P. Dubenskiy, E. A. Boykova, L. Malov, *Izv. AN SSSR, ser. fiz.* **54**, 1833 (1990).
2. N. Tsoneva, Ch. Stoyanov, Yu. P. Gangrsky, et al., *Phys. Rev.* **C61**, 044303 (2000).
3. S. T. Boneva, E. V. Vasilieva, Yu. P. Popov, A. M. Sukhovoj, V. A. Khitrov, *Sov. J. Part. Nucl.* **22**, 479 (1991).
4. V. G. Soloviev, *Theory of atomic nuclei: Quasiparticles and phonons* (Bristol and Philadelphia, Institute of Physics Publishing, 1992).
5. A. I. Vdovin, V. G. Soloviev, *Sov. J. Part. Nucl.* **14**, 99 (1983); V. V. Voronov, V. G. Soloviev, *Sov. J. Part. Nucl.* **14** 583 (1983).
6. A. I. Vdovin, V. V. Voronov, V. G. Soloviev, Ch. Stoyanov, *Sov. J. Part. Nucl.* **16**, 105 (1985).
7. S. Gales, Ch. Stoyanov, A. I. Vdovin, *Phys. Rep.* **166**, 125 (1989).
8. Chan Zuy Khuong, V. G. Soloviev, V. V. Voronov, *J. Phys. G: Nucl. Phys.* **7**, 151 (1985).
9. R. Lombard, A. I. Vdovin, A. V. Sushkov, N. Yu. Shirikova, *Nucl. Phys.* **A720**, 60 (2003).
10. W. Andrejtscheff, A. I. Vdovin, Ch. Stoyanov, *Nucl. Phys.* **A440**, 437 (1985); A. I. Vdovin, W. Andrejtscheff, Ch. Stoyanov, O. Rodriges, *Izv. AN SSSR, ser. fiz.* **49**, 2173 (1985).

KEY TOPICS IN NUCLEAR STRUCTURE
Proceedings of the 8th Int. Spring Seminar on Nuclear Physics
edited by Aldo Covello
© 2005 World Scientific Publishing Co.

COLLECTIVE MODES IN FAST ROTATING NUCLEI

J. KVASIL[1], N. LO IUDICE[2], R. G. NAZMITDINOV[3,4], A. PORRINO[2], F. KNAPP[1]

[1] *Institute of Particle and Nuclear Physics, Charles University, V.Holešovičkách 2, CZ-18000 Praha 8, Czech Republic*
[2] *Dipartimento di Scienze Fisiche, Universitá di Napoli "Federico II" and Istituto Nazionale di Fisica Nucleare, Monte S Angelo, Via Cinthia I-80126 Napoli, Italy*
[3] *Departament de Física, Universitat de les Illes Balears, E-07122 Palma de Mallorca, Spain*
[4] *Bogoliubov Laboratory of Theoretical Physics, Joint Institute for Nuclear Research, 141980 Dubna, Russia*

We report on a microscopic study of the electromagnetic response in fast rotating nuclei undergoing backbending with special attention at the orbital M1 excitations known as scissors mode. We find that the overall strength of the orbital M1 transitions evolves with the rotational frequency in parallel with the nuclear moment of inertia and, eventually, gets enhanced by more than a factor four above the critical backbending region. The physical implications of this result are discussed.

1. Introduction

The effects of fast rotation on the electric giant dipole resonance has been object of several investigations[1,2]. Less explored is its action on monopole and quadrupole resonances. We know only of two investigations, one carried out in cranked random-phase-approximation (CRPA)[3], the other within a phonon plus rotor model[4].

In a recent paper[5], we have completed the above studies by investigating the M1 excitations, with special attention at those of orbital nature generating the scissors mode[6,7,8]. Such a mode is tightly linked to deformation and nuclear rotation. Its properties are therefore expected to change considerably with increasing angular frequencies, especially in nuclei undergoing backbending, whose inertial parameters are strongly affected by fast rotation. We have considered ^{156}Dy and ^{158}Er as an illustrative example. The details of the procedure can be found elsewhere[5].

2. RPA in the rotating frame

We adopt the Hamiltonian

$$H_\Omega = H - \hbar\Omega\hat{I}_1 = H_0 - \sum_{\tau=n,p} \lambda_\tau N_\tau - \hbar\Omega I_1 + V. \quad (1)$$

The unperturbed term H_0 consists of the Nilsson Hamiltonian with a triaxial Harmonic oscillator potential whose frequencies satisfy the volume conserving condition $\omega_1\omega_2\omega_3 = \omega_0^3$. H_0 contains also a second piece which restores the local Galilean invariance broken in the rotating coordinate system[9].

The two-body potential has the following structure

$$V = V_{PP} + V_{QQ} + V_{MM} + W_{\sigma\sigma}. \quad (2)$$

V_{PP} is a proton-proton and neutron-neutron monopole pairing. V_{QQ}, V_{MM}, and $V_{\sigma\sigma}$ are, respectively, separable quadrupole-quadrupole, monopole-monopole, and spin-spin potentials.

All the multipole fields have good isospin T and signature r, according to the definition given in Ref.[10]. Monopole and quadrupole fields are expressed in terms of doubly stretched coordinates $x_i'' = (\omega_i/\omega_0)\, x_i$ [11] so as to fulfill the stability conditions $< \tilde{Q}_\mu >= 0$ $(\mu = 0, 1, 2)$. These ensure the separation of the pure rotational mode from the intrinsic excitations for a crancked harmonic oscillator [12].

By means of a generalized Bogoliubov transformation, we express the Hamiltonian given by Eq. (1) in terms of quasi-particle creation $(\alpha_i^\dagger)$ and annihilation (α_i) operators. We then solve the RPA equations of motion, written in the form[10]

$$[H_\Omega, P_\nu] = i\,\hbar\omega_\nu^2\, X_\nu, \qquad [H_\Omega, X_\nu] = -i\,\hbar\, P_\nu, \qquad [X_\nu, P_{\nu'}] = i\hbar\delta_{\nu\,\nu'}, (3)$$

where X_ν, P_ν are, respectively, the collective coordinates and their conjugate momenta. The solution of the above equations yields the RPA eigenvalues $\hbar\omega_\nu$ and eigenfunctions

$$|\nu >= O_\nu^\dagger|RPA> = \frac{1}{\sqrt{2}}\Big(\sqrt{\frac{\omega_\nu}{\hbar}}\, X_\nu - \frac{i}{\sqrt{\hbar\omega_\nu}}\,\hat{P}_\nu\Big)|RPA>$$
$$= \sum_{ij}\Big(\psi_{ij}^\nu b_{ij}^\dagger - \Phi_{ij}^\nu b_{ij}\Big)|RPA>, \quad (4)$$

where $b_{ij}^\dagger = \alpha_i^\dagger\alpha_j^\dagger$ $(b_{ij} = \alpha_i\alpha_j)$ creates (destroys) a pair of quasi-particles out of the RPA vacuum $| RPA\rangle$. Since the Hamiltonian can be decomposed into the sum of a positive ($H_\Omega(+)$) and a negative signature ($H_\Omega(-)$) terms we solve the eigenvalue equations (3) for each signature, separately.

The symmetry properties of the cranking Hamiltonian yield

$$[H_\Omega\,,\,N_{\tau=n,p}]_{RPA} = 0, \quad [H_\Omega\,,\,I_1]_{RPA} = 0 \tag{5}$$

and

$$[H_\Omega(-),\Gamma^\dagger] = \Omega\Gamma^\dagger, \tag{6}$$

where $\Gamma^\dagger = (I_2 + iI_3)/\sqrt{2\langle I_1\rangle}$ and $\Gamma = (\Gamma^\dagger)^\dagger = (I_2 - iI_3)/\sqrt{2\langle I_1\rangle}$ fulfill the commutation relation $[\Gamma,\Gamma^\dagger] = 1$. According to Eqs. (5), we have two Goldstone modes, one associated with the violation of the particle number operator, the other is a positive signature zero frequency rotational solution associated with the breaking of spherical symmetry. Eq. (6), on the other hand, yields a negative signature redundant solution of energy $\omega_\lambda = \Omega$, which describes a collective rotational mode arising from the symmetries broken by the external rotational field (the cranking term). Eqs. (5) and (6) ensure the separation of the spurious or redundant solutions from the intrinsic ones.

The strength function for an electric ($X = E$) or magnetic ($X = M$) transition of multipolarity λ from a state of the yrast line with angular momentum I is

$$S_{X\lambda}(E) = \sum_{\nu\, I'} B(X\lambda,\, I \to I',\, \nu)\, \delta(E - \hbar\omega_\nu), \tag{7}$$

where ν labels all the excited states with a given I'. In order to compute the reduced strength $B(X\lambda,\, I \to I',\, \nu)$ we should be able to expand the intrinsic RPA state into components with good K quantum numbers, which is practically impossible in the cranking approach. We compute, therefore, the strength in the limits of zero and high angular frequencies. For fast rotating nuclei, we assume a complete alignment of the angular momentum along the rotational x_1-axis.

The strength function method allows to avoid the explicit determination of RPA eigenvalues and eigenfunctions[10]. It also allows to obtain the n-th moments simply as

$$m_n(X\lambda) = \int_0^\infty E^n\, S_{X\lambda}(E)\, dE. \tag{8}$$

The $m_0(X\lambda)$ and $m_1(X\lambda)$ moments give, respectively, the energy unweighted and weighted summed strengths.

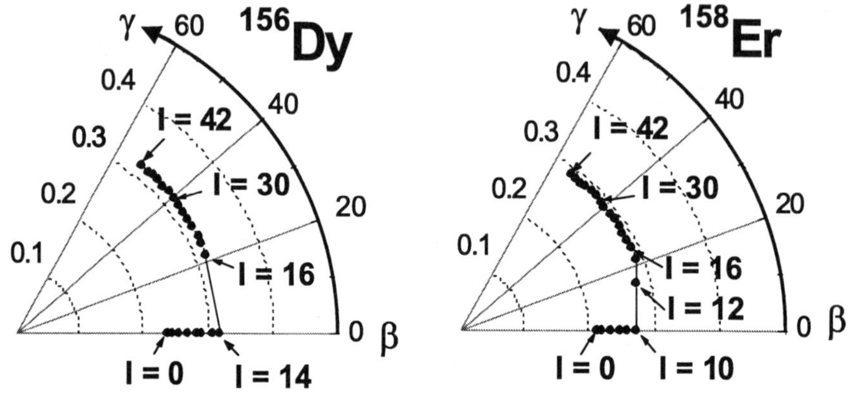

Figure 1. (Color online) Equilibrium deformations in β-γ plane as a function of the angular momentum.

3. Calculations and results

Our approach is not fully selfconsistent. Nonetheless, by using as input for our HB calculations the deformation parameters obtained from the empirical moments of inertia at each Ω [14], we were able to separate the spurious and rotational solutions from the intrinsic modes, to reproduce the experimental dependence of the lowest β and γ bands on Ω and, in particular, to observe the onset of triaxiality (Fig.1), as a result of the crossing of the γ with the ground band in correspondence [15].

Fast rotation does not affect dramatically the $E0$ mode. Its effects get manifest via the suppression of the high energy isovector peak, small in any case, and the appearance of a peak at $\sim 11 \div 12$ MeV, in correspondence with the $K = 0$ branch of the quadrupole resonance, as a result of the stronger coupling with the $K = 0$ quadrupole modes induced by fast rotation.

This has more appreciable effects on the quadrupole transitions (Fig. 2). It broadens considerably the isoscalar quadrupole giant resonance due to the increasing splitting of the different ΔI peaks with increasing Ω and washes out the isovector E2 resonance for the same reason.

At zero rotational frequency, the strength of the magnetic dipole transitions is concentrated in three distinct regions, consistently with the theoretical expectations and the experimental findings[8]. The low-energy interval, ranging from 2 to 4 Mev, is characterized by orbital excitations (scissors mode[6,7]). The high-energy one, located around 24 MeV, consists also of

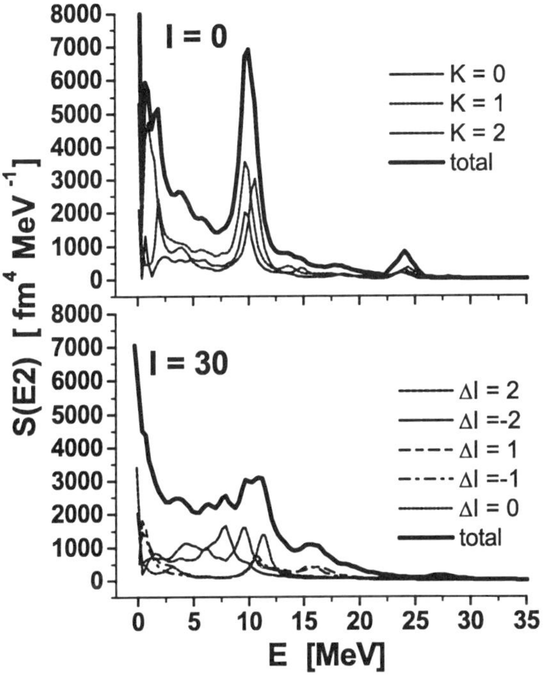

Figure 2. (Color online) E2 strength function at zero and high rotational frequencies in ^{158}Er.

orbital excitations (high energy scissors mode[16]). The intermediate region, ranging from 4 to 12 MeV, is due to spin excitations[17].

We will discuss the spectrum up to 12 MeV, since the effects of rotation are more dramatic in this region. Indeed, as shown in the lower panel of Fig. 3, the distribution of the strength changes considerably as Ω increases, to the point that the dominant peaks shift from 7-8 MeV down to 3 MeV.

The low-lying orbital strength (upper panel of Fig. 3) becomes larger and larger as Ω increases. At $\Omega = 0$, the orbital peaks are small compared to the spin transitions (middle panel) which are dominant in the M1 spectrum. At $I = 30\hbar$, instead, the orbital spectrum covers a wider energy range. Furthermore, it gets magnified, especially in the low-energy sector, where we obtain quite high peaks. The low-lying orbital strength increases by more than a factor six due to fast rotation. One may also observe that the $\Delta I = 0$ transitions, absent at zero frequency ($\Delta I = K = 0$), give a small but nonzero contribution which increases with Ω. This is due to a new

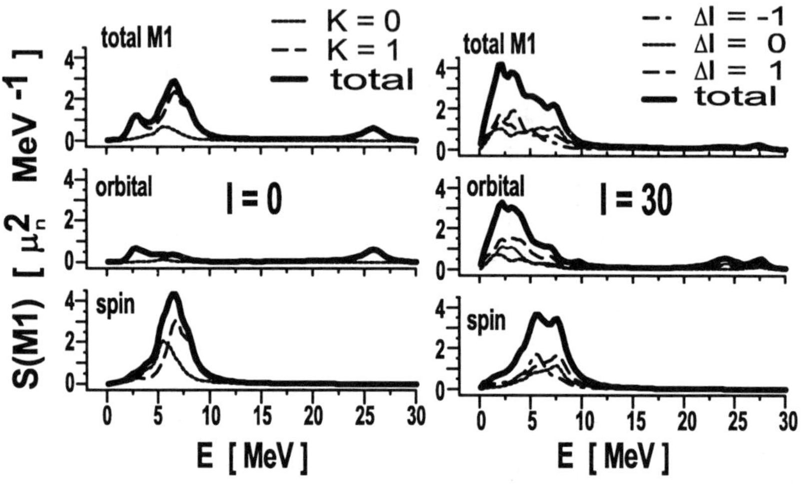

Figure 3. (Color online) Orbital, spin, and total M1 reduced strength distributions at zero (left-hand panels) and high rotational frequencies (right-hand panels) in ^{158}Er.

branch of the scissors mode which arises with the onset of triaxiality[18,19].

We can identify one of the mechanisms responsible for such a large enhancement by comparing (Fig. 4) the Ω behavior of the orbital and total $m_1(M1)$ moments with the corresponding evolution of the kinematical moment of inertia $\Im = I/\Omega$, computed using the cranking method of Ref.[20]. The strikingly similar behavior of the orbital $m_1(M1)$ and the moment of inertia shows that the two quantities are closely correlated at all rotational frequencies. Indeed, at zero frequency, one has the M1 EWSR[21]

$$m_1^{(sc)}(M1) = \frac{3}{8\pi}(1-b)\Im_{rig}\omega_0^2\delta^2, \tag{9}$$

where $b = \kappa(1)/\kappa(0)$ and $\Im_{rig} = 2/3\ mA < r^2 >$. The link is even more explicit in the M1 summed strength. For both low and high energy modes, we have the general form[8]

$$m_0^{(\pm)}(M1) = \sum_{n_\pm} B_{n\pm}^{(\pm)}(M1) \simeq \frac{3}{16\pi}\Im_{sc}^{(\pm)}\bar{E}^{(\pm)}, \tag{10}$$

where $\bar{E}^{(\pm)}$ and $\Im_{sc}^{(\pm)}$ denote the energy centroids and the mass parameters of the high-lying $(+)$ and low-lying $(-)$ scissors modes.

At high energy, protons and neutrons behave as normal irrotational fluids at any rotational frequency.. For the low energy mode, instead, we must distinguish between zero and high rotational frequencies. At zero

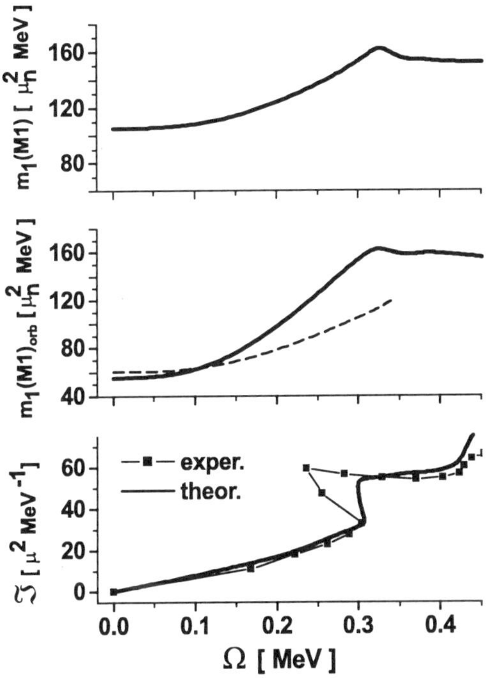

Figure 4. (Color online) Total (top panel), orbital (middle panel) $m_1(M1)$ moments and the kinematical moment of inertia (bottom panel) versus Ω in ^{158}Er. The dashed line in the middle panel displays the M1 EWSR

frequency, protons and neutrons behave as superfluids, so that[22,23] the summed strength $m_0(M1)$ follows the quadratic deformation law found experimentally[24]. At high rotational frequency, instead, the pairing correlations are quenched, so that protons and neutrons behave basically as rigid rotors. We can, therefore, distinguish two different regimes, one below the backbending critical frequency and the other above. Below backbending, while the quasi-particle energy moves downward due to the weakening of pairing, the $M1$ strength increases with Ω due to the increasing axial deformation and the smooth enhancement of the moment of inertia. Above the backbending critical value, when the nucleus undergoes a transition from a superfluid to an almost rigid phase, as a result of the alignment of few quasi-particles with high angular momenta, the $M1$ strength jumps to a plateau, due to a sudden increase of the moment of inertia, while the deformation parameter δ remains practically constant.

Also the onset of triaxiality raises $m_1(M1)$ at high rotational frequency, to a modest extent. A further contribution comes from the changes in the shell structure induced by fast rotation. This, indeed, enhances the number of configurations taking part to the motion over the whole energy range. The new configurations generate new transitions on one hand, and, on the other hand, enhance the amplitudes of collective as well as non collective transitions.

4. Conclusions

Our analysis shows that fast rotation strengthens the coupling between quadrupole and monopole modes, broadens appreciably the isoscalar quadrupole giant resonance and washes out the isovector monopole and quadrupole peaks. These effects are found to be more appreciable than the ones predicted previously[3].

The most meaningful and intriguing result of our calculation concerns the orbital, scissors-like, $M1$ excitations. The enhancement of the overall $M1$ strength at high rotational frequencies emphasizes the dominant role of the scissors mode over spin excitations in fast rotating nuclei and represents an additional signature for superfluid to normal phase transitions in deformed nuclei. If confirmed experimentally, this feature would provide new information on the collective properties of deformed nuclei.

Acknowledgments

This work was partly supported by the Czech grant agency under the contract No. 202/02/0939 and the Italian Ministero dell'Istruzione, Universitá and Ricerca (MIUR).

References

1. K. A. Snover, Ann. Rev. Nucl. Part. Sci. **36**, 545 (1986) and references therein.
2. J.J. Gaardhøje, Ann. Rev. Nucl. Part. Sci. **42**, 483 (1992) and references therein.
3. Y.R. Shimizu and K. Matsuyanagi, Progr. Theor. Phys. **72**, 1017 (1984); *ibid* **75**, 1167 (1986).
4. S. Aberg, Nucl. Phys. A **473**, 1 (1987).
5. J. Kvasil, N. Lo Iudice, R.G. Nazmitdinov, A. Porrino, F. Knapp Phys. Rev. C **69**, 064308 (2004).
6. N. Lo Iudice and F. Palumbo, Phys. Rev. Lett. **41**, 1532 (1978).

7. D. Bohle, A. Richter, W. Steffen, A.E.L. Dieperink, N. Lo Iudice, F. Palumbo, and O. Scholten, Phys. Lett. B **137**, 27 (1984).
8. For an exhaustive list of reference N. Lo Iudice, Rivista Nuovo Cimento **9**, 1 (2000).
9. T. Nakatsukasa, K. Matsuyanagi, S. Mizutori, and Y.R. Shimizu, Phys. Rev. C **53**, 2213 (1996).
10. J. Kvasil, N. Lo Iudice, V.O. Nesterenko, and M. Kopal, Phys. Rev. C **58**, 209 (1998).
11. T. Kishimoto, J. M. Moss, D.H. Youngblood, J.D. Bronson, C.M. Rozsa, D.R. Brown, and A.D. Bacher, Phys. Rev. Lett. **35**, 552 (1975).
12. R.G. Nazmitdinov, D. Almehed, and F. Dönau, Phys. Rev. C **65**, 041307(R) (2002).
13. R. Wyss, W. Satula, W. Nazarewicz, and A. Johnson, Nucl. Phys. A **511**, 324 (1990).
14. R.Ch. Safarov and A.S. Sitdikov, Izv. A.N. **63**, 162 (1999) and references therein.
15. see for instance J. Kvasil and R.G. Nazmitdinov, Phys. Rev. C **69**, 031304 (2004).
16. N. Lo Iudice and A. Richter Phys. Lett. B **228**, 291 (1989).
17. A. Richter, Nucl. Phys. A **553**, 417c (1993)
18. F. Palumbo and A. Richter, Phys. Lett. B **158**, 101 (1985).
19. N. Lo Iudice, E. Lipparini, S. Stringari, F. Palumbo, and A. Richter, Phys. Lett. B **161**, 18 (1985).
20. D. Almehed, F. Dönau, and R.G. Nazmitdinov, J. Phys. G: Nucl. Part. Phys. **29**, 2193 (2003).
21. N. Lo Iudice, Phys. Rev. C **57**, 1246 (1998).
22. N. Lo Iudice and A. Richter Phys. Lett. B **304**, 193 (1993).
23. N. Pietralla, P. von Brentano, R.-D. Herzberg, U. Kneissl, N. Lo Iudice, H. Maser, H. H. Pitz, and A. Zilges, Phys. Rev. C **58**, 184 (1998).
24. W. Ziegler, C. Rangacharyulu, A. Richter, and C. Spieler, Phys. Rev. Lett. **65**, 2515 (1990).

KEY TOPICS IN NUCLEAR STRUCTURE
Proceedings of the 8th Int. Spring Seminar on Nuclear Physics
edited by Aldo Covello
© 2005 World Scientific Publishing Co.

RECENT EXPERIMENTS ON PARTICLE-ACCOMPANIED FISSION

M. MUTTERER [1], Yu.N. KOPATCH [2,3], P. JESINGER [1,4], A.M. GAGARSKI [5], M. SPERANSKY [3], V. TISHCHENKO [6], F. GÖNNENWEIN [4], J. v. KALBEN [1], S.G. KHLEBNIKOV [7], I. KOJOUHAROV [2], E. LUBKIEWICS [8], Z. MEZENZEVA [3], V. NESVISHEVSKY [9], G.A. PETROV [5], H. SCHAFFNER [2], H. SCHARMA [10], D. SCHWALM [11], P. THIROLF [12], W.H. TRZASKA[13], G.P. TYURIN [7], and H.-J. WOLLERSHEIM [2].

[1]*Institut für Kernphysik, Techn. Universität, 64289 Darmstadt, Germany*
[2]*Gesellschaft für Schwerionenforschung mbH, 64291 Darmstadt, Germany*
[3]*Frank Laboratory of Neutron Physics, JINR, 141980 Dubna, Russia*
[4]*Physikalisches Institut der Universität, 72076 Tübingen, Germany*
[5]*Petersburg Nuclear Physics Institute, 188300 Gatchina, Russia*
[6]*Flerov Laboratory of Nuclear Reactions, JINR, 141980 Dubna, Russia*
[7]*V.G. Khlopin Radium Institute, 194021 St. Petersburg, Russia*
[8]*Institute of Physics, Jagiellonian University, 30059 Cracow, Poland*
[9]*Institute Laue-Langevin, 38042 Grenoble-Cedex, France*
[10]*Forschungszentrum Rossendorf, 01314 Dresden, Germany*
[11]*Max-Planck-Institut für Kernphysik, 69115 Heidelberg, Germany*
[12]*Sektion Physik, Universität München, 85748 Garching, Germany*
[13]*Department of Physics, Jyväskylä University, 40351 Jyväskylä, Finland*

The rare ternary fission (TF) process has been studied in a number of correlation experiments that have included registration of neutrons and γ-rays along with the light charged particles (LCPs) or coincident emission of two LCPs. Highly efficient detector setups have permitted, for the fission reactions ^{252}Cf(sf) and ^{233,235}U(n_{th},f), to identify the population of excited states in LCPs, the formation of neutron-unstable nuclei as short-lived intermediated LCPs, as well as the sequential decay of particle-unstable LCP species into charged particle pairs. Quaternary fission with an apparently independent emission of two charged particles has also been observed. The applied technologies are briefly summarized, and particular results are presented and discussed.

1. Introduction

In the ternary fission (TF) process, nucleons from the neck structure evolving right at scission between the forthcoming fission fragments (FF) cluster into light charged particles (LCPs) which are subsequently ejected at about right angle with respect to the fission axis. Although TF is a rare process ($\approx 1/260$ relative to binary fission, for ^{252}Cf), and is consequently difficult to measure, it provides one of the few means to the experimentalist to explore the behaviour of the fissioning system near the instant of scission. In about 87% of TF events so-called long-range α particles (LRA) are present, with $\simeq 16$ MeV mean energy. Helium isotopes (^{4,6,8}He) and hydrogen isotopes (^{1,2,3}H) are responsible for nearly 97% of the total ^{252}Cf TF yield, heavier LCP species being emitted very rarely.

Besides measuring LCP yields, former TF studies concerned angular and energy correlation experiments between the LCPs and FFs, aiming to achieve a complete kinematical description of the three-body break-up. Detailed multiparameter studies, performed during the last 20 years, were concentrated mainly on the relatively abundant α-TF mode and a few fissioning systems, (^{235}U(n_{th},f), ^{239}Pu(n_{th},f)) and, predominantly, ^{252}Cf(sf) (see Refs. [1,2] for an overview). These investigations were inherently limited to the detection of stable or β-radioactive LCP species.

In the present paper, we describe a number of recent more elaborate correlation experiments that either include the registration of neutrons and γ-rays with LCPs and FFs, or the coincident registration of two LCPs. It will be shown that these measurements have permitted to identify several new modes of particle-accompanied fission, such as the population of excited states in LCPs (e.g. in ^{10}Be [3]), the formation of neutron-unstable nuclei (e.g. ^{5}He) as short-lived intermediate LCPs, as well as the sequential decay of particle-unstable LCPs (e.g. ^{8}Be). "Quaternary" fission (QF) with the simultaneous but apparently independent creation of two charged particles right at scission has also been observed [4].

2. Energy Correlations in the Ternary Fission of ^{252}Cf(sf)

In a first experiment, the Darmstadt-Heidelberg 4π-NaI(Tl) Crystal Ball (CB) spectrometer was applied for measuring fission γ- rays [5] (with $\geq 90\%$ efficiency) and neutrons [6] (with $\simeq$ 60% efficiency). The ^{252}Cf sample and the detector system "CODIS" for the FFs and LCPs were mounted at the center of the CB. The set of measured parameters has allowed to determine, for each fission event, the following quantities and their mutual correlations: fragment masses and kinetic energies; multiplicity and angular distribution of fission neutrons; multiplicity, energy and angular distributions of fission γ-rays; energy, nuclear charge (mass), and emission angle of the LCP from ternary fission.

From the kinematical data and the multiplicity of emitted neutrons the fragment total excitation energies TXE could be deduced, for various ternary fission modes with LCPs up to carbon. It turns out that LCP emission proceeds in expense of a considerable amount of TXE (35 MeV, on average, for binary ^{252}Cf fission), with the required energy for particle emission increasing with LCP mass and energy. As an example, the average TXE decreases from 27 MeV to 15 MeV when instead of an α-particle a ternary C-isotope is emitted. In this sense TF with emission of heavier LCPs features a rather cold large-scale rearrangement of nuclear matter. From the sequence of data from α-TF to C-TF there is also clear evidence that there is a pronounced preformation of the FFs right at scission dominated by the well-known double-magic shells, while the TXE of the entire system is spent preferably for excitation of the limited amount of neck nucleons [7]. This observation is in accordance with a recent interpretation of measured LCP yields from TF in terms of nuclear thermometry [8]. In this context also, precise measurements of the yield ratios of excited and ground states for the same LCP could be a useful source of information on the scission-point energetics, e.g. the nuclear temperature. The strong influence especially of the A = 132 (Z = 50, N = 82) double-magic shell on the formation of the heavy mass group of fission fragments, seems to set a natural limit for the LCP mass to about 42 off-shell nucleons available in the supposed neck configuration of ^{252}Cf(sf). The ^{132}Sn shell stabilizations makes also the fragment mass distributions in TF narrower with increasing mass of the LCPs, as is observed in the experiment for ^{10}Be and ^{14}C accompanied fission. At the supposed LCP mass limit around $A_{LCP} = 42$ the FF mass peaks are expected to shrink to quite narrow lines, with the heavy one closely approaching ^{132}Sn.

3. Intermediate ^{5}He, ^{7}He and ^{8}Li* LCPs in ^{252}Cf(sf)

In the CB experiment, the neutron-unstable odd-N isotopes ^{5}He, ^{7}He and ^{8}Li* (in its excited state of $E^* = 2.26$ MeV) were identified to show up as intermediate LCPs in TF of ^{252}Cf [6]. The emergence of the ternary ^{5}He and ^{7}He particles (lifetimes: $1 \times 10^{-21} s$, and $4 \times 10^{-21} s$, respectively) as LCPs was disclosed from the measured angular distributions of their decay neutrons focused by the emission in flight towards the direction of motion of ^{4}He and ^{6}He ternary particles (see Fig. 1). Previously, only ternary ^{5}He emission was observed by analyzing relative neutron intensities measured at forward and backward angles with respect to the α particles [9,10]. In the present work, due to the high counting efficiency, neutrons were also observed to be correlated with the rather rare ^{6}He and Li LCPs. The neutrons peaked around Li-particle motion are attributed to the decay of the second excited state at $E^* = 2.26$ MeV (lifetime: 2×10^{-20} s) in ^{8}Li. The fractional yields of the ^{5}He and ^{7}He TF modes relative to "true" ternary ^{4}He and ^{6}He TF, respectively, were determined to be 0.21(5) for both cases. The surprisingly high yields for these exotic clusters indicate that they are formed inside nuclear matter in a similar manner as the stable species [6]. We note that the formation of ^{5}He in ^{252}Cf fission has the second highest yield among all LCPs, being only superseded (by a factor of $\simeq 5$) by ^{4}He emission, but downgrading ^{3}H (by a factor of $\simeq 2$) to the third most-abundant LCP. In fact, 17(4)% of all long-range α- particles from ^{252}Cf fission are actually residues from the ^{5}He breakup reaction. The population of ^{8}Li* was deduced to be 0.06(2), relative to Li ternary fission, and 0.33(20) relative to the yield of particle stable ^{8}Li.

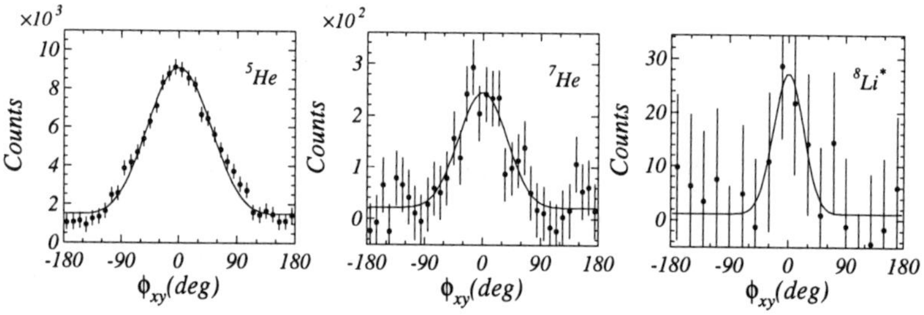

Figure 1. Angular distribution of neutrons from the decay of ^{5}He, ^{7}He and ^{8}Li* LCPs (from left to right) with respect to the direction of motion of ^{4}He, ^{6}He and Li residues. The data represent the projections of the measured angular distributions on the plane perpendicular to the fission axis. The solid lines are the corresponding distributions deduced from trajectory calculations. The more frequent prompt fission neutrons emanating from the fission fragments are already subtracted. (for details see Ref. [6]).

In the analysis the measured neutron angular distributions could be well reproduced by trajectory calculations (see Fig. 1) and, thus, a rather representative picture of the fragment configuration at a very short time after the system's separation could be established [6]. It is worthwhile to note that ternary fission with the emission of neutron-unstable LCPs provides a source of neutrons that are emitted at about right angles to the fission axis, the dominant part coming from ^{5}He, with about one neutron in every 1500 binary fission events. This source of neutrons thus may mimic the search for so-called scission neutrons [11] thought to be related to the binary fission process.

4. γ-Ray Emission in ^{252}Cf(sf) Ternary Fission

The CB experiment has yielded, in spite of the limited energy resolution of NaI(Tl), important information on γ-ray emission in ternary fission. Special issues, partly unexpected ones, concern the so-called "high-energy γ-ray component" in the fission γ-ray spectrum [12], and γ-ray angular anisotropy [5].

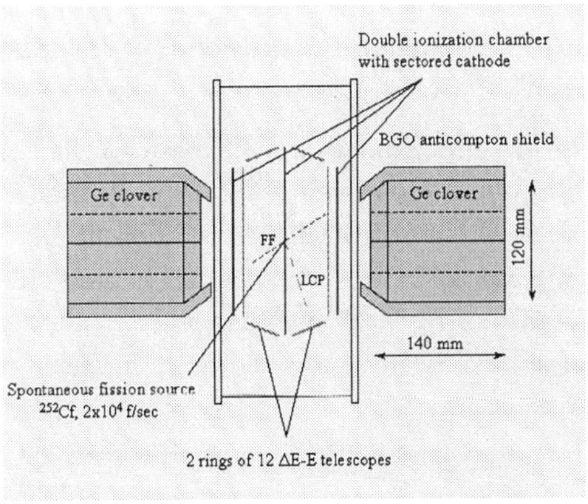

Figure 2. Experimental setup for γ-ray spectroscopy in ^{252}Cf ternary fission. The central part is the FF and LCP detector system CODIS2, contained in a cylindrical vessel filled with 570 torr methane as the counting gas. The two large segmented Super Clover Ge detectors on both sides of CODIS2 are equipped with BGO anticompton shields.

In a second experiment performed at the GSI Darmstadt in summer 2002, two segmented large-volume Super-Clover Ge-detectors combined with the improved detection system "CODIS2" (see Fig. 2) have permitted high-resolution γ-ray spectroscopy of FFs and LCPs in-flight after Doppler correction [13]. CODIS2 is the successor of CODIS, having a similar Frisch-gridded 4π twin ionization chamber (IC) for measuring FF energies, with a cathode divided into sectors for deducing FF emission angles, and two rings of LCP detectors with 12 ΔE-E_{res} telescopes each. Compared to CODIS, several modifications have been made for the FF IC to accept the higher counting rate (2×10^4 fissions/s) and for the LCP telescopes to improve mass and nuclear charge resolution. The VEGA segmented Super Clover

Ge detectors used in the experiment are among the largest Ge detectors in the world consisting of 4 Ge crystals, each one 14 cm in length and 6 cm in diameter.

From this experiment, angular distributions of individual γ-rays in ternary (and binary) fission may be deduced for providing further information on the fragment spins and their alignment. By studying γ-rays from LCPs, the population of their excited states becomes more generally accessible. Furthermore, new data on isotopic LCP yields in ^{252}Cf fission are obtained due to the outstanding resolution of the LCP telescopes in CODIS2 [13].

5. Sequential LCP Decay and Quaternary Fission of ^{252}Cf(sf) and ^{233,235}U(n_{th},f)

Another type of recent experiments was devoted to the coincident registration of two LCPs in one fission event. Quaternary fission in ^{252}Cf(sf) [4,14] and ^{233,235}U(n_{th},f)[15] was studied with two different experimental setups. Not surprisingly QF is still much rarer than TF, and this is probably the reason why in 50 years of research into particle-accompanied fission only a few QF experiments have been conducted, e.g., in Refs. [16,17,18].

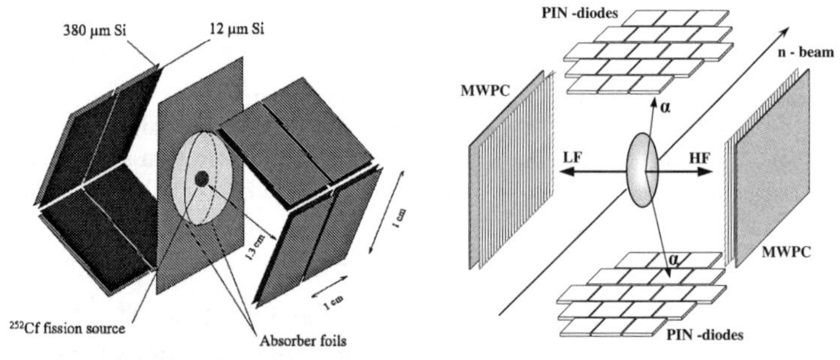

Figure 3. **left:** Experimental setup for measuring α-α and α-t coincidences in ^{252}Cf. LCP identification is performed by a ΔE-E_{rest} measurement. The source was covered with 20 μm kapton for protecting the 8 detector telescopes from ^{252}Cf α's. (Ref. [4]). **right:** Experimental setup for measuring α-α, α-t, and t-t coincidences in ^{233,235}U(n_{th},f). Here, pulse-shape analysis of the current signals from the 38 Si detectors was used to separate H isotopes from α-particles. (Ref. [15])

Figure 3 (left) sketches the setup for the ^{252}Cf(sf) measurement. With a ^{252}Cf source of 5×10^3 fissions/s a total of 255 α-α and 37 α-t coincidences were registered in four weeks of measurements. Figure 3 (right) shows the setup for the measurements on ^{233,235}U(n_{th},f) performed at the cold neutron beam line PF1 located at the high-flux reactor of the ILL Grenoble, France. With a beam intensity of 6×10^8 neutrons/$(s \cdot cm^2)$ a binary fission rate of more than 10^6 fissions/s could be achieved, and, hence, ternary particle rate on the arrays of semiconductor detectors was almost 10^3 /s. Thus, besides α-α and α-t coincidences events with t-t coincidences could be registered (see Fig. 4 (right)). In both experiments, angular distributions and correlations of two light charged particles accompanying the two main fission fragments were measured. Likewise the energy spectra of the LCPs could be taken.

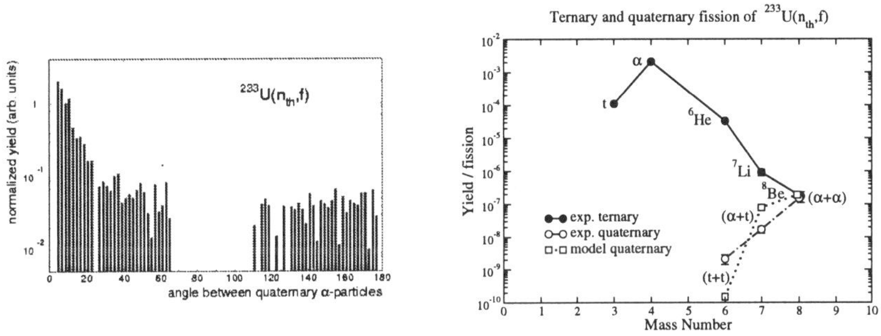

Figure 4. **left:** Distribution of mutual angle between the two α-particles in quaternary fission of ^{233}U(n_{th},f) (preliminary data), measured at an intense cold neutron beam at the ILL Grenoble. The experimental setup is sketched in Fig. 3. **right:** Quaternary yields in ^{233}U(n_{th},f), relative to the corresponding yields of the composite ternary particles. In the model calculation the law of mass action was assumed to be valid for the number of nucleons forming the LCPs. The temperature being requested for applying the law was deduced from the measures yield ratio ^{7}Li*/^{7}Li.

Two LCPs (mainly either two α-particles, or an α and a triton) in one fission event can originate either from an independent emission of the two LCPs ("true" quaternary fission) or from the break-up of particle-unstable species among the heavier LCPs (also called "pseudo" quaternary fission). In the latter case short-lived particle-unstable LCP species are decaying, similar to the neutron-unstable LCPs like ^{5}He, close to the fissioning nucleus and, thus, escape from direct observation. The most prominent example is ^{8}Be which is disintegrating into two α-particles, with $T_{1/2} = 0.07$ fs from its ground state, and with $T_{1/2} = 3 \times 10^{-22}$s from the 3.13 MeV excited state The pseudo α-t QF most probably arises from the sequential decay of ^{7}Li LCPs in the second excited $7/2^-$ state, which compared to the ground state and first excited state decomposes into an α-particle and a triton. In the present experiments, the two varieties of QF have been differentiated from one another by exploiting the different patterns of angular correlations between the two charged LCPs. In the example shown in Fig. 4 (left), isotropically distributed angles are due to true QF, while the events peaking at zero angle are related to pseudo QF. Yields (see Fig.4 (right)) and energy distributions of LCPs for each of the two processes were obtained here for the first time in one and the same experiment [4,15]. As to the yields, it is remarkable that for all types of QF the yields observed for $^{233,235}U(n_{th},f)$ are roughly an order of magnitude lower than for the heavier ^{252}Cf nucleus.

6. Energy Correlations in Ternary and Quaternary Fission of $^{235}U(n_{th},f)$

Very recently a first multi-parameter experiment on $^{235}U(n_{th},f)$ particle-accompanied fission was successfully finished, where a thin Uranium sample at the center of the CODIS2 detection system (Fig. 2) was faced with cold neutron from the PF1 beam at the high-flux reactor of the ILL Grenoble, France [19]. With a beam intensity of 3×10^9 neutrons/$(s \cdot cm^2)$ a binary fission rate of 2.5×10^5 fissions/s could be achieved and, thus, ternary fission studied with a previously unachieved statistical accuracy, while maintaining high resolution for the measurement of both, the LCPs and FFs.

It is of utmost interest to study how the FF mass distribution develops with the emission of heavier LCPs in $^{235}U(n_{th},f)$ where the supposed neck configuration at scission is reduced to $\simeq$ 26 nucleons compared to $\simeq$ 42 in $^{252}Cf(sf)$ (see Sect.2). One thus expects rather narrow ternary FF mass peaks already for medium mass LCPs (around $A_{LCP} = 12$), and the heavy mass peak closely approaching $A = 132$. Compared to $^{252}Cf(sf)$, $^{235}U(n_{th},f)$ TF is, on the one hand, easier to measure because of the low yield of α radioactivity. However, ternary fission yields of heavier LCPs are generally lower by about one order of magnitude, requiring a high source strength as was realized in the experiment. Furthermore, the determination of the yield - TXE relation is very useful for the theory of ternary fission, which establishes the connection between LCP emission probabilities and the configuration (and behaviour) of the fissioning system at scission.

The high fission rate achieved with CODIS2 and the granularity of the LCP detectors have enabled to register, for the first time, fission fragment parameters correlated with the α-α and α-t QF events. It is anticipated that the information on the correlated fragment mass and TXE distributions may help enlighten our understanding of this rare and rather complex particle-accompanied fission mode.

7. Summary and Outlook

The described experimental studies on ^{252}Cf(sf) and ^{233,235}U(n$_{th}$,f) have revealed new aspects of the TF process and, although not being yet fully exploited, may provide new insight into the exit channel of fission, more generally. For ^{252}Cf(sf), the fission fragment data from α-TF to C-TF give a hint of a pronounced preformation of the FFs right at scission due to double-magic shells (mainly A = 132), while the TXE at scission preferably concentrates in the deformable neck. The new measurement on ^{235}U(n$_{th}$,f) has aimed at confirming this view. From the ^{252}Cf experiment on high-resolution γ-ray spectroscopy of FFs and LCPs de-excitation (Sec. 4), angular distributions of individual γ-rays in ternary (and binary) fission may be deduced for providing information on the formation of fragment spins and their alignment. By studying γ-rays from LCPs, the population of their excited states becomes more generally accessible.

Ternary fission of actinides is a source of a large variety of light neutron-rich nuclei not being limited, as demonstrated, to stable or β-radioactive LCP species. LCPs are also born in excited states, and precise measurements of the yield ratios of excited and ground states for the same LCP could be a useful source of information on the scission-point energetics, e.g. the nuclear temperature. Very similar to the neutron-unstable LCPs observed in ^{252}Cf(sf), such as ^{5}He, ^{7}He and ^{8}Li*, decaying before detection into a charged particle and a neutron, there exist also particle-unstable LCPs decaying with short lifetimes into charged particle pairs. The most prominent example for such an LCP is ^{8}Be disintegrating both, from its ground and excited state, into two α-particles. Besides this basically ternary decays turned quaternary in a sequential process, there is also true QF with the independent emission of two charged particles right at scission. In the new experiment on ^{235}U(n$_{th}$,f), fission fragment parameters (fragment mass and TXE distributions) correlated with α-α and α-t QF events could be registered, due to the high fission rate achieved. On the other hand, the observation of the QF yield to be roughly an order of magnitude higher in ^{252}Cf than in ^{233,235}U(n$_{th}$,f) makes ^{252}Cf a promising candidate for going a step further, namely searching a fission mode with the coincident emission of three LCPs, to be called "quinary" fission. In that case, ternary ^{12}C* could be a source for pseudo quinary fission.

Acknowledgements

This work was supported in parts by INTAS (call 99-229) and the German Minister for Education and Research (BMBF) under contracts 06DA461, 06DA913 and 06TU669.

References

1. C. Wagemans, in *The Nuclear Fission Process*, C. Wagemans (Ed.), CRC Press, Boca Raton, Fl., USA, 1991. Chapt. 12.
2. M. Mutterer and J.P. Theobald, in *Nuclear Decay Modes*, D. Poenaru (Ed.), IOP Publ., Bristol, England, 1996, Chapt. 12.
3. P. Singer *et al.*, *Proc. Int. Conf. on Dynamical Aspects of Nuclear Fission, DANF96, Častá Papiernička, Slovakia*, ed. J. Kliman and B.I. Pustylnik, (JINR, Dubna, 1996), p. 262; P. Singer, Ph.D. Thesis, TU Darmstadt (1997).
4. M. Mutterer *et al.*, *Proc. Int. Conf. on Dynamical Aspects of Nuclear Fission, DANF01, Častá Papiernička, Slovakia, 2001*, (World Scientific, Singapore, 2002), p. 326.
5. Yu.N. Kopatch *et al.*, *Phys. Rev. Lett.* **82**, 303 (1999).
6. Yu.N. Kopatch *et al.*, *Phys. Rev.* C **65**, 044614 (2002).
7. M. Mutterer *et al.*, to be published.
8. M.N. Andronenko *et al.*, *Euro. Phys. J.* A **12**, 185 (2001).
9. E. Cheifetz *et al.*, *Phys. Rev. Lett.* **29**, 805 (1972).
10. A.P. Graevskii, and G.E. Solyakin, *Sov. J. Nucl. Phys.* **18**, 369 (1974) .
11. H.H. Knitter *et al.*, in *The Nuclear Fission Process*, ed. C. Wagemans, CRC Press, Boca Raton, FL., USA, 1991, Chap. 11.
12. P. Singer *et al.*, *Z. Phys.* A **359**, 41 (1997).
13. Yu.N. Kopatch *et al.*, *Proc. Symp. on Nuclear Clusters: from Light Exotic to Superheavy Nuclei, Rauischholzhausen, Germany, 2002*, (EP Systema Bt., Debrecen, Hungary, 2002), p. 273.
14. Yu.N. Kopatch *et al.*, *GSI Scientific Report 2000, GSI-2001-1*, (ISSN 0174-0814); URL: http://www.gsi.de/annrep), 23 (2001).
15. F. Gönnenwein *et al.*, *Proc. Int. Symp. New Projects and Lines of Research in Nuclear Physics, Messina, Italy, 2002*, (World Scientific, Singapore, 2003), p. 107.
16. V.N. Andreev *et al.*, *Sov. J. Nucl. Phys.* **18**, 22 (1969).
17. S.K. Kataria *et al.*, *Proc. Conf. Physics and Chemistry of Fission, Rochester 1973*, (IAEA, Vienna, 1973), Vol. II, p. 389.
18. A.S. Fomichev *et al.*, *Nucl. Instr. and Meth. in Phys. Res.* A **384**, 519 (1997).
19. M. Speransky *et al.*, *Proc. Intern. Seminar on the Reaction of Neutrons with Nuclei, ISINN12, Dubna. Russia, June 2004*, (JINR, Dubna, 2004), to be published.

KEY TOPICS IN NUCLEAR STRUCTURE
Proceedings of the 8th Int. Spring Seminar on Nuclear Physics
edited by Aldo Covello
© 2005 World Scientific Publishing Co.

POTENTIAL BARRIERS IN THE QUASI-MOLECULAR DEFORMATION PATH FOR ACTINIDES

G. ROYER, C. BONILLA
Laboratoire Subatech, UMR : 6457, 4 rue A. Kastler, 44307 Nantes, France
E-mail: royer@subatech.in2p3.fr

The deformation energy of actinides in the fusionlike deformation path has been determined from a generalized liquid drop model taking into account both the proximity energy, the asymmetry and an accurate nuclear radius and from the shell and pairing energies. Double and triple-humped potential barriers appear. The second maximum corresponds to the transition from compact and creviced one-body shapes to two touching ellipsoids. Third minimum and peak appear in certain asymmetric exit channels where one fragment is almost a double magic nucleus with a quasi-spherical shape while the other one evolves from oblate to prolate shapes. The heights of the double and triple-humped fission barriers agree with the experimental results. The predicted half-lives follow the experimental data trend.

1. Introduction

The heights of the inner and outer asymmetric fission barriers are almost constant (5-6 MeV) from Th to Am isotopes [1]. Their reproduction is a very difficult task. The fission probability and the angular distribution of the fragments suggest also the presence of hyperdeformed states in a deep third well in several Th and U isotopes [2].

It has been previously shown within a Generalized Liquid Drop Model taking into account both the proximity energy between close opposite surfaces and an accurate radius that the proximity forces strongly lower the deformation energy in the fusionlike shape path and allow to reproduce the experimental fission barrier heights in this exit channel. Recently, the asymmetric fission barrier heights for the Se and Mo nuclei have also been reproduced with the GLDM [3] but not with the RFRM and RLDM. Within the same approach, the α and cluster emission [4,5] as well as the highly deformed state [6] data have been also reproduced.

We focus now on the actinide region taking into account the ellipsoidal deformations of the fission fragments and their associated shell and pairing energies.

2. Potential energy

The GLDM energy includes the volume, surface, Coulomb, proximity and rotational energies. All along the deformation path the proximity energy term E_{prox} allows to take into account the effects of the attractive nuclear forces between nucleons facing each other across a neck or a gap.

3. One and two-body shapes

The one-body shape sequence is described within two joined elliptic lemniscatoids which allow to simulate the development of a deep neck in compact and little elongated shapes with almost spherical ends (see Fig. 1). The proximity energy is very important in this deformation path. For two-body shapes, the coaxial ellipsoidal deformations have been considered [7].

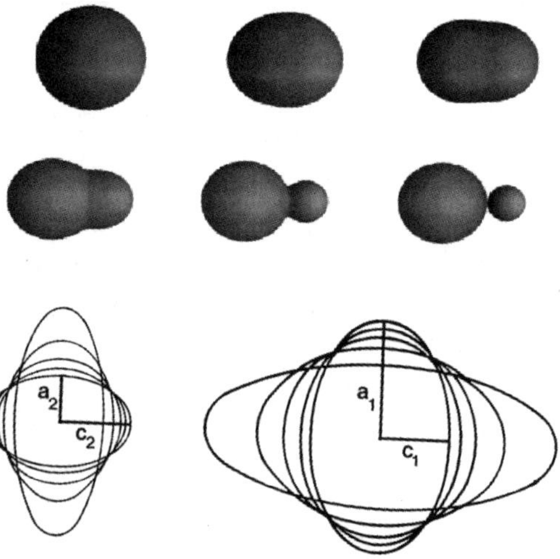

Figure 1. Selected shape sequence to simulate the one-body shape evolution and two coaxial ellipsoid configuration describing the two-body shapes.

The shape-dependent shell corrections have been determined within the Droplet Model expressions [8]. The selected highest proton magic number is 114 while, for the two highest neutron magic numbers, the values 126 and

184 have been retained. For the two-body shapes, the total shell energy is the sum of the shell corrections for each deformed fragment. The pairing energy has been calculated from the expressions proposed by the Thomas-Fermi model.

4. Potential barriers

The dependence of the deformation energy on the shape sequence and introduction of the microscopic corrections is displayed in Fig. 2 for an asymmetric fission path of the ^{230}Th nucleus. The shell effects generate

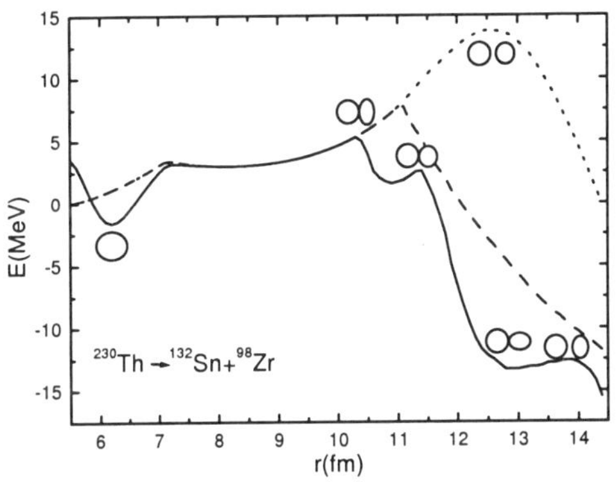

Figure 2. Asymmetric fission barrier of a ^{230}Th nucleus emitting a doubly magic nucleus ^{132}Sn. The dotted and dashed curves give respectively the macroscopic energy within two spheres and the ellipsoidal deformations for the two-body shapes. The solid line includes the shell and pairing energies. r is the distance between mass centres.

the slightly deformed ground state and contribute to the formation of the first peak. The proximity energy flattens the potential energy curve and will explain with the shell effects the formation of a deep second minimum lodging the superdeformed isomeric states for heaviest nuclei. In the exit channel corresponding to the two-sphere approximation the top of the barrier is reached after the rupture of the matter bridge between the two spherical fragments ($r = 11.4$ fm). Then, the top corresponds to two separated fragments maintained in unstable equilibrium by the balance between the attractive nuclear forces and the repulsive Coulomb ones. In this mass range, the introduction of the shell and pairing effects for two-sphere shapes is not sufficient to reproduce the experimental data on the fission barrier heights of actinide nuclei. When the ellipsoidal deformations of the fragments are taken into account, the transition corresponds to the passage (at $r = 11$ fm for ^{230}Th) from a one-body shape with spherical ends and a deep neck to two touching ellipsoidal fragments, one or both of them being slightly oblate. The barrier height is reduced by several MeV. The introduction of the shell effects still lowers the second peak and shifts it to an inner position ($r = 10.3$ fm here). It even leads to a third minimum and third peak in this asymmetric decay path. A plateau appears also at larger distances around 10 MeV below the ground state. It is due to the persistence of the prolate deformation of the lightest fragment. The end of the plateau corresponds to the end of the contact between the two fragments and to a rapid transition from prolate to oblate shapes for the non-magical fragment and the vanishing of the proximity energy. Later on, this second fragment returns to a prolate shape when the interaction Coulomb energy is smaller.

The potential barriers for the ^{235}U, ^{238}Pu and ^{250}Cf nuclei are shown in Fig. 3. For a given mass asymmetry, the charge asymmetry which minimizes the deformation energy has been selected. The proximity energy and the attenuated microscopic effects are responsible for the formation of a second one-body shape minimum. The heights of the two peaks generally increase with the asymmetry but the shell and pairing corrections induce strong variations from this global behaviour. Their main effect is to favour, for the U and Pu isotopes, an asymmetric path where one fragment is close to the doubly magic $^{132}_{50}$Sn nucleus, and, consequently, keeps a spherical shape. This effect is less pronounced for ^{250}Cf since for nuclei with $Z \sim 100$ the symmetric fission gives fragments with a charge of around 50. A third minimum and third peak appear only in the asymmetric decay path and for some specific isotopes.

The calculated and experimental energies of the extrema of the fission barriers are compared in table 1. E_a, E_b and E_c are the first, second and third peak heights while E_{III} is the energy of the third potential minima relatively to the ground state energy (in MeV). There is a very good agreement between the experimental and theoretical heights E_a and E_b of the two peaks. The still sparse but exciting data for the third barrier are correctly reproduced.

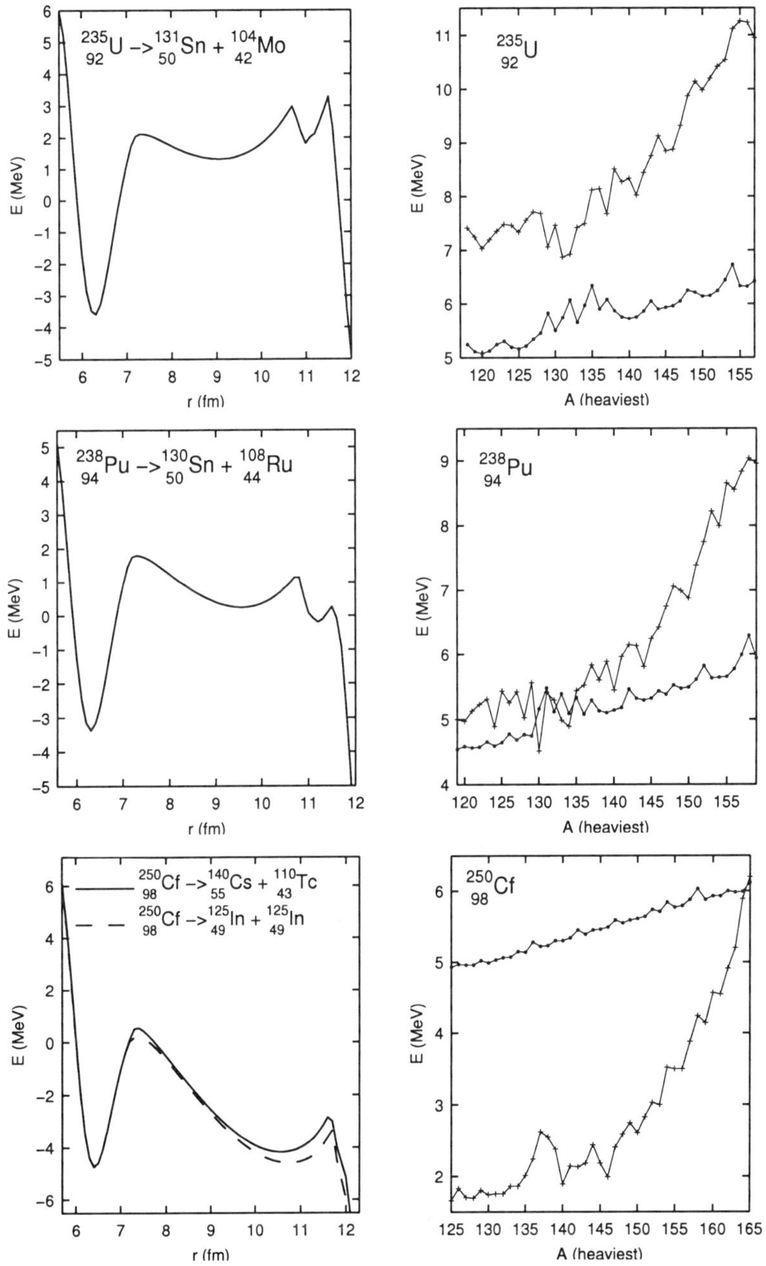

Figure 3. On the left, multiple-humped fission barriers in the mentioned asymmetric fission path for ^{235}U, ^{238}Pu and ^{250}Cf. On the right, inner (full circles) and outer (crosses) fission barrier heights as a function of the mass of the heaviest fragment.

Reaction	$E_a(th)$	$E_a(exp)$	$E_b(th)$	$E_b(exp)$	$E_{III}(th)$	$E_c(th)$
$^{231}_{90}Th \rightarrow ^{132}_{50}Sn + ^{99}_{40}Zr$	5.5	-	7.1	6.5	3.9 exp: 5.6	6.9 exp: 6.3
$^{233}_{90}Th \rightarrow ^{132}_{50}Sn + ^{101}_{40}Zr$	5.6	-	7.0	6.8	5.0 exp: 5.2	7.8 exp: 6.8
$^{232}_{92}U \rightarrow ^{134}_{52}Te + ^{98}_{40}Zr$	4.5	4.9	5.0	5.4	4.2	5.1
$^{234}_{92}U \rightarrow ^{131}_{50}Sn + ^{103}_{42}Mo$	5.0	5.6	5.9	5.5	3.7 exp: 3.1	5.6
$^{236}_{92}U \rightarrow ^{132}_{50}Sn + ^{104}_{42}Mo$	5.5	5.6	6.2	5.5	3.1 exp: 3.15	4.4
$^{238}_{92}U \rightarrow ^{132}_{50}Sn + ^{106}_{42}Mo$	5.5	5.7	5.6	5.7	4.1	5.6
$^{238}_{94}Pu \rightarrow ^{130}_{50}Sn + ^{108}_{44}Ru$	5.2	5.5(5.7)	4.5	5.0	3.2	3.6
$^{240}_{94}Pu \rightarrow ^{130}_{50}Sn + ^{110}_{44}Ru$	5.25	5.6(5.8)	4.6	5.1		
$^{243}_{94}Pu \rightarrow ^{132}_{50}Sn + ^{111}_{44}Ru$	6.3	5.9	5.2	5.2(5.6)	3.2	4.6
$^{242}_{95}Am \rightarrow ^{131}_{50}Sn + ^{111}_{45}Rh$	6.8	6.5	5.7	5.4	4.1	5.1
$^{244}_{95}Am \rightarrow ^{132}_{50}Sn + ^{112}_{45}Rh$	7.0	6.3	5.7	5.4	2.4	4.2
$^{243}_{96}Cm \rightarrow ^{130}_{50}Sn + ^{113}_{46}Pd$	6.0	6.4	4.2	4.2	2.4	2.7
$^{243}_{96}Cm \rightarrow ^{122}_{48}Cd + ^{121}_{48}Cd$	5.5	6.4	4.3	4.2		
$^{250}_{97}Bk \rightarrow ^{130}_{50}Sn + ^{120}_{47}Ag$	6.4	6.1	3.7	4.1		
$^{250}_{98}Cf \rightarrow ^{125}_{49}In + ^{125}_{49}In$	4.9	5.6	1.7	-		
$^{256}_{99}Es \rightarrow ^{128}_{50}Sn + ^{128}_{49}In$	5.9	4.8	2.4	-		

5. Third barrier and half-lives

The origin of the third well in the asymmetric decay path is examined now (see Fig. 4). The dashed line represents the potential for two touching ellipsoids when the one-body shape is still energetically favoured. The second peak (but first on the figure) corresponds to the point where these touching ellipsoids begin to give the lowest energy. The heaviest fragment is a magic nucleus. It therefore preserves its almost spherical shape. The non magic fragment was born in an oblate shape ($s \sim 1.4$), due to the small distance between the mass centres at this step. When this distance increases, the ratio s decreases, because of the proximity energy which tends to keep close the two tips of the fragments. Thus, the lightest fragment remaining in contact with the other spherical fragment approaches the spherical shape and its shell energy increases to reach a maximum which is at the origin of the third peak and which corresponds to two touching different spheres. Before reaching this third peak a third minimum appears. Its shape is hyperdeformed and asymmetric in agreement with the experimental data [2]. Later on, the proximity forces maintain the two fragments in contact and the shape of the smallest one evolves to prolate shapes ($s < 1$) and the shell corrections decrease. The third barrier appears only in the asymmetric decay path and for some specific nuclei. In the symmetric mass exit path, the proximity and Coulomb energies counterbalance the smallest shell effects and induce an asymmetric shape, the two fragments remain in contact but one fragment is oblate while the other one is prolate. With increasing distance between the mass centres the two nuclei become prolate.

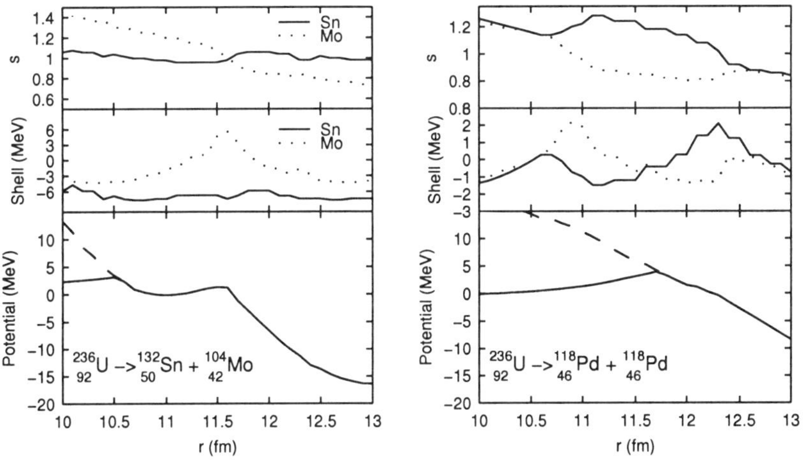

Figure 4. Fission barriers, shell energies and ratio of the semi-axes of the two ellipsoidal fragments for an asymmetric decay channel and the symmetric one for ^{236}U. On the lowest part, the fission barrier is given by the solid line.

The dependence of the fission barrier heights and profiles on the asymmetry for the ^{236}U nucleus, for which experimental data on the third barrier exist, are given in Fig. 5. The position of the second peak in the symmetric decay path corresponds to the position of the third peak in the asymmetric deformation path. Clearly the magicity of some Sn isotopes plays the main role.

Within this asymmetric fission model the decay constant is the product of the assault frequency ν_0 by the barrier penetrability P. The experimental fission half-lives and theoretical predictions for the supposed most probable exit channels have been compared. There is a correct agreement with the experimental data on 20 orders of magnitude.

6. Summary and conclusion

Super and hyperdeformed minima lodging possibly isomeric states appear for the actinide elements in the quasi-molecular shape path within a deformation energy derived from a generalized liquid drop model and including the shell and pairing energies. The second peak corresponds to the tran-

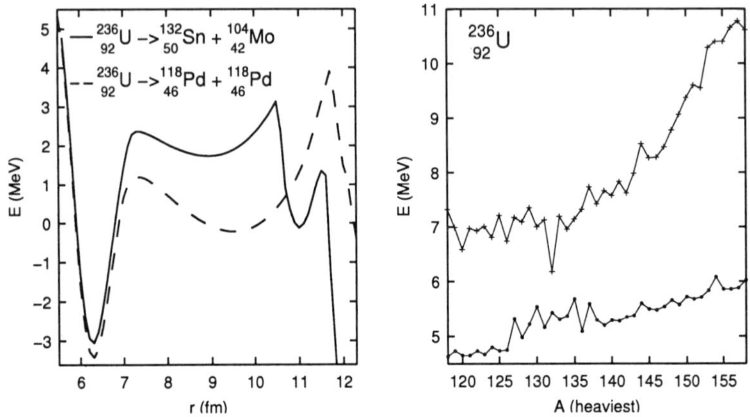

Figure 5. Fission barriers for ^{236}U. The inner and outer fission barrier heights are given, on the right, respectively by the full circles and crosses.

sition from one-body shapes to two touching ellipsoids. The third barrier appears only in the asymmetric decay path and for some specific nuclei. Then, the heaviest fragment is almost a magic nucleus and it preserves its shape close to the sphere. The other fragment evolves from an oblate ellipsoid to a prolate one and the third peak corresponds to the maximum of the shell effects in the non magic fragment and, consequently, to two touching different spheres. The barrier heights agree with the experimental results for the double and triple-humped fission barriers. The predicted half-lives follow the experimental data trend.

References

1. C. Wagemans, *The nuclear fission process* (CRC Press, Boca Raton, 1991).
2. A. Krasznahorkay et al, *Phys. Lett.* **B461**, 15 (1999).
3. G. Royer and K. Zbiri, *Nucl. Phys.* **A697**, 630 (2002).
4. G. Royer, *J. Phys. G: Nucl. Part. Phys.* **26**, 1149 (2000).
5. G. Royer and R. Moustabchir, *Nucl. Phys.* **A683**, 182 (2001).
6. G. Royer, C. Bonilla and R. A. Gherghescu, *Phys. Rev.* **C67**, 34315 (2003).
7. G. Royer and C. Piller, *J. Phys. G: Nucl. Part. Phys.* **18**, 1805 (1992).
8. W.D. Myers, *The droplet model of atomic nuclei* (Plenum, New-York, 1977).

KEY TOPICS IN NUCLEAR STRUCTURE
Proceedings of the 8th Int. Spring Seminar on Nuclear Physics
edited by Aldo Covello
© 2005 World Scientific Publishing Co.

SEMICLASSICAL QUANTIZATION OF THE TRIAXIAL RIGID ROTATOR: DENSITY OF STATES AND SPECTRAL STATISTICS

J.M.G. GÓMEZ[1], V.R. MANFREDI[2], A. RELAÑO[1], L. SALASNICH[2,3]

[1] *Departamento de Fisica Atomica, Molecular y Nuclear, Universidad Complutense de Madrid, Avenida Complutense, E-28040 Madrid, Spain*

[2] *Dipartimento di Fisica "G. Galilei", Università di Padova, Istituto Nazionale di Fisica Nucleare, Sezione di Padova, Via Marzolo 8, I-35131 Padova, Italy*

[3] *Istituto Nazionale per la Fisica della Materia, Unita di Milano Università, Dipartimento di Fisica, Università di Milano, Via Celoria 16, I-20133 Milano*

As is well know, also if the exact energy spectrum of the triaxial rigid rotator cannot be analytically determined, the classical Hamiltonian of the triaxial top is integrable and depends only on two action variables. We use the classical integrability of the triaxial top to analytically calculate the semiclassical energy spectrum via torus quantization. We analyze the global (number of states and density of states) and the local (single energy levels and their fluctuations) properties by using both the semiclassical approximation and the exact diagonalization. The agreement betweeen exact and semiclassical global properties is excellent, while for single energy levels the agreement is not so good. Note that among the spectral fluctuations we have also investigated a new statistic recently proposed in Phys. Rev. Lett. **89**, 244102 (2002).

1. Introduction

In a previous paper we investigated the pathological behaviour of spectral statistics of the triaxial rigid rotator (TRR) [1]. In fact, even if the rotor model is a classically integrable system, the spectral statistics of the energy levels do not follow the prediction of the Poisson statistic. In a more recent paper [2] we have shown that the semiclassical approximation of the TRR gives spectral fluctuations which are in good agreement with exact numerical ones.

In this contribution we analyze in detail the semiclassical approximation of the TRR by studying both global and local properties of the energy spectrum. In order to have a deeper test of the pathology of the TRR spectral fluctuations, we have also used a new spectral statistic. Such a statistic has been recently introduced to characterize quantum integrability and quantum chaos independently of the classical analog [3].

2. Semiclassical Quantization

As mentioned in the introduction, the classical Hamiltonian of the TRR is integrable [4,5] and it is given by

$$H = \frac{1}{2}\left(aJ_1^2 + bJ_2^2 + cJ_3^2\right) , \tag{1}$$

where $\mathbf{J} = (J_1, J_2, J_3)$ is the angular momentum of the rotation, and $a = 1/I_1$, $b = 1/I_2$, $c = 1/I_3$ are three parameters such that I_1, I_2 and I_3 are the principal momenta of inertia of the top. Here we choose $a < b < c$. We use the classical integrability of our triaxial top to calculate the semiclassical energy spectrum via torus quantization. The first step is to find out the action variables of the Hamiltonian. By using the Deprit transformation [5]

$$J_1 = \sqrt{J^2 - J_3^2}\, \sin\theta , \qquad J_2 = \sqrt{J^2 - J_3^2}\, \cos\theta , \tag{2}$$

where $J^2 = J_1^2 + J_2^2 + J_3^2$, the classical Hamiltonian can be written as

$$H = \frac{1}{2}aJ^2 + \frac{1}{2}\left[(c-a) - (b-a)\cos^2\theta\right] J_3^2 + \frac{1}{2}(b-a)J^2\cos^2\theta , \tag{3}$$

where the dynamical variable J_3 is the action variable of the angle θ. By performing another canonical transformation, the angular dependence of the Hamiltonian $H = H(J, J_3, \theta)$ can be removed. Thus, the Hamiltonian can be written as a function of two action variables: the total angular momentum J and a new action variable I given by

$$I = \frac{J}{\pi}\int_{\theta_1}^{\theta_2} \sqrt{\frac{\frac{2H}{J^2} - a - (b-a)\cos^2\theta}{(c-a) - (b-a)\cos^2\theta}}\, d\theta . \tag{4}$$

The action variable I is a function of the Hamiltonian H and of the total angular momentum J, namely $I = I(H, J)$. It follows that the Hamiltonian H is now an implicit function of two action variables: $H = H(J, I)$. It means that the system is integrable and the classical trajectories of the 6-dimensional phase-space are restricted to a 2-dimensional torus. The classical energy E of the system is given by $E = H(J, I)$. For a fixed value of J, the Hamiltonian (4) describes a sort of nonlinear pendulum where the extrema of integration are

$$\theta_1 = arcos\left[\sqrt{\frac{2H - aJ^2}{2(b-a)}}\right] , \qquad \theta_2 = arcos\left[-\sqrt{\frac{2H - aJ^2}{2(b-a)}}\right] , \tag{5}$$

for $0 \leq H < (b/2)J^2$ and it corresponds to librational motion, while they are

$$\theta_1 = 0 \, , \qquad \theta_2 = \pi \, , \tag{6}$$

for $(b/2)J^2 < H \leq (c/2)J^2$ and it corresponds to rotational motion. Note that $H = (b/2)J^2$ is the energy of the separatrix between librational and rotational motions. The semiclassical (torus) quantization of the energy is performed by setting

$$J = (j + \frac{1}{2})\hbar, \quad I = k\hbar \, , \tag{7}$$

where j and k are integer quantum numbers such that $k = -j, -j+1, ..., j-1, j$ and $j \geq 0$. Thus, the semiclassical energy $E^{sc}_{j,k}$ depends on the two quantum number, j and k. For a fixed value of j, there is a set of $2j+1$ energy levels $E^{sc}_{j,k}$ given by the implicit equation

$$I\left(E^{sc}_{j,k} \, , \, (j+\frac{1}{2})\hbar\right) = k\hbar \, , \tag{8}$$

where $I(H, J)$ is given by Eq. (4). Note that the lowest energy level of the set is $E^{sc}_{j,0} = a\hbar^2(j+1/2)^2/2$, while the higher is $E^{sc}_{j,j} = c\hbar^2(j+1/2)^2/2$.

3. Global spectral properties

Our semiclassical quantization gives immediately the semiclassical number of states $\bar{N}_j(E)$ up to the energy E with a fixed angular quantum number j. In fact, from Eq. (4) one finds

$$\bar{N}_j(E) = \frac{1}{\hbar} I\left(E \, , \, (j+\frac{1}{2})\hbar\right) \, . \tag{9}$$

The semiclassical density of states $\bar{\rho}_j(E)$ can be obtained from the number of states $\bar{N}_j(E)$ by using the formula

$$\bar{\rho}_j(E) = \frac{d}{dE}\bar{N}_j(E) \, . \tag{10}$$

In Figure 1 we plot the "exact" density of states $\rho_j(E)$ of the TRR and compare it with the semiclassical one $\bar{\rho}_j(E)$. Note that the "exact" energy spectrum is obtained with the numerical diagonalization of the quantized version of the Hamiltonian shown in Eq. (1). The quantization method and the numerical procedure are discussed in detail in [1]. Here we point out that, fixing the quantum number j of the total angular momentum, the quantum Hamiltonian can be divided into 4 submatrices. These submatrices correspond to 4 classes of symmetey of the quantum states [1,7].

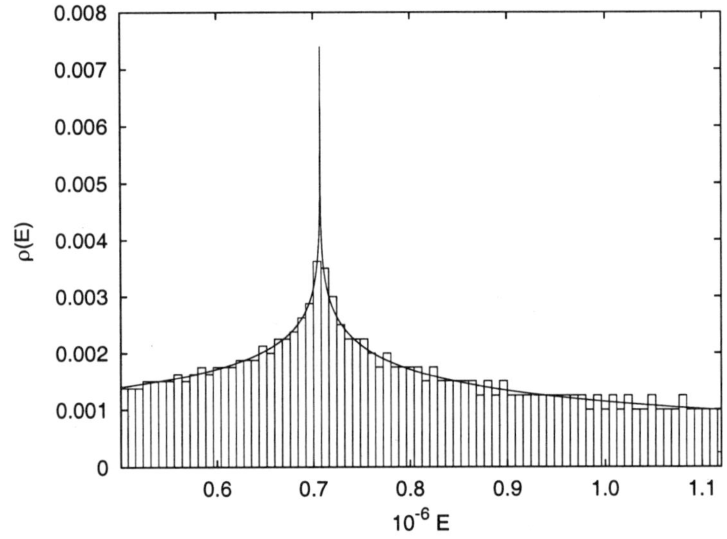

Figure 1. Density of states $\rho(E)$ of the TRR. States with even parity and total quantum number $j = 1000$. Parameters of the TRR: $a = 1$, $b = \sqrt{2}$, $c = \sqrt{5}$ and $\quad = 1$. The solid line is the semiclassical prediction.

Figure 1 shows that the semiclassical density of states is in remarkable very good agreement with the histograms of the "exact" density of states apart very close to the separatrix. These results suggest that the global properties of the TRR, like the number of states and the density of states, are very well reproduced by the semiclassical approximation.

4. Local spectral properties

The semiclassical approximation can be also used to derive the individual energy levels of the TRR. In Figure 2 we compare the "exact" energy spectrum, obtained with the numerical diagonalization of the quantum TRR Hamiltonian, with the semiclassical energy spectrum, calculated using Eqs. (4,5,8).

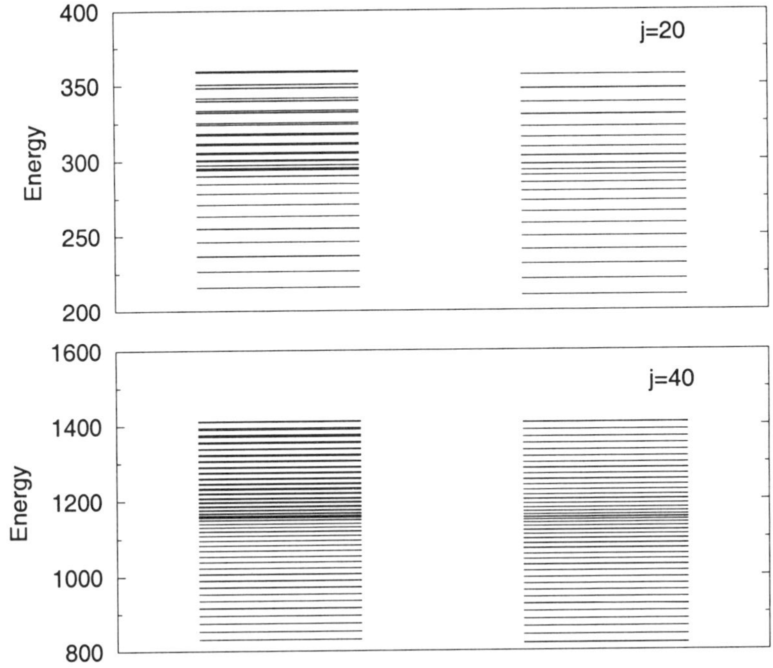

Figure 2. "Exact" (left) vs semiclassical (right) energy spectrum of the TRR with a fixed value of the angular quantum number j. Parameters as in Fig. 1. Figure adapted from Ref. 2.

The figure 2 shows that the semiclassical quantization cannot take into account the broken degeneracy of pairs of "exact" energy levels. Moreover, one observes that the semiclassical energy levels are more accurate in the higher part of the spectrum (rotational levels, see Eqs. (4-6)).

Having obtained the individual energy levels, we can calculate their fluctuation properties. The most used spectral statistic of fluctuations is the $P(s)$ function. $P(s)$ is the distribution of nearest-neighbor spacings $s_i = (\tilde{E}_{i+1} - \tilde{E}_i)$ of the unfolded levels $\tilde{E}_i$. It is obtained by accumulating the number of spacings that lie within the bin $(s, s + \Delta s)$ and then normalizing $P(s)$ to unit. As shown by Berry and Tabor, for quantum systems whose classical analogs are integrable, $P(s)$ is expected to follow the Poisson distribution

$$P(s) = \exp(-s) \ . \tag{11}$$

In Figure 3 the spectral statistic $P(s)$ is plotted for the four submatrices of $\hat{H}$ and $J = 1000$. Note that the level spectrum is mapped into unfolded

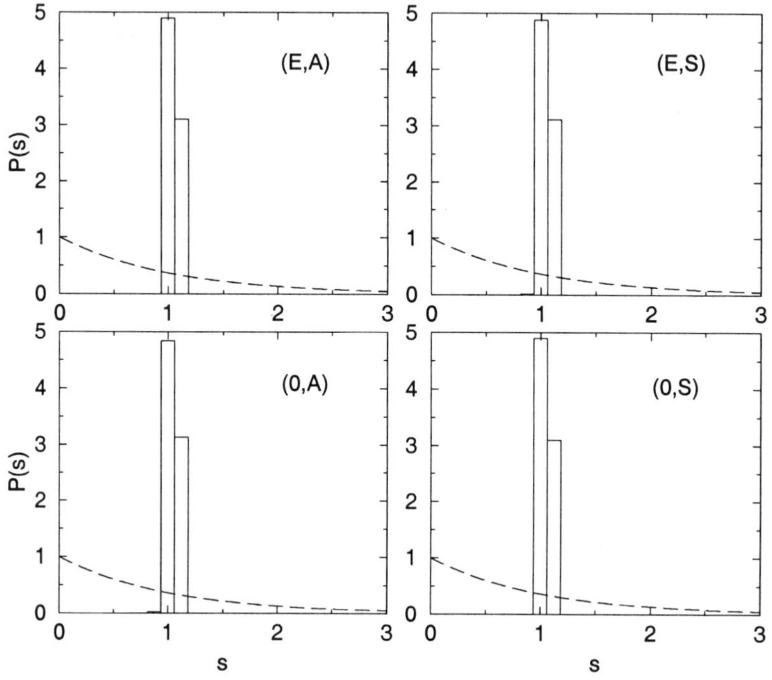

Figure 3. Nearest neighbor spacing distribution $P(s)$ of the TRR. Quantum states with angular quantum number $j = 1000$. States with 4 classes of symmetry: even (E) and odd (O) parity, symmetric (S) and anti-symmetric (A) (see Ref. 1 and Ref. 7). The dashed line is the Poisson prediction $P(s) = \exp(-s)$. Parameters as in Fig. 1. Figure adapted from Ref. 1.

levels with quasi-uniform level density by using a standard procedure described in [8]. $P(s)$ is practically the same for the four classes of symmetry. Moreover, $P(s)$ has a pathological behavior: a peak near $s = 1$ and nothing elsewhere. The same pathological behavior has been found by calculating, again for a fixed j, the $P(s)$ distribution of the semiclassical energy levels [2]. It is important to stress that the pathological behavior of the semiclassical $P(s)$ distribution is a direct consequence of the existence of the constant of motion I of Eq. (4): this action variable I fixes the second quantum number k of the semiclassical energy spectrum. Instead, in the pure quantum case, it is not (yet) found a hermitian operator $\hat{I}$, that will be the quantum analog of the classical function I, such that $[\hat{H}, \hat{I}] = [\hat{J}, \hat{I}] = 0$. Nevertheless, the behavior of the $P(s)$ distribution support the conjecture that the quantum constant of motion $\hat{I}$ should exist.

An interesting new statistic of energy levels has been recently introduced [3]. It is defined as

$$\delta_n = N^{av}(E_{n+1}) - N(E_{n+1}), \tag{12}$$

and it measures the deviations of the "exact" number of states $N(E)$ with respect to the average (secular) number of states $N^{av}(E)$. Note that, when the semiclassical quantization is possible, $N^{av}(E)$ is identified with the semiclassical number of states $\bar{N}(E)$. Usually, the δ_n statistic is strongly fluctuating and its power spectrum follows a power law $1/f^{\alpha}$, where the exponent α strictly depends on the chaoticity of the system [3,9].

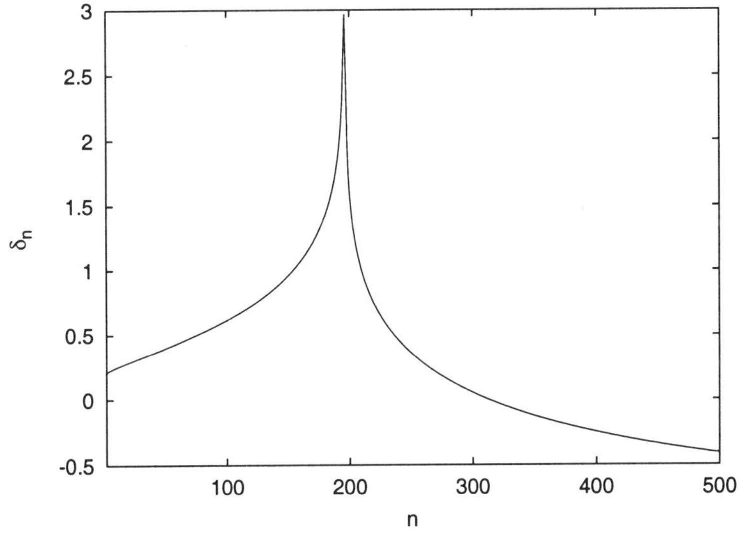

Figure 4. Statistic δ_n of the TRR for odd and antisymmetric states with angular quantum number $j = 1000$. Parameters as in Fig. 1.

In Figure 4 we plot the δ_n statistic obtained with the "exact" odd and antisymmetric levels of the TRR. We can see that δ_n is not a fluctuating but a smooth signal, whose shape is similar to the density of states. This result is quite strange because, as previously stressed, it is expected that δ_n fluctuates around zero in a stochastic way. Note that the δ_n statistic obtained with the semiclassical levels is instead identically zero if one takes $\bar{N}(E)$ as average number of states in Eq. (12). The δ_n statistics of the TRR shows that the semiclassical number of states $\bar{N}(E)$ is quite accurate. In fact, the differences between $\bar{N}(E)$ and $N(E)$, that are exactly expressed by δ_n, become relevant only near the classical sepatratrix, where there is a clustering of energy levels.

5. Conclusions

Our analysis of the triaxial rigid body has shown that the semiclassical quantization we have introduced is remarkably accurate in the description of the global properties of the energy spectrum, like the average number of states and the average density of states. Only very close to the separatrix, where there is the level clustering, the semiclassical approximation is less accurate. This conclusion is strongly supported by the study of the recently introduced δ_n statistic, that explicitly measures the deviations of the semiclassical number of states with respect to the exact one.

References

1. V.R. Manfredi and L. Salasnich, Phys. Rev. E **64**, 066201 (2001).
2. V.R. Manfredi, V. Penna, L. Salasnich, Mod. Phys. Lett. B **17**, 803 (2003); V.R. Manfredi, V. Penna, L. Salasnich, Theoretical Nuclear Physics in Italy, Cortona, Ed. by S. Boffi et al., 291 (World Scientific, Singapore, 2002).
3. A. Relano, J.M.G. Gomez, R.A. Molina, J. Retamosa, E. Faleiro, Phys. Rev. Lett. **89**, 244102 (2002).
4. V.I. Arnold, Mathematical Methods in Classical Mechanics (Springer, New York, 1974).
5. A.R.P. Rau, Rev. Mod. Phys. **64**, 623 (1992); A. Turchetti, Dinamica Classica dei Sistemi Fisici (Zanichelli, Bologna, 1998).
6. M. Tabor, Chaos and Integrability in Nonlinear Dynamics (Wiley, New York, 1989).
7. L. Landau and E. Lifshitz, *Course in Theoretical Physics*, vol. 3: Quantum Mechanics (Pergamon, 1977).
8. V.R. Manfredi, Lett. Nuovo Cimento **40**, 135 (1984).
9. J.M.G. Gomez, A. Relano, J. Retamosa, E. Faleiro, L. Salasnich, M. Vranicar, M. Robnik, "$1/f^{\alpha}$ noise in spectral fluctuations of quantum systems", submitted for publication (2004).

SECTION V

SPECIAL TOPICS

KEY TOPICS IN NUCLEAR STRUCTURE
Proceedings of the 8th Int. Spring Seminar on Nuclear Physics
edited by Aldo Covello
© 2005 World Scientific Publishing Co.

CHAOS AND 1/f NOISE IN NUCLEAR SPECTRA

J. M. G. GÓMEZ, A. RELAÑO, AND J. RETAMOSA
Departamento de Física Atómica, Molecular y Nuclear, Universidad Complutense de Madrid, E-28040 Madrid, Spain

R. A. MOLINA
CEA/DSM, Service de Physique de l'Etat Condensé, 91191 Gif-sur-Yvette, France.

E. FALEIRO
Departamento de Física Aplicada, E. U. I. T. Industrial, Universidad Politécnica de Madrid, E-28012 Madrid, Spain

Many complex systems in nature and in human society exhibit time fluctuations characterized by a power spectrum which is a power function of the frequency f. We show that the energy spectrum fluctuations of quantum systems can be formally considered as a discrete time series. The power spectrum behavior of such a signal for paradigmatic quantum systems suggests the following conjecture: The energy spectra of chaotic quantum systems are characterized by $1/f$ noise.

1. Introduction

One of the most ubiquitous features of complex physical systems is the appearance of so called $1/f^{\alpha}$ noise in fluctuating physical variables, meaning that the Fourier power spectrum $S(f)$ behaves as $1/f^{\alpha}$ in terms of the frequency f. Examples of such systems are electronic devices, sun spot activity, human heartbeat and the DNA sequence, but there are also quite different cases, like the music of J. S. Bach [1,2]. Why this type of fluctuations are so very ubiquitous is not yet well understood. But surprisingly or not, we shall show in this talk that to the long list of known systems with $1/f^{\alpha}$ noise we can further add all the Hamiltonian quantum systems with purely regular or purely chaotic motion.

During the last twenty years, the classical concept of chaos has been extended to quantum systems, and great progress has been achieved in understanding the universal properties of chaotic quantum systems. It is well known that there is a clear relationship between the energy level fluctuation properties of a quantum system and the large time scale behavior of its classical analogue. The pioneering work of Berry and Tabor [3] showed that the spectral fluctuations of a quantum system whose classical analogue is fully integrable are well described by Poisson statistics, i. e. the successive energy levels are not correlated. Some years later, Bohigas *et al.* [4] conjectured that the fluctuation properties of generic quantum systems, which in the classical limit are fully chaotic, coincide with those of random matrix theory (RMT). This conjecture is strongly supported by experimental data, many numerical calculations, and analytical work based on semiclassical arguments. A review of later developments can be found in [5].

Here we present a recently discovered, very different approach to quantum chaos [6], which is based on traditional methods of time series analysis. The essential feature of chaotic energy spectra in quantum systems is the existence of level repulsion and correlations. To study these correlations, we can consider the energy spectrum as a discrete signal, and the sequence of energy levels as a time series where the energy plays the role of time. We shall see that examination of the power spectrum of energy level fluctuations reveals very accurate power laws for completely regular or completely chaotic Hamiltonian quantum systems. It turns out that chaotic systems have $1/f$ noise, in contrast to the $1/f^2$ Brown noise of regular systems.

2. Outline of the power spectrum approach

In order to observe the universal features of spectral fluctuations in quantum systems, it is necessary to unfold the energy spectrum. Since level fluctuation amplitudes are modulated by the value of the mean level density $\bar{\rho}(E)$, to compare the fluctuations of different systems we have to remove the smooth behavior of the level density. The unfolding consists in locally mapping the real spectrum into another with mean level density equal to one. The actual energy levels E_i are mapped into new dimensionless levels ϵ_i [7]. The nearest neighbor spacing sequence is defined by $s_i = \epsilon_{i+1} - \epsilon_i$, $i = 1, \cdots, N-1$. For the unfolded levels, the mean level density is equal to 1 and $\langle s \rangle = 1$. In practical cases the unfolding procedure can be a difficult task for systems where there is no analytical expression for the mean level density [7].

Generally, the nearest neighbor spacing distribution $P(s)$ and the $\Delta_3(L)$ statistic are used to characterize spectral fluctuations, but in this paper we use the statistic δ_n [8] defined by

$$\delta_n = \sum_{i=1}^{n}(s_i - \langle s \rangle) = \sum_{i=1}^{n} w_i, \tag{1}$$

where the index n runs from 1 to $N-1$. The quantity w_i gives the fluctuation of the i-th spacing from its mean value $< s >= 1$.

The statistic δ_n represents the deviation of the unfolded excitation energy from its mean value n. The function δ_n has a formal similarity with a time series. For example, we may compare the energy level spectrum with the diffusion process of a particle. The analogy is clear if the index i of the nearest level spacings is considered as a discrete time, and the spacing fluctuation w_i as the analogue of the particle displacement d_i from the collision at time i to the next collision. Certainly, there are some differences between the two systems, but the essential point is that the function δ_n is the analogue of the particle total displacement at time n.

Our aim is to study the δ_n signal of chaotic quantum systems. We can analyze their spectral statistics with numerical techniques normally used in the study of complex systems and try to relate the emerging properties with some universal features that appear in many other branches of physics. One of those techniques is the calculation of the power spectrum $S(k)$ of a discrete and finite series δ_n given by $S(k) = \left|\hat{\delta}_k\right|^2$, where $\hat{\delta}_k$ is the discrete Fourier transform of δ_n, and N is the size of the series [6].

3. Results and discussion

As an example of a very chaotic system, we take the atomic nucleus at high excitation energy, where the level density is very large. To obtain the energy spectrum, shell-model calculations for selected nuclei are performed, using realistic interactions that reproduce well experimental data of nuclei in a mass region. The Hamiltonian matrices for different angular momenta, parity and isospin are fully diagonalized using the shell-model code Nathan [9], and careful global unfolding is performed. Then, sets of 256 consecutive levels of the same $J^\pi T$, from the high level density region, are selected. To characterize the statistical properties of the δ_n signal, we calculate an ensemble average of its power spectrum, in order to reduce statistical fluctuations and clarify its main trend. The average $\langle S(k) \rangle$ is calculated with 25 sets.

Fig. 1 shows the results for a typical stable *sd* shell nucleus, ^{24}Mg, with matrix dimensionalities up to about 2000, using the effective W interaction

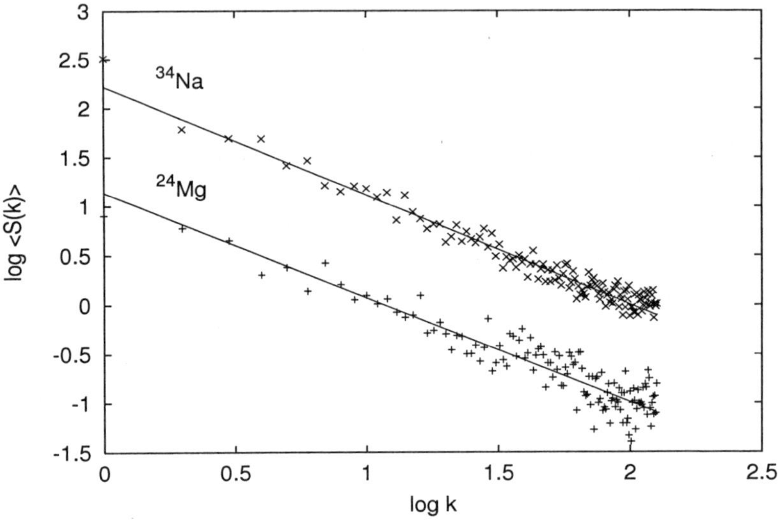

Figure 1. Average power spectrum of the δ_n function for ^{24}Mg and ^{34}Na, using 25 sets of 256 levels from the high level density region. The plots are displaced to avoid overlapping.

[10]; and for a very exotic nucleus, ^{34}Na, with dimensions up to about 5000, in the *sd* proton and *pf* neutron shells, using a realistic interaction [11]. Clearly, the power spectrum of δ_n follows closely a power law. We may assume the simple functional form $\langle S(k)\rangle \sim 1/k^{\alpha}$. A least squares fit to the data of Fig. 1 gives $\alpha = 1.11 \pm 0.03$ for ^{34}Na, and $\alpha = 1.06 \pm 0.05$ for ^{24}Mg. These results rise the question of whether there is a general relationship between quantum chaos and the power spectrum of the δ_n fluctuations of the system.

Probably, the simplest and most reliable way to clarify this issue is to compare δ_n and $\langle S(k)\rangle$ for Poisson energy levels and random matrix spectra. Random matrix theory plays a predominant role in the description of chaotic quantum systems [5]. It deals with three basic Hamiltonian matrix ensembles: The Gaussian orthogonal ensemble (GOE) of N-dimensional matrices, the Gaussian unitary ensemble (GUE), and the Gaussian symplectic ensemble (GSE), which apply to different systems, depending on the integer or half-integer spin, the time-reversal and the rotational symmetries of the system. As an additional collectivity, we introduce here the ensemble of diagonal matrices whose elements are random Gaussian variables, and call it the *Gaussian diagonal ensemble* (GDE).

Fig. 2 shows the signal δ_n for a GDE (Poisson) and a GOE spectrum of dimension 1000. Clearly the two signals are very different. It is also illustrative to compare those signals with a discrete time series $x(t)$, with $1/k$ and $1/k^2$ power laws, generated with the random-phase approximation procedure used in Ref. [12]. The similarity of the $\alpha = 2$ time series with the Poisson spectrum, and the $\alpha = 1$ time series with the GOE spectrum is obvious.

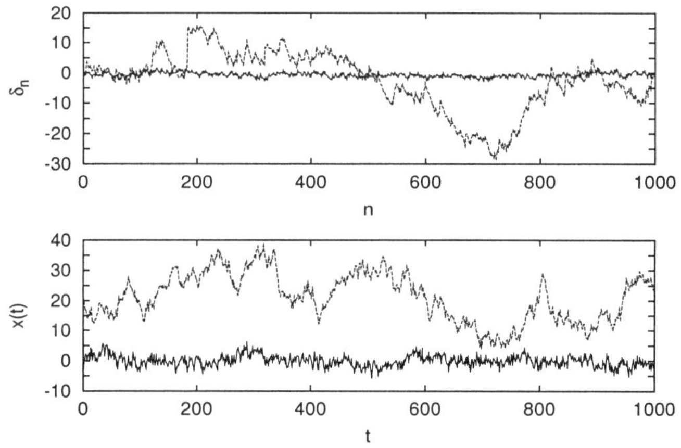

Figure 2. Comparison of the δ_n function for Poisson (dashed line) and GOE spectra (solid line), with a standard time series $x(t)$ with $1/k^{\alpha}$ power spectrum, for $\alpha = 2$ (dashed line) and $\alpha = 1$ (solid line).

To compute the average $\langle S(k) \rangle$, we generate 30 different matrices of dimension 1000 for each type of random matrix ensemble. Fig. 3 shows the results of these calculations in a decimal log-log scale. In all the cases the main trend is essentially linear, except for very high frequencies, where some deviation is observed, probably due to finite size effects.

Ignoring frequencies greater than $\log k = 2.2$, the fit gives $\alpha_{\text{GDE}} = 1.99$ with an uncertainty near 2%. The spectrum of any matrix pertaining to GDE consists of N uncorrelated levels. This is due to the diagonal character of the matrix and to the fact that its matrix elements are independent random variables. Consequently, the nearest level spacings are also uncorrelated and δ_n is just a sum of $N - 1$ independent random variables. The power spectrum of such a signal is well known to present $1/k^2$ behavior, and that is in full agreement with our numerical value for α in the Poisson spectrum. Furthermore, Berry and Tabor [3] showed that in a semiclassical integrable system, the spacings s_i are random independent variables for $i >> 1$. As a consequence their δ_n power spectrum behaves as $1/k^2$. However this behavior may be modified by the levels of the ground state region.

By contrast, the spectrum of any GOE member of large dimension is generally considered the paradigm of chaotic quantum spectra. It presents level correlations at all scales. The same applies to GUE and GSE, in increasing order of level repulsion. As is well known, the nearest neighbor spacing distribution for these three ensembles behaves as $P(s) \sim s^\beta$ for small s, where β is known as the level repulsion parameter. For our diagonal ensemble with Poisson statistics, $\beta = 0$, while $\beta = 1, 2, 4$ for GOE, GUE and GSE, respectively [5].

The power spectrum of δ_n for these three ensembles is displayed in Fig. 3. The fit of $\langle S(k) \rangle$ to the $1/f^\alpha$ power law is excellent. For the exponents we obtain $\alpha_{\rm GOE} = 1.08$, $\alpha_{\rm GUE} = 1.02$ and $\alpha_{\rm GSE} = 1.00$. In all the cases the error of the linear regression is about 2%. The three ensembles yield the same power law, with $\alpha \simeq 1$. Clearly, the the power spectrum $\langle S(k) \rangle$ behaves as $1/k^\alpha$ both in regular and chaotic energy spectra, but level correlations decrease the exponent from the $\alpha = 2$ limit for uncorrelated spectra to apparently a minimum value $\alpha = 1$ for chaotic quantum systems.

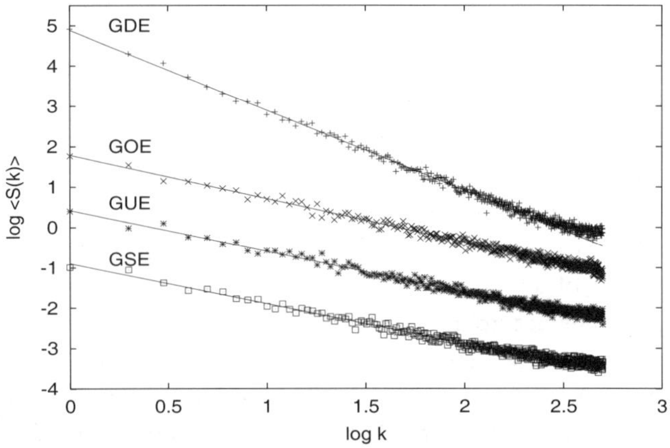

Figure 3. Power spectrum of the δ_n function for GDE (Poisson) energy levels, compared to GOE, GUE and GSE. The plots are displaced to avoid overlapping.

4. The power spectrum conjecture

The concept of quantum chaos has no precise definition as yet. Quantum systems with classical analogues are considered chaotic when their classical analogues are chaotic. Quantum systems without classical analogues may be called chaotic if they show the same kind of fluctuations as chaotic quantum systems with classical analogues. In practice, the Bohigas-Gianoni-Schmit RMT conjecture [4] is generally used as a criterion. But the results obtained above for the power spectrum of the δ_n statistic suggest a new conjecture: *The energy spectra of chaotic quantum systems are characterized by $1/f$ noise* [7].

This conjecture has several appealing features. It is a property characterizing the chaotic spectrum by itself, without any reference to the properties of other systems like GOE. It is universal for all kinds of chaotic quantum systems, either time reversal invariant or not, either of integer or half-integer spin. Furthermore, the $1/f$ noise characterization of quantum chaos includes these physical systems into a widely spread kind of systems appearing in many fields of science, which display $1/f$ fluctuations. Thus, the energy spectrum of chaotic quantum systems exhibits the same kind of fluctuations as many other complex systems.

5. Conclusions

Chaotic quantum energy spectra are characterized by strong level repulsion and strong spectral rigidity. In terms of the conventional $\Delta_3(L)$ statistic, strong rigidity corresponds to small values of $\Delta_3(L)$ and slow, logarithmic dependence on L. We can try to interpret what spectral rigidity means in terms of the δ_n function.

Rigidity of the energy spectrum means that the deviations of the energy spacings s_i from their mean value $< s >= 1$ are generally small, and that the spectrum is organized in such a way that a deviation of a spacing from the mean tends to be balanced by neighboring spacings. Therefore it is unlikely to find a long series of consecutive spacings all above or below the mean spacing.

In a time series, antipersistence means that an increasing or decreasing trend in the past makes the opposite trend in the future probable. In the present approach, where a quantum energy spectrum is considered as a time series, the spectral rigidity is analogous to antipersistence. We have seen that a rigid energy spectrum gives rise to a δ_n power spectrum of $1/k^\alpha$ type with $\alpha = 1$. On the other hand, as is well known, a time series with a $1/k^{\alpha}$ power spectrum where $\alpha \simeq 1$ is very antipersistent. Therefore, the interpretation of spectral rigidity as the analogue of antipersistence is consistent with the behavior of the power spectrum.

Summarizing, we have seen that for quantum systems the δ_n function can be considered as a time series, where the level order index n plays the role of a discrete time. The power spectrum $\langle S(k)\rangle$ of δ_n has been studied for representative energy spectra of regular and chaotic quantum systems. Neat power laws $\langle S(k)\rangle \sim 1/k^{\alpha}$ have been found in all cases. For Poisson spectra, we get $\alpha = 2$, as expected for independent random variables. For spectra of atomic nuclei at higher energies, in regions of high level density, and for the GOE, GUE and GSE ensembles, we obtain $\alpha = 1$.

These results suggest the conjecture that chaotic quantum systems are characterized by $1/f$ noise in the energy spectrum fluctuations. This property is not a mere statistic to measure the chaoticity of the system. It provides an intrinsic characterization of quantum chaotic systems without any reference to the properties of RMT ensembles. As is well known $1/f$ noise is quite ubiquitous. It characterizes sunspot activity, music, and chronic illness [1]. And we believe that it characterizes quantum chaos as well.

This work is supported in part by Spanish Government grants BFM2003-04147-C02 and FTN2003-08337-C04-04.

References

1. B. B. Mandelbrot, *Multifractals and 1/f noise* (Springer, New York, 1999).
2. M. Schroeder, *Fractals, Chaos and Power Laws* (Freeman, New York, 1992).
3. M. V. Berry and M. Tabor, Proc. R. Soc. London A **356**, 375 (1977).
4. O. Bohigas, M. J. Giannoni and C. Schmit, Phys. Rev. Lett. **52**, 1 (1984).
5. T. Guhr, A. Müller-Groeling and H. A. Weidenmüller, Phys. Rep. **299**, 189 (1998); H. J. Stöckmann, *Quantum Chaos*, (Cambridge U. P., Cambridge, 1999).
6. A. Relaño, J. M. G. Gómez, R. A. Molina, J. Retamosa and E. Faleiro, Phys. Rev. Lett. **89**, 244102 (2002).
7. J. M. G. Gómez, R. A. Molina, A. Relaño and J. Retamosa, Phys. Rev. E **66**, 036209 (2002).
8. M. L. Mehta, *Random Matrices*, (Academic Press, 1991)
9. E. Caurier, *et al.* Phys. Rev. C **59**, 2033 (1999).
10. B. H. Wildenthal, in *Progress in Particle and Nuclear Physics*, Ed. D. H. Wilkinson, Vol. **11** (Pergamon, Oxford 1984).
11. E. Caurier, *et al.* Phys. Rev. C **58**, 2033 (1998).
12. N. P. Greis and H. S. Greenside, Phys. Rev. A **44**, 2324 (1991).

KEY TOPICS IN NUCLEAR STRUCTURE
Proceedings of the 8th Int. Spring Seminar on Nuclear Physics
edited by Aldo Covello
© 2005 World Scientific Publishing Co.

SUPER-RADIANCE: FROM NUCLEAR PHYSICS TO PENTAQUARKS *

VLADIMIR ZELEVINSKY

NSCL, Michigan State University, East Lansing, MI 48824-1321, USA

ALEXANDER VOLYA

Department of Physics, Florida State University, Tallahassee, FL 32306-4350, USA

The phenomenon of super-radiance in quantum optics predicted by Dicke 50 years ago and observed experimentally has its counterparts in many-body systems on the borderline between discrete spectrum and continuum. The interaction of overlapping resonances through the continuum leads to the redistribution of widths and creation of broad super-radiant states and long-lived compound states. We explain the physics of super-radiance and discuss applications to weakly bound nuclei, giant resonances and widths of exotic baryons.

1. Introduction

Traditionally nuclear theory is divided into nuclear structure and nuclear reactions. Being naturally related by physics of nuclei as the subject of research, these fields are still significantly different in their methods, quality and justification of approximations and level of understanding. Moving away from the line of nuclear stability, we have to overcome this barrier. Only the consistent consideration of microscopic structure together with the response of the system to external fields, including various reaction amplitudes, is capable of developing the general picture of loosely bound nuclei. Such systems which become *open* even under weak excitations have their specifics in extreme sensitivity of internal properties to the proximity of continuum. Similar problems emerge in atomic and molecular physics as well as in mesoscopic condensed matter physics and quantum optics.

*The collaboration with N. Auerbach is highly appreciated. The work was supported by the NSF grant PHY-0244453 and in part by a grant from the US-Israel BSF.

In a weakly bound quantum system, couplings of intrinsic dynamics with reaction channels can crucially determine many important physical properties. Above threshold, intrinsic energy levels become resonances embedded in the continuum [1] with the lifetime typically shortening as excitation energy increases. The relevant physical parameter here is $\kappa = \gamma/D$, the ratio of the characteristic partial width (for a given channel) of a resonance to the spacing between the resonances. The presence of overlapping resonances, $\kappa > 1$, is usually associated with a very complicated and unpredictable pattern of Ericson fluctuations that can only be considered in statistical terms of random amplitudes. It turns out however that different, less known, dynamics can take place in the region of a relatively small number of open channels. New collective phenomena are possible leading to the redistribution of the widths (and of corresponding time scales) and creation of short-lived (*super-radiant*) as well as long-lived (*trapped*) structures. Being a consequence of the unitarity of the dynamics in the continuum, this restructuring appears in any consistent theory that considers the bound states and continuum on equal footing. In particular, it follows from classical Fano theory [2] widely used in quantum optics [3].

The term *super-radiance* (SR) refers to the discovery by Dicke [4] who 50 years ago has shown that, among 2^N states of a system of N two-level atoms confined to the volume of size smaller than the radiation wavelength between the two levels, one state exists that radiates very fast and coherently so that its width is close to $\Gamma = N\gamma$, where γ is the width of an individual isolated atom, and the intensity is proportional to N^2. The remaining states live for a long time having very small widths. The SR is observed [5] in the laser pulse transmission through a resonant medium. It is important that the coherent coupling of the atoms is reached due to their interaction through the common radiation field, independently of the direct atom-atom interaction.

The analog of the SR emerges in many-body quantum systems, Fig. 1, where N intrinsic states of the same symmetry are coupled to common decay channels as was seen in numerical simulations for the nuclear continuum shell model [6]. At $\kappa \sim 1$, the system undergoes a phase transition from the separated narrow resonances to the width accumulation by the SR states; their number correlates with that of the open channels. The theory of the phenomenon was given in [7], where the analogy to the Dicke SR state was pointed out; for various aspects of theory see [8,9,10,11].

Due to the very general character of the SR, it was observed in many different situations in atomic, molecular, condensed matter and nuclear physics. We present a brief review of the theory and discuss selected applications to nuclear physics of low and intermediate energies, including the suggestions for the SR as a reason for the narrow width of exotic baryons.

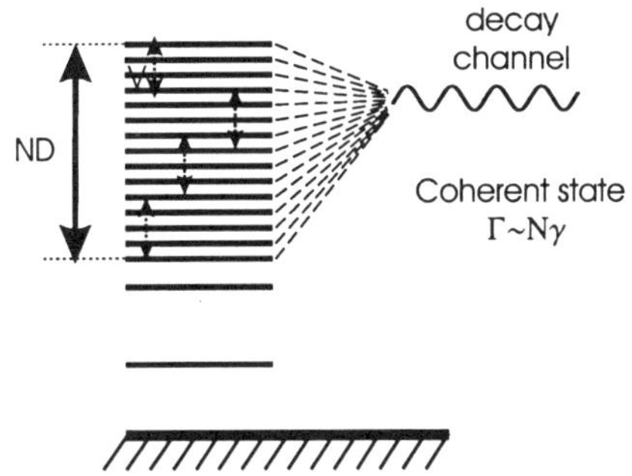

Figure 1. Schematic figure showing the SR mechanism.

2. Ingredients of the theory

1. The convenient approach to the unified theory of intrinsic states coupled to the continuum is provided by the Feshbach projection techniques [12]. The full system is described by the Hermitian Hamiltonian H. The Hilbert space is decomposed into two classes, internal, $Q \equiv \{|1\rangle\}$, and external, $P = \{|c; E\rangle\}$, where c marks the continuum channels. The total eigenfunction at energy E contains the contributions of both classes; they are coupled by the matrix elements $\langle c; E|H|1\rangle$. Eliminating the external states by the projection *at given energy*, one comes to the effective eigenvalue problem in the intrinsic space, where the effective Hamiltonian is given by

$$\mathcal{H}_{QQ}(E) = H_{QQ} + H_{QP} \frac{1}{E - H_{PP} + i0} H_{PQ}. \tag{1}$$

2. The internal energy-independent Hamiltonian H_{QQ} determines the eigenvalues ϵ_n of the states which would be bound without the continuum coupling. This coupling converts at least some of them into resonances with *complex energies* $\mathcal{E}_j = E_j - (i/2)\Gamma_j$.

3. The effective Hamiltonian (1) depends on running energy E. The continua c are started at threshold energies E_c. At $E > E_c$, the denominator of the propagator in eq. (1) contains singular terms corresponding to the real (*on-shell*) decay into the channel c with energy conservation. The real part Δ of the propagator corresponds to the principal value of the singular terms and describes the virtual (*off-shell*) processes of coupling through all (closed and open) channels, while only open channels contribute to the imaginary part $(-i/2)W$ that makes the effective Hamiltonian *non-Hermitian*.

4. The anti-Hermitian part of the effective Hamiltonian is factorized into a number of terms equal to the number k of channels open at energy E. Thus, the general form of $\mathcal{H}$ is

$$\mathcal{H}_{12} = H_{12} + \Delta_{12}(E) - \frac{i}{2} W_{12}(E), \quad W_{12} = \sum_{c;\,\text{open}} A_1^c A_2^{c*}, \tag{2}$$

where the amplitudes A_1^c for the coupling of the intrinsic state $|1\rangle$ to the channel c are proportional to the original matrix elements $\langle 1|H|c\rangle$.

5. The principal part Δ renormalizes the intrinsic Hamiltonian H_{QQ}. In many situations it is of minor significance being small and weakly dependent on energy. The anti-Hermitian part is the driving force for new physics. The amplitudes A_1^c should vanish at threshold energy E_c, and this energy dependence is crucial for weakly bound systems.

6. The diagonalization of $\mathcal{H}$, eq. (2), determines the complex eigenvalues $\mathcal{E}_j = \tilde{E}_j - (i/2)\Gamma_j$ and the (biorthogonal) set of the eigenfunctions $|j\rangle$ of quasistationary states depending on running energy E. The resonance centroid E_j on the real axis is self-consistently determined by $\tilde{E}_j(E = E_j) = E_j$.

7. The same formalism determines the reaction cross sections at a given energy with the aid of the scattering matrix

$$S^{ab}(E) = s_a^{1/2} \left\{ \delta^{ab} - \sum_{12} A_1^{a*} \left(\frac{1}{E - \mathcal{H}} \right)_{12} A_2^b \right\} s_b^{1/2}. \tag{3}$$

Here the "potential" phases $s_a = \exp(2i\delta_a(E))$ describe the contributions of remote resonances which usually are not taken into account explicitly. The full effective Hamiltonian $\mathcal{H}$, eq. (2) in the denominator in eq. (3) includes the same amplitudes A_1^c as in the numerator, and this makes the S-matrix explicitly unitary.

3. SR phase transition

At energy below the lowest threshold, we have the Hermitian Hamiltonian $H + \Delta$. This determines the discrete spectrum of bound states ϵ_n and the stationary wave functions $|\psi_n\rangle$. The unitary transformation to this basis of the *internal representation* keeps the anti-Hermitian part factorized so that at energy in the continuum

$$\mathcal{H}_{nn'} = \epsilon_n \delta_{nn'} - \frac{i}{2} W_{nn'}, \quad W_{nn'} = \sum_{c;\,\text{open}} A_n^c A_{n'}^{c*}. \tag{4}$$

The strength of the continuum coupling in channel c is measured by the ratio $\kappa_c = \gamma_c/D$ of the typical partial width $\gamma_c = |A^c|^2$ to the mean level spacing D (we consider the class of intrinsic states with the same exact quantum numbers). At small κ_c, we have in this channel isolated narrow resonances with the widths $\sim \gamma_c$.

As κ_c increases, the role of the *off-diagonal damping* [3] increases. When κ_c is getting close to 1, the *percolation* happens: decay of a level n with return to the overlapping level n' becomes likely, and all levels are coherently coupled through the continuum, similarly to the Dicke coherent state. Now we are close to another limit where the dynamics are defined by $W_{nn'}$. At strong coupling one can start with the *doorway representation* of the eigenstates of W. Due to the factorization of W, the rank of this matrix is equal to the number k of open channels, and there are k doorway states with non-zero widths. The remaining $N-k$ states are weakly coupled to the continuum only through the doorways. The eigenstates of $\mathcal{H}$ are divided into two groups: the SR states sharing almost all summed width of the N states, and the trapped states which are very long-lived.

In the simplest limit of a single open channel and degenerate intrinsic spectrum, $\epsilon_n \equiv \epsilon$, the whole width $\Gamma = \mathrm{Tr}\, W$ is concentrated in one SR state. If the intrinsic levels ϵ_n are not degenerate but their spread $\delta\epsilon \sim ND \ll \Gamma$, the sum $\tilde{\Gamma}$ of all small $N-1$ widths is [13]

$$\tilde{\Gamma} \approx 4\,\frac{(\Delta\epsilon)^2}{\Gamma}. \tag{5}$$

The continuum coupling in the limit of $\kappa \sim (\Gamma/ND) \gg 1$ implements the segregation of distinct time scales in the reaction process:

the shortest time $\tau_d \sim \hbar/\Gamma$ corresponds to the fast direct reaction through the SR intermediate state;

fragmentation time $\tau_f \sim \hbar/\Delta\epsilon \sim \kappa\tau_d$ is what is necessary for the internal damping of the original excitation of one of the intrinsic states;

Weisskopf time $\tau_W \sim \hbar/D \sim N\kappa\tau_d$ is the recurrence time of the wave packet (intrinsic equilibration);

compound lifetime of trapped states $\tau_c \sim \hbar/(\tilde{\Gamma}/N) \sim \kappa\tau_W$ that allows for the full exploration of intrinsic space.

4. Some applications

The universality of the SR mechanism guarantees its appearance in any situation with a not very large number of open channels (otherwise the off-diagonal damping is quenched by the Ericson fluctuations of amplitudes corresponding to different channels) and the necessary ingredients present. There are many examples of the manifestation of the SR in molecular, atomic and condensed matter physics, see for instance [14,15,16]. Below we shortly discuss some examples from nuclear physics.

4.1. *Two collectivities*

Giant resonances (GR) in the response function of the nucleus to the multipole excitation appear as a result of the coherent coupling of simple particle-hole excitations. By our terminology, they are generated by the intrinsic interaction that accumulates the *multipole strength* at some energy shifted from the interval $\Delta\epsilon$ of the unperturbed intrinsic states. The strength collectivization, however, is not what is seen in the reaction. The observed cross sections are determined by the partial *widths* of unstable intrinsic states with respect to a given channel.

The simple model [17,18] with the effective Hamiltonian

$$\mathcal{H}_{nn'} = \epsilon_n \delta_{nn'} + \lambda d_n d_{n'} - \frac{i}{2} A_n A_{n'}, \tag{6}$$

contains unperturbed energies, real multipole interaction and interaction through the continuum. The dynamics here are determined by two multidimensional vectors $\mathbf{d} = \{d_n\}$ and $\mathbf{A} = \{A_n\}$. The multipole interaction creates the GR shifted by $\Omega = \lambda \mathbf{d}^2$ *along the real energy axis* from the unperturbed centroid $\bar{\epsilon}$ and accumulating the multipole strength. The continuum interaction creates the SR state shifted by $\Gamma = \mathbf{A}^2$ *along the imaginary energy axis* to the lower part of the complex plane and accumulating almost all available decay width.

The interplay of the collective effects depends on the "angle" ϕ between the vectors $\mathbf{d}$ and $\mathbf{A}$. In the degenerate case, $\epsilon_n = \epsilon$, the reaction amplitude looks like

$$T(E) = \frac{(E - \epsilon - \Omega)\Gamma + \lambda(\mathbf{A}\cdot\mathbf{d})^2}{(E - \epsilon - \Omega)[E - \epsilon + (i/2)\Gamma] + (i/2)\lambda(\mathbf{A}\cdot\mathbf{d})^2}. \tag{7}$$

In the limits of parallel internal and external couplings, $\phi = 0°$, we have

$$T(E) = \frac{\Gamma}{E - \epsilon - \Omega + (i/2)\Gamma}. \tag{8}$$

Here the SR and GR collectivization are combined, and the experiment will reveal the shifted "*Giant Dicke resonance*" with full strength and full width. This is what we expect to see in gamma scattering, where both intrinsic excitation and decay have the same multipole nature. In the opposite case of orthogonal couplings, $\phi = 90°$, the result is $T(E) = \Gamma/[E - \epsilon - (i/2)\Gamma]$. Now the experiment would show only the unshifted SR resonance with vanishing multipole strength but broad width (the decay channel has another nature, for example results from evaporation). The GR state is *dark* with collective strength but no access to the continuum. Fig. 2 shows a more realistic case of non-degenerate intrinsic states when both the strength and the width are shared between the displaced GR and the SR in the region of unperturbed excitation energies. As the coupling with continuum increases, the low-energy branch is expected to develop into what is called *pigmy giant resonance.* We hope to present the study of the pigmy resonance elsewhere but the mechanism shown here is quite universal. Because of complicated interference between the two types of collectivity, the observed picture should differ in different channels.

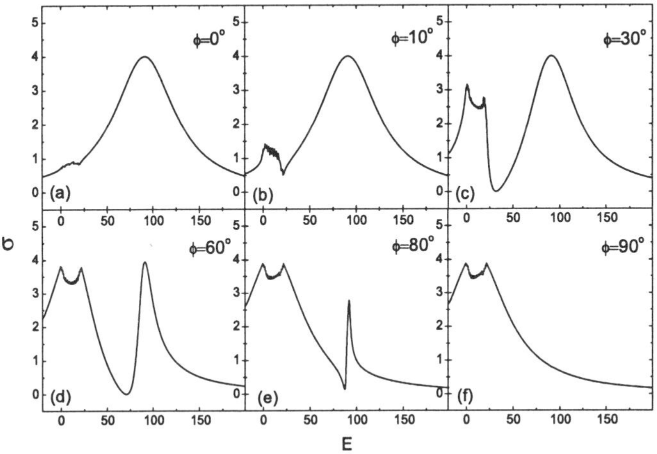

Figure 2. The scattering amplitude in the model of eq. (6) with 20 states with spacing between ϵ_n set as a unit of energy; the parameters are $\Omega = \Gamma = 80$. The panels show the evolution of scattering as a function of increasing angle between the multidimensional vectors **d** and **A**.

4.2. *Loosely bound nuclei*

These applications stimulated the renewed interest to the SR physics. As we have discussed at the previous Seminar [19], in the proximity of thresholds the coupling of intrinsic states with and through the continuum dominates the dynamics. The shell model machinery can be generalized [20] to include the effective non-Hermitian energy-dependent Hamiltonian. In particular, one can consistently consider systems like ^{11}Li where the single-particle states are unbound but additional correlations bring paired states back from the continuum. In all such cases the correct energy dependence of the amplitudes A_1^c near threshold is crucial. However, this dependence is sensitive to the absolute position of thresholds. To solve the problem in a nucleus A with one- and two-body decay channels, we need to know the spectrum of the $(A-1)$ and $(A-2)$ nuclei, and the whole chain of daughter nuclei. The plausibility of such a program was demonstrated for the chain of oxygen isotopes [20]. Fig. 3 shows full shell model calculation for the few oxygen nuclei above the ^{16}O core and comparison with data; this will be further discussed in a forthcoming publication.

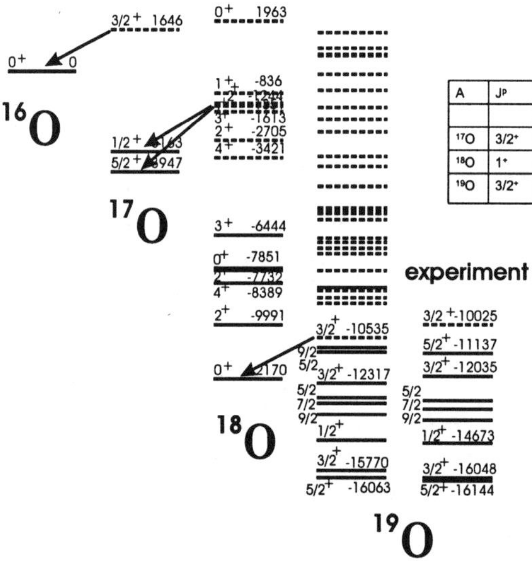

A	J^P	Experiment		CSM calculation	
		E(MeV)	Γ (keV)	E(MeV)	Γ (keV)
^{17}O	3/2⁺	1.646	96	1.646	102
^{18}O	1⁺	8.817	70	10.820	86
^{19}O	3/2⁺	6.119	110	5.530	90

Figure 3. Continuum shell model calculation for the oxygen isotopes with $A = 16$ to 19. One and two-neutron emission processes are considered. On a single energy scale stable nuclear states are shown with solid lines and neutron-unstable states by dashed lines. The experimentally observed states of ^{19}O are shown to the right on the same energy scale. The scarce experimental data for energies and neutron decay widths are listed in the table along with theory predictions. Correspondingly, in the level scheme the arrows indicate the decay transitions of these states.

5. Widths of exotic baryons

The SR mechanism works in hadron physics. A baryon resonance R created, for example, in a photonuclear experiment mixes with the RN^{-1} states of R-(nucleon hole) nature. The baryon resonances have excitation energies starting from few hundred MeV and typical vacuum widths exceeding 100 MeV. The continuum mixing with the RN^{-1} states of the same symmetry produces an SR state and few trapped states which are observed as narrow resonances on the broad SR background. This pattern was seen in the $^{12}C(e, e'p\pi^-)^{11}C$ Mainz experiment [21] at the Δ-isobar region and attributed [13] to the SR mechanism. In terms of two collectivities, the SR state here is a good candidate for the parallel case. As shown by old RPA calculations [22], the pionic coherent state (similar to the nuclear GR) accumulates the pion decay width as well.

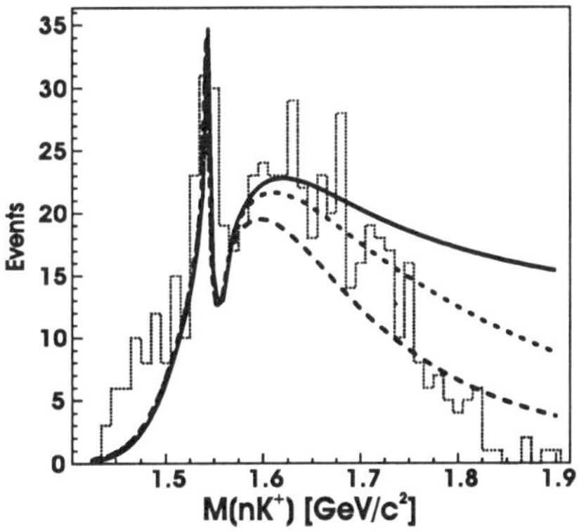

Figure 4. The two-state model for the pentaquark in comparison with the results of the CLAS experiment $\gamma d \to pK^-(K^+n)$ [24]. The p-wave (K^+n) continuum channel is considered with kinematical dependence $A(E) \sim E^{3/4}$ near threshold, solid curve. For two other curves, the amplitude $A(E)$ at high energy is damped as $A(E) \sim [E^3/(1 + E/\Lambda)^3]^{1/4}$ and the cutoff parameters Λ are 300 MeV, dashed curve, and 500 MeV, dotted curve.

We have applied [23] the same idea to recent observations by different groups of the Θ^+ resonance with strangeness +1. The resonance coined as *pentaquark* has a very narrow width smaller than few MeV. In the experiments on nuclei the situation should be quite similar to that in the Δ-case. In [23] the arguments are given that in the deuteron and proton experiments the same mechanism can still be at work because the intrinsic QCD dynamics create few states of different nature but the same symmetry which are coupled through the continuum. It is sufficient to have just two such states, for example (but not necessarily) a standard quark bag and a large size quasimolecular state. Fig. 4 shows the result for a simple two-state model with naturally chosen parameter values where the narrow resonance on the broad background emerges due to the SR mechanism.

6. Conclusion

The phenomenon of SR is very general; it has to appear, provided all necessary ingredients are in place, in any theory that describes physics on the borderline of discrete and continuum spectrum and agrees with the unitarity requirements. The segregation of physical time scales is accompanied by the appearance of long-lived states on a broad background. One can expect the most spectacular manifestations in many-body systems (autoionizing atomic states, chemical reactions, loosely bound nuclei, collective dynamics in the continuum and analogous physics on a quark level). There are important ramifications for quantum chaos and its experimental tests, mesoscopic condensed matter physics, studies of entanglement, decoherence and quantum information.

References

1. C. Mahaux and H.A. Weidenmüller, *Shell Model Approach to Nuclear Reactions* (North Holland, Amsterdam, 1969).
2. U. Fano, Nuovo Cim. **12** (1935) 156; Phys. Rev. **124** (1961) 1866.
3. S.M. Barnett and P.M. Radmore, *Methods in Theoretical Quantum Optics* (Clarendon Press, Oxford, 2002).
4. R.H. Dicke, Phys. Rev. **93** (1054) 99.
5. N. Scribanowitz, I.P. Herman, J.C. MacGillivray, and M.S. Fields, Phys. Rev. Lett. **30** (1973) 309.
6. P. Kleinwächter and I. Rotter, Phys. Rev. C **32** (1985) 1742.
7. V.V. Sokolov and V.G. Zelevinsky, Phys. Lett. B **202** (1988) 10; Nucl. Phys. **A504** (1989) 562.
8. I. Rotter, Rep. Prog. Phys. **54** (19910 635.
9. V.V. Sokolov and V.G. Zelevinsky, Ann. Phys. (N.Y.) **216** (1992) 323.
10. F.M. Izrailev, D. Saher, and V.V. Sokolov, Phys. Rev. E **49** (1994) 130.
11. A. Volya and V. Zelevinsky, J. Opt. B **5** (2003) 450.
12. H. Feshbach, Ann. Phys. (N.Y.) **5** (1958) 357; **19** (1962) 287.
13. N. Auerbach and V. Zelevinsky, Phys. Rev. C **65** (2002) 034601.
14. V.B. Pavlov-Verevkin, Phys. Lett. A **129** (1988) 168.
15. V.V. Flambaum, A.A. Gribakina, and G.F. Gribakin, Phys. Rev. A **54** (1996) 2066.
16. I. Rotter, E. Persson, K. Pichugin, and P. Seba, Phys. Rev. E **62** (2000) 450.
17. V.V. Sokolov and V.G. Zelevinsky, Physika (Zagreb) **22** (1990) 303.
18. V.V. Sokolov, I. Rotter, D.V. Savin, and M. Müller, Phys. Rev. C **56** (1997) 1031, 1044.
19. V. Zelevinsky and A. Volya, Challenges of Nuclear Structure, Proceedings of the 7th International Spring Seminar on Nuclear Physics, ed. A. Covello (World Scientific, Singapore, 2002) p. 261.
20. A. Volya and V. Zelevinsky, Phys. Rev. C **67** (2003) 054322.
21. P. Bartsch *et al.*, Eur. Phys. J. A **4** (1999) 209.
22. M. Hirata, J.H. Koch, F. Lenz, and E.J. Moniz, Ann. Phys. (N.Y.) **120** (1979) 205.
23. N. Auerbach, V. Zelevinsky, and A. Volya. Phys. Lett. B **590** (2004) 45.
24. S. Stepanyan *et al.* (CLAS collaboration), Phys. Rev. Lett. **91** (2003) 252001.

KEY TOPICS IN NUCLEAR STRUCTURE
Proceedings of the 8th Int. Spring Seminar on Nuclear Physics
edited by Aldo Covello
© 2005 World Scientific Publishing Co.

THE PHYSICS OF PROTEIN FOLDING AND OF DRUG DESIGN

R. A. BROGLIA

Department of Physics, University of Milano,
INFN, Sez. di Milano, Via Celoria 16 20133 Milano, Italy, and
The Niels Bohr Institute, Bledgamsvej 16, 2100 Copenhagen, Denmark

G. TIANA

Department of Physics, University of Milano,
INFN, Sez. di Milano, Via Celoria 16 20133 Milano, Italy

The problem of protein folding consists in understanding how the aminoacid sequence of a protein (primary structure) determines its unique, biological active equilibrium conformation (tertiary structure). By mean of simplified models, we explore the dynamical processes which are at the basis of the folding of model proteins and find a simple hierarchical mechanism which governs the folding phenomenon. Exploiting this result, it is possible not only to develop an algorithm to determine the equilibrium conformation of a protein from its sequence, that is to solve the protein folding problem provided one knows the interaction among the amino acids, but also to design a novel class of drugs which interfere with the folding mechanism and whose inhibitor effect cannot be neutralized through mutations, as it is the case with standard drugs acting, as a rule, on the active site of enzymes.

1. Introduction

The problem of protein folding is to understand *how a protein molecule of specified amino acid sequence ends up in a unique configuration which, among other things, determines its biological function* [1]. In physical terms, the problem is how the one–dimensional information provided by the sequence of twenty types of amino acids encodes for a unique and stable three–dimensional equilibrium conformation.

This problem has a self–evident biological and medical importance. The sequencing of the human genome [2,3], that is the identification of the way the thousands of milions of basis follow each other in the human DNA, provides information on the sequence of amino acids forming each of the tens of thousands of proteins which build our cells and catalize the chemical reactions which make them function. The acquisition of sequence data by DNA sequencing is relatively quick, and vast quantities of data have become available through international efforts. But the knowledge of the sequence alone is of little help in understanding the function of the corresponding protein, in manipulating its function and in designing drugs to act on it. For that, one needs the three dimensional equilibrium conformation. On the other hand, the acquisition of three-dimensional data is still slow and is limited to proteins that either crystallize in a suitable form or are sufficiently small and soluble to be solved by NMR in solution [4]. In fact, while at present data banks contain information concerning the linear sequence of about 10^5 proteins, atomic coordinates of only 10^4 native structures are available [5]. Algorithms are thus required to translate the linear information into spatial information.

During the last twenty years a remarkable developement of protein models has taken place, ranging from simple two–state models of the kind used by chemists to describe chemical reactions, to all–atom models which take advantage of the power achieved by modern computers which allow to carry out simulations of the folding of proteins over periods of time which, in spite of being a small fraction of the full folding time, are still not negligible, at least for the case of small proteins.

A particularly interesting model describing the protein as a chain of beads on a cubic lattice, seems to represent an appropriate balance between solvability and realism (cf. e.g. ref. [6] and refs. therein). Studying in detail this model, one can find some remarkable simplicities in the folding of proteinlike chains. The folding process is controlled, within the framework of this model, by a small subset of the amino acids of the protein. As we shall see in more detail in the next Section, these "hot" amino acids [6] build very early in the folding process few local elementary structures (LES), which diffuse as essentially rigid entities. When the local elementary structures, which display a high affinity for each other, find their correct partners, they build the folding nucleus (FN), the minimum set of native contacts needed to overcome the main barrier of the free energy associated with the entire folding process [7,8].

The point of view of folding in terms of assembly of local elementary structures into the folding nucleus not only accounts for known experimental facts, but also opens the way to predictions and manipulations. In fact, while the direct prediction of the native conformation of a protein from the amino acid sequence is difficult. On the other hand, the localization of the local elementary structures is much easier, elementary structures are known, it is not impossible to determine the folding nucleus, and from it the native conformation. Furthermore, the knowledge of local elementary structures can be used to design drugs able to inhibit the folding, and consequently the biological activity, of selected proteins [9].

2. The model

An important ingredient which is at the basis of the folding of proteins is the heterogeneity of the interaction arising from the presence of twenty kinds of different amino acids. It is known that physical systems displaying such an heterogeneity display, as a rule, a rough energy landscape with many competing low–energy states. This is a picture incompatible with that of proteins, which must display a unique ground state, well separated from the others, and as few metastable states as possible. Consequently, the purpose of these models is to understand what makes a protein, characterized by a well defined amino acid sequence, different from a generic heterogeneous system, whose paradigm is found in a random sequence of amino acids.

The simplest choice for a heterogeneous potential is that of a contact potential, in the form

$$U(\{r_i\},\{\sigma(i)\}) = \sum_{ij} B_{\sigma(i)\sigma(j)}\Delta(r_i - r_j), \tag{1}$$

where r_i and $\sigma(i)$ are the position and kind of the ith amino acid, $\Delta(r_i - r_j)$ is a contact function which assumes the value 1 if $|r_i - r_j| \leq 1$ and zero otherwise, while $B_{\sigma\tau}$ is the element of the 20×20 interaction matrix which defines the iteraction energy between amino acids of kind σ and τ. A widely used interaction matrix has been calculated by Miyazawa and Jernigan (MJ) [10] from the statistical analysis of the contacts of a large database of known proteins, assuming that the more often a given contact appears in the database, the more attractive it is. This is done by calculating the probability $p_{\sigma\tau}$ of appearence of the contact between the amino acids of kind σ and τ, and assuming a Boltzmann–like relationship of the kind $B_{\sigma\tau} \sim -\log p_{\sigma\tau}$.

The second approximation used, consisting in locating the beads representing the amino acids on the vertices of a cubic lattice of unitary side length, implies that the conformational degrees of freedom are discrete. This is very convenient from a computational point of view and makes conformational entropy easy to handle.

Good–folder sequences are characterized by a large gap $\delta = E_c - E_n$ (compared to the standard deviation σ of the contact energies) between the energy of the sequence in the native conformation E_n, and the lowest energy (threshold energy) of the conformations structurally dissimilar to the native conformation [11,12]. The quantity E_c is the lowest energy a random sequence can achieve in the process of compacting, and is a quantity which is solely determined by the composition of the protein. In other words, good folders are associated with an normalized gap $\xi = \delta/\sigma \gg 1$, quantity closely related to the z-score [13]. Furthermore, starting from a designed sequence which displays a large gap, all mutated sequences which preserve (to some extent) the gap fold into the native conformation [14]. For the sake of definitess, we will consider in the following a particular sequence made out of 36 amino acids called S_{36} and folding to the native structure shown in Fig. 3(d), which can be seen as prototype of folding model sequence [6].

3. Folding of small proteins

A striking result which emerges from studying the inverse folding approach is that the stabilization energy of a protein is note distributed evenly across its amino acid, but is concentrated in few "hot" residues [6]. Locating "hot" amino acids is quite simple. In fact, for this purpose one introduces point mutations in each site of the native structure, that is, one replaces each of the amino acids of the designed (low energy) sequence by all of the possible 19 amino acids and study whether the resulting sequence still folds or not. It is found that mutations in only few sites denaturate (i.e., impedes its folding) as well as destabilizes (strongly reduces the native state occupation probability) the protein. To be quantitative, we find that only $8\% \pm 2\%$ (Fig. 3(d)) of the amino acids of a designed sequence are highly conserved, strongly interacting and occupy a hot site in the native conformation, in general well protected inside the protein, as it will suit an hydrophobic residue. Mutations of the amino acids occupying the hot sites denaturate the protein, that is block the unfolded (denaturated) $\rightarrow$ native ($D \rightarrow N$) phase transition. Mutations of amino acids occupying the other sites have little effect on the ability the resulting sequence has to fold onto the native conformation, but lead to sequences which, in the native conformation, still display an energy lower than E_c, thus qualifying as good folders. The resulting families of (homologous) proteins (folding to the same native structures) display in common essentially only the few amino acids which occupy the hot sites.

The hot amino acids not only determine the stability of the protein but also the hierarchy of native contacts formation through which the protein, starting from an elongated phase reaches the native conformation (cf. Fig. 2): 1) formation, almost instantaneously of few local elementary structures (LES, i.e. hidden intermediates corresponding to incipient α–helices and β–sheets, the secondary structures of proteins) stabilized by the interaction between the hot amino acids, b) formation of the minimum set of native contacts which brings the system over the major free energy barrier of the whole folding process resulting from the docking of the LES (i.e., formation of the post–critical folding nucleus (FN)), c) relaxation of the remaining amino acids onto the native structure shortly after the formation of the FN giving rise to a unique system with an energy below E_c [7,8].

Summing up, the folding of proteins is controlled by the corresponding hot amino acids through the LES, ultimate building blocks of this molecular LEGO [15]. In other words, the simple, most important feature common to all designed sequences folding to the same native structure is the presence of few, highly conserved, strongly interacting, hot, amino acids which stabilize the LES and which are buried inside the folding nucleus of the protein in its native conformation.

4. Drug Design

LES elementary structures are also at the basis of a protocol for non–conventional drug design recently proposed by us [9]. Conventional drugs perform their activity either by activating or by inhibiting some target component of the cell. In particular, many inhibitory drugs bind to an enzyme and deplete its function by preventing the binding of the substrate. This is done by either capping the active site of the enzyme (competitive inhibition) or, binding to some other part of the enzyme, by provoking structural changes which make the enzyme unfit to bind the substrate (allosteric inhibition). The two main features that inhibitory drugs must display are efficiency and specificity. In fact, it is not sufficient that the drug binds to the substrate and reduces efficiently its activity. It is also important that it does not interfere with other cellular processes, binding only to the protein it was designed for. These features are usually accomplished designing drugs which mimick the molecular properties of the natural substrate. In fact, the pair enzyme/substrate have undergone milions of years of evolution in order to display the required features. Consequently, the more similar the drug is to the substrate, the lower is the probability that it interferes

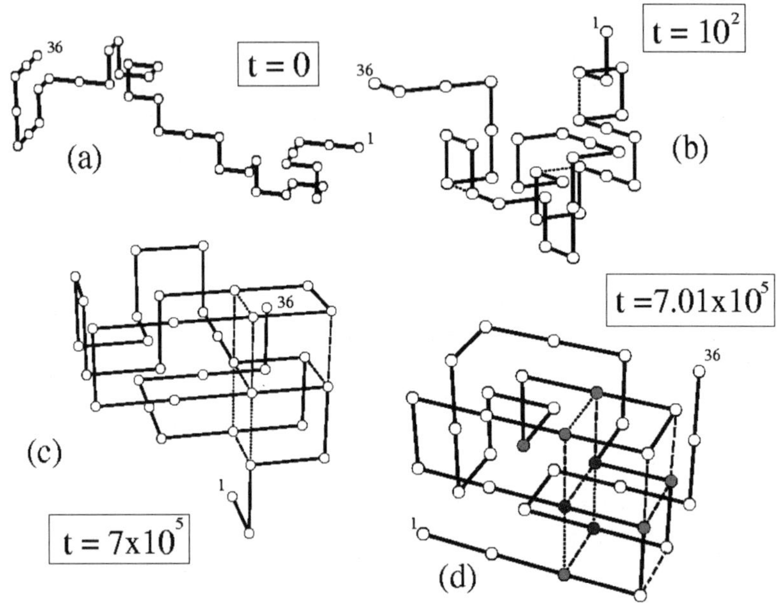

Figure 1. Dynamics of contact formation for a MC simulation of the folding of the model sequence sequence S_{36}. With a dashed line we label the contacts 3–6, 27–30 and 11–14 stabilizing the LES S_4^1, S_4^2 and S_4^3 (cf. Fig. 3(d)). With solid dot lines along the vertical axis we label (from top to bottom) the contacts: 5–28, 3–30, 14–27, 6–11, 13–28, 6–27, 12–5, 4–29 forming the folding nucleus.

with other cellular processes. Something that this kind of inhibitory drugs are not able to do is to avoid the development of resistance, a phenomenon which is typically related to viral protein targets. Under the selective pressure of the drug, the target is often able to either mutate the amino acids at the active site or at sites controlling its conformation in such a way that the activity of the enzyme is essentially retained, while the drug is no longer able to bind to it. An important example of drug-resistance is connected with AIDS. In this case, one of the main target proteins, HIV-protease, a dimer formed out of two identical chains each containing 99 residues and folding according to the LES paradigm discussed above (cf. e.g. [17]), is able to mutate its active site so as to avoid the effects of drug action within a period of time of 6-8 months. In keeping with this result and with the central role played by LES in the folding process of proteins, we suggest the use of short peptides with the same sequence as LES (p–LES) as non–conventional drugs which interfere with the folding mechanism of the target protein, destabilizing it and making it prone to proteolisis. These drugs are efficient, specific and do not suffer from the upraise of resistance.

In fact, the very reason why LES make single domain proteins fold fast confers p–LES the required features to act as effective drugs, that is, efficiency and specificity. They are efficient because they bind as strongly as LES do. Since LES are responsible for the stability of the protein, their stabilization energy must be of the order of several times kT. These peptides are also as specific as LES are. In fact LES have evolved over millions of years so as to prevent the upraise of metastable states and to avoid aggregation, aside from securing that the protein to fold fast. The possibility of developing non–conventional drugs for actual situations is tantamount to being able to determine the LES for a given protein. This can be done either experimentally (e.g. through molecular engineering [18]), or extending the algorithm discussed in ref. [16] making use of a realistic force field. The resulting peptides can be used either directly as drugs, or as templates to build mimetic molecules, which eventually do not display side effects connected with digestion or allergies. A feature which makes, in principle, these drugs quite promising as compared to conventional ones is to be found in the fact that the target protein cannot evolve through mutations to escape the drug, as happens in particular in the case of viral proteins in response to conventional drugs, because the mutation of residues in the LES would, anyway, lead to protein denaturation.

References

1. J. Maddox, Does folding determine protein configuration?, Nature, **370** (1994) 13
2. D. D. Shoemaker *et al.*, Experimental annotation of the human genome using microarray technology, Nature **409** (2001) 922
3. J. C. Venter *et al.*, The sequence of the human genome, Science **291** (2001) 1304
4. R. F. Service, Tapping DNA structures produces a trickle, New Focus, Science **298** (2002) 948
5. Protein Data Bank, on the website http://www.rcsb.org
6. G. Tiana, R. A. Broglia, H. E. Roman, E. Vigezzi and E. I. Shakhnovich, Folding and misfolding of designed protein–like chains with mutations, J. Chem. Phys. **108** (1998) 757
7. R. A. Broglia and G. Tiana, Hierarchy of Events in the folding of model proteins, J. Chem. Phys. **114** (2001) 7267
8. G. Tiana and R. A. Broglia, Statistical Analysis of Native Contact Formation in the Folding of Designed Model Proteins, J. Chem. Phys. **114** (2001) 2503
9. R. A. Broglia, G. Tiana and R. Berera, Resistance proof, folding-inhibitor drugs, J. Chem. Phys. **118** (2003) 4754

10. S. Miyazawa and R. Jernigan, Estimation of effective interresidue contact energies from protein crystal structures, Macromolecules **18** (1985) 534
11. E. I. Shakhnovich, Proteins with selected sequences fold into unique native conformation, Phys. Rev. Lett. **72** (1994) 3907
12. E. I. Shakhnovich and A. Gutin, Enumeration of all compact conformations of copolymers with random sequence of links, J. Chem. Phys. **93** (1989) 5967
13. V. I. Abkkevich, A. M. Gutin and E. I. Shakhnovich, Specific nucleus as the transition state for protein folding, Biochemistry **33** (1994) 10026
14. R. A. Broglia, G. Tiana, H. E. Roman, E. Vigezzi and E. Shakhnovich, Stability of Designed Proteins against Mutations, Phys. Rev. Lett., **82** (1999) 4727
15. R.A. Broglia, G. Tiana, S. Pasquali, H. E. Roman, E. Vigezzi, Folding and Aggregation of Designed Protein Chains, Proc. Natl. Acad. Sci. USA, **95** (1998) 12930
16. R. A. Broglia and G. Tiana, Reading the three–dimensional structure of a protein from its amino acid sequence, Proteins **45** (2001) 421
17. G. Tiana and R. A. Broglia, Folding and design of dimeric proteins, Proteins **49** (2002) 82
18. A. Fersht, *Structure and Mechanism in protein science*, W. H. Freeman and Co., New York (1999)

LIST OF PARTICIPANTS

V. Abrosimov — Institute for Nuclear Research
47 Prospect Nauki, 03028 Kiev, UKRAINE
abrosim@kinr.kiev.ua

F. Andreozzi — Dipartimento di Scienze Fisiche, Università di Napoli Federico II
Complesso Universitario di Monte S. Angelo
Via Cintia, 80126 Napoli, ITALY
andreozzi@na.infn.it

G. de Angelis — Laboratori Nazionali di Legnaro
Via Romea 4, 35020 Legnaro (Padova), ITALY
deangelis@lnl.infn.it

N. Aoi — Heavy Ion Nuclear Physics Lab., RIKEN
2-1 Hirosawa, Wako-shi, Saitama-ken, 351-0198 JAPAN
aoi@riken.jp

T. Aumann — GSI, Gesellschfat für Schwerionenforschung mbH (GSI)
Planckstrasse 1, D-64291, Darmstadt, GERMANY
t.aumann@gsi.de

C. Baktash — Oak Ridge National Laboratory, Physics Division
BLDG 6000, Oak Ridge, TN 37831-6371, USA
baktashc@ornl.gov

S. Baroni — Dipartimento di Fisica, Università di Milano
Via Celoria 16, 20133 Milano, ITALY
baroni@mi.infn.it

B. R. Barrett — Department of Physics, P.O.Box 210081, University of Arizona
1118 East 4th Street, Tucson, AZ 85721-0081, USA
bbarrett@physics.arizona.edu

N. Benczer-Koller — Department of Physics and Astronomy, Rutgers University
New Brunswick, NJ 08903, USA
nkoller@physics.rutgers.edu

D. Bonatsos — Institute of Nuclear Physics, N.C.S.R. "Demokritos"
GR-15310, Aghia Paraskevi, Attiki, GREECE
bonat@inp.demokritos.gr

G. C. Bonsignori — Dipartimento di Fisica "A. Righi", Università di Bologna
Via Irnerio 46, 40127 Bologna, ITALY
bonsignori@bo.infn.it

R. Broda — Niewodniczański Institute of Nuclear Physics PAN
Radzikowskiego 152, 31-342 Kraków, POLAND
rafal.broda@ifj.edu.pl

R. A. Broglia — Dipartimento di Fisica, Università di Milano
Via Celoria 16, 20133 Milano, ITALY
broglia@mi.infn.it

A. Brondi — Dipartimento di Scienze Fisiche, Università di Napoli Federico II
Complesso Universitario di Monte S. Angelo
Via Cintia, 80126 Napoli, ITALY
brondi@na.infn.it

R. F. Casten — Wright Nuclear Structure Laboratory, Physics Department
Yale University, New Haven, CT 06520-8124, USA
rick@riviera.phyisics.yale.edu

F. Catara — Dipartimento di Fisica, Università di Catania
Corso Italia 57, 95129 Catania, ITALY
catara@ct.infn.it

L. Coraggio — Istituto Nazionale di Fisica Nucleare, Sezione di Napoli
Complesso Universitario di Monte S. Angelo
Via Cintia, 80126 Napoli, ITALY
coraggio@na.infn.it

L. Corradi — Laboratori Nazionali di Legnaro
Via Romea 4, 35020 Legnaro (Padova), ITALY
corradi@lnl.infn.it

A. Covello — Dipartimento di Scienze Fisiche, Università di Napoli Federico II
Complesso Universitario di Monte S. Angelo
Via Cintia, 80126 Napoli, ITALY
covello@na.infn.it

P. J. Daly — Chemistry Department, Purdue University
West Lafayette, IN 47907, USA
glshively@chem.purdue.edu

N. D. Dang
RI-Beam Factory Project Office, RIKEN
2-1 Hirosawa, Wako-City 351-0198, Saitama, JAPAN
dang@postman.riken.go.jp

J. M. Daugas
CEA/DIF/DPTA/SPN
BP 12, 91680 Bruyères le Châtel, FRANCE
jean-michel.daugas@cea.fr

A. Dellafiore
Istituto Nazionale di Fisica Nucleare, Sezione di Firenze
Largo E. Fermi 2, 50125 Firenze, ITALY
della@fi.infn.it

J. P. Draayer
Department of Physics and Astronomy, Lousiana State University
Baton Rouge, Lousiana, 70803-4001, USA
draayer@lsu.edu

C. Fahlander
Division of Cosmic and Subatomic Physics, Lund University
Lund, SWEDEN
claes.fahlander@kosufy.lu.se

L. S. Ferreira
CFIF and Departamento de Física, Instituto Superior Técnico
Av. Rovisco Pais, 1049-001, Lisboa, PORTUGAL
flidia@ist.utl.pt

B. Fornal
Niewodniczanski Institute of Nuclear Physics
Radzikowskiego 152, 31-342 Kraków, POLAND
bogdan.fornal@ifj.edu.pl

S. Fujii
Department of Physics, University of Tokyo
7-3-1 Hongo, Bunkyo-ku, Tokyo 113-0033, JAPAN
sfujii@nt.phys.s.u-tokyo.ac.jp

M. Gai
Department of Physics, Yale University
272 Whitney Avenue, New Haven, CT 06520-8124, USA
gai@uconn.edu

A. Gargano
Istituto Nazionale di Fisica Nucleare, Sezione di Napoli
Complesso Universitario di Monte S. Angelo,
Via Cintia, 80126 Napoli, ITALY
gargano@na.infn.it

K. Gelbke
National Superconducting Cyclotron Laboratory
and Department of Physics
Michigan State University, East Lansing, MI 48824-1321, USA
gelbke@nscl.msu.edu

J. Genevey | Laboratoire de Physique Subatomique et de Cosmologie
53 Avenue des Martyrs, F-038026 Grenoble, FRANCE
genevey@lpsc.in2p3.fr

A. I. Georgieva | Institute of Nuclear Research and Nuclear Energy
Bulgarian Academy of Sciences
Bul. Tzarigradsko Chaussee 72, Sofia 1784, BULGARIA
anageorg@inrne.bas.bg

J. Ginocchio | Los Alamos National Laboratory, MS B283
Los Alamos, NM 87545, USA
gino@lanl.gov

M. Girod | Service de Physique Nucleaire
BP 12, 91680 Bruyères le Châtel, FRANCE
michel-g.girod@cea.fr

J. M. Gómez | Departamento de Fisica Atómica, Molecular y Nuclear
Facultad de Ciencias Fisicas
Universidad Complutense, 28040 Madrid, SPAIN
gomezk@nucl.fis.ucm.es

P. M. Gore | Department of Physics and Astronomy, Vanderbilt University
VU Station B 351807, Nashville, TN 37235, USA
philip.m.gore@vanderbilt.edu

M. Górska | GSI Darmstadt
Planckstrasse 1, 64291, Darmstadt, GERMANY
M.Gorska@gsi.de

D. Goutte | GANIL
BP 55027, 14076 Caen CEDEX 5, FRANCE
goutte@ganil.fr

G. Graw | Sektion Physik, Ludwig-Maximilians-Universität
Am Coulombwall 1, D 85748 Garching, GERMANY
Gerhard.Graw@lmu.de

P. Guazzoni | Dipartimento di Fisica, Università di Milano
Via Celoria 16, 20133 Milano, ITALY
paolo.guazzoni@mi.infn.it

J. H. Hamilton | Department of Physics and Astronomy, Vanderbilt University
VU Station B 351807, Nashville, TN 37235, USA
j.h.hamilton@vanderbilt.edu

M. Hjorth-Jensen — Department of Physics, University of Oslo
POB 1048 Blindern, N-0316 Oslo, NORWAY
mhjensen@fys.uio.no

F. Iachello — Center for Theoretical Physics, Sloane Laboratory
Yale University, New Haven, CT 06520-8120, USA
francesco.iachello@yale.edu

R. Id Betan — Royal Institute of Technology, AlbaNova University Center
SE-10691, Stockholm, SWEDEN
idbetan@nuclear.kth.se

N. Itaco — Dipartimento di Scienze Fisiche, Università di Napoli Federico II
Complesso Universitario di Monte S. Angelo
Via Cintia, 80126 Napoli, ITALY
itaco@na.infn.it

E. F. Jones — Department of Physics and Astronomy, Vanderbilt University
VU Station B 351807, Nashville, TN 37235, USA
elizabethfj@earthlink.net

R. Julin — Department of Physics, University of Jyväskylä
40351 Jyväskylä, FINLAND
Rauno.Julin@phys.jyu.fi

K. Kemper — Physics Department, Florida State University
Tallahassee, FL 32306-4350, USA
Kirby@phy.fsu.edu

U. Kneissl — Institut für Strahlenphysik der Universität Stuttgart
Allmandring 3, 70569 Stuttgart, GERMANY
kneissl@ifs.physik.uni-stuttgart.de

T. T. S. Kuo — Physics Department, State University of New York at Stony Brook
Stony Brook, NY 11794-3800, USA
kuo@nuclear.physics.sunysb.edu

J. Kvasil — Institute of Particle and Nuclear Physics, Charles University
V Holesovickach 2, 180 00 Prague 8, CZECH REPUBLIC
kvasil@ipnp.troja.mff.cuni.cz

E. G. Lanza — Istituto Nazionale di Fisica Nucleare, Sezione di Catania
via S. Sofia 64, 95123 Catania, ITALY
lanza@ct.infn.it

G. La Rana — Dipartimento di Scienze Fisiche, Università di Napoli Federico II
Complesso Universitario di Monte S. Angelo
Via Cintia, 80126 Napoli, ITALY
giovanni.larana@na.infn.it

D. Lenis — Institute of Nuclear Physics, N.C.S.R. "Demokritos"
GR-15310 Aghia Paraskevi, Attiki, GREECE
lenis@inp.demokritos.gr

S. M. Lenzi — Dipartimento di Fisica, Università di Padova
Via Marzolo 8, 35131 Padova, ITALY
lenzi@pd.infn.it

A. Leviatan — Racah Institute of Physics, The Hebrew University
Jerusalem 91904, ISRAEL
ami@vms.huji.ac.il

K.-P. Lieb — II Physikalisches Institut, Universität Göttingen
Tammannstrasse 1, D-37073 Göttingen, GERMANY
plieb@gwdg.de

N. Lo Iudice — Dipartimento di Scienze Fisiche, Università di Napoli Federico II
Complesso Universitario di Monte S. Angelo
Via Cintia, 80126 Napoli, ITALY
loiudice@na.infn.it

S. Lunardi — Dipartimento di Fisica, Università di Padova
Via Marzolo 8, 35131 Padova, ITALY
lunardi@pd.infn.it

H. Mach — Department of Radiation Sciences, Uppsala University
61182 Nykoping, SWEDEN
henryk.mach@studsvik.uu.se

A. Maj — Niewodniczanski Institute of Nuclear Physics
ul Radzikowskiego 152, 31-342 Krakow, POLAND
Adam.Maj@ifj.edu.pl

V. R. Manfredi — Dipartimento di Fisica, Università di Padova
Via Marzolo 8, 35131 Padova, ITALY
manfredi@pd.infn.it

P. Mason — Dipartimento di Fisica, Università di Milano
Via Celoria 16, 20133 Milano, ITALY
Paolo.Mason@mi.infn.it

J. Meng	School of Phyiscs, Peking University Beijing 100871, CHINA mengj@pku.edu.cn
R. Moro	Dipartimento di Scienze Fisiche, Università di Napoli Federico II Complesso Universitario di Monte S.Angelo Via Cintia, 80126 Napoli, ITALY moro@na.infn.it
M. Mutterer	Institut für Kernphysik, Technische Universität Darmstadt Schlossgartenstrasse 9, D-64289 Darmstadt, GERMANY mutterer@ikp.tu-darmstadt.de
D. R. Napoli	Laboratori Nazionali di Legnaro Via Romea 4, 35020 Legnaro (Padova), ITALY napoli@lnl.infn.it
T. Otsuka	Department of Physics, University of Tokyo Hongo, Bunkyo-ku, Tokyo 113-0033, JAPAN otsuka@phys.s.u-tokyo.ac.jp
F. Palumbo	Laboratori Nazionali di Frascati Istituto Nazionale di Fisica Nucleare P. O.Box 13, 00044 Frascati, ITALY fabrizio.Palumbo@lnf.infn.it
S. C. Pieper	Physics Division, Argonne National Laboratory Building 203, Argonne, IL 60439, USA spieper@anl.gov
J. A. Pinston	Laboratoire de Physique Subatomique et de Cosmologie 53 Avenue des Martyrs, F-038026 Grenoble, FRANCE pinston@lpsc.in2p3.fr
G. Pisent	Dipartimento di Fisica, Università di Padova Via Marzolo 8, 35131 Padova, ITALY pisent@pd.infn.it
A. Porrino	Dipartimento di Scienze Fisiche, Università di Napoli Federico II Complesso Universitario di Monte S. Angelo Via Cintia, 80126 Napoli, ITALY porrino@na.infn.it

A. V. Ramayya — Department of Physics and Astronomy, Vanderbilt University
VU Station B 351807, Nashville, TN 37235, USA
a.v.ramayya@vanderbilt.edu

K. E. Rehm — Physics Division, Argonne National Laboratory
9700 South Cass Av., Argonne, IL 60439, USA
rehm@anl.gov

P. Ring — Physykdepartment der Technischen Universität München
D-85748 Garching, GERMANY
Peter_Ring@ph.tum.de

D. J. Rowe — Department of Physics, University of Toronto
ON M5S1A7, CANADA
rowe@physics.utoronto.ca

G. Royer — Laboratoire Subatech
4 rue A. Kastler, La Chantrerie BP 20722
44307 Nantes Cedex 03, FRANCE
royer@subatech.in2p3.fr

D. Schwalm — Max Planck Institut für Kernphysik
Postfach 103980, D 69029 Heidelberg, GERMANY
schwalm@mpi-hd.mpg.de

M. Serra — Department of Physics, University of Tokyo
Hongo, Bunkyo-ku, Tokyo 113-0033, JAPAN
mserra@tkyntm.phys.s.u-tokyo.ac.jp

A. Shotter — TRIUMF
4004 Wesbrook Mall, Vancouver, B.C. V6T 2A3, CANADA
ashotter@triumf.ca

P. Sona — Dipartimento di Fisica, Università di Firenze
Largo E. Fermi 2, 50125 Firenze, ITALY
sona@fi.infn.it

G. S. Stoitcheva — Physics Division, Oak Ridge National Laboratory
Oak Ridge, TN 37831, USA
stoitchevags@ornl.gov

M. Stoitsov — Department of Physics and Astronomy
University of Tennessee
Knoxville, Tennessee 37996, USA
stoitsov@ornl.gov

Ch. Stoyanov — Bulgarian Academy of Sciences
Institute for Nuclear Research and Nuclear Energy
72 Bvd. Tzarigradsko Chaussee, Sofia 1784, BULGARIA
stoyanov@inrne.bas.bg

I. Talmi — Department of Physics, The Weizmann Institute of Science
Rehovot 76100, ISRAEL
igal.talmi@weizmann.ac.il

M. Tomaselli — GSI
Planckstrasse 1, 64291, Darmstadt, GERMANY
m.tomaselli@gsi.de

D. Tonev — Laboratori Nazionali di Legnaro
Via Romea 4, 35020 Legnaro (Padova), ITALY
mitko@lnl.infn.it

A. Vdovin — Bogoliubov Laboratory of Theoretical Physics
Joint Institute for Nuclear Research
141980 Dubna, Moscow Region, RUSSIA
vdovin@thsun1.jinr.dubna.ru

E. Vigezzi — Istituto Nazionale di Fisica Nucleare, Sezione di Milano
Via Celoria 16, 20133 Milano, ITALY
vigezzi@mi.infn.it

V. Voronov — Bogoliubov Laboratory of Theoretical Physics
Joint Institute for Nuclear Research
141980 Dubna, Moscow Region, RUSSIA
voronov@thsun1.jinr.ru

R. Wyss — Royal Institute of Technology, Alba Nova University Center
SE-10691, Stockholm, SWEDEN
wyss@kth.se

N. Yoshida — Faculty of Informatics, Kansai University
Takatsuki 569-1095, JAPAN
yoshida@res.kutc.kansai-u.ac.jp

L. Zamick — Department of Physics and Astronomy, Rutgers University
Busch Campus, Piscataway, NJ 08854-8019, USA
zamick@physics.rutgers.edu

V. Zelevinsky | National Superconducting Cyclotron Laboratory
Michigan State University, East Lansing, MI 48824-1321, USA
zelevinski@nscl.msu.edu

L. Zetta | Dipartimento di Fisica, Università di Milano
Via Celoria 16, 20133 Milano, ITALY
luisa.zetta@mi.infn.it

L. Zuffi | Dipartimento di Fisica, Università di Milano
Via Celoria 16, 20133 Milano, ITALY
zuffi@mi.infn.it

A. Zuker | Institut de Recherches Subatomiques, IN2P3-CNRS
Université Louis Pasteur, F-67037, Strasbourg, FRANCE
andres.zuker@ires.in2p3.fr

AUTHOR INDEX